# The Nuclear Shell Model 70 Years after Its Advent: Achievements and Prospects

Editors

**Angela Gargano**
**Giovanni De Gregorio**
**Silvia Monica Lenzi**

Basel • Beijing • Wuhan • Barcelona • Belgrade • Novi Sad • Cluj • Manchester

*Editors*
Angela Gargano
Istituto Nazionale di Fisica Nucleare
Napoli, Italy

Giovanni De Gregorio
Università degli Studi della Campania "Luigi Vanvitelli"
Caserta, Italy

Silvia Monica Lenzi
Università degli Studi di Padova
Padova, Italy

*Editorial Office*
MDPI
St. Alban-Anlage 66
4052 Basel, Switzerland

This is a reprint of articles from the Special Issue published online in the open access journal *Physics* (ISSN 2624-8174) (available at: https://www.mdpi.com/journal/physics/special_issues/NuclearShell70).

For citation purposes, cite each article independently as indicated on the article page online and as indicated below:

| Lastname, A.A.; Lastname, B.B. Article Title. *Journal Name* **Year**, *Volume Number*, Page Range. |
| --- |

**ISBN 978-3-0365-9504-7 (Hbk)**
**ISBN 978-3-0365-9505-4 (PDF)**
**doi.org/10.3390/books978-3-0365-9505-4**

© 2023 by the authors. Articles in this book are Open Access and distributed under the Creative Commons Attribution (CC BY) license. The book as a whole is distributed by MDPI under the terms and conditions of the Creative Commons Attribution-NonCommercial-NoDerivs (CC BY-NC-ND) license.

# Contents

# About the Editors

**Angela Gargano**

Angela Gargano is a First Researcher at the Italian Institute for Nuclear Physics (INFN-Napoli). Her research studies nuclear structure physics within the shell-model framework by employing effective interactions derived from realistic potentials. She has published around 200 articles in peer-reviewed journals, including five review papers, and contributed to more than 20 plenary invited talks at international conferences. She has served on national/international committees, as a referee for peer-reviewed journals, and participated in the organization of many workshops and conferences. In particular, she is one of the promoters of the topical meeting "International Spring Seminar on Nuclear Physics," held in the Napoli area since 1987. Her dissemination activity concentrates on gender issues, especially integrating gender perspectives in science and academic/research organizations.

**Giovanni De Gregorio**

Giovanni De Gregorio is a Researcher at Università degli studi della Campania "Luigi Vanvitelli". His research focuses on developing many-body methods and related computational tools. He has published around 60 articles in peer-reviewed journals, including two review papers, and contributed to more than 20 plenary talks at national and international conferences. He was also a member of the organizing committee of different international conferences and was a referee of different peer-reviewed journals.

**Silvia Monica Lenzi**

Silvia Monica Lenzi is a Professor at the Department of Physics and Astronomy of the University of Padua. Her research covers different theoretical and experimental aspects of the structure of atomic nuclei. The main research lines regard the study of isospin symmetry along the N=Z line and the structure of neutron-rich nuclei. The experimental activity consists of gamma-ray spectroscopy performed with large gamma-ray spectrometers; this is complemented by theoretical developments in the framework of the nuclear shell model. She has published over 300 articles in international scientific journals and contributed to more than 90 invited talks at conferences and seminars. She serves as a coordinator and manager for several international networks and collaborations and is the Director of the School of Specialization in Medical Physics at the University of Padua.

# Preface

The shell model (SM) entered nuclear physics almost 70 years ago after Maria Goeppert Mayer and Hans Jensen—who shared the Nobel prize in 1963 for their work—were able to explain the mystery behind the magic numbers associated with the large stability in the ground state of some nuclides.

The SM is widely considered the basic scheme for the microscopic description of the nucleus, and, starting from its introduction; it has been successfully applied for investigating a variety of nuclear structure phenomena, which have important implications in our understanding of both astrophysics and physics beyond the standard model.

In the last two decades, thanks to high-performance computing facilities and the implementation of very efficient codes and methods, large-scale SM calculations have become a well-established approach to investigating medium- and heavy-mass nuclei whose description involves many valence nucleons in large model space. Today, it is possible to deal with huge matrices of dimension $10^{11}$ - $10^{12}$, which was unthinkable until a few years ago.

Over about the same period, the extraordinary improvements in sensitivity and efficiency of the experimental tools and the development of a large variety of radioactive beams have allowed the exploration of new regions of the nuclide chart towards the drip lines. The richness of data emerging from these experimental studies has probed the reliability and robustness of the SM in describing the new phenomena observed far from stability, such as the onset of collectivity at the historical magic numbers, the development of islands of inversion and the appearance of new magic numbers, the evidence of shape coexistence. These studies have revealed modifications in the shell structure as a function of proton and neutron numbers, leading to a paradigm shift away from the universality of the magic numbers. Large-scale SM calculations are required to investigate these phenomena and to understand the role of the different components of the nuclear force in determining the evolution of the shell structure towards the driplines.

While the SM has been used predominantly with empirical effective Hamiltonians, substantial progress has been achieved in recent years in deriving the SM Hamiltonians from realistic bare interactions, including two and three-body forces based on chiral effective field theory; this has given a further impulse towards a fully microscopic description of atomic nuclei starting from the quantum chromodynamics degrees of freedom. Within this context, valence-space Hamiltonians can be derived using many-body perturbation theory, and only very recently, nonperturbative approaches have been introduced.

This Special Issue collects 14 contributions of leading experts in the field, intending to provide, starting from the historical setting, a clear overview of the status and future developments of the nuclear shell model, including its applications in describing various nuclear structure phenomena. The book is primarily addressed to the nuclear physics community, but it may also interest other branches of physics.

In closing, we would like to take this opportunity to express our deep gratitude to all authors for their valuable contributions. We also warmly acknowledge the MDPI Book staff and the Atomic Physics Section Editorial team. Thanks to Ms. Ling Yang, the Managing Editor, for her assistance during the Volume preparation.

**Angela Gargano, Giovanni De Gregorio, and Silvia Monica Lenzi**
*Editors*

Article

# The Very Long Lifetime of $^{14}$C in the Shell Model

**Igal Talmi**

Department of Particle Physics and Astrophysics, The Weizmann Institute of Science, 234 Herzl Street, Rehovot 7610001, Israel; igal.talmi@weizmann.ac.il

**Abstract:** This is a fitting memory for our late friend and colleague Aldo Covello. For many years, he was our host in the series of Spring Seminars which he organized. In these conferences, the shell model was a central subject which was taken very seriously. This paper is written after 70 years of successful shell model calculations of nuclear energies and also various transitions. The beta decay of $^{14}$C has been an enigma. The history and present situation are described. The importance check of any theory to yield the strength of the mirror transition of $^{14}$O is pointed out.

**Keywords:** the shell model; $^{14}$C beta decay; mirror decay of $^{14}$O

**Citation:** Talmi, I. The Very Long Lifetime of $^{14}$C in the Shell Model. *Physics* **2022**, *4*, 940–947. https://doi.org/10.3390/physics4030062

Received: 17 June 2022
Accepted: 22 July 2022
Published: 24 August 2022

**Publisher's Note:** MDPI stays neutral with regard to jurisdictional claims in published maps and institutional affiliations.

**Copyright:** © 2022 by the authors. Licensee MDPI, Basel, Switzerland. This article is an open access article distributed under the terms and conditions of the Creative Commons Attribution (CC BY) license (https://creativecommons.org/licenses/by/4.0/).

In this paper, I look back at 70 years of the nuclear shell model. Actually, there is a prehistoric part, which is close to 100 years old. Atomic nuclei were discovered by Rutherford in 1911. For several years, their composition was a mystery until the neutron was discovered by Chadwick in 1932. In the same year, Heisenberg published a paper in which he showed that nuclei are composed of protons and neutrons.

In those early days before mass-spectroscopy was used to measure nuclear binding energies, physicists used various transitions and reactions to determine that certain nuclei are more stable than others. In atoms, extra stability is associated with closed shells. In the same year, 1932, Bartlett suggested a similar structure for some nuclei [1]. He suggested that, in $^{4}$He, there are closed $1s$ shells of protons and neutrons and in $^{16}$O there was also a closed $1p$ shell. Not many papers followed Bartlett's idea. Most thorough and systematic ones were written by Elsasser [2–6]. By studying experimental data, he discovered several nuclei with extra stability. It was difficult for most physicists to understand how a system with a rather large number of particles interacting by strong short-range forces may be described by an independent particle model. In addition, the nature of the magic numbers was baffling. The lowest of them, 2, 8 and 20, could somehow make sense. Higher magic numbers, discovered by Elsasser, 50, 82 and 126, could be obtained only from very strange central potentials.

In a comprehensive review article, Bethe and Bacher [7] gave a description of nuclear physics in 1936. They present the shell model, arguments against it but also a case where only it seems to explain the data. A very devastating paper against any shell model was published in the same year by no lesser person than Niels Bohr [8]. He wrote: "In the atom and in the nucleus we have indeed to do with two extreme cases of mechanical many-body problems for which a procedure of approximation resting on a combination of one-body problems, so effective for the former case, loses any validity in the latter".

During World War II, nuclear physicists in major countries were occupied with work on nuclear weapons. In 1948, Maria Mayer published a detailed study [9] in which she showed that nuclei whose proton and/or neutron numbers were found by Elsasser to be magic have indeed extra stability. Mayer's paper revived the interest in the shell model. Feenberg and Hammack [10] and Nordheim [11] published detailed papers in which they tried to reproduce the new data in models similar to that of Elsasser. The shells which proposed contained certain orbits characterized by the orbital angular momenta $l$ of the nucleons.

Maria Mayer [12] and, independently, Jensen et al. [13] introduced a novel idea. The shell structure, taken to be that of a harmonic oscillator central potential, is modified by a

strong spin–orbit interaction. The $l$ orbit is split into a lower $j = l + 1/2$ orbit and a higher $j = l - 1/2$ one. Both orbits remain in the same major shell with an important exception.

The spin–orbit interaction,

$$2a(\mathbf{s}.\mathbf{l}) = a[(\mathbf{s}+\mathbf{l}).(\mathbf{s}+\mathbf{l}) - \mathbf{l}.\mathbf{l} - \mathbf{s}.\mathbf{s}] = a[j(j+1) - l(l+1) - 3/4], \tag{1}$$

is equal to $al$ for $j = l + 1/2$ and to $-a(l+1)$ for $j = l - 1/2$. The $j = l + 1/2$ orbit with the highest value of $l$ in a shell is mostly affected. If its energy is sufficiently pushed down (by a negative value of $a$), it may join the shell below it. This effect leads to the observed magic numbers. For example, the closed shells $1s, 1p, 1d2s, 1f2p$ contain 40 protons or neutrons but only when joined by the 10 protons (or neutrons) in the $1g_{9/2}(l = 4, j = 9/2)$ orbit is the magic number 50 reached. The simple Mayer–Jensen shell model has been widely accepted and used by experimentalists and theorists. Wigner, who made seminal contributions to nuclear physics, remained skeptical. He could not understand the origin of the interaction (1).

Apart from the radial dependence, wave functions in the shell model are well defined for ground states of closed shells nuclei. They remain well defined also if a single nucleon is added or removed from such nuclei. If there are several nucleons outside closed shells (valence nucleons), they may couple in several states (with the exception of two identical $j = 1/2$ nucleons or holes). To calculate energies and wave functions, it is necessary to calculate eigenvalues and eigenstates of the sub-matrix of the Hamiltonian, which is defined by states of the shell. Maria Mayer was aware of this situation and stated coupling rules for the spins of ground states. They were $J = 0$ for even-even nuclei of which there are no exceptions and $J$ equal to $j$ of one of the orbits in the shell for odd-even (or even-odd) nuclei. There are some exceptions to the second rule.

Mayer tried to find some theoretical basis for her rules. She looked at some $j^n$ configurations of neutrons and of protons and "for simplicity", calculated their energy levels with a delta interaction [14]. The calculated ground state spins agreed with her rules! In addition, the pairing energy emerged from the calculation. She thought that the exceptions to her rule are due to the finite range of the interaction. Her student Dieter Kurath wrote a short paper [15] on this subject and so did I, a student of Pauli in Zurich [16].

In spite of the schematic nature of the zero range delta interaction, Mayer's novel approach made an important impact. In earlier calculations, it was assumed that the interaction between nucleons may be approximated by a constant over the nucleus. Short range interactions were not easy to handle in standard spectroscopy. Matrix elements of two-body interactions were expanded in terms of Slater integrals, each obtained from a term in the expansion of the integrand in terms of the particle coordinates $r_1$ and $r_2$ . If the interaction is constant where the wave functions do not vanish, only the $k = 0$ Slater integral does not vanish. If, however, the interaction is a delta function, all Slater integrals need not vanish; each is proportional to $2k + 1$.

Kurath in Ref. [15] showed that an argument of Racah against a ground state spin calculated in the shell model [17] is based on the $k = 0$ Slater integral. Racah created modern spectrometry for the earlier version of the nuclear shell model, but, when it became unpopular, he moved to atomic spectroscopy.

I had been looking for an expansion in which matrix elements of the delta interaction will have only one term. Instead of the Slater expansion of the interaction, the product of wave functions could be expanded in terms of the center-of-mass $\mathbf{R} = (\mathbf{r}_1 + \mathbf{r}_2)/2$ and relative coordinate $\mathbf{r} = \mathbf{r}_1 - \mathbf{r}_2$. A simple and finite expansion occurs only for the kinetic energy and for the harmonic oscillator potential. The results should be independent of $\mathbf{R}$ since the interaction is translationally invariant [18]. This transformation is used now everywhere.

As mentioned above, for an efficient use of the shell model, it is necessary to know the two-body interaction. In the early days, people tried to use the interaction between free nucleons with or without some theoretical modifications. Even if agreement with some experimental data was obtained, no agreement with other ones could be reached. The

standard excuse was "configuration mixing"—disagreements were blamed on the effect of shell model configurations not included. Such a procedure may lead to good agreement, but it has no meaning. An amusing example of failure of such a calculation is offered by dealing with calcium isotopes. In 1955, Ford and Levinson published a series of three papers on levels of $^{43}$Ca. They started from levels of $^{42}$Ca of which the first excited level, with $J = 2$, was known. Above it, two higher levels were measured, but their spins were unknown. Ford and Levinson took those spins to be $J = 4$ and $J = 6$, as in some other nuclei and, using their energies with those of the measured states, calculated positions of $^{43}$Ca levels—so far, so good. However, they did not get correctly the position of the first excited $J = 5/2$ state. This situation was known to several authors who left this problem. Ford and Levinson did not give up but called for help with the mixing of other configurations with the $(f_{7/2})^n$ one. As if by miracle, the calculated position of the $J = 5/2$ state came out very close to the measured value and so was the case with some other observables. In the fall of 1956, there was an international conference on theoretical physics in Seattle, WA, USA. Jensen gave a talk on the nuclear shell model and the Ford–Levinson work seemed to be one of its successes. Not much later, the spins of the 2 levels, taken by Ford and Levinson to have $J = 4$ and $J = 6$, were measured to have spins $J = 0$ and $J = 2\ldots$ They do not belong to the $(f_{7/2})^2$ configuration. They are "intruder states" from some other configuration.

Slowly, the realization that the operators which should be used in a model may be very different from the real ones dawned on nuclear physicists. This was pointed out very clearly by Keith Brueckner and his collaborators. The problem was how to determine the effective interaction for the shell model in nuclei where the shell model seems to give a reasonable description. Clearly, no theoretical derivation seems easy for such a complex system. I have been looking for a case where the shell model description will indicate a simple configuration. I chose a case which seemed simple and in summer 1954 asked a student to study it and determine the effective two-body interaction. The system we considered was the four low lying states of $^{40}$K which have spins obtained by coupling a $1d_{3/2}$ proton hole with a $1f_{7/2}$ neutron ($J = 2, 3, 4, 5$).

To check the consistency of our calculation, we calculated energies of states of another simple related shell model configuration. We chose the energy levels of the $1d_{3/2}$ proton—$1f_{7/2}$ neutron configuration ($J = 2, 3, 4, 5$) expected in $^{38}$Cl. We looked at the levels published by 1954, and the only agreement was in the spin $J = 2$ of the ground state. We were disappointed but not surprised. We used pure $jj$-coupling wave functions which may have been rather extreme. In addition, it was not clear that it is a good approximation to use matrix elements from one nucleus in another one. Only in 1956 were accurate measurements of $^{38}$Cl levels published, and they were in very good agreement with our calculated ones [19]. A few days after Ref. [19] was published, the case considered there appeared in a paper by Pandya [20]. He derived the Pandya relation expressing analytically particle–hole interaction as a linear combination of particle–particle interactions. He looked for an example and found the $^{38}$Cl—$^{40}$K case.

The work and results of Ref. [19] started a new period in shell model calculations. I gave a talk on this work in a meeting of the Israel Physical Society and Racah said that this is the beginning of nuclear spectroscopy. Matrix elements of the effective interaction, diagonal and non-diagonal, were extracted from complex nuclei [21]. It took some time, but the successful calculations were convincing. When I was advocating this method, Arthur Kerman told me that, while he was a student at Caltech (California Institute of Technology, CA, USA), Richard Feynman was trying to use this method on light nuclei (probably in the $1p$ shell). In 1962, I gave a colloquium talk in Caltech. After the talk, Feynman told me that he tried this approach until, for a certain nucleus, his prediction was that four low lying levels are almost degenerate. He thought that this is very unlikely and dropped this approach. Perhaps Feynman talked about $^{16}$N in which measured energies of the four lowest states are below 0.4 MeV.

Eigenvalues of the Hamiltonian obey the variational principle and hence may be calculated even with approximate wave functions. This method has been widely applied in calculations of energies and other observables [21]. Only one of the earlier papers and books, where this approach is discussed, is mentioned here. It referred to the specific case which is discussed in the following [22]. It is the $^{14}$C beta decay which should be an allowed Gamow–Teller beta decay with a half life of a few days. Its observed lifetime is more than 5000 years. It has been a great gift to archaeology and a great enigma to nuclear structure physics.

The $J = 0$ ground state of $^{14}$C, according to the shell model, is due to the two proton holes in the $1p$ orbit. This configuration has two possible $J = 0$ states, one with $S = 0$, $L = 0(^1S_0)$ and the other with $S = 1, L = 1(^3P_0)$. The single nucleon spin–orbit interaction Equation (1), which gives rise to the shell model, leads to a linear combination of these two states:

$$x|^1S_0 > +y|^3P_0 > . \tag{2}$$

This is an isospin $T = 1$ level. It decays by emitting an electron and neutrino to the $T = 0$ ground state of $^{14}$N which has $J = 1$. In the $p^{-2}$ configuration, there are three independent $T = 0$ states in which any state with $J = 1$ may be expressed as a linear combination of them. Thus, the $^{14}$N ground state may be expressed as

$$\alpha|^3S_1 > +\beta|^1P_1 > +\gamma|^3D_1 > . \tag{3}$$

The operator of the allowed beta decay is $\sigma = 2\ \mathbf{s}$, which has non-vanishing matrix elements only between states with the same part which depends on $\mathbf{r}$. Thus, in the case considered here, it is equal to a linear combination of $\alpha x$ and $\beta y$. Its precise value is

$$\sqrt{6}(\alpha x - \beta y/\sqrt{3}) . \tag{4}$$

Inglis noted that, if the effective forces are central and even in the presence of spin-orbit interaction, the matrix element (4) cannot vanish [23]. He argued, as shown below, that the coefficients x and y in Equation (2) have the same signs. In the case of central-and spin-orbit interactions, the $L = 2$ $D$ state has a non-vanishing non-diagonal matrix element only with the $^1P_1$ state. This is due to the single particle spin–orbit interaction. The $\alpha$ and $\beta$ coefficients in Equation (3) have opposite signs. As a result, the matrix element (4) cannot vanish. Inglis suggested that the near vanishing of the matrix element is due to mixing of higher configurations into states of the $1p$ shell.

This is the situation, in the extreme case of $jj$-coupling. The coefficients of the $(1p_{1/2})^{-2}$ state with $J = 0, T = 1$ are $x = \sqrt{1/3}$ and $y = \sqrt{2/3}$. The coefficients of the $J = 1, T = 0$ state are $\alpha = -\sqrt{1/27}, \beta = \sqrt{6/27}$ and $\gamma = \sqrt{20/27}$ as in Ref. [23].

In 1954, Jancovici, a graduate student in Princeton and I, a visitor there, noticed that the situation is changed if tensor forces are included in the shell model interactions. The $^3D_1$ state may then strongly interact with the $^3S_1$ state in addition to its spin–orbit interaction with the $^1P_1$ state. If this interaction is sufficiently strong, the signs of the $\alpha$ and $\beta$ become equal and cancellation or near cancellation may be possible [24]. That work was carried out in the period when various interactions were used taken from attempts to extract them from analysis of nucleon–nucleon reactions. Ref. [24] was no exception, and this was the case with the several publications which followed it. One conclusion was clear: tensor forces may play an important role in $^{14}$C beta decay.

Cohen and Kurath [22] carried out shell model calculations in the $1p$ shell. Following our paper [24] and its followers, they determined the effective interaction by attempting to achieve the best fit to measured energy levels. No attempt was made to obtain the longevity of $^{14}$C, but they refer to it. They write in their comments on their results on $^{14}$N: "The major point of interest here concerns the beta decay of $^{14}$C which has been difficult for the shell model . . . actually, changing the $^{14}$N ground state to one which has an over-lap of 0.998 with the state from the latter case would produce the nearly exact cancellation indicated by

experiment". They mention a possible effect of mixing with configurations from a higher shell. The experimental situation is still complex and the results of Ref. [22] support the view that the longevity of $^{14}$C may have a simple explanation in the shell model.

The wave-function [2] is an eigenstate of the sub-matrix of the Hamiltonian

$$\begin{matrix} ^3P_0 \\ ^1S_0 \end{matrix}\begin{vmatrix} E(^3P_0) & -2a\sqrt{2} \\ -2a\sqrt{2} & E(^1S) \end{vmatrix} = \begin{vmatrix} E(^3P_0)-E' & -2a\sqrt{2} \\ -2a\sqrt{2} & E(^1S)-E' \end{vmatrix} + \begin{vmatrix} E' & 0 \\ 0 & E' \end{vmatrix} \tag{5}$$

In the matrices (5), $E(^3P_0)$ and $E(^1S_0)$ are the energies of the states before diagonalization. $E''$ is the lower $J = 0$ eigenvalue of Hamiltonian (5), either of the left or right side of it. The higher eigenvalue, $E'$, lies 13.75 MeV above $E''$. To diagonalize the matrix on the left, it is sufficient to diagonalize the left matrix on the right. Diagonalization of the matrix on the left yields the two eigenvalues $E''$ and $E'$. Hence, diagonalization of the matrix on the right yields the corresponding eigenvalues 0 and $E''$–$E'$ . This difference is taken here to be –13.75 MeV.

The difference between the two eigenvalues of the spin–orbit interaction (1) of a single nucleon is

$$E_{l+1/2} - E_{l-1/2} = a(2l+1)\,. \tag{6}$$

The value of $a$ in the case considered here can be obtained from the difference of the single hole states in $^{15}$N. The ground state of $^{15}$N is a $J = 1/2^-$ state, and a $J = 3/2^-$ state is 6.324 MeV above it. Thus, the value of $a$ is 2.108 MeV, the coefficient in Equation (1) is 4.216 and the non-diagonal matrix element in Hamiltonian (5) is $-4.216\sqrt{2} = -5.962$ MeV.

The matrix to be diagonalized:

$$\begin{matrix} ^3P_0 \\ ^1S_0 \end{matrix}\begin{vmatrix} U & -4.216\sqrt{2} \\ -4.216\sqrt{2} & W \end{vmatrix} \tag{7}$$

Since one of its eigenvalues is 0, its determinant vanishes so that $UW = 8 \times 2.108^2 = 8 \times 4.442 = 35.54$. The trace of matrix (7) is equal to the trace of the diagonalized matrix. Thus, $U + W = (E'' - E') + (E' - E') = -13.75$ MeV. From these values follows $U = -10.3$ and $W = -3.45$ MeV. The derivations above may be summarized by a simple expression of $U$ and $W$. The matrix (7) has an eingenvalue 0 and hence its determinant vanishes:

$$UW - 8a^2 = 0\,. \tag{8}$$

The other eigenvalue, $N$, is equal to the trace of the matrix, $E = U + W$. Thus, Equation (8) may be rewritten as

$$U(N-U) - 8a^2 = 0 = U^2 - UN + 8a^2 = (U - N/2)^2 - N^2/4 + 8a^2\,. \tag{9}$$

From Equation (9), the explicit expression for the lower diagonal matrix element follows:

$$U = \frac{1}{2}(E - \sqrt{(E^2 - 4(8a^2))}\,. \tag{10}$$

The higher element is given by

$$W = \frac{1}{2}(E + \sqrt{(E^2 - 4(8a^2))}\,. \tag{11}$$

The sum of Equations (10) and (11) is $U + W = N$ and their product is

$$UW = (E - \sqrt{(E^2 - 4(8a^2))})(E + \sqrt{(E^2 - 4(8a^2))})/4 = (E^2 - (E^2 - 4(8a^2)))/4 = 8a^2\,. \tag{12}$$

The coefficients of the eigenstate (3) can be obtained from the matrix (5) by

$$\left|\begin{array}{c} -10.3 - 5.962 \\ \\ -5.962 - 3.45 \end{array}\right| \begin{array}{c} y \\ \\ x \end{array} = -13.75 \left| \begin{array}{c} y \\ \\ x \end{array} \right. \quad (13)$$

The relation (8) is satisfied by

$$x = (13.75 - 10.3)y/5.962 = 0.579y \text{ or } x = 5.962y/(13.75 - 3.45) = 0.579y. \quad (14)$$

The normalized coefficients are equal, to a good approximation, to $x = 0.5$ and $y = \sqrt{0.75}$. The wave function (2) with these coefficients has an overlap larger than 0.995 with the $J = 0$ state of two $p_{1/2}$ holes.

The amplitudes in the $^{14}$N ground state, the coefficients of Equation (4), may be obtained by diagonalization of the shell model sub-matrix

$$\begin{array}{c|lll} ^3S_1 & V(S) & a\sqrt{2/3} & V_T \\ ^1P_1 & a\sqrt{2/3} & V(P) & -a\sqrt{5/6} \\ ^3D_1 & V_T & -a\sqrt{5/6} & V(D) - a(3/2) \end{array} \quad (15)$$

As shown in Ref. [24], due to the tensor forces, the coefficients $\alpha$ and $\beta$ may have the same sign. If it is absent, they have different signs due to the positive non-diagonal element of the spin–orbit interaction between the $^3S_1$ and the $^1P_1$ states.

The matrix element of the Gamow–Teller beta decay is given by Equation (4) as

$$\sqrt{6}(x\alpha - y\beta/\sqrt{3})\,. \quad (16)$$

Due to the values of $x$ and $y$ obtained above, $\sqrt{0.75}/\sqrt{3} = \sqrt{0.25} = 0.5$ and this matrix element vanishes for any equal coefficients, $\alpha = \beta$. The experimental situation in $^{14}$N is more complicated than in $^{14}$C. In the following, we assume that the matrix element (11) is strongly reduced. Using plausible coefficients, it is possible to check this mechanism by looking at the mirror beta decay. The faster decay of $^{14}$O to the $^{14}$N ground state should be due to the difference in Coulomb energies. The values of $\alpha$ and $\beta$ in obtaining 0 or close to them may occur in the actual nuclei. The measured matrix element in the $^{14}$C is actually very small but not exactly zero.

As stated above, the near cancellation of the beta decay matrix element in $^{14}$C, should not occur for the mirror transition of $^{14}$O. In the case considered above, states are of two proton holes interacting also by the Coulomb interaction. In $^{14}$O, there are neutron holes and the difference [10] is slightly smaller, 6.176 MeV. Hence, $a$ = 2.059, the coefficient in Equation (1) is $2a$ = 4.118 and the non-diagonal matrix element in Hamiltonian (5), in the present case, is $-4.118\sqrt{2}$ = –5.824 MeV.

In matrix (5), E($^3P_0$) includes -2$\alpha$, and its energy should be increased by 0.1 MeV. Using Equation (10) in this case yields $U = \frac{1}{2}(-13.65 - \sqrt{13.65^2 - 4 \times 5.824^2})$ = –10.38 MeV and $W$ =-3.27 MeV. In the case considered now, $x = 3.27y/5.824 = 0.561$ and $x = 10.38y/13.65 =$ 0.561. The normalized coefficients are $x$ = 0.489 and $y$ = 0.872. To calculate the beta decay matrix element (9), it is not sufficient to take $\alpha = \beta$. A plausible choice, consistent with a large value of $\gamma$, is $\alpha = \beta = 0.3$. With this value, the square of the element (9) becomes equal to $0.09 \times 6(0.489 - 0.872/\sqrt{3})^2 = 0.000104$. The $ft$ value is obtained by dividing 5300 by this number [25]. Thus, log$ft$ = log(5300/0.000104) = 7.7, which is in the region of experimental results.

The importance of $^{14}$C dating in archaeology is demonstrated by the University of Cambridge which publishes a journal called *Radicarbon*. Most of the articles in it deal with applications of $^{14}$C, but, several years ago, an issue was devoted to the nuclear physics behind the phenomenon. The paper "The half-life of $^{14}$C—Why is it so long?" by Kutschera was published online by Cambridge University Press [26]. It contains a detailed bibliography on this subject—Ref. [24]—as well as recent large scale computations. No

results on beta decay of $^{14}$O are presented. In the Introduction, Kutschera states that, in the approach of Ref. [24], using the simple shell model, some reduction of the rate of $^{14}$C beta decay is obtained but not sufficiently. No reference is quoted. This approach may turn out to be not the correct one but not for the reason given in *Radiocarbon*.

Only three of the many papers listed in Ref. [27] are mentioned here. Fayache, Zamick, and Muther considered central, spin–orbit and tensor forces [28]. They considered, however, also mixing of nearby configurations and various values of the interactions. They found a set of values which fits the data. They quote a theoretical derivation of these values (if "the enhancement of the small component of the Dirac spinors of the nucleons is taken into account").

Negret et al. (38 authors, all listed in the References) present experimental results relevant to beta decay from which information on A = 14 nuclei may be deduced. The theoretical analysis is based on calculations in which shell model wave functions were used, but no central potential (core) is assumed, NCSM. Not all low levels appear and the big reduction of $^{14}$C beta decay is not reproduced. The important result is that the main component of the $J = 1$, $T = 0$ ground state of $^{14}$N is indeed $^3D_1$ . The authors, like some others, believe that clustering is an important ingredient that should be included. The authors of the next paper do not share this opinion.

Maris, Vary, Navratil, Ormand, Nam, and Dean use "ab initio no-core shell model" in their calculation [29]. The Hamiltonian is taken from "the chiral effective field theory including three-nucleon force terms". They find that the latter have a large effect leading to the large reduction of the beta decay rate of $^{14}$C. If this sounds simple, the order of the matrix with which they deal is 872,999,912. The number of non-vanishing, diagonal and non-diagonal 3NF (3-nucleon-forces) matrix elements is about $2 \times 10^{13}$. The shell model is supposed to give some simplicity from the complexity of the nuclear many body system. The results of this paper are far from simple.

To find if the shell model is sufficiently detailed to yield the results of beta decay of $^{14}$C and $^{14}$O, it may be necessary to understand the level structure of these nuclei and of $^{14}$N. It may be necessary to consider possible mixing between levels of the $1f2$p configuration and some $1g2$d levels. Even small admixtures may affect the values of the small decay rates. At this time, it is too early to give up the hope.

**Funding:** This research received no external funding.

**Data Availability Statement:** Not applicable.

**Conflicts of Interest:** The author declares no conflict of interest.

## References

1. Bartlett, J.N. Nuclear structure. *Nature* **1932**, *130*, 165. [CrossRef]
2. Elsasser, W.M. Sur le principe de Pauli dans les noiyaux. *J. Phys. Radium* **1933**, *4*, 549–556. [CrossRef]
3. Elsasser, W.M. Remarques sur la constitution des noyaux atomiques. I. *J. Phys. Radium* **1934**, *5*, 253–256. [CrossRef]
4. Elsasser, W.M. Sur le principe de Pauli dans les noiyaux. II. *J. Phys. Radium* **1934**, *5*, 389–397. [CrossRef]
5. Elsasser, W.M. Remarques sur la constitution des noyaux atomiques. II. *J. Phys. Radium* **1934**, *5*, 475–485. [CrossRef]
6. Elsasser, W.M. Sur le principe de Pauli dans les noiyaux. III. *J. Phys. Radium* **1934**, *5*, 635–639. [CrossRef]
7. Bethe, H.A.; Bacher, R.F. Nuclear Physics. A. Stationary states of nuclei. *Rev. Mod. Phys.* **1936**, *8*, 82–229. [CrossRef]
8. Bohr, N. Neutron capture and nuclear constitution. *Nature* **1936**, *137*, 344–348. [CrossRef]
9. Mayer, M.G. On closed shells in nuclei. *Phys. Rev.* **1948**, *74*, 235–239. [CrossRef]
10. Feenberg, E.; Hammack, K.C. Nuclear shell structure. *Phys. Rev.* **1949**, *75*, 1877–1893. [CrossRef]
11. Nordheim, L. On spins, moments, and shells in nuclei. *Phys. Rev.* **1949**, *75*, 1894–1901. [CrossRef]
12. Mayer, M.G. On closed shells in nuclei. II. *Phys. Rev.* **1949**, *75*, 1969–1970. [CrossRef]
13. Haxel, O.; Jensen, J.H.D.; Suess, H.E. Zur interpretation der ausgezeichneten Nucleonzahlen im Bau der Atomkerne. *Naturwiss* **1948**, *5*, 376. (Received on 12 February 1949). [CrossRef]
14. Mayer, M.G. Nuclear configurations in the spin-orbit coupling model II. Theoretical considerations. *Phys. Rev.* **1950**, *78*, 22–23. [CrossRef]
15. Kurath, D. Effect of finite range interactions in the *jj* coupling shell model. *Phys. Rev.* **1950**, *80*, 98–99. [CrossRef]
16. Talmi, I. On the spin of the nuclear ground state in the *jj*-coupling scheme. *Phys. Rev.* **1951**, *82*, 101–102. [CrossRef]

17. Racah, G. Nuclear coupling and shell model. *Phys. Rev.* **1950**, *78*, 622–623. [CrossRef]
18. Talmi, I. Nuclear spectroscopy with harmonic oscillator wave-functions. *Helv. Phys. Acta* **1952**, *25*, 185–234. Available online: https://www.e-periodica.ch/digbib/view?pid=hpa-001 (accessed on 3 August 2022)
19. Goldstein, S.; Talmi, I. Related *jj*-coupling configurations in $K^{40}$ and $Cl^{38}$. *Phys. Rev.* **1956**, *102*, 589–590. [CrossRef]
20. Pandya, S.P. Nucleon–hole interaction in *jj* coupling. *Phys. Rev.* **1956**, *103*, 956–957. [CrossRef]
21. Talmi, I. Effective interactions and coupling schemes in nuclei. *Rev. Mod. Phys.* **1962**, *34*, 704–722. [CrossRef]
22. Cohen, S.; Kurath, D. Effective interactions for the 1p shell. *Nucl. Phys.* **1965**, *73*, 1–24. [CrossRef]
23. Inglis, D.R. The energy levels and the structure of light nuclei. *Rev. Mod. Phys.* **1953**, *25*, 390–450. [CrossRef]
24. Jancovici, B.; Talmi, I. Tensor forces and the $\beta$ decay of $^{14}C$ and $^{14}O$. *Phys. Rev.* **1954**, *95*, 289–291. [CrossRef]
25. Winther, A.; Kofoed-Hansen, O. On the coupling constants of $\beta$-decay. Evidence from allowed transitions. *Mat. Fys. Medd. Dan. Vid. Selsk.* **1953**, *27/14*, 1–22. Available online: http://gymarkiv.sdu.dk/MFM/kdvs/mfm (accessed on 3 August 2022).
26. Kutschera, W. The half-life of $^{14}C$—Why is it so long? *Radiocarbon* **2019**, *61*, 1135–1142. [CrossRef]
27. Fayache, M.S.; Zamick, L.; Müther, H. Vanishing Gamow-Teller rate for $A = 14$ and the nucleon–nucleon interaction in the medium. *Phys. Rev. C* **1999**, *60*, 067305. [CrossRef]
28. Negret, A.; Adachi, T.; Barrett, B.R.; Baumer, C.; van den Berg, A.M.; Berg, G.P.A.; Von Brentano, P.; Frekers, D.; De Frenne, D.; Fujita, H.; et al. Gamow-Teller strengths in the $A = 14$ multiplet: A challenge to the shell model. *Phys. Rev. Lett.* **2006**, *97*, 062502. [CrossRef]
29. Maris, P.; Vary, J.P.; Navratil, P.; Ormand, W.E.; Nam, H.; Dean, D.J. Origin of the anomalous long lifetime of $^{14}C$. *Phys. Rev. Lett.* **2011**, *106*, 202502. [CrossRef]

MDPI

*Opinion*

# To Shell Model, or Not to Shell Model, That Is the Question

**Andrew E. Stuchbery** [1,*] **and John L. Wood** [2]

[1] Department of Nuclear Physics and Accelerator Applications, Research School of Physics, The Australian National University, Canberra, ACT 2600, Australia
[2] School of Physics, Georgia Institute of Technology, Atlanta, GA 30332, USA; john.wood@physics.gatech.edu
* Correspondence: andrew.stuchbery@anu.edu.au

**Abstract:** The present review takes steps from the domain of the shell model into open shell nuclei. The question posed in the title is to dramatize how far shell model approaches, i.e., many nucleons occupying independent-particle configurations and interacting through two-body forces (a configuration interaction problem) can provide a description of nuclei as one explores the structure observed where neither proton nor neutron numbers match closed shells. Features of doubly closed and singly closed shell nuclei and adjacent nuclei are sketched, together with the roles played by seniority, shape coexistence, triaxial shapes and particle–core coupling in organizing data. An illuminating step is taken here to provide a detailed study the reduced transition rates, $B(E2; 2_1^+ \to 0_1^+)$, in the singly closed shell nuclei with doubly closed shell plus or minus a pair of identical nucleons, and the confrontation between such data and state-of-the-art shell model calculations: this amounts to a review of the effective charge problem. The results raise many questions and point to the need for much further work. Some guidance on criteria for sharpening the division between the domain of the shell model and that of deformation-based descriptions of nuclei are provided. The paper is closed with a sketch of a promising direction in terms of the algebraic structure embodied in the symplectic shell model.

**Keywords:** nuclear structure; shell model; seniority; shape coexistence; effective charge; emergent structure

**Citation:** Stuchbery, A.E.; Wood, J.L. To Shell Model, or Not to Shell Model, That Is the Question. *Physics* **2022**, *4*, 697–773. https://doi.org/10.3390/physics4030048

Received: 30 November 2021
Accepted: 8 March 2022
Published: 29 June 2022

**Publisher's Note:** MDPI stays neutral with regard to jurisdictional claims in published maps and institutional affiliations.

**Copyright:** © 2022 by the authors. Licensee MDPI, Basel, Switzerland. This article is an open access article distributed under the terms and conditions of the Creative Commons Attribution (CC BY) license (https://creativecommons.org/licenses/by/4.0/).

## 1. Introduction

The shell model has served as the most fundamental view one possesses when looking at the structure of nuclei. With its inception, at the hands of Maria Goeppert-Mayer [1] and Hans Jensen and colleagues [2] in 1949, at "three-score years and ten", it is not going to die. It is based on the premise of independent-particle motion in a spherical mean field with strong spin–orbit coupling. The quantum mechanical solution, at the level of independent-particle motion in a harmonic-oscillator potential, can be obtained using methods that all senior-year undergraduate students should be able to handle. It provides a far-reaching language for talking about nuclear structure. With the "gift" of the harmonic oscillator potential to the mathematics of quantum physics, the symmetries that emerge are without equal in the quantum domain. Thus, why question "shell model" in its verbal (i.e., operative) form?

The problem is correlations. Correlations are the antithesis of independent-particle motion in quantum many-body systems. The problem in nuclei is: Just how deeply do correlations influence what we are studying? A shell modeler must start by assuming a correlation-free basis: a complete set of states, which are many copies of single-particle states each labelled by a principal quantum number ($N$), an angular momentum quantum number ($l$), a directional component of angular momentum ($m_l$), and spin plus direction-of-spin quantum numbers ($s$, $m_s$). (Spin–orbit coupling favors a $j$-coupled basis, $|N, j, l, m_j\rangle$, where $j$ and $m_j$ are the total angular momentum and its projection.) However, pairing correlations immediately dominate singly closed-shell nuclei; and most nuclei are deformed

in their ground states while many that have spherical ground states exhibit low-energy deformed states. A simple extension of the shell model philosophy to a deformed mean field, the Nilsson model, augmented with adiabatic rotational degrees of freedom, provides an enormously powerful organizing principle for handling large amounts of data, in the guise of the unified or Bohr–Mottelson model. However, a very large number of nuclei do not separate into this simple adiabatic factorization. Such nuclei are often called "transitional nuclei". Herein lies the biggest challenge that remains in order to achieve a unified view of nuclear structure. Transitional nuclei are "sandwiched" between the shell model [3] and the unified model [4], and correlations are dominant. How do we develop theories applicable to such nuclei? To shell model or not to shell model?

The use of the term "to shell model" here is in reference to the time-honoured theoretical approach to nuclear structure which uses a basis of spherical independent-particle states, truncated at a small number of shell model energy shells, and a residual two-body interaction. The shell model is therefore a configuration interaction problem. The question then is which correlations are important, and how can one ensure that the relevant correlations emerge in the calculations.

The shell model approach is straightforward for handling all nuclei: start by introducing two-body interactions. Indeed, at the level of pairing interactions, this leads to the quasispin and seniority concepts. Quasispin is a formulation that manifestly illustrates what is meant by correlations in a quantum mechanical many-body system. With a simple approximation (by use of quasispin coherent states) this leads to the Bardeen–Cooper–Schrieffer (BCS) theory of superconductivity (see Section 4.5.3 in [5]). In finite many-body systems, as applied to nuclei, the language only needs some simple constraints to accommodate shell structure. Seniority, and its implied quasispin structure, dominates excitation patterns in singly closed shell nuclei. However, seniority breaks down immediately, at low spin, when both protons and neutrons are active. This is again due to correlations, but these correlations are not yet well understood: this is the point where nuclear deformation emerges. This nexus is the focus of the present review.

The shell model provides the most fundamental language one possesses for discussion of nuclear structure. This conceptual basis is often called the "shell model". Here, as defined, the term "shell model" is adopted in its more restricted usage as a computational model, where a Hamiltonian defined by residual interactions is diagonalized in a spherical independent-particle basis. Our view is that, with sufficient computing power, a suitable basis, and appropriate interactions, all structural details of nuclei would likely emerge. The issue, apart from the magnitude and complexity of the problem, is whether the structures in the output would be evident and intelligible. Here, the task of discussing the emergent structures in nuclei and the use of algebraic models to understand them is adopted in the context of the nuclear shell model. Therefore, the experimental data are broadly reviewed and the cases where simple models based on phenomenology and algebraic models give insights that would not be evident in a complex large-scale shell model approach are highlighted.

## 2. Nuclei with Closed Shells: An Experimental Perspective

Nuclei with closed shells, both singly and doubly closed, have been the base upon which the shell model has been built. However, such nuclei are neither manifestations of nor a sound basis for the shell model in its extreme independent-particle form. Such nuclei (i.e., closed shell) can usefully be classified into three types: doubly closed shell nuclei with equal numbers of protons ($Z$) and neutrons ($N$), i.e., $N = Z$; doubly closed shell nuclei with $N > Z$; and singly closed shell nuclei.

The distinction of doubly closed shell nuclei with $N = Z$ is that they exhibit shape coexistence at low energy, even at the level of the first excited states in $^{16}$O and $^{40}$Ca, as shown in Figure 1. In doubly even nuclei with $N > Z$, shown in Figure 2, shape coexistence has not yet been observed. The simple explanation is that, for $N = Z$, spatial overlap of the

proton and neutron configurations is maximal, and it is proton–neutron correlations that are deformation producing.

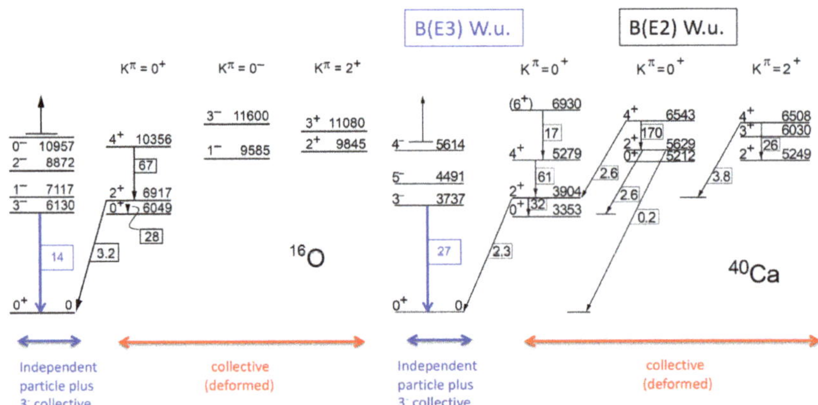


**Figure 1.** Excited states in the $N = Z$ doubly closed shell nuclei $^{16}$O and $^{40}$Ca. Collectivity associated with the $2_1^+$ and $3_1^-$ states is shown. Collectivity involving deformation is supported by large electric-quadrupole transition rates, as indicated by the $B(E2)$ values in Weisskopf units (W.u.). Inferred $K$ quantum numbers for collective bands are indicated. The horizontal bars with upward pointing arrows indicate excitation energies above which states are omitted. Adapted from [6].

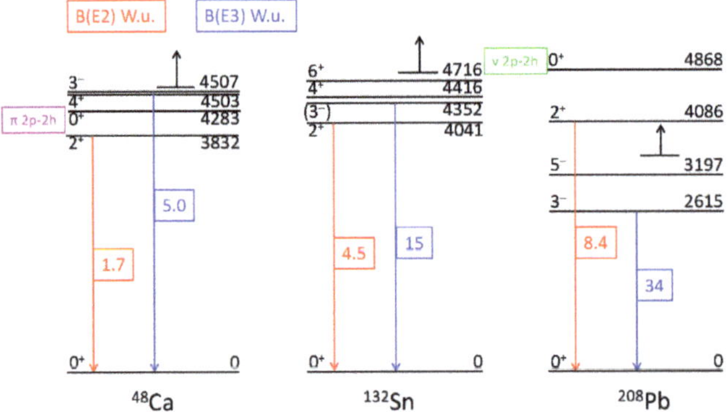


**Figure 2.** Excited states in the $N > Z$ doubly closed shell nuclei. Collectivity associated with the $2_1^+$ and $3_1^-$ states is indicated by $B(E\lambda)$ values. The lowest known pair excitations are labelled. The horizontal bars with upward pointing arrows indicate excitation energies above which states are omitted. Electromagnetic decay strengths for $^{132}$Sn are calculated from data appearing in [7]. Adapted from [8].

The distinction of singly closed shell nuclei is that they are dominated by the emergence of pairing correlations. Pairing correlations are concisely formulated using the concept of the seniority quantum number, $v$, i.e., the number of unpaired nucleons. This was first recognized by Maria Goeppert-Mayer [9,10]. The quantum mechanics of pairing correlations is concisely, even elegantly, described using quasispin, as introduced by Arthur Kerman [11]. The basic features of quasispin, as applied to a series of $(j = 7/2)^n$ configurations, where $n$ denotes the occupation of the orbit, are shown in Figure 3; a view which complements that in Figure 3 is shown for a series of $(j = 9/2)^n$ configurations in Figure 4. The quasispin algebra is developed in detail in Chapter 6 of [6]. That Chapter includes a thorough treatment of the origins of the key ideas from Racah's seniority [12–14] through

Flowers' handling of $j-j$ coupling [15], Helmers' unitary symplectic invariants [16], Lawson and Macfarlane's identification of the rank-1/2 quasispin $\mathrm{su}(2)$ tensorial character of one-body annihilation and creation operators [17], to Kerman's simple formulation [11]. Furthermore, it can be noted that there is a profound duality structure residing in these algebras [18], which shows how algebraic structure provides insight into the complexity of many-body quantum systems. A pedagogical treatment of the quasispin algebra is presented in Chapter 4 in [5]. That Chapter illustrates how P.W. Anderson's idea [19] provided the first conceptual recognition of quasispin as the essential algebraic structure underlying many-fermion systems with Cooper pairs [20].

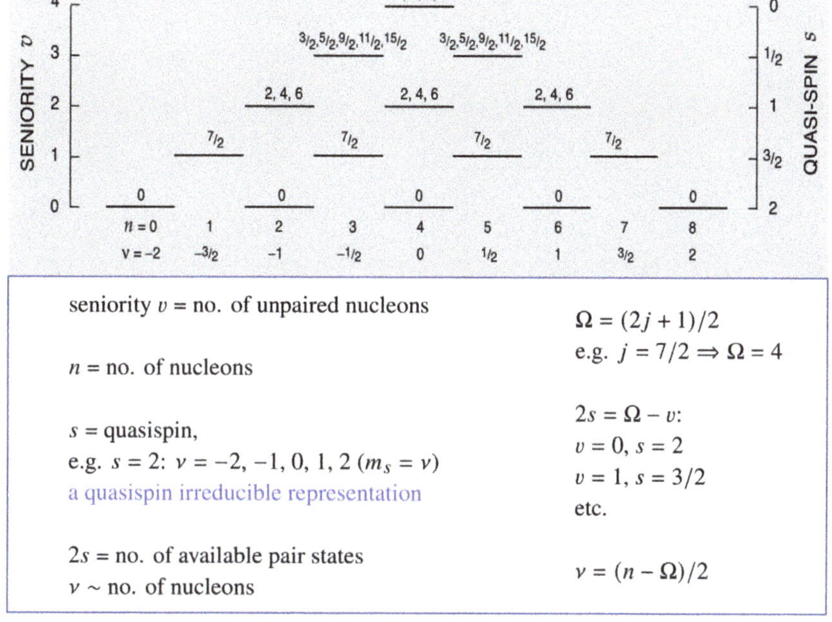


**Figure 3.** A schematic view of basic features possessed by a seniority-dominated $j = 7/2$ shell with a many-proton or many-neutron structure. The excitation patterns and associated spins are shown relative to the seniority zero, $v = 0$ states across the filling of the shell, where the filling is designated by the particle number, $n$. The quasispin quantum numbers, $s$ and $\nu$ are $\mathrm{su}(2)$ quantum numbers and their relationship to shell model quantum numbers is shown in the box. Adapted from [6].

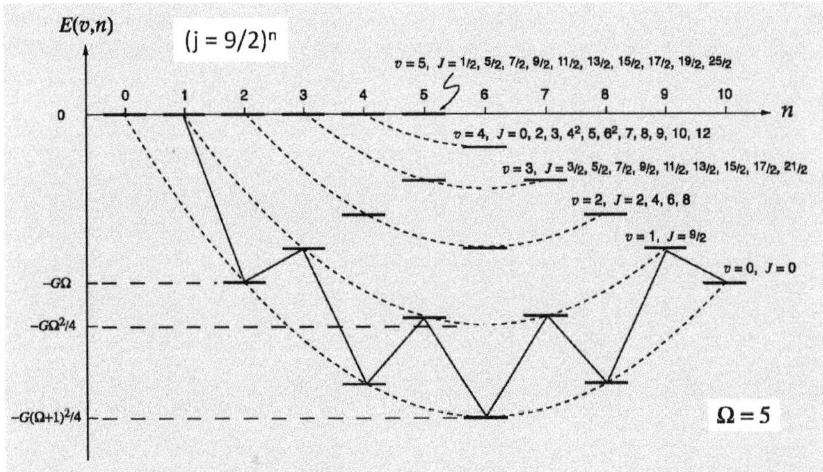


**Figure 4.** A schematic view of basic features of a seniority-dominated $j = 9/2$ shell. The excitation patterns and associated spins $J$ are shown relative to the uncorrelated, $v$ = max. states across the filling of the shell, where the filling is designated by the particle number, $n$. For other details, see Figure 3. Taken from [6].

Experimentally, the seniority coupling scheme is realized essentially exactly when the low-energy structure of singly closed shell nuclei is dominated by a high-$j$ orbital. This is shown in Figure 5 for $j = 11/2$ neutron subshell filling in the Sn isotopes and in Figure 6 for $j = 11/2$ proton subshell filling in the $N = 82$ isotones. The patterns are almost indistinguishable. The domination of seniority extends into patterns of electric quadrupole, $E2$ transition probabilities: this is shown in Figure 7 for $j = 9/2$ configurations in even-Cd and even-Pd nuclei with $N = 50$ and $N = 82$. The pattern of $E2$ matrix elements in nuclei dominated by seniority coupling shows a smoothly changing character which is well described by the following relationship for the reduced transition strength [6]:

$$B(E2; s\nu J_i \to s\nu J_f) \propto \langle s\nu 10|s\nu\rangle^2 = \frac{\nu^2}{s(s+1)} = \frac{(n-\Omega)^2}{4s(s+1)}, \quad (1)$$

where $J_i$ and $J_f$ are spins of initial and final states, $s$, $\nu$ are quasispin quantum numbers, details of which appear in Figure 3; $\langle s\nu 10|s\nu\rangle$ is an $\mathrm{su}(2)$ Clebsch–Gordan coefficient and $\Omega = (2j+1)/2$, e.g., $\Omega = 6$ for $j = 11/2$. This Clebsch–Gordan coefficient emerges from the quasispin $\mathrm{su}(2)$ algebra when applying the Wigner–Eckart theorem to the $E2$ operator: this operator is a rank-1 quasispin tensor. Details are beyond the present discussion and are given in [6]. (Note: $\nu$ (designated by the Greek letter nu) is distinct from the seniority quantum number, $v$ (designated by the Latin letter vee).) This relationship is illustrated in Figures 8 and 9 for the $j = 11/2$ configurations in the even-mass Sn isotopes and $N = 82$ isotones, respectively. Indeed, these patterns are one of the best signatures of structure unique to singly closed shell nuclei. However, the clarity and interpretation of these structures are dictated by quantum mechanics that is beyond that of the independent-particle shell model in that correlations in the form of Cooper pairs have emerged. Pairing Hamiltonians can be derived as a simplification of the nucleon–nucleon residual interaction; however, the focus here is on the empirical simplicity of the seniority structures that persist toward mid-shell where the number of valence nucleons is large, in contrast with the connection between pairing correlations and the two-body residual interactions in a large-basis shell model calculation, which is not obvious. Stated in rhetorical terms: Could one ascertain the algebraic structure of Cooper pairs, in the guise of quasispin, and manifestly controlling structure in all singly closed shell nuclei, based on a shell model computational

program? Once the quasispin structure is recognized, its implications for the residual interactions required in the shell model can be explored so that the structure emerges from the calculations.

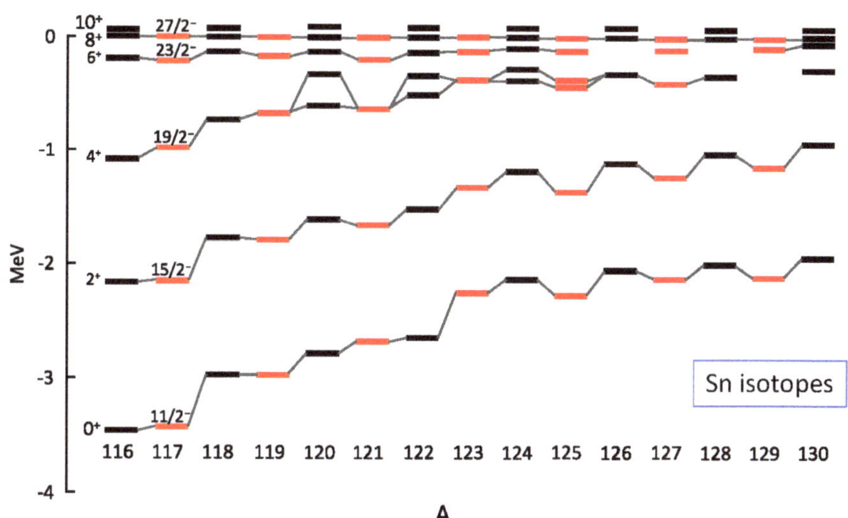


**Figure 5.** The seniority-dominated spectra versus the atomic mass number, $A$, in the neutron-rich tin isotopes, shown relative to the highest spin state in each multiplet (note the $J = 27/2$ state in the odd-mass isotopes is set at the same level as the $J = 8$ state in the even mass isotopes). These structures are dominated by neutrons filling the $1h_{11/2}$ orbital. Note: multiple $J = 4$ states are seen in $^{120,122,124}$Sn and multiple $J = 19/2$ states are seen in $^{125}$Sn. Reproduced from [8].

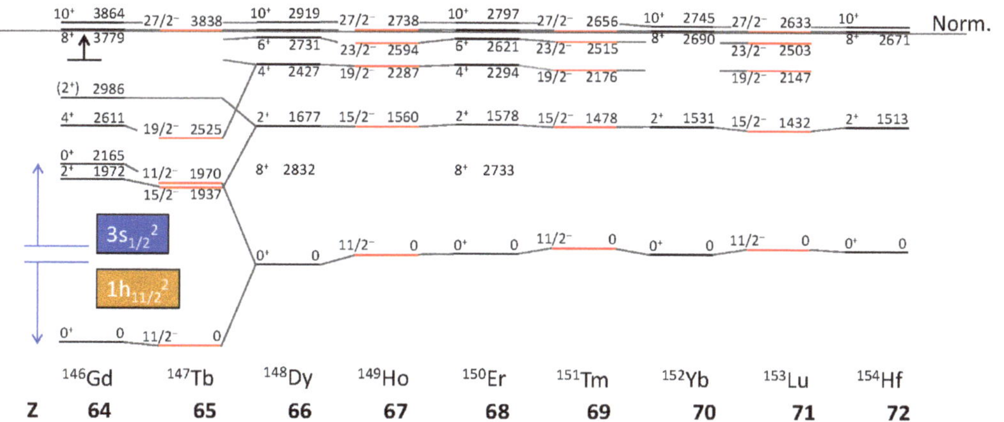


**Figure 6.** The seniority-dominated spectrum in the proton-rich $N = 82$ isotones, shown relative to the highest spin state in each multiplet (note the $J = 27/2$ state in the odd-mass isotones is set at the same level as the $J = 8$ state in the even-mass isotones). These structures are dominated by protons filling the $1h_{11/2}$ orbital. The structure of $^{146}$Gd and $^{147}$Tb involves two-state mixing, as depicted schematically. Reproduced from [8].

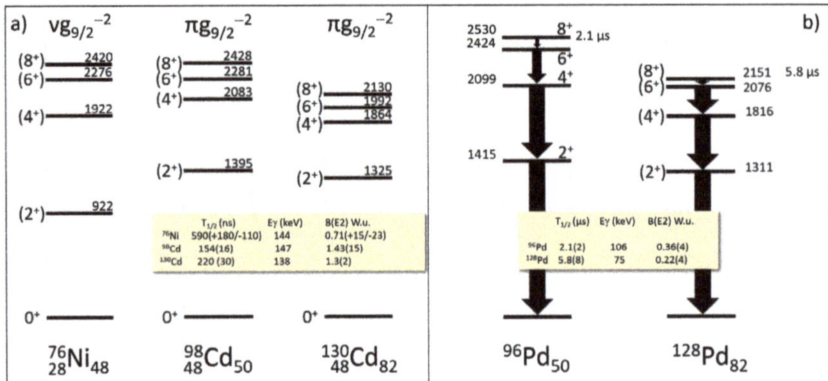


| | $T_{1/2}$ (ns) | Eγ (keV) | B(E2) W.u. |
|---|---|---|---|
| $^{76}$Ni | 590(+180/-110) | 144 | 0.71(+15/-23) |
| $^{98}$Cd | 154(16) | 147 | 1.43(15) |
| $^{130}$Cd | 220 (30) | 138 | 1.3(2) |

| | $T_{1/2}$ (μs) | Eγ (keV) | B(E2) W.u. |
|---|---|---|---|
| $^{96}$Pd | 2.1(2) | 106 | 0.36(4) |
| $^{128}$Pd | 5.8(8) | 75 | 0.22(4) |

**Figure 7.** (**a**) Seniority isomers involving $j = 9/2$ structures. The inset shows the half lives of the states with spin 8, the corresponding $8^+ \rightarrow 6^+$ transition energies, and the deduced $B(E2)$ values for these transitions. The constancy of the $B(E2)$ values, independent of mass, is remarkable and shows the simple nature of seniority structures. The figure is adapted from one appearing in [21]. Data are from the Evaluated Nuclear Structure Data File (ENSDF) [22]. The $6^+$-state energy in $^{130}$Cd, which is uncertain in ENSDF, is from [23]. (**b**) Seniority isomers involving the proton $(1g_{9/2})^{-4}$ configurations in the palladium isotopes at the $N = 50$ and $N = 82$ shell closures. The inset shows the deduced $B(E2)$ values. The $^{96}$Pd scheme is adapted from one appearing in [24] and the $^{128}$Pd scheme is from [25]. The tabulated half lives and $B(E2)$ values are taken from ENSDF. There are more recent published values [26,27], but the conclusions do not change.

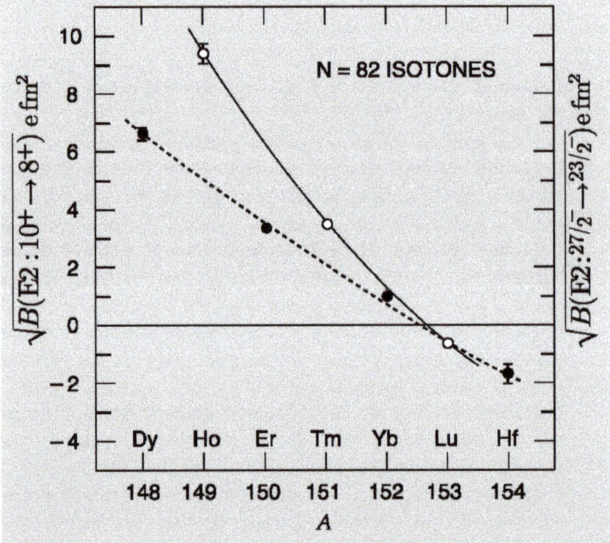


**Figure 8.** Illustration of Equation (1), expressed in square-root form, for the proton $1h_{11/2}$ configurations in the $N = 82$ isotones. The $B(E2)$ data shown are for the $10^+ \rightarrow 8^+$ transitions in the even-mass nuclei and for the $27/2^- \rightarrow 23/2^-$ transitions in the odd-mass nuclei, cf. Figure 6. The sign of the square root is allowed to change to match the matrix element changing from positive to negative as depicted. If the proton number is counted with reference to $^{146}$Gd as $n = 0$: with $\Omega = 6$, according to Equation (1), the $B(E2)$ value should vanish at $^{152}$Yb. Note that this is an effect emerging from the Wigner–Eckart theorem for su(2), applied to reduction of the $E2$ matrix elements with respect to their quasispin tensor structure. Redrawn from [28].

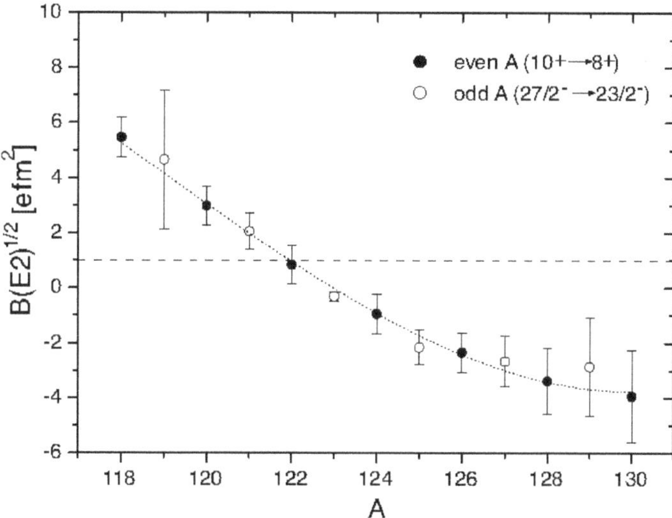


**Figure 9.** A pattern of $B(E2)$ values, similar to that shown in Figure 8, for the even-mass and odd-mass Sn isotopes. These data suggest that the half-filled shell, where the $B(E2)$ value goes to zero, is at $A \sim 122$, i.e., that the $1h_{11/2}$ orbital is not at the highest energy within the $50 < N < 82$ shell: this is consistent with $^{129}$Sn (and likely $^{131}$Sn) exhibiting a ground-state spin–parity of $3/2^+$. Note: there is a scale factor of 0.514 applied between the even and odd-mass values, which accommodates the $v = 2$ and $v = 3$ seniorities involved via the Clebsch–Gordan coefficient in Equation (1). Reprinted with permission from [29]. Copyright (2008) by the American Physical Society.

In the remainder of this Section, some observations are made with respect to the mathematical structure on which quasispin is based, in order to place this shell model view into perspective.

The arrival at the concept of quasispin as a degree of freedom in nuclei requires the recognition of mathematical structures that are not obvious. A brief sketch of the essential details is given here in words. Full details are given by Rowe and Wood [6] and, at an introductory level, by Heyde and Wood [5]. Specifically, the quasispin algebra is recognized by expressing the Hamiltonian and the interaction using second quantization. The mathematics emerge by taking bilinear combinations of the elements (one-body fermionic creation and annihilation operators) of a Jordan algebra (anticommutator brackets of the creation and annihilation operators). These bilinear combinations obey a Lie algebra (commutator brackets). This is impossible to see until one works out the Lie bracket values of the bilinear combinations, which is done by expanding them using anticommutator bracket relations so as to express everything in terms of Jordan algebra elements in "normal order"; see Equation (4.93) in Ref. [5]. Normal order means annihilation operators all to the right and creation operators all to the left. Furthermore, the Lie bracket algebra for a Jordan algebra element (single creation or annihilation operator) with quasispin algebra elements (bilinear combinations of creation and annihilation operators) reveals that the creation and annihilation operators are rank-1/2 quasispin tensors. This is also impossible to see until one works out the Lie bracket values. Indeed, rank-1/2 tensors are unknown in spin-angular momentum theory; see p. 423 in Ref. [6] for additional details.

Spectroscopy of low-spin and medium-spin states is beginning to provide a comprehensive (near-complete) view of excited states in doubly even nuclei at and near closed shells. Consequently, seniority coupling has been shown to apply in nuclei where the structure is dominated by two medium-spin $j$ shells. This is illustrated in Figures 10 and 11 for the $N = 82$ isotones with $Z < 64$. The $v = 2$ structures in $^{134}$Te, $^{136}$Xe, $^{138}$Ba, and $^{140}$Ce are labelled in Figure 10: these include the $1g_{7/2}$ structures, with $J = 2, 4,$ and $6$, and

the $1g_{7/2}$-$2d_{5/2}$ structures, with $J = 1, 2, 3, 4, 5$, and 6. In $^{136}$Xe only, as expected, $v = 4$ structures are observed with the allowed spins, $J = 2, 4, 5$, and 8, cf. Figure 3. The comprehensive view of $^{136}$Xe is the result of an $(n, n'\gamma)$ study [30]. Note that this seniority-based organization of data is essentially complete; for example, there is no excited $7/2^+$ state observed, as might be expected from a $1g_{7/2} \otimes 2_1^+$ coupling—such a coupling is forbidden by the Pauli exclusion principle if the $2_1^+$ states are seniority-dominated structures. The $B(E2; 2_1^+ \rightarrow 0_1^+) = B_{20}$ values and the magnetic moments, $\mu(2_1^+)$, are shown for reference and discussed further in Figure 12 as the $g$ factors, where $g(2_1^+) = \mu(2_1^+)/2$.

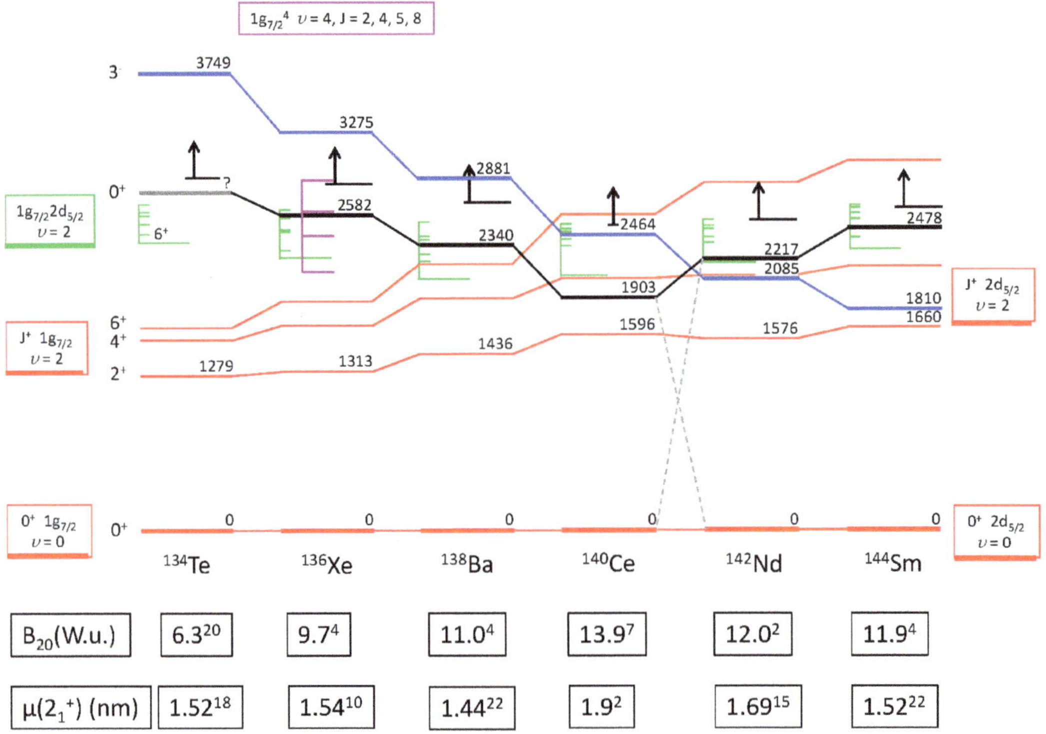


**Figure 10.** A view of the systematics of the even-mass $N = 82$ isotones with $50 < Z < 64$. The low-energy structure of these isotones is dominated by occupancy of the $\pi 1g_{7/2}$ and $\pi 2d_{5/2}$ shell model configurations: the Fermi surface progressing from the $\pi 1g_{7/2}$ to the $\pi 2d_{5/2}$ orbit is schematically indicated by dashed lines between $^{140}$Ce and $^{142}$Nd. The seniority structures are identified. The $3^-$ states are shown for reference. Horizontal bars with vertical arrows indicate excitations above which states are omitted from the figure.

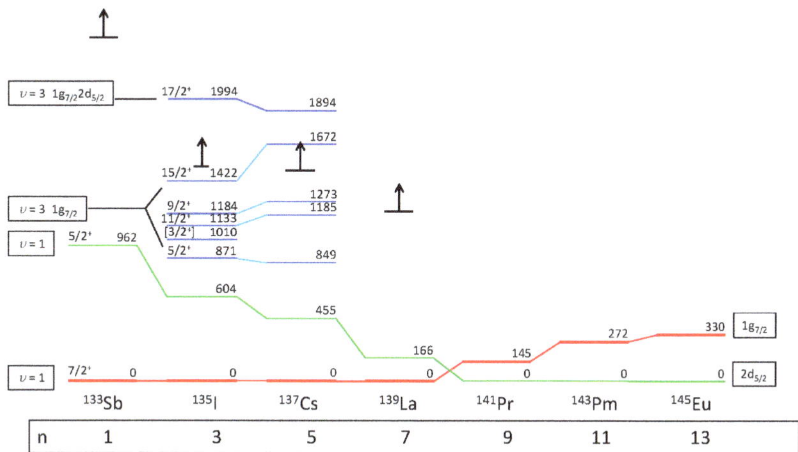


**Figure 11.** A view of the systematics of the odd-mass $N = 82$ isotones with $50 < Z < 64$. The low-energy structure of these isotones, as noted in Figure 10, is dominated by occupancy of the $1g_{7/2}$ and $2d_{5/2}$ shell model configurations: here, the completion of the filling of the $1g_{7/2}$ orbital at $Z = 58$ ($^{140}$Ce) is manifest in the change in ground-state spins between $^{139}$La and $^{141}$Pr. In $^{135}$I and $^{137}$Cs only, as expected, $v = 3$ structures are observed with the allowed spins, $J = 3/2, 5/2, 9/2, 11/2$, and $15/2$, cf. Figure 3. Note: the spin of 1010 keV state in $^{135}$I is not known but is consistent with $3/2^+$. A state with spin–parity $3/2^+$ is predicted at about 1 MeV excitation energy in $^{137}$Cs. Horizontal bars with vertical arrows indicate excitations above which states are omitted from the figure. Additional data for $^{141}$Pr, $^{143}$Pm, and $^{145}$Eu are not shown because they are not part of the present focus. Taken from [8].

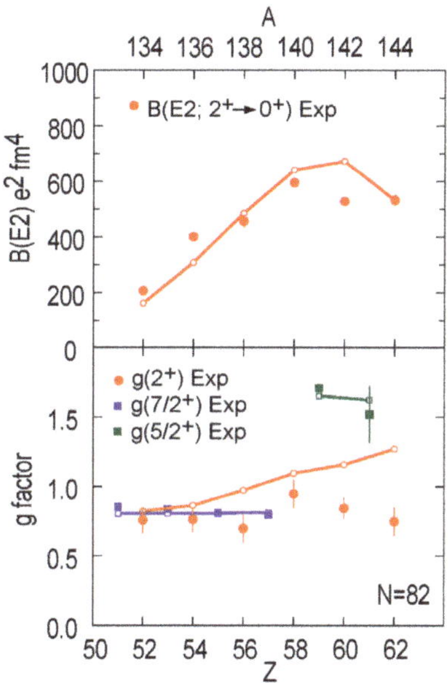


**Figure 12.** Shell model calculations of $B(E2)$ and $g$ factors in the $N = 82$ isotones with $50 < Z < 64$. Reproduced from [31], with the permission of AIP Publishing.

The seniority structure of the $N = 82$ isotones and its breakdown is an issue for future detailed study. However, shell model calculations affirm the dominant seniority structures. The case of $^{136}$Xe has been studied comprehensively [30,32]. Table 1 shows experimental $B(E2)$ values between low-excitation states in $^{136}$Xe in comparison to the $(1g_{7/2}2d_{5/2})$ seniority model, as well as several shell model calculations that include all orbits in the $50 \leq Z \leq 82$ major shell but use alternative interactions. The $B(E2)$ data indeed demonstrate the pattern predicted by the seniority scheme. It should be noted that $^{136}$Xe represents the mid-shell for the $\pi 1g_{7/2}$ orbit, for which several $E2$ transitions are forbidden. In such cases, the observed transition strengths result from small components of the wavefunction, which can lead to considerable variations in the shell model predictions, despite the calculations agreeing on the dominant structure of the states. It was noted in [32] that the large-basis shell model calculations support the dominant configurations assigned in the $(1g_{7/2}2d_{5/2})$ seniority model up to the $4_2^+$ state at 2.1 MeV excitation, although there is considerable configuration mixing. The $(1g_{7/2}2d_{5/2})$ model accounts for all states up to about 2.8 MeV, with the exception of the $0_2^+$ state (more on the $0_2^+$ state below in this Section). However, above the 2.1-MeV $4_2^+$ state, where the level of density increases, the correspondence between the two-level and full basis is less clear.

The $0_2^+$ states are consistent with a multi-pair structure distributed over the $1g_{7/2}$ and $2d_{5/2}$ orbitals. For example, the `jj55` model with `sn100` interactions [33] has dominant configurations of $\pi(2d_{5/2})^2$ (76%) [$^{134}$Te], $\pi(1g_{7/2})^2(2d_{5/2})^2$ (45%) [$^{136}$Xe], and $\pi(1g_{7/2})^6$ (51%) [$^{138}$Ba], for the $0_2^+$ states.

**Table 1.** Electric quadrupole transition rates in $^{136}$Xe. The seniority model in the $(1g_{7/2}2d_{5/2})$ space is described in [30] and in the text. The shell model calculations from [30,32,34] use alternative interactions in the model space $1g_{7/2}, 2d_{5/2}, 3s_{1/2}, 2d_{3/2}, 1h_{11/2}$, which covers the $50 \leq Z \leq 82$ major shell.

| **Transition** | $B(E2)$ ($e^2$fm$^4$) | | | | |
|---|---|---|---|---|---|
| | **Exp. [30,35]** | **Seniority** | **Ref. [34]** | **N82K [30]** | **jj55 [32]** |
| $2_1^+ \to 0_1^+$ | 415(12) | 415 $^{(a)}$ | 357 | 400 | 398 |
| $4_1^+ \to 2_1^+$ | 53.2(7) | 0 | 63.6 | 86 | 48 |
| $6_1^+ \to 4_1^+$ | 0.55(2) | 0 | 0.088 | 0.12 | 4.8 |
| $2_2^+ \to 0_1^+$ | 23(3) | 0 | – | 12 | 0.7 $^{(b)}$ |
| $2_3^+ \to 0_1^+$ | 38(3) | 22.2 | – | 12 | 48 $^{(b)}$ |
| $2_2^+ \to 2_1^+$ | 299(71) | 419 | – | 103 | 308 $^{(b)}$ |
| $2_3^+ \to 2_1^+$ | $21^{+58}_{-21}$ | 0.63 | – | 117 | 8 $^{(b)}$ |

$^{(a)}$ The seniority model uses proton effective charge $e_p = 1.81$, set to reproduce the experimental $B(E2; 2_1^+ \to 0_1^+)$.
$^{(b)}$ The calculated $2_2^+$ state is identified with the experimental $2_3^+$ state and vice versa.

Figure 12 shows the experimental $g$ factors of the $2^+$ states and the $B(E2; 2_1^+ \to 0_1^+)$ values of the even-even $N = 82$ isotones with $50 < Z < 64$, and compares them with large-basis shell model calculations. In addition, the ground-state $g$ factors of the interleaving odd-$A$ isotones are shown, which indicate that the Fermi surface moves from the $1g_{7/2}$ orbit into the $2d_{5/2}$ orbit at $Z = 59$. The $B(E2)$ trend is quite well described, but the $g(2_1^+)$ trend is not well described, particularly when the Fermi surface moves into the $2d_{5/2}$ orbit. In contrast, the odd-$A$ isotopes are well described. Focusing on the range $51 \leq Z \leq 57$, the $g$ factor data in Figure 12, for both odd and even-$A$ isotones, are near constant and thus consistent with a simple $\pi 1g^n_{7/2}$ structure in both the ground states (odd-$Z$) and $2_1^+$ states (even-$Z$). The lowered experimental $g(2_1^+)$ values for $^{140}$Ce, $^{142}$Nd and $^{144}$Sm have been attributed to increasing contributions from $\nu(1h^{-1}_{11/2}2f_{7/2})$ excitations [36]. Nevertheless, the basic seniority structure appears to persist in these nuclei.

The complete pattern of excitations in odd-mass, singly closed shell nuclei is somewhat more complex than for even-mass singly closed shell nuclei. This is shown for $j = 11/2$ in

the tin isotopes in Figure 13. Note that the states expected for seniority $v = 3$ range over 14 spin values for $j = 11/2$, viz. $2J = 3, 5, 7, 9, 9, 11, 13, 15, 15, 17, 19, 21, 23$, and 27 (see, e.g., [37]). The experimental view is incomplete, but there is sufficient detail to conclude that the seniority scheme provides a reliable basis for understanding the low-energy excitations in these isotopes. This perspective is supported by a more global view of odd-mass nuclei shown in Figure 14, wherein patterns for seniority-three multiplets in selected nuclides and selected spin couplings are visible for $j = 7/2, 9/2$, and $11/2$. This global behavior appears not to have been recognized. We conjecture that there may be a geometric interpretation of this pattern, similar to the geometrical interpretation of two-body interactions for a pair of identical nucleons in a moderate to high $j$ orbit, as introduced by Schiffer and True [38]. An angle between the two spins can be defined, which gives a measure of the overlap of the two orbits for different resultant spins; see discussions in Refs. [3,8].

One can conclude that seniority likely provides a complete description of the lowest-energy excited states in singly closed shell nuclei—with one proviso: singly closed shell nuclei exhibit low-energy deformed structures that "coexist" with the low-excitation seniority-dominated structures.

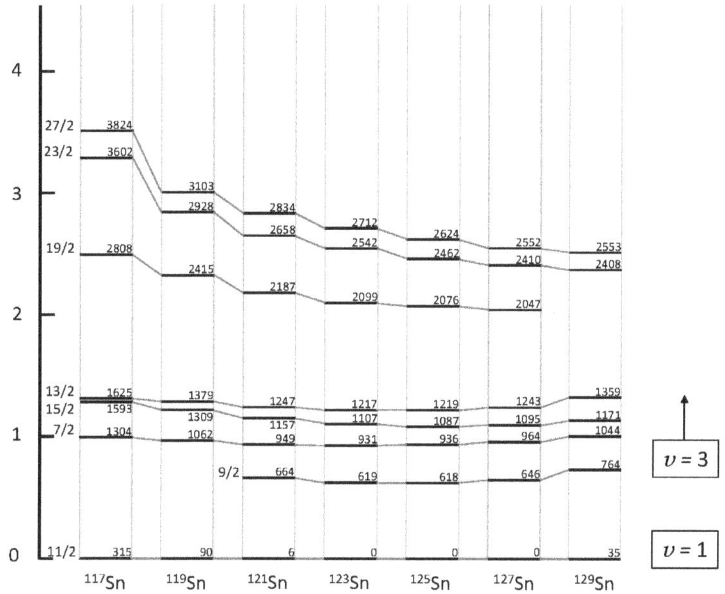


**Figure 13.** A view of the systematics of the seniority-three $\nu h^n_{11/2}$ states in the neutron-rich odd-mass tin isotopes. There are some states missing, according to seniority-dominated coupling; the full set contains: $2J = 3, 5, 7, 9, 9, 11, 13, 15, 15, 17, 19, 21, 23$, and 27 (see, e.g., [37]). Because of ambiguities in some parity assignments, other potential candidate states are omitted. Note there are "second" $19/2^-$ states observed in $^{123,125,127}$Sn.

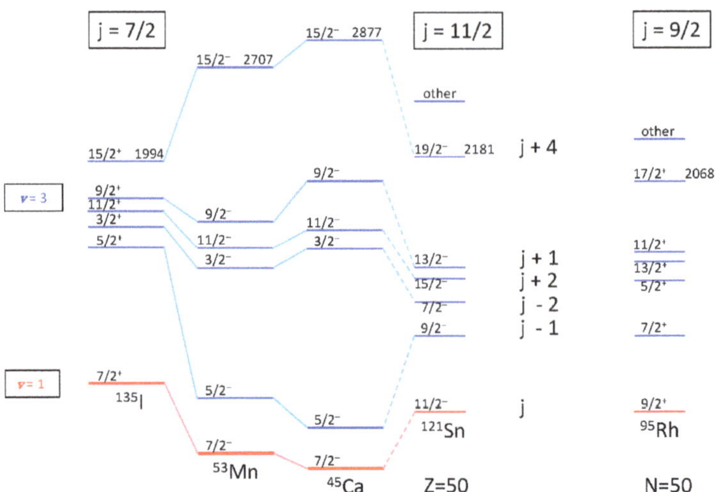


**Figure 14.** A global view of seniority-three multiplets in selected nuclides and selected spin couplings, for $j = 7/2, 9/2$, and $11/2$. Energies are omitted to avoid cluttering the figure; energies are also relative, per isotope. To our knowledge, this universal behaviour has not been recognized. Note that the $j = 7/2$ multiplets (with the proviso made for $^{135}$I in Figure 11) are complete; the $j = 9/2$ and $11/2$ multiplets contain more states than shown here, cf. Figures 4 and 13.

The manifestation of shape coexistence in singly closed shell nuclei was recognized already forty years ago [39] and was reviewed thirty years ago [40]. It is well established for $Z = 20$, 50, and 82 and for $N = 20$ and 28; there are hints to its presence for $Z = 8$ and 28, and for $N = 8$, 50, and 82. Details can be found in the most recent review [41], together with some details in the earlier review [40]. The emerging view is that shape coexistence likely occurs in all nuclei; including that spherical states occur in nuclei with deformed ground states [42]. A concise perspective of the occurrence of deformation in nuclei as compared to atoms can be encapsulated in: "The difference between atoms and nuclei is that atoms are a manifestation of many-fermion quantum mechanics with one type of fermion, which repel, whereas nuclei involve two types of nucleon, which attract. By deforming, the system can lower its energy via relaxing the constraints of the Pauli exclusion principle in such a manner that more spatially symmetric configurations become accessible, which leads to a lowering of the energy of the system". (It can be noted that the emerging view of baryons may signal correlated, even deformed structures, especially the recent realization [43] that the proton contains more (virtual) anti-down quarks than anti-up quarks: this is simply a manifestation of correlations that involve "particle–hole" excitations, i.e., quark–antiquark pairs, and the Pauli principle.)

## 3. Hints of Correlations, beyond Pairing and Seniority, at Closed Shells

The dominance of seniority, with intruding shape coexistence, in singly closed shell nuclei is not quite "the whole story". The following analysis of effective charges implied by the observed $B(E2; 2_1^+ \rightarrow 0_1^+)$ in even-even nuclei adjacent to doubly closed shells demonstrates what can be encapsulated in the term "the effective charge problem".

Electric multipole transition rates in the shell model are usually evaluated using harmonic oscillator wavefunctions. For a single-particle transition $j_\beta \rightarrow j_\alpha$, the reduced matrix element $\langle j_\alpha || T(E2) || j_\beta \rangle$ can be evaluated from

$$\langle j_\alpha || T(E\lambda) || j_\beta \rangle = \frac{e}{\sqrt{4\pi}} (-1)^{j_\beta + \lambda - \frac{1}{2}} \frac{1 + (-1)^{l_\alpha + l_\beta + \lambda}}{2} \hat{\lambda} \hat{j}_\alpha \hat{j}_\beta \begin{pmatrix} j_\alpha & j_\beta & \lambda \\ \frac{1}{2} & -\frac{1}{2} & 0 \end{pmatrix} b^\lambda \tilde{R}^{(\lambda)}_{\alpha\beta}, \quad (2)$$

where $\hat{j} \equiv \sqrt{(2j+1)}$ and $\tilde{R}^{(\lambda)}_{\alpha\beta}$ is the dimensionless radial integral that can be evaluated in closed form with harmonic oscillator wavefunctions. The oscillator length $b$ is defined as

$$b = \sqrt{\frac{\hbar}{m_N \omega}}, \tag{3}$$

where $\hbar$ is the reduced Planck constant, $m_N$ is the nucleon mass, and $\hbar\omega$ can be evaluated as a function of the mass number $A$ as

$$\hbar\omega = (45A^{-1/3} - 25A^{-2/3}) \text{ MeV}, \tag{4}$$

which has been found to give satisfactory agreement with observed charge radii. In general,

$$B(E\lambda; J_i \to J_f) = |\langle J_f || T(E2) || J_i \rangle|^2 / (2J_i + 1). \tag{5}$$

For transitions between the states of the pure $j^2$ configuration, the $B(E2)$ values are related to the single-particle matrix element $\langle j || T(E2) || j \rangle$, by

$$B(E2; J_i \to J_i - 2) = 4(2J_i - 3) \left\{ \begin{array}{ccc} j & J_i - 2 & j \\ J_i & j & 2 \end{array} \right\}^2 |\langle j || T(E2) || j \rangle|^2. \tag{6}$$

It is instructive to begin with the textbook cases of $^{17}$O and $^{18}$O, which can be considered as adding one and two neutrons, respectively, to a $^{16}$O core. Identifying the first-excited state to ground, $1/2^+_1 \to 5/2^+_1$, transition in$^{17}$O as due to the neutron transition from the $2s_{1/2}$ to $1d_{5/2}$ orbits, the experimental value of $B(E2) = 2.39(3)$ W.u. (Weisskopf units) requires an effective neutron charge of $e_n = 0.534(3)e$. This value is close to $e_n = 0.5e$, which is the default often adopted for shell model calculations. However, turning to $^{18}$O, and identifying the $2^+_1 \to 0^+_1$ transition with $\nu(d_{5/2})^2_{2^+} \to \nu(d_{5/2})^2_{0^+}$, requires $e_n = 1.054(14)e$ to explain the observed transition strength of 3.32(9) W.u. One might hope that this discrepancy between $^{17}$O and $^{18}$O would be resolved by a shell model calculation in the full `sd` model space with one of the "universal" `sd` interactions, but it is not. Such shell model calculations describe $^{17}$O well. The same calculations, however, fall short of explaining the $B(E2 : 2^+_1 \to 0^+_1)$ in $^{18}$ O by a factor of nearly 3. It is worth noting that the experimental $B(E2)$ for $^{18}$O is based on about 20 independent measurements by four independent techniques, all in reasonable agreement. The conclusion must be that the effective charge handles $^{17}$O, but fails for $^{18}$O due to additional correlations.

Table 2 shows shell model calculations for the reduced transition rate, $B(E2; 2^+_1 \to 0^+_1)$, in doubly magic nuclides plus or minus two like nucleons. The shell model calculations were performed with NuShellX [44] and generally use a contemporary set of interactions for the relevant basis space, and either the recommended effective charges for the selected basis space, or the default $e_p = 1.5e$ and $e_n = 0.5e$, for protons and neutrons, respectively. The effective charges required to bring the shell model calculations into agreement with experiment are shown. For those nuclides adjacent to $^{48}$Ca and $^{56}$Ni, calculations were run in a basis that treats these nuclei as doubly magic, as well as in the full `fp` shell, which allows for excitations from the $1f_{7/2}$ shell across the $N, Z = 28$ shell gap into the $2p_{3/2}$, $1f_{5/2}$, and $1p_{1/2}$ orbits. These calculations account for the neutron core excitation in $^{48}$Ca, including the $\nu(2p-2h)\ 0^+$ state at 5.46 MeV, but cannot describe the $\pi(2p-2h)\ 0^+$ state at 4.28 MeV; see Figure 2, and cf. Figure 60.

**Table 2.** Effective charges, $e_p$ and $e_n$, in units of the elementary charge $e$, for $B(E2; 2_1^+ \rightarrow 0_1^+)$ in doubly magic nuclides plus or minus two like nucleons. The experimental values are from the Evaluated Nuclear Structure Data File (ENSDF) [22], with the following exceptions: $^{46}$Ar [45], $^{130,134}$Sn [46,47], $^{134}$Te [48], $^{210}$Po [49].

| Nuclide | Basis $^a$ | Interaction | $B(E2)$ (W.u.) | | $e_p$ | $e_n$ |
|---|---|---|---|---|---|---|
| | | | Experiment | Shell Model $^b$ | | |
| $^{16}$O core: | | | | | | |
| $^{14}$C | p | pewt [50,51] | $1.8 \pm 0.3$ | 5.42 | $0.86 \pm 0.07$ | – |
| $^{18}$O | sd | usdb [52] | $3.32 \pm 0.09$ | 1.16 $^c$ | – | $0.76 \pm 0.01$ |
| $^{18}$Ne | sd | usdb [52] | $17.7 \pm 1.8$ | 10.64 $^c$ | $1.75 \pm 0.09$ | – |
| $^{40}$Ca core: | | | | | | |
| $^{38}$Ar | sd | usdb [52] | $3.4 \pm 0.16$ | 3.36 $^c$ | $1.37 \pm 0.03$ | – |
| $^{38}$Ca | sd | usdb [52] | $2.5 \pm 0.6$ | 0.37 $^c$ | – | $1.17 \pm 0.14$ |
| $^{42}$Ca | f7 | f7cdpn [53] | $9.5 \pm 0.4$ | 0.64 | – | $1.92 \pm 0.04$ |
| | fp | gx1a [54,55] | $9.5 \pm 0.4$ | 0.77 | – | $1.76 \pm 0.04$ |
| $^{42}$Ti | f7 | f7cdpn [53] | $16 \pm 4$ | 5.80 | $2.49 \pm 0.31$ | – |
| | fp | gx1a [54,55] | $16 \pm 4$ | 6.94 | $2.28 \pm 0.28$ | – |
| $^{48}$Ca core: | | | | | | |
| $^{46}$Ar | sdpf | sdpfmu [56] | $4.4 \pm 0.4$ | 7.77 $^d$ | – | – |
| $^{46}$Ca | f7 | f7cdpn [53] | $3.63 \pm 0.3$ | 0.60 | – | $1.23 \pm 0.05$ |
| | fp | gx1a [54,55] | $3.63 \pm 0.3$ | 0.94 | – | $0.98 \pm 0.04$ |
| $^{50}$Ca | ho | ho [57] | $0.68 \pm 0.02$ | 0.83 | – | $0.45 \pm 0.01$ |
| | fp | gx1a [54,55] | $0.68 \pm 0.02$ | 0.84 | – | $0.45 \pm 0.01$ |
| $^{50}$Ti | ho | ho [57] | $5.46 \pm 0.19$ | 5.05 | $1.56 \pm 0.03$ | – |
| $^{50}$Ti | fp | gx1a [54,55] | $5.46 \pm 0.19$ | 9.19 | – | – |
| $^{56}$Ni core: | | | | | | |
| $^{54}$Fe | f7 | f7cdpn [53] | $11.1 \pm 0.3$ | 4.76 | $2.29 \pm 0.03$ | – |
| | fp | gx1a [54,55] | $11.1 \pm 0.3$ | 13.08 | – | – |
| $^{54}$Ni | f7 | f7cdpn [53] | $10 \pm 2$ | 0.53 | – | $2.17 \pm 0.22$ |
| | fp | gx1a [54,55] | $10 \pm 2$ | 6.69 | – | – |
| $^{58}$Ni | ho | ho [57] | $10.0 \pm 0.4$ | 0.83 | – | $1.73 \pm 0.03$ |
| | fp | gx1a [54,55] | $10.0 \pm 0.4$ | 9.28 | – | – |
| $^{132}$Sn core: | | | | | | |
| $^{130}$Sn | jj55 | sn100 [33] | $1.18 \pm 0.25$ | 0.76 | – | $0.62 \pm 0.06$ |
| $^{134}$Sn | jj56 | jj56cdb [33] | $1.42 \pm 0.25$ | 0.94 | – | $0.62 \pm 0.05$ |
| $^{134}$Te | jj55 | sn100 [33] | $5.12 \pm 0.21$ | 4.00 | $1.70 \pm 0.03$ | – |
| $^{208}$Pb core: | | | | | | |
| $^{206}$Pb | jj56 | khhe [58] | $2.8 \pm 0.09$ | 0.79 | – | $0.94 \pm 0.02$ |
| $^{210}$Pb | jj67 | khpe [58] | $1.4 \pm 0.4$ | 0.55 | – | $0.80 \pm 0.11$ |
| $^{210}$Po | jj67 | khpe [58] | $1.83 \pm 0.28$ $^e$ | 3.51 | $1.08 \pm 0.08$ | – |

$^a$ Model basis spaces:
p: $\pi$ & $\nu$ $(1p_{3/2}, 1p_{1/2})$
sd: $\pi$ & $\nu$ $(1d_{5/2}, 2s_{1/2}, 1d_{3/2})$
f7: $\pi$ & $\nu$ $(1f_{7/2})$
fp: $\pi$ & $\nu$ $(1f_{7/2}, 2p_{3/2}, 1f_{5/2}, 2p_{1/2})$
sdpf: $\pi$ $(1d_{5/2}, 1d_{3/2}, 2s_{1/2})$; $\nu$ $(1f_{7/2}, 2p_{3/2}, 1f_{5/2}, 2p_{1/2})$
ho: $\pi$ $(1f_{7/2})$; $\nu$ $(2p_{3/2}, 1f_{5/2}, 2p_{1/2})$
jj55: $\pi$ & $\nu$ $(1g_{7/2}, 2d_{5/2}, 2d_{3/2}, 3s_{1/2}, 1h_{11/2})$
jj56: $\pi$ $(1g_{7/2}, 2d_{5/2}, 2d_{3/2}, 3s_{1/2}, 1h_{11/2})$; $\nu$ $(1h_{9/2}, 2f_{7/2}, 2f_{5/2}, 3p_{3/2}, 3p_{1/2}, 1i_{13/2})$
jj67: $\pi$ $(1h_{9/2}, 2f_{7/2}, 2f_{5/2}, 3p_{3/2}, 3p_{1/2}, 1i_{13/2})$; $\nu$ $(1i_{11/2}, 2g_{9/2}, 2g_{7/2}, 3d_{5/2}, 3d_{3/2}, 4s_{1/2}, 1j_{15/2})$
$^b$ The default charges are $e_p = 1.5$ and $e_n = 0.5$, unless otherwise indicated.
$^c$ For usdb the recommended values $e_p = 1.36$, $e_n = 0.45$ were used.
$^d$ For sdpgmu the recommended values $e_p = 1.35$, $e_n = 0.35$ were used.
$^e$ This experimental value has been questioned; see text.

There is no overall pattern in the effective charges shown in Table 2. Most of the shell model $B(E2)$ values are within a factor of 2 to 3 of the experiment; however, those for the calcium isotopes, $^{38}$Ca and $^{42}$Ca, are underestimated by an order of magnitude. The

experimental $B(E2)$ value for $^{46}$Ar is almost a factor of two smaller than theory. While a lifetime measurement [59] gave a $B(E2)$ value consistent with theory, the weight of evidence from independent Coulomb excitation measurements [45,60,61] makes the adopted value in Table 2 firm and in tension with theory.

Good agreement in the `fp`-shell calculation is obtained for $^{50}$Ca and $^{54}$Fe. As noted above, in these cases, $^{48}$Ca and $^{56}$Ni are not doubly magic cores but part of the `fp` model space. It is puzzling that the calculation for $^{50}$Ti in the same model space is twice the experiment, but the restricted $f_{7/2}$ model space agrees with experiment.

Moving to heavier nuclei, the effective charges in the $^{132}$Sn region are near the default values [62], although most recent calculations adopt $e_p \approx 1.7e$ and $e_n \approx 0.8e$ [32,34,63,64]. The measured $B(E2)$ for $^{130}$Sn [46,47] is lower than theory and the experimental systematics (see [65]); the experiment should be repeated.

In the $^{208}$Pb region, $e_n$ approaches $+e$. The experimental result for $^{210}$Po is problematic. As shown below in this Section, an analysis of higher-excited states in $^{210}$Po corresponding nominally to the $\pi 1h_{9/2}^2$ configuration implies $e_p \approx 1.5e$. The experimental $B(E2)$ in Table 2 for $^{210}$Po is deduced from a recent lifetime measurement by the Doppler shift attenuation method following the $^{208}$Pb($^{12}$C,$^{10}$Be)$^{210}$Po reaction, which gave $\tau = 2.6 \pm 0.4$ ps [49]. This new result is certainly an improvement on the previous measurement which used (d,d′) above the Coulomb barrier to excite a $^{210}$Po target [66]. However, it is difficult to measure such a short lifetime below the longer-lived $4^+$, $6^+$ and $8^+$ states that tend to also be populated in heavy ion reactions; Kocheva et al. [49] recommend additional experiments. Coulomb excitation of the radioactive beam (e.g., at ISOLDE where $^{210}$Po activity remains in used ion sources) would be a possibility, avoiding the problem of feeding from the longer-lived higher excited states.

In several cases in Table 2, a $j^2$ approximation is (at least at face value) a reasonable starting point. For the case of $^{14}$C, it is not: holes in $^{16}$O nominally occupy the $1p_{1/2}$ orbit which must couple with $1p_{3/2}$ to form a $2^+$ state. In other cases, like $^{130}$Sn, the $2d_{3/2}$, $3s_{1/2}$, and $1h_{11/2}$ single-particle orbits are so close in energy that a single-$j^2$ approximation cannot be applicable.

In some respects, the comparison of effective charges from the $2_1^+ \to 0_1^+$ transitions alone may be considered selective and not altogether fair. However, as discussed in this Section, it fits our purpose, which is to examine the emergence of collectivity in nuclei. To explore further the successes and limitations of the shell model approach, comparisons of $E2$ strengths and $g$ factors are now made for a selection of the semimagic nuclides in Table 2 that can be approximated as a single-$j^2$ configuration adjacent to a doubly magic core. Later in this section and again in Section 8, we argue that the properties of $2_1^+$ states, especially their electromagnetic properties, play an important part in developing an understanding of the emergence of collectivity in nuclei.

Table 3 shows the effective charges required to explain $B(E2)$ values between low-excitation states associated with nominal $j^2$ configurations in doubly magic nuclides plus or minus two like nucleons. For most cases, only protons or neutrons are active in the basis space. For $^{50}$Ti and $^{54}$Fe, calculations were performed in the `fp` model space which allows neutron excitations across $N = 28$; thus, both protons and neutrons contribute to the transition rate. In these cases, the proton effective charge required by experiment was evaluated assuming that $e_n = 0.5e$. The uncertainty given is due to the uncertainty in the experimental $B(E2)$ alone. Concerning the uncertainty in the assumed value of $e_n$, it can be noted that $e_p + e_n$ is near constant for $^{50}$Ti, so a decrease in $e_n$ by say $0.1e$ leads to an increase in $e_p$ of approximately 0.1. For $^{54}$Fe, the value of $e_p$ is less sensitive to the assumed value of $e_n$.

As expected, the effective charge is generally reduced when the basis space is enlarged; the $j^2$ model is obviously an oversimplification. However, it is a better approximation for the nuclei adjacent to the $N > Z$ doubly magic $^{132}$Sn and $^{208}$Pb. One reason is that, for nuclei adjacent to $N = Z$ doubly closed shells, intruder configurations are present at

low energy and these place the active nucleons in a much larger Hilbert space than can be handled by the shell model.

From Table 3, one can conclude that the effective charge required to describe the $B(E2; 2_1^+ \rightarrow 0_1^+)$ transition is often greater than that required to explain the transitions between the higher spins in the $j^2$ multiplet (i.e., the $E2$ decays of the states with $J^\pi = 4^+$, $6^+$, ... $(2j)^+$), particularly for the $j^2$ model. One can also see that the effective charges exceed the bare nucleon values, even in the large basis shell model calculations. The effective proton charges are reduced significantly for $^{50}$Ti and $^{54}$Fe when the basis space is expanded to include the whole `fp` shell. The proton charge deduced for $^{50}$Ti even approaches unity, but this assumes that $e_n = 0.5e$.

**Table 3.** Effective charges for nominal $j^2$ configurations in selected doubly magic nuclides plus or minus two like nucleons. The effective charges are evaluated assuming a pure $j^2$ configuration and for the mixed configurations of the (large basis) shell model (SM) calculations in Table 2. The experimental transition rates, $B(E2)_{\text{exp}}$, are from ENSDF [22] and from the references in Table 2.

| Nuclide | Config. | Transition | $B(E2)_{\text{exp}}$ (W.u.) | $e^{\text{eff}}$ | |
|---|---|---|---|---|---|
| | | | | $j^2$ | SM |
| Protons: | | | | | |
| $^{18}$Ne | $\pi 1d_{5/2}^2$ | $2_1^+ \rightarrow 0_1^+$ | $17.7 \pm 1.8$ | $2.43 \pm 0.12$ | $1.75 \pm 0.09$ |
| | | $4_1^+ \rightarrow 2_1^+$ | $8.9 \pm 1.2$ | $2.08 \pm 0.14$ | $1.34 \pm 0.09$ |
| $^{42}$Ti | $\pi 1f_{7/2}^2$ | $2_1^+ \rightarrow 0_1^+$ | $16 \pm 4$ | $2.5 \pm 0.3$ | $2.28 \pm 0.28$ |
| | | $6_1^+ \rightarrow 4_1^+$ | $3.2 \pm 0.2$ | $1.65 \pm 0.05$ | $0.95 \pm 0.03$ |
| $^{50}$Ti | $\pi 1f_{7/2}^2$ | $2_1^+ \rightarrow 0_1^+$ | $5.46 \pm 0.19$ | $1.56 \pm 0.03$ | $1.06 \pm 0.03$ [a] |
| | | $4_1^+ \rightarrow 2_1^+$ | $5.5 \pm 1.5$ | $1.57 \pm 0.21$ | $1.1 \pm 0.2$ [a] |
| | | $6_1^+ \rightarrow 4_1^+$ | $3.14 \pm 0.13$ | $1.76 \pm 0.04$ | $1.24 \pm 0.03$ [a] |
| $^{54}$Fe | $\pi 1f_{7/2}^{-2}$ | $2_1^+ \rightarrow 0_1^+$ | $11.1 \pm 0.3$ | $2.29 \pm 0.03$ | $1.36 \pm 0.02$ [a] |
| | | $4_1^+ \rightarrow 2_1^+$ | $6.3 \pm 1.3$ | $1.73 \pm 0.18$ | $1.60 \pm 0.18$ [a] |
| | | $6_1^+ \rightarrow 4_1^+$ | $3.24 \pm 0.06$ | $1.84 \pm 0.02$ | $1.36 \pm 0.01$ [a] |
| $^{134}$Te | $\pi 1g_{7/2}^2$ | $2_1^+ \rightarrow 0_1^+$ | $5.12 \pm 0.21$ | $1.80 \pm 0.04$ | $1.70 \pm 0.03$ |
| | | $4_1^+ \rightarrow 2_1^+$ | $4.3 \pm 0.4$ | $1.65 \pm 0.08$ | $1.58 \pm 0.07$ |
| | | $6_1^+ \rightarrow 4_1^+$ | $2.05 \pm 0.04$ | $1.69 \pm 0.02$ | $1.54 \pm 0.02$ |
| $^{210}$Po | $\pi 1h_{9/2}^2$ | $2_1^+ \rightarrow 0_1^+$ | $1.83 \pm 0.28$ [b] | $1.07 \pm 0.08$ | $1.08 \pm 0.08$ |
| | | $4_1^+ \rightarrow 2_1^+$ | $4.46 \pm 0.18$ | $1.56 \pm 0.03$ | $1.50 \pm 0.03$ |
| | | $6_1^+ \rightarrow 4_1^+$ | $3.05 \pm 0.09$ | $1.55 \pm 0.02$ | $1.50 \pm 0.02$ |
| | | $8_1^+ \rightarrow 6_1^+$ | $1.12 \pm 0.04$ | $1.48 \pm 0.03$ | $1.44 \pm 0.03$ |
| Neutrons: | | | | | |
| $^{18}$O | $\nu 1d_{5/2}^2$ | $2_1^+ \rightarrow 0_1^+$ | $3.32 \pm 0.09$ | $1.054 \pm 0.014$ | $0.76 \pm 0.01$ |
| | | $4_1^+ \rightarrow 2_1^+$ | $1.19 \pm 0.06$ | $0.76 \pm 0.02$ | $0.51 \pm 0.01$ |
| $^{42}$Ca | $\nu 1f_{7/2}^2$ | $2_1^+ \rightarrow 0_1^+$ | $9.5 \pm 0.4$ | $1.92 \pm 0.04$ | $1.76 \pm 0.04$ |
| | | $4_1^+ \rightarrow 2_1^+$ | $8.3 \pm 1.2$ | $1.80 \pm 0.13$ | $1.6 \pm 0.1$ |
| | | $6_1^+ \rightarrow 4_1^+$ | $0.777 \pm 0.022$ | $0.82 \pm 0.01$ | $0.47 \pm 0.01$ |
| $^{134}$Sn | $\nu 2f_{7/2}^2$ | $2_1^+ \rightarrow 0_1^+$ | $1.42 \pm 0.25$ | $0.80 \pm 0.07$ | $0.62 \pm 0.05$ |
| | | $6_1^+ \rightarrow 4_1^+$ | $0.89 \pm 0.17$ | $0.94 \pm 0.09$ | $0.52 \pm 0.05$ |
| $^{210}$Pb | $\nu 2g_{9/2}^2$ | $2_1^+ \rightarrow 0_1^+$ | $1.4 \pm 0.4$ | $0.81 \pm 0.12$ | $0.80 \pm 0.11$ |
| | | $4_1^+ \rightarrow 2_1^+$ | $4.8 \pm 0.9$ | $1.40 \pm 0.13$ | $1.28 \pm 0.12$ |
| | | $6_1^+ \rightarrow 4_1^+$ | $2.1 \pm 0.8$ | $1.11 \pm 0.21$ | $1.0 \pm 0.2$ |
| | | $8_1^+ \rightarrow 6_1^+$ | $0.7 \pm 0.3$ | $1.02 \pm 0.22$ | $0.90 \pm 0.19$ |

[a] Evaluated in the `fp` basis with `gx1a` interactions and $e_n = 0.5e$. See text for details. [b] This experimental value has been questioned; see text.

There are broadly two scenarios to explain the effective charge. First, and universally applicable, is the coupling of the valence nucleons to collective excitations of the core, including the giant resonances, in such fashion that the concept of an effective charge as a renormalization procedure has some operational justification. Second, and specific to particular cases, is the coupling between the valence nucleons and low-excitation configurations outside the shell model basis. This later scenario means that the shell model configuration is wrong in a more fundamental way.

An examination of the magnetic moments, or rather the $g$ factors ($g = \mu/J$), can distinguish between these scenarios. To this end, Table 4 shows an evaluation of the $g$ factors for those nominal $j^2$ configuration cases in Table 3 for which there are experimental data. It is useful to make use of the fact that $g(j^n) = g(j)$; that is, the $g$ factor of any number of nucleons in a single-particle orbit is equal to the $g$ factor of the single-particle orbit, independent of the number of nucleons ($n$) and the resultant spin.

The empirical $g$ factor of the $j^2$ configuration was evaluated as the average of the $g$ factors of the ground-states of the neighbouring nuclei with $A \pm 1$ and odd-$Z$ or odd-$N$, as appropriate. The shell model calculations in the `sd` and `fp` spaces use the default effective $M1$ operator for those basis spaces. For the `jj55` space, the $M1$ operator is as in Refs. [48,64,65,67,68]. For $^{210}$Pb and $^{210}$Po (`jj67`), the effective $g_s$ was set to 70% of the free nucleon value and $g_l$ adjusted to reproduce the ground state $g$ factors of $^{209}$Pb ($\nu 2g_{9/2}$) and $^{209}$Po ($\pi 1h_{9/2}$). The values so obtained conform to expectations ($g_l(\pi) \approx 1.1$ and $g_l(\nu) \lesssim 0$). It is important to note that the renormalization of the $M1$ operator is due to processes quite distinct from those that give rise to the effective charge, namely meson exchange currents, and core polarization. Here, the core polarization involves particle–hole excitations between spin–orbit partners, which couple strongly to the $M1$ operator. It thus differs in a fundamental way from the core polarization associated with the $E2$ effective charge.

It is convenient to discuss the results in Table 4 beginning with the heavier nuclei, $^{210}$Pb and $^{210}$Po. For these nuclei adjacent to $^{208}$Pb, there is good agreement between the experimental $g$ factors of the $6_1^+$ and $8_1^+$ states, and both the empirical $j^2$ estimate and the shell model. These can be considered text book examples. It is unfortunate that there are no data for the $2_1^+$ and $4_1^+$ states, which, as the following discussion in this Section suggests, might show additional collectivity.

Turning to $^{134}$Te, the $E2$ and $g$ factor data for the $\pi(1g_{7/2})^2$ multiplet are complete, and there is reasonable agreement with both the $j^2$ model and the shell model calculations. A detailed analysis has been given in Ref. [48], wherein it is shown that there is additional quadrupole collectivity in the $2_1^+$ state of $^{134}$Te that is not accounted for by large-basis shell model calculations that assume an inert $^{132}$Sn core. It was demonstrated that coupling the valence $\pi g_{7/2}^2$ configuration to a core vibration with the properties of the first-excited state in $^{132}$Sn can readily account for the observed $2_1^+ \to 0_1^+$ transition strength in $^{134}$Te, and that the wavefunctions of the $2_1^+$, $4_1^+$ and $6_1^+$ states of $^{134}$Te nevertheless remain dominated by the $\pi g_{7/2}^2$ configuration. It can be concluded that $^{132}$Sn is a relatively inert shell-model core. The caveat, however, is that the shell model calculations still require relatively large effective charges.

In the `fp` shell, $^{50}$Ti shows quite good agreement with both the $j^2$ model and the shell model. For $^{54}$Fe, the experimental $g$ factors show better agreement with the large-basis shell model than the $j^2$ model. The shell model calculations in the `fp` basis with the `gx1a` interactions do a reasonable job of describing the different behaviour of the $g$ factors in $^{50}$Ti and $^{54}$Fe.

The isotopes with two neutrons outside the $N = Z$ cores $^{16}$O and $^{40}$Ca show similar behaviour: $g(2_1^+)$ is reduced significantly in magnitude compared to both the $j^2$ model and the shell model calculation, whereas the higher excited states, $4^+$ in $^{18}$O, and $6^+$ in $^{42}$Ca, have $g$ factors in agreement with both the $j^2$ model and the larger-basis shell model. In

these isotopes, both the $E2$ transition strengths and the $g$ factors indicate that the $2^+$ state must contain collective admixtures. Writing the $2_1^+$ wavefunction in the form

$$|2_1^+\rangle = \alpha|\mathrm{SM}\rangle + \sqrt{1-\alpha^2}|\mathrm{coll}\rangle, \tag{7}$$

where SM denotes the part from the shell model basis space and "coll" denotes the collective part (from multiparticle-multihole excitations), implies that

$$g(2_1^+) = \alpha^2 g_{\mathrm{SM}} + (1-\alpha^2) g_{\mathrm{coll}}. \tag{8}$$

Assuming that the collective $g$ factor is $g_{\mathrm{coll}} \approx Z/A \approx 0.5$, and taking the shell model $g$ factor from Table 4 implies that there is a collective contribution of $\alpha^2 = 20 \pm 2\%$ in the first excited state of $^{18}$O, and a huge $59 \pm 5\%$ collective contribution in the first-excited state of $^{42}$Ca. This mixing in $^{42}$Ca is in excellent agreement with a 50% collective contribution deduced from Coulomb excitation data and one-neutron transfer reaction data (see Figures 41 and 42 for full details). To explain the observed $g$ factor in $^{42}$Ca, Ref. [69] requires that the basis space be expanded to include the `sd` as well as `fp` orbits for both protons and neutrons. This strongly collective structure of the $2_1^+$ state is in stark contrast with the near pure $\nu(f_{7/2})^2$ structure of the $6_1^+$ state.

**Table 4.** $g$ factors for nominal $j^2$ configurations in doubly magic nuclides plus or minus two like nucleons. Data are from [70] (with a correction for $^{54}$Fe $g(2_1^+)$ from [71]).

| Nuclide | Config. | $J_i^\pi$ | $g$ (exp) | $g$ (emp $j^2$) | $g$ (SM) |
|---|---|---|---|---|---|
| $^{18}$O | $\nu 1d_{5/2}^2$ | $2_1^+$ | $-0.285 \pm 0.015$ | $-0.685$ | $-0.476$ |
| | | $4_1^+$ | $-0.63 \pm 0.10$ | $-0.685$ | $-0.603$ |
| $^{42}$Ca | $\nu 1f_{7/2}^2$ | $2_1^+$ | $0.04 \pm 0.06$ | $-0.416$ | $-0.615$ |
| | | $6_1^+$ | $-0.415 \pm 0.015$ | $-0.416$ | $-0.538$ |
| $^{50}$Ti | $\pi 1f_{7/2}^2$ | $2_1^+$ | $1.45 \pm 0.08$ | $1.538$ | $1.235$ |
| | | $6_1^+$ | $1.55 \pm 0.17$ | $1.538$ | $1.379$ |
| $^{54}$Fe | $\pi 1f_{7/2}^{-2}$ | $2_1^+$ | $0.95 \pm 0.11$ | $1.407$ | $1.091$ |
| | | $6_1^+$ | $1.37 \pm 0.03$ | $1.407$ | $1.354$ |
| $^{134}$Te | $\pi 1g_{7/2}^2$ | $2_1^+$ | $0.76 \pm 0.09$ | $0.833$ | $0.837$ |
| | | $4_1^+$ | $0.75 \pm 0.50$ | $0.833$ | $0.833$ |
| | | $6_1^+$ | $0.847 \pm 0.025$ | $0.833$ | $0.842$ |
| $^{210}$Pb | $\nu 2g_{9/2}^2$ | $6_1^+$ | $-0.312 \pm 0.015$ | $-0.320$ | $-0.304$ [a] |
| | | $8_1^+$ | $-0.312 \pm 0.007$ | $-0.320$ | $-0.307$ [a] |
| $^{210}$Po | $\pi 1h_{9/2}^2$ | $6_1^+$ | $0.913 \pm 0.008$ | $0.913$ | $0.912$ [b] |
| | | $8_1^+$ | $0.919 \pm 0.006$ | $0.913$ | $0.911$ [b] |

[a] $g_s(\nu) = 0.7 g_s^{\mathrm{free}}(\nu) = -2.678$ and $g_l(\nu) = -0.033$ set to reproduce the ground-state moment of $^{209}$Pb.
[b] $g_s(\pi) = 0.7 g_s^{\mathrm{free}}(\pi) = 3.910$ and $g_l(\pi) = 1.16$ set to reproduce the g.s. moment of $^{209}$Bi.

To sum up, for the nuclei with $N = Z$ cores, the $2_1^+$ structure is apparently affected by mixing with low-excitation deformed multparticle-multihole states, whereas the higher-spin states are closer to the naïve $j^2$ structure. For $N > Z$ cores, the low-spin states are better approximated by the empirical $j^2$ model and quite well described by the shell model. However, in all cases, a substantial effective charge is required to explain the $E2$ strength, even when the $g$ factor suggests a relatively pure shell model configuration.

Although a first assessment of the effective charges required to explain the $B(E2;2^+_1 \rightarrow 0^+_1)$ data adjacent to closed shells may appear to show no pattern, some features can be identified: (i) shape coexistence and mixing must be taken into account when the doubly magic core has $N = Z$, (ii) there are always non-zero corrections to the nucleonic charges. Defining $\delta e_p$ and $\delta e_n$, where $e_p^{\text{eff}} = (1 + \delta e_p)e$ and $e_n^{\text{eff}} = \delta e_n\, e$, the common assumption that $\delta e_p \approx \delta e_n \approx 0.5$ is seen to be valid in many cases. However, $\delta e_n$ appears to increase in heavier nuclei.

The above data and discussion shows that, for $E2$ transition strengths, the bare electric charges, $e_p = +1e$ and $e_n = 0$, do not work for configurations confined to a valence shell. A correction to the effective charges $\delta e_{p(n)} \gtrsim 0.5$ is usually required, even when the low-lying core excitations are taken into account. Certainly, the use of effective charges has provided a means for exploring nuclear structure using the shell model applied to nuclei that do not have closed shells. However, such practice buries important aspects of the origin of quadrupole collectivity in nuclei; one cannot learn the whole story about the origin of nuclear collectivity using such theories. We suggest that the path forward is two-fold: first, to develop models that obviate the need for effective charges, and second, where the use of effective charges is unavoidable, to formulate appropriate strategies to understand and manage their use.

There are "standard" approaches to evaluate effective charge—often conceptually based on the particle-vibration model of Bohr and Mottelson for nuclei with a single valence nucleon. The vibration can be described microscopically by particle–hole excitations in a Random Phase Approximation (RPA)-type approach [72–76]. There is then some choice of—and sensitivity to—the interaction used in the RPA calculation [76]. This procedure, based on single particle–hole excitations, will not account for the effects of mixing between the valence configurations and low-excitation multiparticle-multihole configurations, which will particularly affect the $E2$ effective charge. The procedure to generalize from one valence nucleon to many is less often discussed. The effective charge must vary to some extent with the number of valence nucleons, but, in practice, it is usually held constant.

Some further comments on the path forward are made in Section 10.

A wider view of what one means by the shell model as an independent-particle model is provided by quasi-elastic electron scattering knockout of protons from closed shell nuclei. A summary view is provided in Figure 15. Quasielastic electron-scattering knockout of protons is a probe of independent-particle behaviour in nuclei that is distinct from the more familiar one-nucleon transfer reaction spectroscopy such as (d, $^3$He). First, the interaction is purely electromagnetic; second, entrance and exit channel effects are limited to the outgoing (high-momentum) proton. Thus, confidence can be placed in the extracted spectroscopic factors for $(e, e'p)$ reactions and the revelation that the single-particle view is "incomplete". The important insight is that one is never dealing with independent particles in a quantum many-body system such as the atomic nucleus: correlations are ubiquitous. Indeed, there are severe warnings of this in the theoretical literature, e.g., [77,78]. These correlations go much deeper than pairing correlations. The subject of nucleon correlations in nuclei is broad. Reference to them in the narrative here is minimal because our focus is on systematics of low-energy phenomenology. For the interested reader, a useful entry point is Ref. [79]. For recent access to the topic, a useful source is Ref. [80].

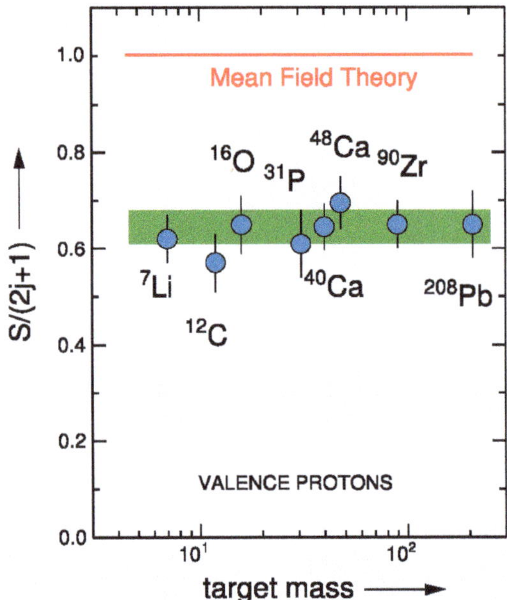


**Figure 15.** Spectroscopic strengths from quasielastic electron scattering knockout of valence protons, $A(e, e'p)$. Adapted from [81] (taken from [82]). The conclusion is, relative to a mean-field view, that never more than 70% of independent-particle strength is manifested in valence nucleon structure, even at doubly closed shells, i.e., other degrees of freedom are contributing to these structures.

The dilemma presented by the data in Figure 15 is a direct confrontation of the shell model approach to nuclear structure, so it can be viewed as a restatement of the question that is used for the title of this review. The data raise two questions: (1) Where has the single-particle strength gone? (2) What has replaced the single-particle strength? We do not attempt to answer these questions. Note that we are in good (bad?) company with the Standard Model of particles and fields. The Standard Model has a plethora of parameters, and nobody knows where they come from. There is one difference in our favour: we believe that protons and neutrons underlie the low-energy degrees of freedom in nuclei, but to employ their bare parameters requires much larger model spaces. Let us note the subtle point regarding correlations: it is primarily the number of configurations involved, not the number of particles, that is relevant. Shell model computations are only tractable in (relatively) small Hilbert spaces: the accumulating evidence is that these spaces are too small. There is an exponential growth in matrix dimensions as the shell model space is increased. However, "symmetry guided" approaches are beginning to circumvent this limitation [83]. A few details are given in Section 10.

It is relevant to note here that the missing strength in $(e, e'p)$ knockout and the effective charge problem must be related at a fundamental level because the $T(E\lambda)$ matrix elements for mass $A$ can be expanded in terms of one-body spectroscopic factors connecting $A$ and $A - 1$. Whether the general missing strength in transfer reactions [84] is associated with short-range [85] or long-range [86] correlations is crucial for the question of emerging collectivity. Moreover, the role of this missing strength in the emergence of quadrupole collectivity in nuclei could possibly be illuminated by examining how the effective charges for higher multipolarities, particularly $E4$ and $E6$, compared to those for $E2$ transitions. The negative polarization charge required for the $E6$ transition in $^{53}$Fe remains a puzzle; see, e.g., [74,76]. Experimental verification of this sole example of an $E6$ transition is clearly important.

A useful tool that has been used to explore independent-particle degrees of freedom in nuclei has been one-nucleon transfer reactions. However, the so-called spectroscopic strengths extracted from such data must be treated with great caution. This was recognized long ago by Baranger [87], and even earlier by Macfarlane and French [88]. These issues have received renewed attention; see, e.g., [89,90] and references therein for a discussion of the problem. The key issue is: Which nuclei provide the best view of independent-particle degrees of freedom? The approach of looking at how degrees of freedom, which manifestly are not independent-particle degrees of freedom, "intrude" into nuclei where independent-particle degrees of freedom have the best chance of dominating (and are widely assumed to do so [91]) is explored here.

By now, it is recognized that structures, even highly deformed structures, "intrude" into the low-energy excitations of spherical nuclei [41]. However, there are subtleties in the mechanism by which such intruder states appear at low excitation energy. An example is shown in Figure 16 for low-energy excited states in $^{47,49}$Ca. The naïve interpretation of the low energies of the $3/2^-$ state in $^{47}$Ca and the $7/2^-$ state in $^{49}$Ca would be that the $N = 28$ shell gap has broken down; but, with an understanding of the manifestation of pairing correlations, the reality is that the $N = 28$ shell gap is strongly present. The persistence of the shell gap can be seen on the right side of Figure 16 where the difference between the observed excitation energies of the first-excited states in $^{47}$Ca and $^{49}$Ca (which correspond to excitation of a neutron across $N = 28$) and the shell gap energy of $\approx 5.1$ MeV is very close to the pairing energy determined from the odd-even staggering in the neutron separation energy, $S_n$. However, one reads about "collapse of shells" and "dissolving of shells". This would be true if there were no correlations present; but correlations *are* present.

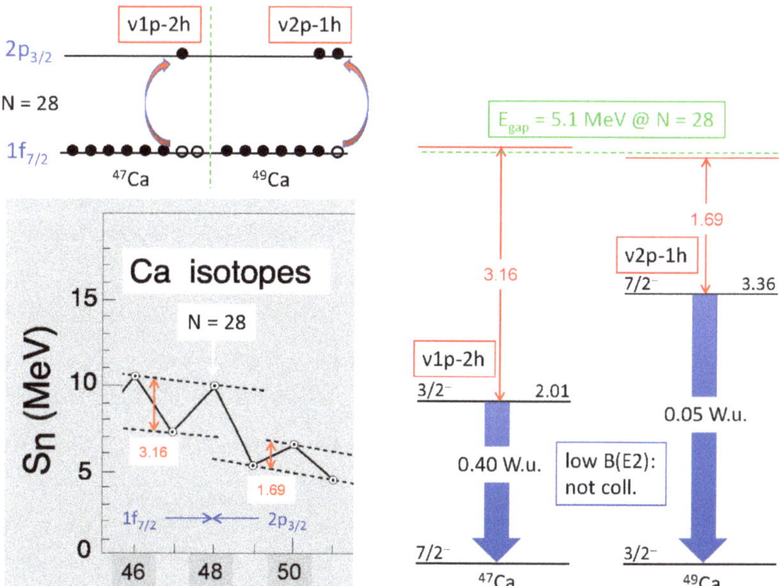


**Figure 16.** Intruder states in $^{47,49}$Ca. The low energy of the $3/2^-$ state in $^{47}$Ca and the $7/2^-$ state in $^{49}$Ca result from pairing correlations. The low $B(E2)$ values associated with these states indicate little or no collective core excitation is involved. The left-hand side of the figure illustrates how a simple estimate of the pairing correlation energy can be made. This analysis shows that the energy gap for $N = 28$ at $Z = 20$ is 5.1 MeV, in line with a well-defined shell gap. The data are taken from ENSDF [22], AME2020 [92], and [90]. Reproduced from [8].

A classic example of intruder states that illustrate the role of deformation is shown in Figure 17 for the odd-mass thallium isotopes. The first hints of these deformed intruder states were recognized long ago [93]; the thallium isotopes were a prime motivational origin of the first review of shape coexistence [94]. The spectroscopic evidence resides in the hindrances of the isomeric transitions and in the band structures associated with the isomer ($9/2^-$ states). The key excitation is a proton across $Z = 82$ to leave a hole pair below $Z = 82$; this hole pair correlates with the valence neutron pairs. These correlations result in near-identical "parabolas" in Bi and Pb isotopes, scaled by the number of proton pairs (see Figure 17 in [41]) and the parabolas exhibit a near collinearity when plotted versus neutron number. The $9/2^-$ intruder structure is the oblate Nilsson $9/2^-[505]$ configuration. There are extensive band structures which are well-described by the Meyer-ter-Vehn model [95,96]. The cores are ${}^{A-1}$Hg; the parameters are the same as for odd-Hg $1i_{13/2}$ bands and odd-Au $1h_{11/2}$ bands (viz. $\beta = 0.15$, $\gamma = 37°$). However, these details raise serious questions about using simple shell model configurations when interpreting excited states even in nuclei with one nucleon coupled to a singly closed shell.

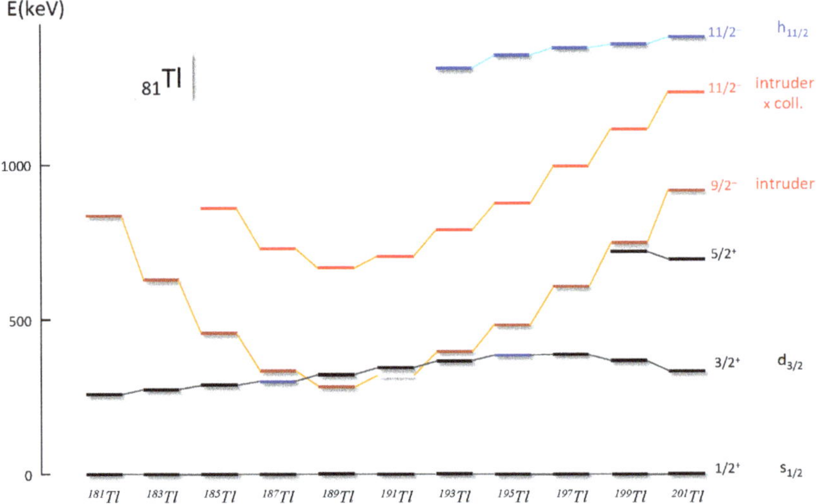


**Figure 17.** The lowest-energy intruder states in the odd-mass thallium isotopes. A naïve interpretation (from a spherical shell model perspective) would lead to the conclusion that because a $9/2^-$ state is below an $11/2^-$ state in excitation energy, spin–orbit coupling has "broken down" or "collapsed". In reality, the $9/2^-$ states shown are dominated by proton $1p - 2h$ excitations and are deformed structures: the first collective excitation on these $9/2^-$ states is shown. [41]. Further details are given in the text. Reproduced from [8].

## 4. Nuclei with Open Shells; Emergence of Collectivity

Nuclear structure is dominated by open-shell nuclei. With excitation of nucleons across shell gaps, and the resulting correlations, "open-shell" configurations intrude to low energy, even to the ground state, in some closed-shell nuclei. Thus, one must understand open-shell nuclei from a microscopic perspective. There are excellent limiting cases for nuclear behavior in open-shell nuclei: these are the strongly deformed nuclei, but a detailed microscopic understanding is lacking. Some perspectives on the current situation are presented here. This is the main focal point of this paper.

The key criterion for this exploration is to identify signatures of shell model structure in open-shell nuclei. Doubly even nuclei obscure shell model structure because of the correlations of pairs of nucleons. Odd-mass nuclei manifestly provide a view, via the unpaired nucleon. However, correlations are still an issue because there can be mixing

of the configurations with different $j$ values within a given shell. However, spin–orbit coupling provides a way forward: each shell has a unique-parity orbital and configurations involving this orbital will be the least mixed of any structures observed.

The power of the systematics of unique-parity states is illustrated in Figures 18 and 19. These figures show the systematics of the positive-parity states in the odd-mass yttrium isotopes across two shells, Figure 18, and of the negative-parity states in the odd-mass $N = 63$ isotones across two shells, Figure 19. Noting that the "parent" $j$ configurations are $1g_{9/2}$ and $1h_{11/2}$, i.e., they differ by one unit of spin, the patterns are similar to the point that they are close to identical. (We recognize that, in the yttrium isotopes, there is a "delayed" onset of collectivity in $^{91,93,95,97}$Y, an issue which does not concern us here). These patterns suggest that there is an underlying coupling scheme that is defined by just a few simple basic features. Since multi-$j$ shell structures (as manifested in, e.g., the negative-parity states in the $28 < Z < 50$ shell, involving the configurations $1f_{5/2}$, $2p_{3/2}$, and $2p_{1/2}$) are dominated by mixing of these configurations, the unique-parity states may provide a basic guide, via recognized single-$j$ shell dominated patterns, for a mixed $j$-shell description across all open-shell odd-mass nuclear structure. Thus, we point to patterns that are independent of specific open shells; and to the implication that "shell-specific" interactions may be unnecessarily complex and intricate.

At present, the best description of experimental data for odd-mass nuclei in regions where deformation is not large is: "incomplete". However, a small number of such nuclei have been sufficiently well studied that they can provide guidance to likely a more complete view of the structure of unique-parity states. The best experimental example of the structure we draw attention to is shown in Figure 20, i.e., the nucleus $^{125}$Xe. This is a pattern of organization related to the studies of a single-$j$ particle coupled to a rigid triaxial rotor, by Juergen Meyer-ter-Vehn [95,96]; indeed, he suggested such a pattern in $^{187}$Ir, long before detailed spectroscopic information was available: an up-to-date view of $^{187}$Ir is shown in Figures 21 and 22, and these strongly support the view. Further note that a very similar pattern appeared in a weak coupling description [97].

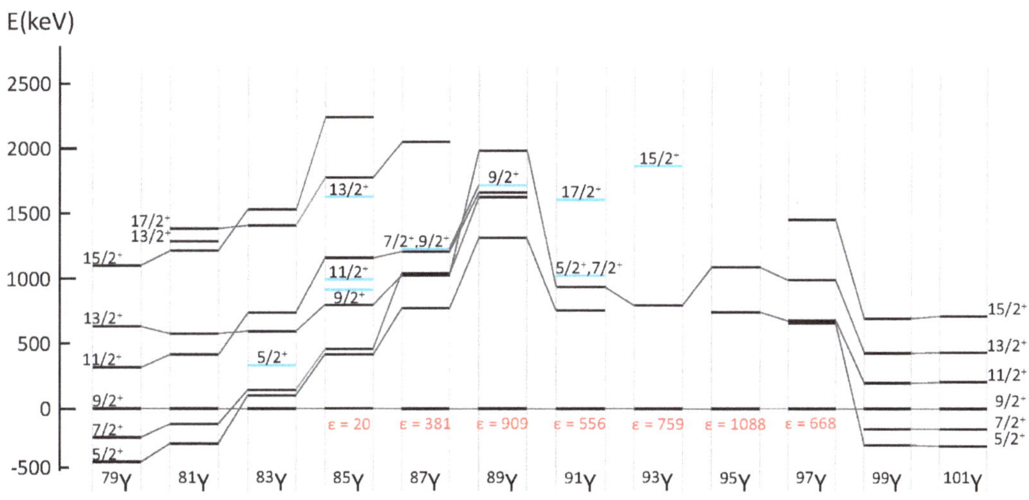


**Figure 18.** Systematics of the positive-parity states across the yttrium isotopes. These states are the unique-parity states for protons in the $28 < Z < 50$ open shell. Note the emergence of near-identical excitation patterns at the extreme mass numbers (these are the Nilsson configurations $\Omega^{\pi}[Nn_z\Lambda] = 5/2^+[422]$, which differ only in rotational energy parameters).

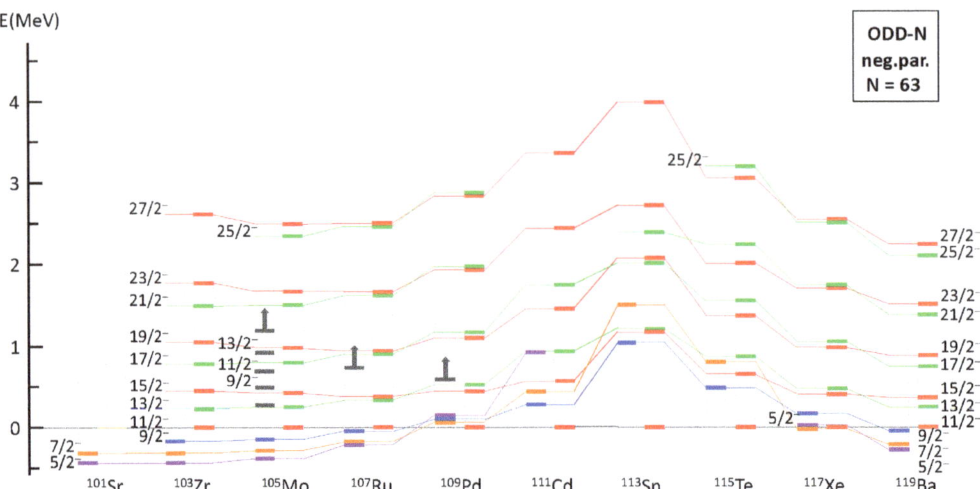


**Figure 19.** Systematics of negative-parity states across the $N = 63$ isotones. These states are the unique-parity states for neutrons in the $50 < N < 82$ open shell. Note the emergence of near-identical excitation patterns at the extreme mass numbers (these are the Nilsson configurations $\Omega^{\pi}[Nn_z\Lambda] = 5/2^{-}[532]$, which differ only in rotational energy parameters and, slightly, in "staggering" or signature splitting).

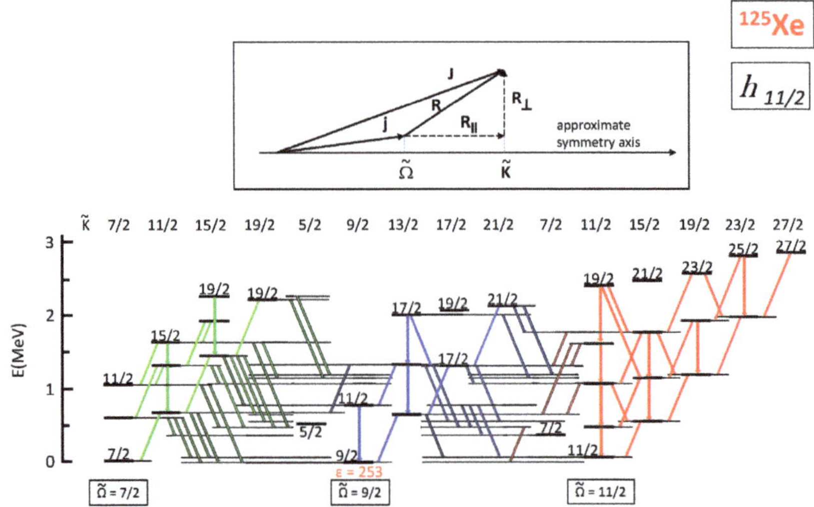


**Figure 20.** Organization of the unique-parity states in $^{125}$Xe, associated with $j = 11/2$ into a "hyperband" pattern due to Meyer-ter-Vehn [95,96]. The inset shows the quantum numbers used to define this pattern [98]. The data are taken from [99], but the pattern was not recognized there. Note that "vertical" $\Delta I = 2$ (i.e., $E2$) transitions are almost totally absent from the observed data. See the text for more details.

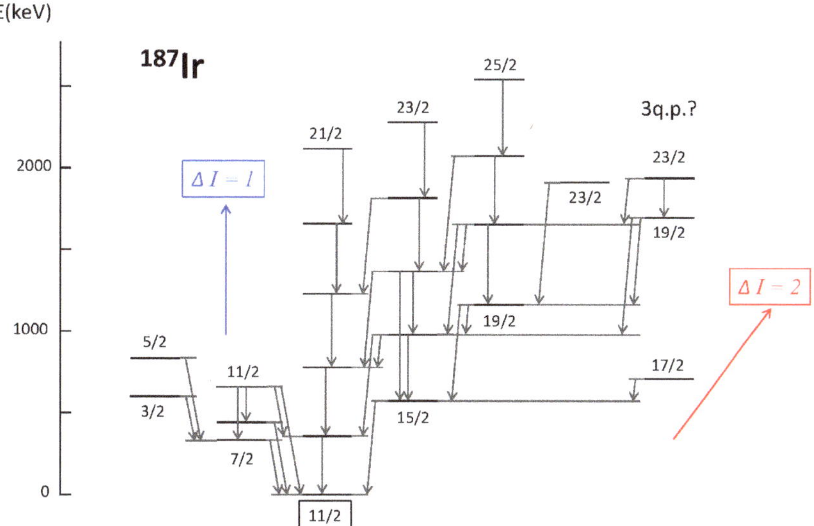


**Figure 21.** Organization of the unique-parity states in $^{187}$Ir associated with $j = 11/2$ into a "hyper-band" pattern due to Meyer-ter-Vehn [95,96]. The data are taken from [100], but the pattern was not recognized there. The states to the right may be associated with three-quasiparticle (3 q.p.) excitations.

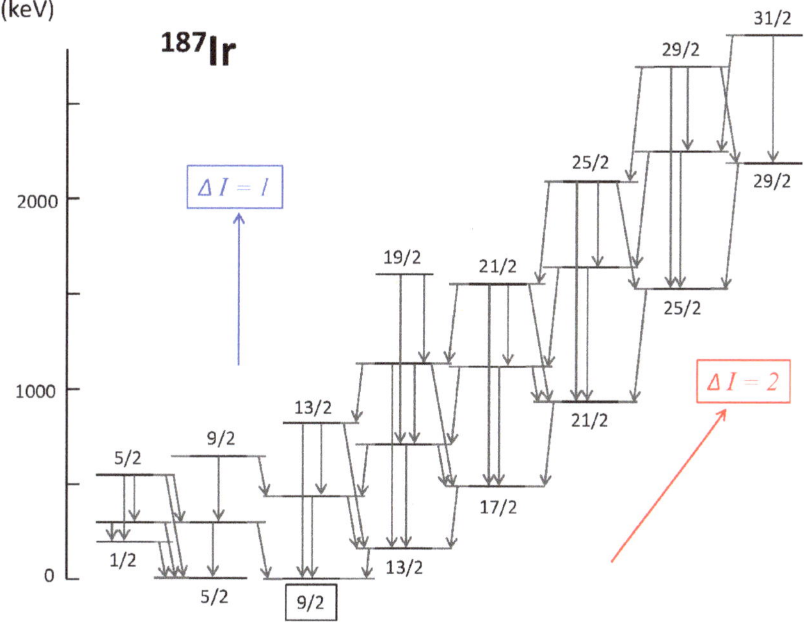


**Figure 22.** Organization of the unique-parity states in $^{187}$Ir associated with $j = 9/2$ intruder configuration into a "hyper-band" pattern due to Meyer-ter-Vehn [95,96]. The data are taken from [100], but the pattern was not recognized there.

The pattern shown in Figure 20 is an organization of experimental data to reflect the dominance of so-called "rotational-aligned" coupling, which occurs in odd-mass nuclei that are not strongly deformed. The leading rotationally aligned set of states is highlighted

in red and extends from the lowest $I = 11/2$ state, diagonally upwards to the right. The spin 11/2 originates from the $1h_{11/2}$ spherical shell model state, which dominates all low-energy negative-parity states in $^{125}$Xe. The lowest tier of states in this set has the spin sequence 11/2, 15/2, 19/2, 23/2, 27/2, ...; the tier just above has the spin sequence 13/2, 17/2, 21/2, 25/2, ... However, as shown, this basic pattern is "repeated", within the set of states highlighted in red, with tiers possessing spin sequences 15/2, 19/2, 23/2, ..., 17/2, 21/2, ... Furthermore, with sets of states, highlighted in blue and green, the red pattern is repeated built on states of $I = 9/2$ and 7/2, respectively. The tiers of $\Delta I = 2$ spin sequences, beyond the first two, result from axial asymmetry and the coupling of the $j = 11/2$ particle to an axially asymmetric rotor. The repeated sets of states, coded with different colours, identified as $I = 9/2$ and $I = 7/2$, arise from alignment of the $j = 11/2$ particle in the deformed quadrupole field of $^{125}$Xe (such as occurs in the Nilsson model) and rotations about the unfavored axis of the triaxial rotor. These multiple tiers have been considered in some nuclei, by some authors, as candidates for so-called "wobbling": such wobbling, however, requires strong $E2$ transitions between tiers of states, i.e., decays appearing as vertical arrows; in $^{125}$Xe, these transitions appear to be dominated by $M1$ transitions, which might be termed "magnetic" rotation; for further details, see Refs. [95,96]. There is controversy regarding $E2$ admixtures in $\Delta I = 1$ transitions; see the general remarks in [101].

A perusal of the literature over recent decades suggests that the view of Meyer-ter-Vehn has been "forgotten". The question that arises from consideration of Figures 18–20 (and Figures 21 and 22) is: How small a deformation is meaningful in weakly deformed nuclei? We consider this question but do not reach a final answer. An important outcome of the Meyer-ter-Vehn model [95,96] has been a multi-$j$ version of the model, which is usually described as the particle-triaxial-rotor model (PTRM) [102]. Indeed, it was applied to a description of $^{125}$Xe [103,104] before the more recent detailed data set [99]. Thus, the focus here is on a deeper look at the basics of these models, especially near their weak deformation limit.

A major factor in particle-rotor models, both axially symmetric and axially asymmetric, when the deformation is not large, is so-called "Coriolis" or "rotational" alignment. A milestone paper that pointed to this effect was by Frank Stephens and coworkers [105], based on observations in the odd-mass lanthanum isotopes; an up-to-date view of their perspective is shown in Figure 23. An up-to-date view of all known negative-parity states in the odd-mass lanthanum isotopes is shown in Figure 24. Except for $^{133}$La, the low-spin couplings are not yet observed. The coupling to low spin is addressed shortly in this Section. The pattern in Figure 23 is referred to as "rotation-aligned" coupling. A simple explanation is given in Figure 25. The essential mechanism is the competition between "rotation alignment" and "deformation alignment", where deformation alignment is embodied in the basic quantum mechanics of the Nilsson model. The quantum mechanics of rotation alignment is described by the $\boldsymbol{I} \cdot \boldsymbol{j}$ term of the particle-rotor model: Figure 25 is a semi-classical view of this term. A naïve view of the weak-coupling limit of this term is that it dominates the coupling, and the total spin, $I$ and $j$ become collinear. This already appears to happen in the odd-mass lanthanum isotopes for the $I = j + 2$ states, but this does not address the question for the other possible couplings of $j$ (to the even-even core) to yield a resultant total spin $I$.

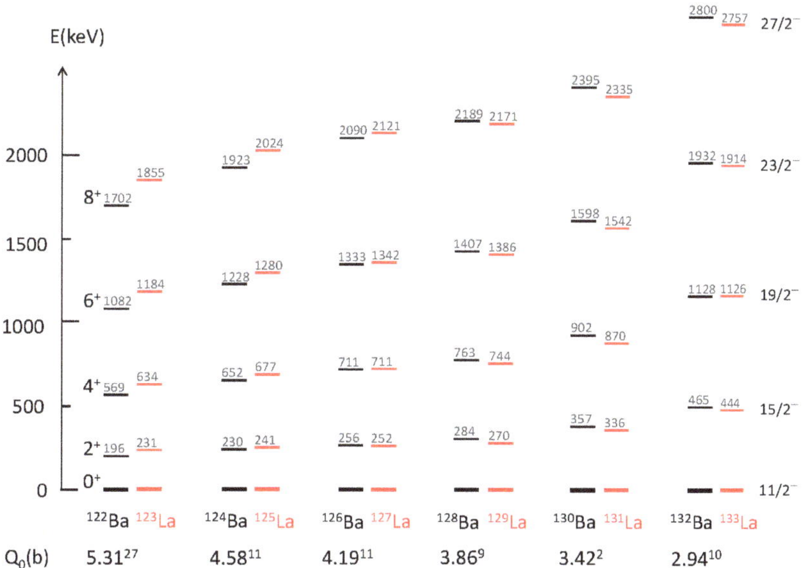


**Figure 23.** Energy pattern of the high-spin unique-parity states in the odd-mass lanthanum isotopes, compared to the ground-state bands of the $(A-1)$ even-mass barium isotopes. This figure is an up-to-date view of one first proposed by Stephens et al. [105], where the term "rotation-aligned coupling scheme" was introduced.

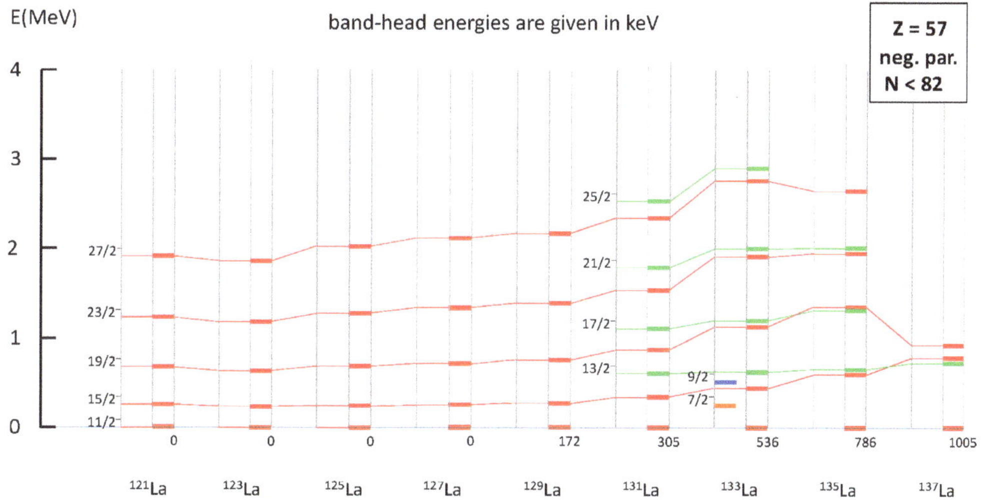


**Figure 24.** Systematics of unique-parity states in the odd-mass lanthanum isotopes for $N \leq 82$. Note the sparse information on low-spin states. The data are taken from ENSDF [22].

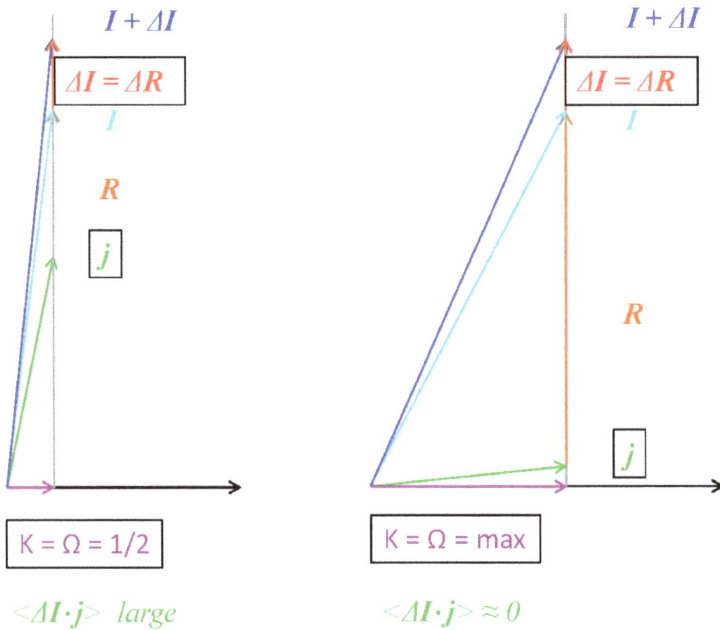


**Figure 25.** Semi-classical view of the rotation-alignment or Coriolis term in the particle-rotor model. The two coupling schemes depict the two extremes of the Nilsson model for a single-$j$ state in a spheroidally deformed mean-field, labelled by $\Omega = 1/2$ and $\Omega = \max$. For the particle-rotor coupling, $\boldsymbol{I} = \boldsymbol{R} + \boldsymbol{j}$ and very different alignments of $\boldsymbol{I}$ and $\boldsymbol{j}$ are possible. Recognizing that, e.g., Figure 23 focuses on energy differences, one sees from these diagrams that *differences* in $\boldsymbol{I}$, i.e., $\boldsymbol{\Delta I}$, a vector quantity, result in very different values for $\boldsymbol{\Delta I} \cdot \boldsymbol{j}$ and hence for expectation values of this quantity. Reproduced from [8].

Coupling of $j$ to an even-even core to yield low-spin states with unique parity is sparsely characterized in weakly deformed nuclei, as already noted. An extreme "weak coupling" example is shown in Figure 26. By weak coupling, one means that a set of states, resulting from coupling an odd-nucleon of spin $j$ to the core $2_1^+$ excitation, $j \otimes 2_1^+$, with spins $|j-2| \leq J \leq |j+2|$, appears as a closely spaced multiplet, at an excitation centred on the $2_1^+$ energy of the even-even core, connected by unfragmented $E2$ strength to the spin-$j$ ground state, is observed. This simple view is approximately realized in Figure 26: Coulomb excitation strongly populates five states with $J = 5/2, 7/2, 9/2, 11/2$, and $13/2$; it also weakly populates a $5/2^+$ state at 941 keV and a $9/2^+$ state at 1461 keV. These two states are due to a shape coexisting or intruder band (a Nilsson $1/2^+[431]$ decoupled rotational band) details of which are not important to the present focus. It is sufficient to note that the weakly coupled multiplet is identifiable, with the provision that the spin 5/2 and 9/2 members of the multiplet are manifested with some configuration mixing due to near degeneracies with intruder band configurations. This would suggest that the weak coupling limit is a familiar pattern, and the quest is nearly complete, pending filling in some minor details. However, a recent result [64] shows that the situation is far from being the weak-coupling limit: this is illustrated in Figure 27. Even though the energies appear to approximate the weak-coupling pattern, significant collective $E2$ strength has been "acquired" by the addition of a single extra-core proton. More specifically, the odd-$A$ nucleus $^{129}$Sb shows additional collectivity in Coulomb excitation from the ground state, above that of the $^{128}$Sn core. A shell model description with effective charges of $e_p = 1.7e$ and $e_n = 0.8e$ set from the $B(E2; 0_1^+ \to 2_1^+)$ values of the semimagic neighbours $^{130}$Te for protons and $^{128}$Sn for neutrons, goes some way towards describing this additional collectivity. This simple

particle–core coupling situation therefore gives evidence of emerging collectivity over and above that implied by the significant effective charges associated with the individual proton and neutron contributions.

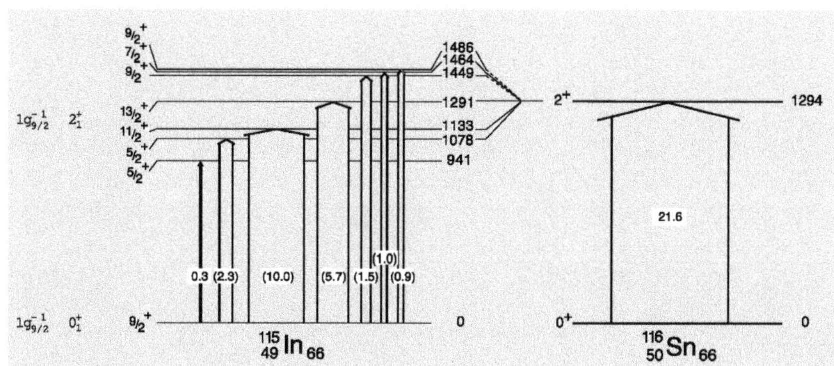


**Figure 26.** Example of near-weak coupling in $^{115}$In, observed by Coulomb excitation and shown in comparison to Coulomb excitation for the neighbouring even-even "core" nucleus, $^{116}$Sn. These states are due to the proton coupling $1g_{9/2}^{-1} \otimes 2_1^+$. There is some fragmentation of strength for specific spin-parities: this results from mixing with intruder states. The states at 941 keV, $5/2^+$ and 1449 keV, $9/2^+$ are members of a decoupled rotational band built on the $1/2^+[431]$ Nilsson configuration [106]. This configuration has $1g_{7/2}$ parentage and results in a rotational band with decoupling parameter, $a \simeq -2$, which puts the $3/2^+$ state below the $1/2^+$ state. Note that the negative parity states are not shown; the lowest negative parity states in the odd-In isotopes are shown in Figure 28. The numbers in parentheses are $B(E2)$ values for the excitation process, in units of $e^2\text{fm}^4 \times 10^2$ ($100\ e^2\text{fm}^4 = 60$ W.u. for $A = 115$.) The data are taken from [107]. Reproduced from [6].

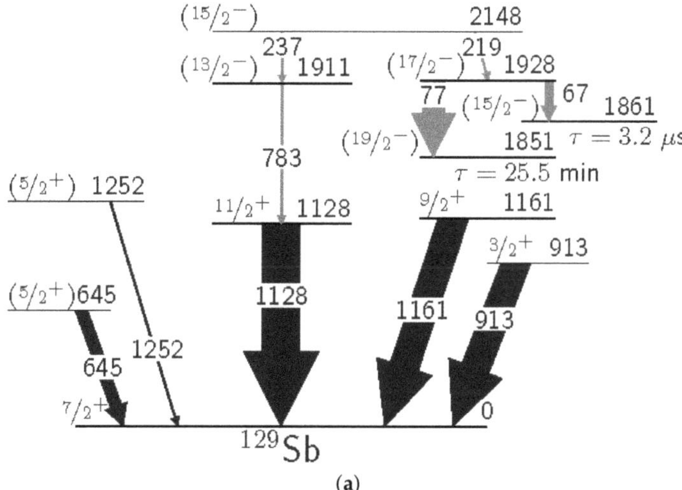


**Figure 27.** *Cont.*

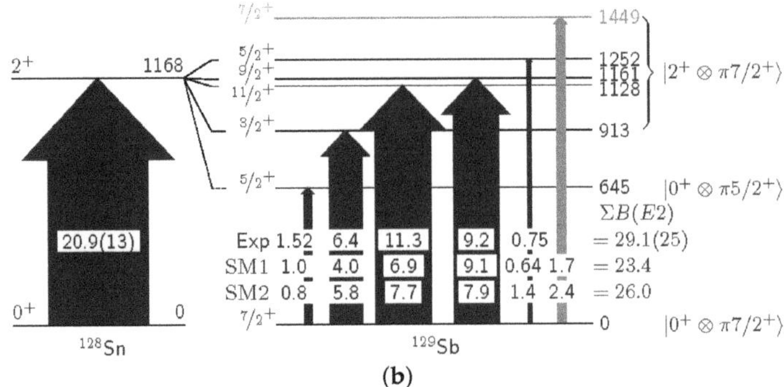


**Figure 27.** (**a**) Partial level scheme for $^{129}$Sb. Grey transitions result from excitation of the 1851-keV isomer present in the beam. (**b**): Fragmentation of the $E2$ strength in W.u. over the $2^+ \otimes \pi g_{7/2}$ multiplet members and candidate $\pi d_{5/2}$ state of $^{129}$Sb and enhancement of total strength as compared to the $^{128}$Sn core. The grey colored transition was not experimentally observed. The figure is reproduced from [64]. See also Ref. [64] for details of the large-basis shell model calculations SM1 and SM2, which employ the same basis space but alternative contemporary residual interactions.

The status of particle–core coupling presented above, and in additional calculations by Gray et al. [108], suggests that there is not a good understanding with respect to the $Z = 50$ closed shell and the odd-mass In and Sb isotopes. The issue extends across the entire mass surface due to a severe lack of critical data. The systematics of the low-lying states in the odd mass In and Sb isotopes are shown in Figures 28 and 29, respectively. The pattern of the In isotopes suggests that, for the negative-parity states, there may be important collective effects which would explain the energy minimum at mid shell. Weak deformation is supported by laser hyperfine spectroscopy studies [109] and is shown in Figure 30. Note that two views of deformation for the In isotopes are presented in Figure 30: a direct view via spectroscopic quadrupole moments—the lower sequence of data points centred on $\beta \sim 0.1$, and an indirect view via isotope shifts—the upper sequence of data points. The latter view can be inferred to contain a dynamical contribution, but this aspect lies beyond the present discussion. The observed pattern for the Sb isotopes suggests a "crossing" of the $2d_{5/2}$ and $1g_{7/2}$ configurations. However, at present, the question of the collectivity associated with low-lying states in the odd-Sb isotopes suggests caution is needed in making the interpretation of the lowest $5/2^+$ and $7/2^+$ states as resulting from pure shell model configurations.

Skyrme Hartree–Fock calculations with the SKX interaction [110] correctly track the nominal $2d_{5/2}$ vs. $1g_{7/2}$ level ordering in the Sb isotopes, but the location of the $3s_{1/2}$ orbit does not track with the behaviour of the observed $J = 1/2^+$ state with its shift in energy across the observed $7/2^+$ state. In the indium isotopes, the single-particle levels in the potential generated by SKX are more separated in energy and roughly track with the observed levels of the relevant spin–parity. It appears that the indium levels remain quite regular because the parent orbits are well separated in energy in the mean field and the observed states are less affected by residual interactions; however, one sees clear evidence from the quadrupole moments in Figure 30 that deformation develops at mid-shell. In contrast, the Sb isotopes have the $2d_{5/2}$ and $1g_{7/2}$ single-particle states quite close in the mean field calculation. Thus, the observed level ordering can be sensitive not only to changes in the mean field, but also to residual interactions and deformation effects.

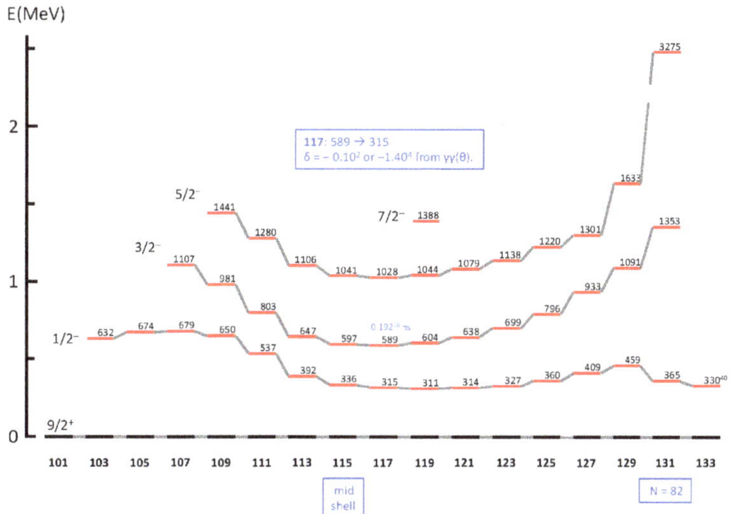


**Figure 28.** Systematics of the negative-parity states in the odd-mass indium isotopes relative to the $9/2^+$ ground states. Naïvely, these states could be interpreted as the shell model single proton–hole configurations $2p_{1/2}$, $2p_{3/2}$, and $1f_{5/2}$; but $E2$ transition strengths would be desirable before such an interpretation is made. The "parabolic" energy trend suggests interactions with neutrons across the shell with a characteristic energy minimum near the mid-shell point ($N = 66$), as indicated. Note the severe deficiency of data for electromagnetic decay strengths: there is one half life, for the 589 keV state in $^{117}$In, and the $E2/M1$ mixing ratio for the decay of this state is ambiguous. Data for $^{107}$In are from [111] and for $^{131}$In are from [112]; other data are taken from ENSDF [22].

**Figure 29.** Systematics of selected states in the odd-mass antimony isotopes. The short blue lines show the energies of the $2_1^+$ states in the even-even $^{A-1}$Sn isotopes with respect to which weak coupling in the $^A$Sb isotopes can be assessed. The ground states of $^{125-135}$Sb are not shown; they all have spin–parity $7/2^+$.

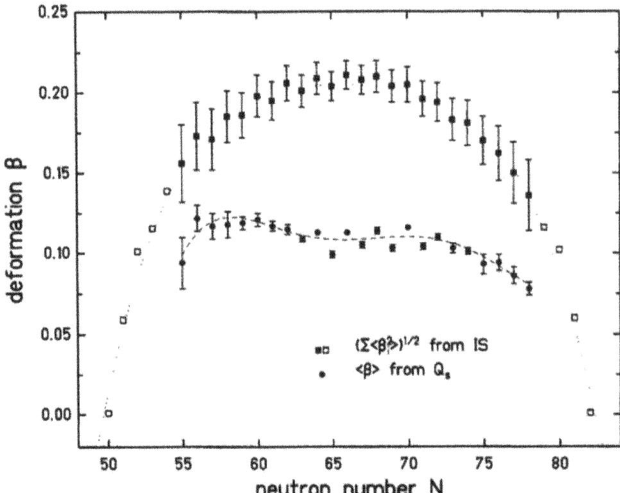


**Figure 30.** Deformations of the odd-In isotopes deduced from spectroscopic quadrupole moments and isotope shifts following laser hyperfine spectroscopy [109]. Reprinted from [109], Copyright (1987), with permission from Elsevier.

Mass regions where the issue of emergent collectivity needs detailed spectroscopic study are addressed in Section 5 through Section 9. In particular, Sections 5 and 6 focus on the Ni and Ca isotopes, respectively.

Let us emphasize that there is a substantial body of evidence for the role of triaxial shapes in nuclei that are of moderate deformation. This is supported by the observation of "too many low-energy states for axial symmetry" in unique-parity excitations, such as shown in Figures 20–22. It is also supported by the application of the Kumar–Cline sum rules [113,114] to shell model electromagnetic strengths, as summarized for calculations of the Bohr-model deformation parameters derived from the shape invariants for the tellurium isotopes in Figure 31. These features do not imply that $^{128-134}$Te can be modeled as weakly deformed triaxial rotors in their low-lying states up to spin $6^+$. Scrutiny of the wave functions and predicted $g$ factors, for example, indicates that the structures of the lowest few states are very different, despite their apparently similar shape parameters. These excitations are not rotations of a single intrinsic structure as is supposed in the triaxial rotor model. Although the magnetic moments indicate that the Te isotopes near the $N = 82$ shell closure cannot be accurately modelled as weakly deformed triaxial rotors, a triaxial rotor description may prove appropriate as the number of neutron holes increases. The fact that the excited-state shapes in Figure 31 are all triaxial with $\gamma \approx 30^\circ$ may suggest that the pathway of emerging collectivity in this region progresses from near-spherical nuclei near $^{132}$Sn, to weakly-deformed triaxial rotors as an intermediate step, before finally reaching more strongly deformed prolate rotors near mid-shell. Further data and calculations across an extended range of Te and Xe isotopes would help to assess this conjecture.

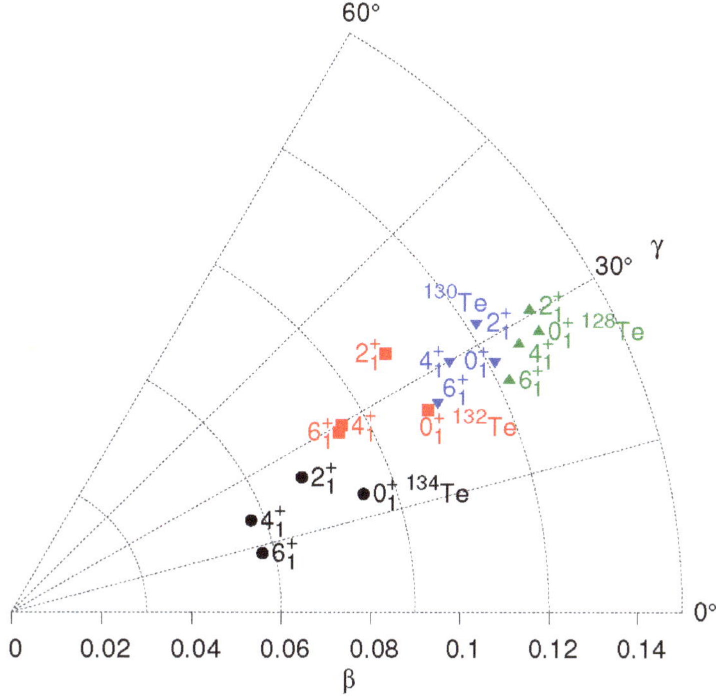


**Figure 31.** Average Bohr-model deformation parameters for yrast states in $^{128,130,132,134}$Te, assuming an ellipsoidal deformed nucleus and determined from shell-model calculations using the Kumar–Cline sum rules. For clarity, the fluctuations are not plotted. They are similar in magnitude for all cases, and by happenstance, the "softness" or fluctuation associated with each point is comparable to the scatter in the plotted points. Reproduced from [115].

## 5. Emergent Collectivity in the Nickel Isotopes

Currently, there is a high interest in neutron-rich nuclei. This is because of unprecedented access to completely new mass regions, and soon to come facilities that may "reach" even further. In particular, the neutron-rich Ni isotopes and the adjacent open-shell isotopes have received much attention. The systematic features of the low-energy excited states in the even-mass Ni isotopes are shown in Figures 32 and 33. A naïve interpretation of $^{58-66}$Ni (Figure 32) is that they are vibrational; however, the error of using only energies to make structural interpretations of weakly deformed nuclei has now been substantially demonstrated [116]. The structure of $^{58-66}$Ni is addressed in detail in this Section, with attention to seniority and shape coexistence. An unequivocal interpretation of $^{70-76}$Ni (Figure 33) is that these isotopes are dominated by seniority coupling. However, this is an incomplete view, as details in Figure 33 imply; the structure of $^{70-76}$Ni is also addressed in detail in this Section.

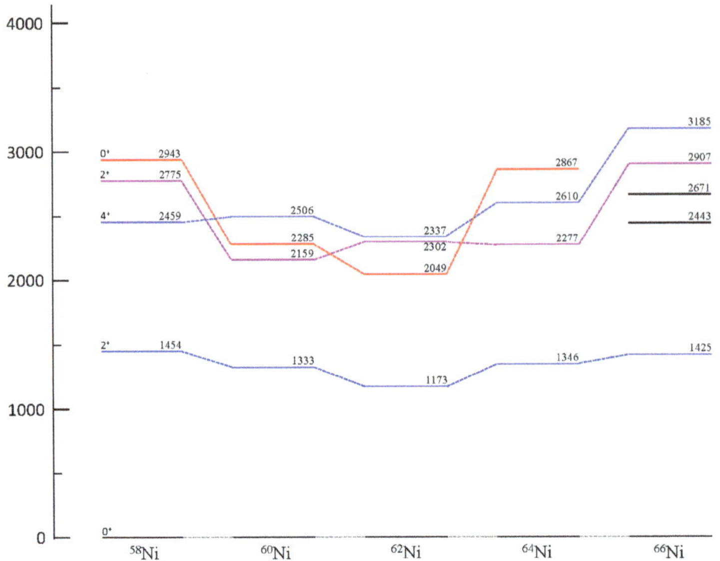


**Figure 32.** Systematics of the states with spin 0, 2, 4 and positive parity in $^{58-66}$Ni. The states in $^{66}$Ni at 2443 and 2671 keV are assigned spin–parity $0^+$ and are taken from [117]; other data are taken from ENSDF [22].

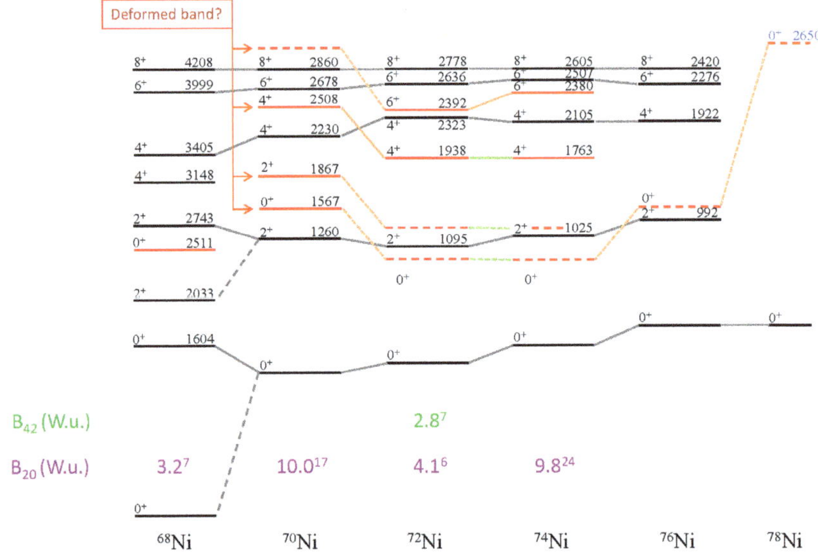


**Figure 33.** Systematics of the lowest positive-parity excited states in $^{68-78}$Ni. Data are taken from: [118,119] ($^{68}$Ni); [120–122] ($^{70}$Ni); [123] ($^{72}$Ni); [124] ($^{74}$Ni); [125] ($^{76}$Ni). The $B(E2)$ data are from ENSDF [22] for $^{68,70}$Ni, from [126] for $^{72}$Ni and from [127] for $^{74}$Ni. The dashed red lines are a conjecture regarding the possibility of deformed intruder bands based on the interpretation of $2_2^+$ and $4_2^+$ states being members of such bands, and would be consistent with the interpretation of such a structure in $^{70}$Ni, suggested by Chiara et al. [122]; they are interpreted as seniority-four states by Morales et al. [124].

Recently, a study of conversion electrons following $(p, p')$ excitation of $^{58,60,62}$Ni was made by Evitts et al. [128,129]. A notable result was the observation of strong $E0$ decay branches from second-excited $2^+$ states to the first-excited $2^+$ states. The details for $^{62}$Ni are shown in Figure 34. Previously, strong $E0$ decays had been established for a series of excited $0^+$ states in $^{58,60,62}$Ni [130]. However, an unresolved puzzle was that, whereas this strength was associated with proton–pair excitations in $^{58,60}$Ni, this was not the situation in $^{62}$Ni. The paper of Evitts et al. [128,129] points to a possible resolution: Figure 34 suggests that the $0_2^+$ state at 2049 keV is the head of a strongly deformed band and the 2302-keV $2^+$ state is the first-excited band member. The proton pair–excitation at 3524 keV is shown. Our interpretation of the 2049 keV $0^+$ state in $^{62}$Ni is a "4p-4h" excitation of the $^{56}$Ni core. Such a structure would not be populated in ($^3$He, n), ($^{16}$O, $^{14}$C), ($^6$Li, d) or ($^{16}$O, $^{12}$C) reactions, which were the spectroscopic probes used to identify the proton–pair excitations in $^{58,60,62}$Ni [131–134].

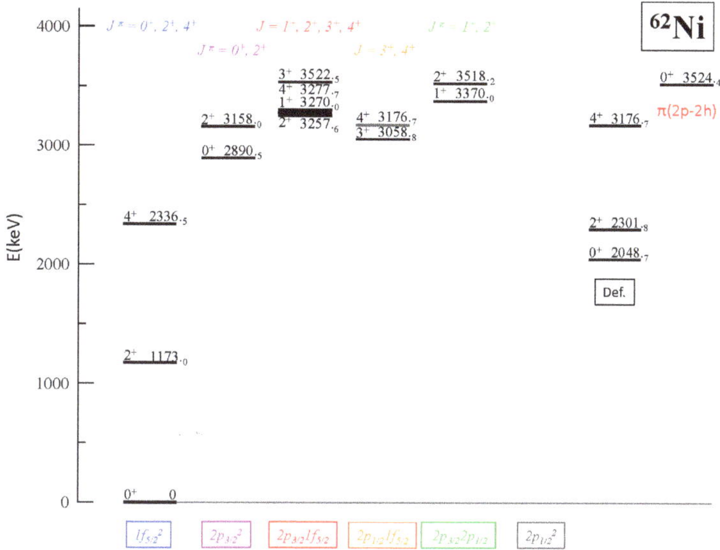


**Figure 34.** Organization of the lowest excited states in $^{62}$Ni into seniority-dominated structures and a new strongly deformed band. At present, this view must be considered a conjecture. The proposed seniority-two structures are labelled by the shell model configurations with their associated spins and parities. The deformed band is discussed in the text. The $0^+$ state at 3524 keV is assigned as a proton–pair excitation based on two-proton [134] and $\alpha$ [132] transfer reaction spectroscopic studies.

The seniority-dominated structure of $^{70,72,74,76}$Ni has an unusual complication. While it is simple in $^{76}$Ni, as established by direct observation of a cascade of four gamma rays from an isomer with half-life 590 ns [125], this isomerism has disappeared in the lighter even-mass nickel isotopes. The situation is now resolved at the level of the multiple decay branches from the candidate spin–parity $8^+$ states characteristic of a $(j = 9/2)^2$ seniority, $v = 2$ multiplet; but an open question is the nature of the low-lying states that facilitate these "fast" decays. Two possibilities exist: the "extra" states are seniority four, $v = 4$ states or, the "extra" states are members of coexisting deformed bands. It is possible for $v = 4$ states to appear lower in energy than $v = 2$ states in the manner manifested in $^{72,74}$Ni [124]. It is also plausible that shape coexistence is occurring at low energy in these nuclei. In favor of the latter interpretation is that shape coexistence has been suggested to occur at low energy in $^{70}$Ni [121,122]. Furthermore, a near identical structure in the $N = 50$ isotones involving the proton $1g_{9/2}$ subshell exhibits robust seniority isomerism

with no involvement of $v = 4$ states producing fast decays (see, e.g., [135], although one has to note seniority breakdown at low spin inferred from lifetime measurements [136]).

## 6. Collectivity in the Calcium Isotopes

The calcium isotopes hold a unique position in the study of nuclear structure. With $Z = 20$ and a reach to either side of $N = 20$ and 28, they should be a perfect illustration of closed-shell behaviour in nuclei, except that they are not. Figures 1 and 2 open the focus of this contribution, with a perspective on $^{40}$Ca as an $N = Z$ doubly closed-shell nucleus and on $^{48}$Ca as an $N > Z$ doubly closed-shell nucleus: $^{48}$Ca conforms to expectations; $^{40}$Ca does not. Indeed, recently, the time-honored view that closed shells only occur at 2, 8, 20, 28, 50, 82, 126 has been questioned due to unusual systematic features in $^{52,54}$Ca: this is of high interest with respect to forthcoming prospects for new facilities which will provide access to very neutron-rich nuclei, and the calcium isotopes in particular. (The current "reach" into the neutron-rich calcium isotopes is two events in 39 h of beam time, assigned to $^{60}$Ca [137]).

A highly attractive feature of the calcium isotopes between $N = 20$ and 28 is that they should be dominated by a single $j$ shell, the $1f_{7/2}$ shell. Figure 35 shows data that support this view. The $j = 7/2$ seniority $v = 2$ states ($J = 2, 4, 6$) are highlighted in red; the $j = 7/2$ seniority $v = 4$ states ($J = 2, 4, 5, 8$) in $^{44}$Ca are highlighted in orange. Note that the $J = 4$, $v = 4$ configuration mixes with the $J = 4$, $v = 2$ configuration. Further note that the $J = 5$ state has not been observed. Figure 35 also shows that other states appear at low energy in $^{42,44,46}$Ca: these are discussed with reference to the following, Figures 36–40 and Table 5.

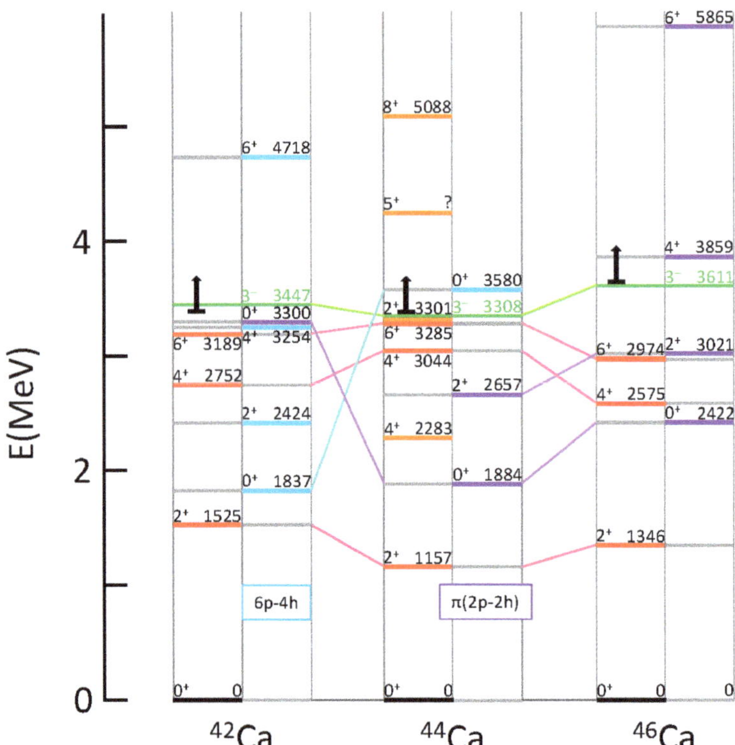


**Figure 35.** Seniority and shape coexistence view of $^{42-46}$Ca. Data for $^{46}$Ca include recent results of Pore et al. [138] and Ash et al. [139]. The deformed band in $^{42}$Ca (shown in blue) is observed up to

spin 12 [140]. The deformed bands in $^{44}$Ca and $^{46}$Ca are indicated in purple. The seniority-two states are indicated in red (but see details below in this caption); the seniority-four states, unique to $^{44}$Ca, are indicated in orange (but, again, see details below in this caption). The distinction between the deformed bands is based on multi-nucleon transfer reactions: see Table 5. The seniority-2 and seniority-4 structures in $^{44}$Ca, the $4^+$ states at 2.28 and 3.04 MeV, and the $2^+$ states at 1.16, 2.66 and (probably) 3.30 MeV are mixed; see [141]. For the seniority structures associated with the $\nu 1f_{7/2}$ configuration, see Figure 3. Note that, while there is a high-spin study of $^{44}$Ca (Lach et al. [142]), the deformed band has not been characterized. For a complementary perspective of the calcium isotopes, see also details in Figures 36 and 41. The $3^-$ states, which are not part of the present discussion, are shown in green. Horizontal bars with vertical arrows indicate excitation energies above which states are omitted.

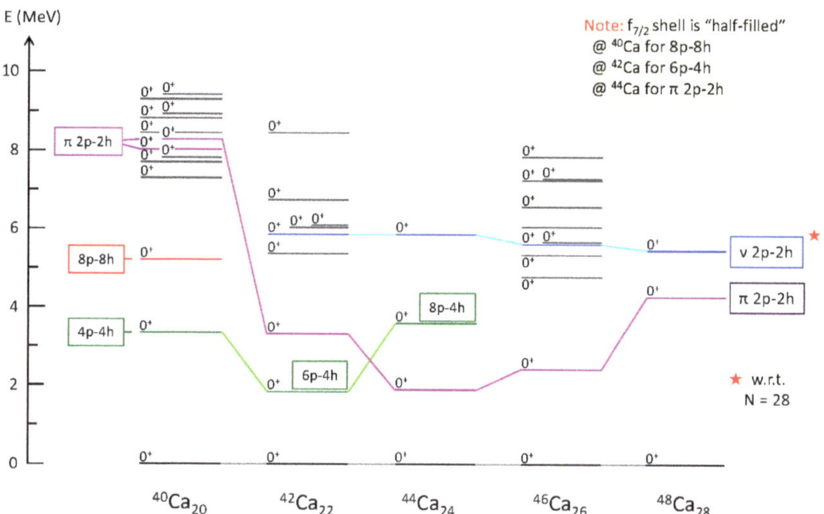


**Figure 36.** Excited $0^+$ states in $^{40-48}$Ca. All known $0^+$ states up to 10 MeV are shown. Assignments to particle–hole configurations are indicated where known and further details are given in Figure 35 and 37, and Table 5. Note the inset box which indicates when the $\nu 1f_{7/2}$ shell is half filled. Further note that the $\nu 2p-2h$ configurations are with respect to $N = 28$, and are identified by the neutron–pair–addition reaction (t,p).

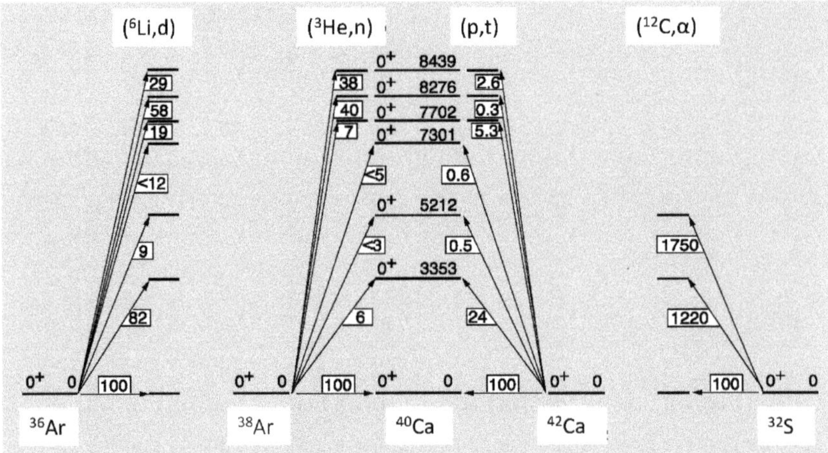


**Figure 37.** Excited $0^+$ states in $^{40}$Ca viewed from the perspective of multiparticle transfer reactions. The first excited $0^+$ state at 3353 keV is usually labelled as "4p-4h" on the basis of its strong population in the $^{36}$Ar($^6$Li,d) reaction, but note the strong population of $0^+$ states, particularly around 8.3 MeV. The second excited $0^+$ state at 5212 keV is usually labelled as "8p-8h" on the basis of its strong population in the $^{32}$S($^{12}$C,$\alpha$) reaction (the population of the 3353 keV state could involve partial filling of the hole states in $^{32}$S, and does not necessarily imply an 8p-8h admixture to the 3353 keV state). Figure 1 depicts the deformed and superdeformed bands built on the 3353 and 5212 keV states, respectively. Proton–pair configurations appear to dominate $0^+$ states around 8 MeV, as supported by the $^{38}$Ar($^3$He,n)$^{40}$Ca reaction. Based on the $^{42}$Ca(p,t)$^{40}$Ca reaction, neutron–pair configurations do not dominate below 8.5 MeV. This leaves the 7301 keV excited $0^+$ state as the leading structure of interest for an interpretation: possibly it is a "6p-6h" state (cf. Figure 40). Taken from [6].

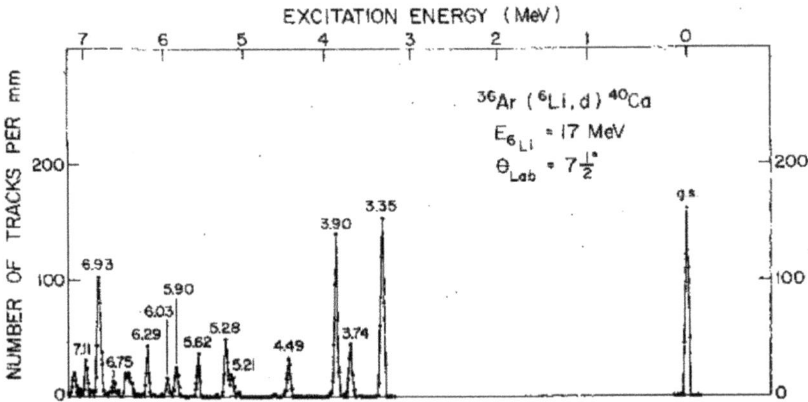


**Figure 38.** Spectrum of deuterons following the reaction ($^6$Li,d) on a $^{36}$Ar target. The most strongly populated excited states are members of the deformed band with $E_x$ ($J^\pi$): 3353 ($0^+$), 3904 ($2^+$), 5279 ($4^+$), 6930 ($6^+$), cf. Figure 1. Note that the peaks at 5.28 and 6.93 MeV are multiplets. Reprinted with permission from [143]. Copyright (1979) by the American Physical Society.

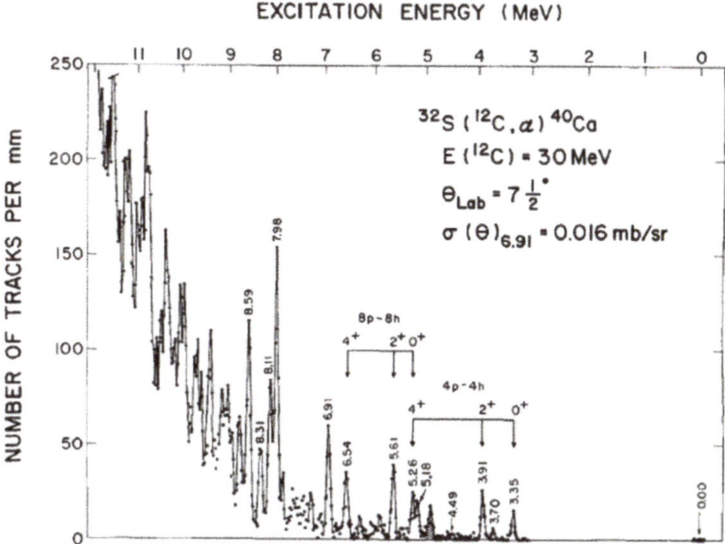


**Figure 39.** Spectrum of alphas following the reaction ($^{12}$C, $\alpha$ ) on a $^{32}$S target. States in both the 4p-4h and the 8p-8h deformed bands are populated. The population of the 4p-4h band may involve a partial filling of the "eight holes" in the target. Reprinted from [144], Copyright (1972), with permission from Elsevier.

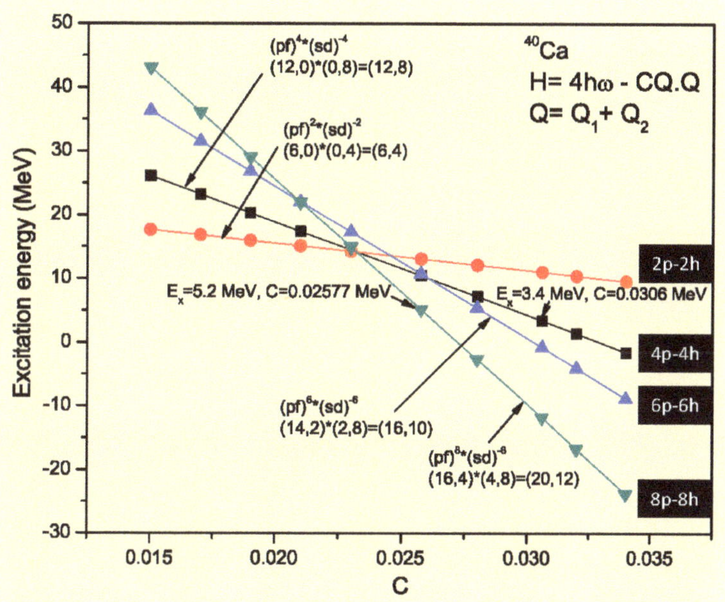


**Figure 40.** Estimate of the multiparticle-multihole basis state energies for $^{40}$Ca using a schematic $su(3)_{\text{particle}} \otimes su(3)_{\text{hole}}$ model with a $Q \cdot Q$ interaction of strength $C$, where $Q = Q_1 + Q_2$ and $Q_1$ and $Q_2$ act on the $N_p$ $(\lambda_1, \mu_1)$ and $N_h$ $(\lambda_2, \mu_2)$, $pf$ and $sd$ irreps, respectively. The figure is from a collaboration between one of us (JLW) and the late David Rowe.

**Table 5.** Tabulation of multi-nucleon transfer reaction spectroscopic data that provide (limited) information on the multiparticle-multihole structures of excited $0^+$ states in $^{42,44,46}$Ca. The terms "strong" and "weak" refer to strengths of population of states in these transfer reactions. Note that the inference of "particle" and "hole" structure depends on the transfer nucleons and the target configuration with respect to closed shells.

| | $0_1^+$ | $0_2^+$ | $0_3^+$ | $0_4^+$ | $0_5^+$ | $0_6^+$ | **reference** |
|---|---|---|---|---|---|---|---|
| $^{42}$Ca | | | | | | | |
| $E_x$ (MeV): | 0.00 | 1.84 [a] | 3.30 [b] | 5.35 | 5.86 [c] | 6.02 | |
| $^{40}$Ca(t,p)$^{42}$Ca | | | | | strong | | [145] |
| $^{40}$Ar($^3$He,n)$^{42}$Ca | | weak | | | | | [146] |
| $^{38}$Ar($^6$Li,d)$^{42}$Ca | | | strong | | | | [147] |
| $^{44}$Ca | | | | | | | |
| $E_x$ (MeV): | 0.00 | 1.88 [b] | 3.58 [a] | 5.86 [c] | | | |
| $^{48}$Ti(d,$^6$Li)$^{44}$Ca | | | strong | | | | [148] |
| $^{46}$Ti($^{14}$C,$^{16}$O)$^{44}$Ca | | strong | | | | | [149] |
| $^{42}$Ca(t,p)$^{44}$Ca | | | | strong | | | [145] |
| $^{46}$Ca | | | | | | | |
| $E_x$ (MeV): | 0.00 | 2.42 [b] | 4.76 | 5.32 | 5.60 [c] | 5.63 [c] | |
| $^{48}$Ti($^{14}$C,$^{16}$O)$^{46}$Ca | | strong | | | | | [149] |
| $^{44}$Ca(t,p)$^{46}$Ca | | | | | strong | strong | [145] |

[a] $4p-4h \otimes \nu 1f_{7/2}^n$, [b] $\pi(2p-2h)$, [c] $\nu(2p-2h)$.

Figure 36 shows the problem of the simple $1f_{7/2}$ shell-based view of $^{42,44,46}$Ca: there are "too many $0^+$ states" at low energy in the even-mass calcium isotopes. A single-shell-seniority view does not possess any excited $0^+$ states; but the second excited state in $^{42,44,46}$Ca is a $0^+$ state. Furthermore, the first-excited $0^+$ states in $^{42,44,46}$Ca are not configurations due to a common origin. The evidence for this is presented in the following paragraphs.

The key to the structure of the calcium isotopes with $N = 20 - 28$ is manifested in multi-nucleon transfer reaction spectroscopy for $^{40}$Ca, summarized in Figure 37. The details are complex and counterintuitive. Indeed, the evidence "has to be seen to be believed"; Figures 38 and 39 show the evidence. It is important to recognize the role of the target nuclei in that they define "hole" structures with respect to which the transferred multi-nucleon "clusters" are "added". Added nucleons can completely fill the holes (ground-state population), or partially fill the holes, or not fill the holes at all. There is a dominance of transfer to states that involve the target holes remaining completely unfilled, i.e., a 4p-4h configuration in the ($^6$Li,d) reaction and an 8p-8h configuration in the ($^{12}$C,$\alpha$) reaction. Note that the ($^6$Li,d) reaction does not populate the states of the "8p-8h" band strongly, but band mixing is suggested. Further note that the ($^{12}$C,$\alpha$) reaction can populate the states of the "4p-4h" band by "partially filling" the holes.

A partial guide to the multiparticle-multihole structure of $^{42,44,46}$Ca is presented in Table 5. This view is only partial because of a lack of stable isotope targets. The view leaves open many questions, but overriding all questions is the clear view that the shell model, as a simple model view, catastrophically breaks down in these isotopes. A guide to a likely interpretation is provided by the schematic-model view presented in Figure 40. This treats particle "clusters" and hole "clusters" as distinct entities that interact. This is tractable using an $su(3)$ algebra with a quadrupole–quadrupole, $Q \cdot Q$ interaction (see [150,151] for details). This schematic view suggests a viable "coupling scheme", which serves much like the Nilsson scheme serves in nuclei with deformed ground states, but here serving to handle multiparticle-multihole excitations at low energy in the calcium isotopes.

Figure 40 reveals the enormous energy shifts associated with interactions that produce nuclear deformation. In Figure 1, $B(E2)$ values in association with the deformed bands in $^{40}$Ca are given: if the $B(E2; 4^+ \rightarrow 2^+)$ strength of 170 W.u. (in the 8p-8h band) is scaled to $A = 172$ ($A^{4/3}$ dependence), it has a strength of 1200 W.u., cf. the $4^+ \rightarrow 2^+$ transition in the ground-state band of $^{172}$Yb, where the strength is 300 W.u. Note that the structural interpretations of the deformed bands in $^{40}$Ca are not model based; they are mandated by the transfer reaction data shown in Figures 38 and 39. Most importantly, Figure 40 and specifically the interaction strength of the quadrupole-quadrupole interaction, $C$, illustrates how configurations that are spread over $\sim$90 MeV in a spherical mean-field, i.e., $8\hbar\omega$, can appear almost degenerate in energy. Indeed, it is worthy of comment that Nature could be said to have "barely realized a spherical ground state for $^{40}$Ca" (as also for $^{16}$O). In the spirit of "islands of inversion", there is a veritable archipelago of islands of inversion, multiple inversions, within one-oscillator shell of excitation energy in $^{40}$Ca. Unfortunately, detailed spectroscopy of multiparticle-multihole states in nuclei is confined to this local mass region. A few more details are given before closing this Section.

Let us make some further comments on Figure 40. Just as one arrives at a shell model basis using an oscillator potential with spin–orbit coupling (and further, by deforming the oscillator potential, as a Nilsson model basis), so in Figure 40 one arrives at a multi-shell basis. The justification of invoking this basis is the observation of the 4p-4h and 8p-8h states in $^{40}$Ca at 3.5 and 5.2 MeV, respectively. The excitations of the 2p-2h and 6p-6h states in $^{40}$Ca are not characterized, but there are many $0^+$ excited states known above 7 MeV, as presented in Figure 37. Note that a 5% change in the interaction strength corresponds to a 5 MeV shift in energy at $C = 0.025$ for the 8p-8h configuration. From this perspective, the very existence of a shell model description of nuclei is a "just-so" story, i.e., for a small change in this $su(3)$-model interaction, spherical states in nuclei would have only been encountered as rare, exotic excitations. An example of this is realized in $^{44}$Ti, as depicted in Figure 43. Details behind these schematic estimates are given in [150–152] (see also [153,154]).

A useful spectroscopic view of the persistence of multiparticle-multihole excitations in this mass region is provided by $^{42}$Ca. This nuclide is accessible to transfer reaction spectroscopy and to Coulomb excitation. A view which combines such spectroscopic data is presented in Figure 41. Figure 42 shows a simple view of the structure using "two-state mixing", applied to the lowest states with spins 0, 2, and 4; this description should be compared with shell model and collective model views summarized in Table 6. An important point to note is that, while these coexisting configurations mix, the mixing is sufficiently weak that the underlying dominant structures can be identified: they are a spherical (valence) neutron particle pair and a "6p-4h" structure resulting from the $^{40}$Ca core 4p-4h structure, cf. Figure 37.

The description presented in Figure 42 correctly reproduces the largest $E2$ transition strength, that between the $2_1^+$ and $0_2^+$ states, i.e., this strength is entirely due to mixing with zero contribution from intrinsic strength. The description fails for the $E2$ transition strength between the $2_2^+$ and $0_1^+$ states, and there is a serious failure for the diagonal matrix elements of the $2_1^+$ and $2_2^+$ states. The conclusion is that two-state mixing for spin 2 is inadequate: three (four)-state mixing is necessary. Experimentally, third and fourth $2^+$ states are known at 3392 and 3654 keV (cf. ENSDF [22]); both are populated in the one-neutron addition reaction: spectroscopic characterization of these states is lacking.

**Table 6.** Comparison of *E*2 matrix elements in $^{42}$Ca with shell model (SM) and beyond mean-field (BMF) calculations. The differences between theory (th) and experiment (ex) are shown as $[(\langle E2\rangle_{\text{ex}} - \langle E2\rangle_{\text{th}})/\langle E2\rangle_{\text{ex}}] \times 100\%$. Details of these calculations are given in [155]: the shell model calculations follow details similar to those employed in [156]. Comparison of the two-state mixing results, shown in Figure 42, suggests serious deficiencies in these two models, which are state of the art. Note that all three calculations obtain an incorrect sign for the $2_2^+ \to 0_1^+$ E2 matrix element, and they all seriously fail for the diagonal matrix elements. Adapted from [155].

| $I_i \to I_f$ | $\langle I_i\|\|E2\|\|I_f\rangle$ $e$ fm$^2$ | | | % Difference | |
|---|---|---|---|---|---|
| | **Experiment** | **SM** | **BMF** | **SM** | **BMF** |
| $2_1^+ \to 0_1^+$ | $20.5 \pm 0.6$ | 11.5 | 9.14 | 78 | 124 |
| $4_1^+ \to 2_1^+$ | $24.3 \pm 1.2$ | 11.3 | 12.2 | 115 | 99 |
| $6_1^+ \to 4_1^+$ | $9.3 \pm 0.2$ | 8.2 | 14.3 | | |
| $0_2^+ \to 2_1^+$ | $22.2 \pm 1.1$ | 11.9 | 6.1 | 87 | 264 |
| $2_2^+ \to 0_1^+$ | $-6.4 \pm 0.3$ | 9.4 | 4.4 | 32 [a] | 45 [a] |
| $2_2^+ \to 2_1^+$ | $-23.7^{+2.3}_{-2.7}$ | −13.6 | −7.7 | 74 | 208 |
| $4_2^+ \to 2_1^+$ | $42^{+3}_{-4}$ | 21.9 | 10.1 | 92 | 316 |
| $2_2^+ \to 0_2^+$ | $26^{+5}_{-3}$ | 32 | 42 | 19 | 38 |
| $4_2^+ \to 2_2^+$ | $46^{+3}_{-6}$ | 52 | 70 | 12 | 34 |
| $2_1^+ \to 2_1^+$ | $-16^{+9}_{-3}$ | −4.3 | 0.1 | | |
| $2_2^+ \to 2_2^+$ | $-55 \pm 15$ | −31 | −42 | | |

[a] Wrong sign.

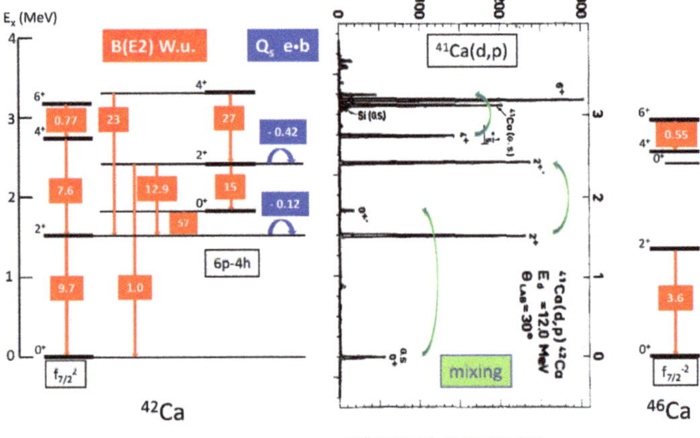


**Figure 41.** Two key spectroscopic views of $^{42}$Ca and a comparison with $^{46}$Ca. On the left, the lowest positive-parity states are shown together with the *E*2 transition strengths in W.u. between these states and diagonal values for the *E*2 matrix elements in *e*b of the $2_1^+$ and $2_2^+$ states, as determined by Coulomb excitation [155]. In the centre of the figure, the population of these states by the $^{41}$Ca(d,p)$^{42}$Ca reaction is shown (the spectrum is reproduced from [157], Copyright (1972), with permission from Elsevier.). This reaction should only populate the $0^+$ ground state and one each for states of spin–parity $2^+, 4^+$, and $6^+$ corresponding to a seniority $v = 2$ multiplet in association with the expected $1f_{7/2}$ orbital, which is the only shell model subshell for $20 < N < 28$: these data provide evidence that there is mixing between these seniority configurations and other structure which is intruding to low energy. On the right, for comparison, the lowest positive-parity states in $^{46}$Ca together with known *E*2 transition strengths are shown. In addition to cited sources, data are taken from ENSDF [22].

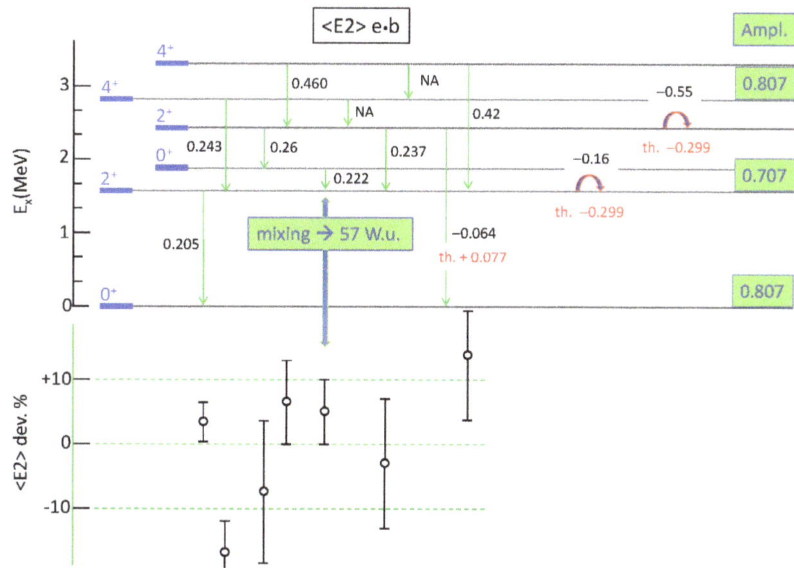


**Figure 42.** Two-state mixing for the lowest pairs of states with spin–parity $0^+$, $2^+$, and $4^+$ in $^{42}$Ca. The $E2$ matrix elements are shown for transitions and level quadrupole moments. The mixing amplitudes are fixed from the fragmentation of the one-neutron addition spectroscopy shown in Figure 41, viz. 0.807 in the ground state, 0.707 in the $2_1^+$ state and 0.807 in the $4_1^+$ for the $1f_{7/2}$ components of these states. There are two fitted parameters: $Q_0 = 10\ e\ \text{fm}^2$ for the $1f_{7/2}$ configurations and $Q_0 = 40\ e\ \text{fm}^2$ for the intruder configurations. The $Q_0$ value for intruder configurations is multiplied by a rotor model Clebsch–Gordan coefficient for the respective spin values, i.e., $-1.195Q_0$ for $\langle 2_1||E2||2_1\rangle$ and $1.604Q_0$ for $\langle 4_1||E2||2_1\rangle$. Differences between theory (th) and experiment (ex) are shown in the lower part of the figure as $[(\langle E2\rangle_{ex} - \langle E2\rangle_{th})/\langle E2\rangle_{ex}] \times 100\%$. Other theoretical views are tabulated in Table 6.

This conclusion that two-state mixing is inadequate is in line with recent experimental observations of $E0$ decays from the normal-deformed and superdeformed $0^+$ states to the nominally spherical ground state in $^{40}$Ca where it is found that two-state mixing cannot explain the observed monopole decay strengths. Rather, three-state mixing is needed [158]. In this case large basis shell model calculations, which include multinucleon excitations of both protons and neutrons across the $Z = N = 20$ shell gap, are able to describe the $E0$ data. Of relevance for the present discussion is that these data confirm that the naïve spherical ground-state configuration of $^{40}$Ca is mixed with deformed intruder structures. This mixing contributes to the shortfall in single-particle strength displayed for valence proton knockout in Figure 15.

A rare view of a deformed nucleus, where a non-intruder spherical excited state has been identified, is $^{44}$Ti as depicted in Figure 43. The double-charge exchange reaction identifies the double-isobaric analog state of the $^{44}$Ca ground state, which is manifestly a spherical state as characterized by its seniority-dominated low-energy structure. This highlights the role of the many-body symmetrization in dictating deformation. Recall, the nucleon–nucleon interaction is short-ranged and attractive, and the total "space $\otimes$ spin $\otimes$ isospin" wave function is antisymmetric: thus, for maximum binding of nucleons in a nucleus, the space-part of the wave function must be as symmetric as possible (a spatially antisymmetric wave function results in "cancellations" in the many-body energy correlations). It also reveals, via the excitation energy of 9.3 MeV, why the identification of spherical states in nuclei with deformed ground states is extremely difficult and therefore essentially never discussed, but such states are present. Identification of deformed states in

nuclei with spherical ground states is usually achieved via the distinctive rotational bands associated with deformed structures in nuclei; spherical states do not exhibit such easily identified patterns.

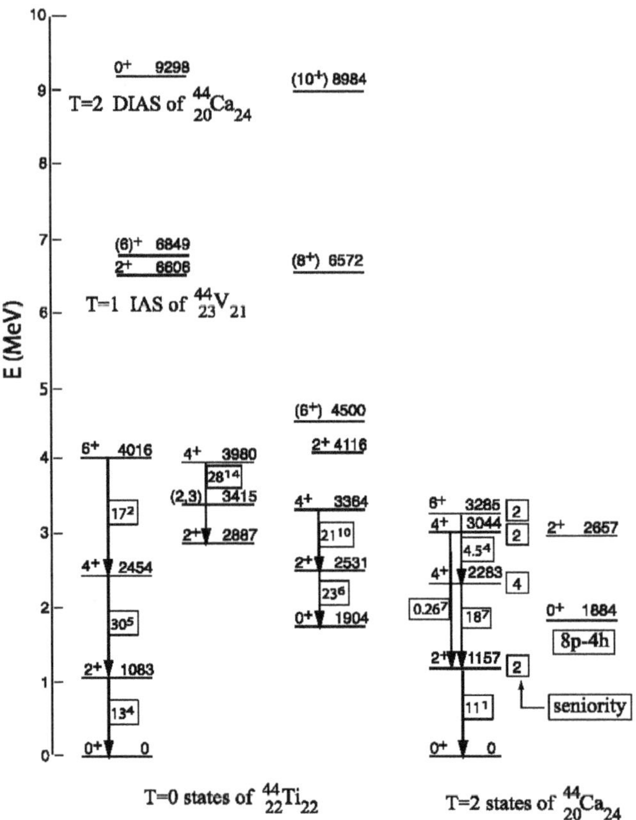


**Figure 43.** Low-energy $T = 0$, 1 and 2 states of $^{44}$Ti and lowest-energy states of $^{44}$Ca which are states of isospin $T = 2$. $E2$ transition rates between states are given by $B(E2)$ values (in W.u.). The key feature of the figure is the state at 9298 keV in $^{44}$Ti. This state is the double-isobaric analog state of the $^{44}$Ca ground state and so, manifestly, it is a "spherical" state. See text for details. Reproduced from [42]. Note: the 1884 keV state in $^{44}$Ca is incorrectly identified. As per Table 5 and Figures 35 and 36, it should be labelled as a $\pi(2p - 2h)$ configuration.

The region of $^{40}$Ca could be regarded as the confrontational meeting point between shell model descriptions of nuclei and the true nature of the structure of nuclei. Multi-shell configurations manifestly dominate the low-energy structure. This is a proven spectroscopically based interpretation, i.e., it is not a model-inspired interpretation. Beyond this mass region, spectroscopic data that reveal the role of multiparticle-multihole configurations become sparse. Indeed, this region is a key meeting point, not only for shell-model based and multi-shell descriptions of nuclear structure, but also for incorporation of configurations from the continuum, as pointed out in [159].

## 7. Intruder States, the Shell Model, and Nuclei Adjacent to Closed Shells

### 7.1. Intruder States

In the lexicon of nuclear structure study, the term "intruder states" has become established. "Intruder" means a state that is observed where it is not expected or "does

not belong". These are states that appear to be shell model states which are observed at low energy on the "wrong side" of a shell model energy gap. Examples are presented in Figures 16 ($^{47,49}$Ca) and 17 (Tl isotopes), and implicitly for the $9/2^-$ intruder structure in $^{187}$Ir in Figure 22. However, intruder states are not simple shell model states because they have underlying correlations in their structure. Indeed, the reason they intrude is because of these correlations. Thus, in Figure 16, an example with important pairing correlations is shown, namely that a simple addition, deduced from one-neutron separation energies, "restores" the low energies of the intruder states to their uncorrelated energies, which reflect the 5 MeV shell gap. In Figure 17, an example with important pairing and deformation correlations is shown, notably that, in addition to the appearance at low energy (pairing correlations), there is a systematic "parabolic" trend in the excitation energies as a function of neutron number, with a minimum near the mid-shell point ($N = 104$), which is where the greatest number of neutrons are active. Much confusion exists in the literature regarding intruder state structures: it appears that they are often viewed as part of a shell model picture. Let us emphasize that the shell model is an independent-particle model based on a spherical mean field. Intruder states are usually strongly deformed and so they are of completely different character to "shell model" structures. For example, they can exhibit rotational bands which can expose their distinctly different character. In effect, the normal and deformed states largely exist in different basis spaces.

It is not implied that rotational bands cannot emerge from shell model calculations. If one could conduct shell model calculations in a sufficiently large space, intruders and their deformation should emerge, but, at present, such calculations are not generally feasible. Consequently, operationally, we have the "coexistence" of shell model descriptions and the Nilsson model plus rotations where nuclei with intruder states are concerned; and we observe actual structures characterized by different $E2$ properties, i.e., quadrupole moments and $B(E2)$ values.

Having noted that rotational structures can begin to emerge in current shell model calculations, we also draw attention to the on-going challenge: the emergence of rotational bands in a finite many-nucleon system calculation is arguably the most profound challenge in a nuclear structure. From everything we understand by the term "a shell model calculation", it is fair to say that this must be a future reality. We look beyond the use of model interactions, such as employed in the Elliott model [160] (where the emergence of rotational bands is guaranteed), and we look to this question using the best effective interactions available. Note that the Elliott model is a single-shell description of states; intruder states demand a multi-shell view. The Elliott model bands are not the generally observed rotational bands in nuclei. It is noteworthy, however, that the Elliott model re-emerges as a submodel of the symplectic shell model, which is discussed in Section 10. By way of shape coexistence, especially in nuclei such as $^{40}$Ca, a convergence on a multi-shell view appears most promising (see, e.g., [158]). However, let us note that the key observables that characterize nuclear rotations are specific $E2$ properties: transition and diagonal $E2$ matrix elements with ratios given by rotor-model Clebsch–Gordan coefficients—these are observed to high precision in some nuclei. To be convincing, such a calculation must use the bare charges of the proton and neutron, $e_p = 1e$ and $e_n = 0$ and match the precision of these observed properties.

The most dramatic examples of intruder states are where they appear as the ground-state structures of nuclei. This only occurs in a few local mass regions (such as the $^{32}$Mg region, the $^{42}$Si region, and perhaps near $^{68}$Ni): the term "island of inversion" has become popular for the description of such an occurrence. There is a tendency to place an unphysical emphasis on this terminology: one reads about mapping the borders (or shores) of islands of inversion. However, the structures are not islands; they persist across the entire mass surface, albeit mostly as excited states. A leading example is shown in Figure 44, which depicts systematics in the even-mass $N = 20$ isotones. This is a celebrated historical example. The first clues came from mass measurements [161] and isotope shift measurements [162] in the sodium isotopes. This was shortly followed by a measurement

of $E(2_1^+)$ in $^{32}$Mg [163]. A suggested unified view of these observations, from earlier times, is reproduced in Figure 45. However, it took thirty years to establish the lowest spherical state in $^{32}$Mg [164], and to explore the structure as an excited state in the neighboring $^{34}$Si [165]. It appears that nobody has yet shown an interest in looking at the underlying structure in $^{36}$S, but it has been known for a long time in $^{38}$Ar. Note that the deformed band in $^{38}$Ar is nearly identical in energy spacing to the ground-state band in $^{32}$Mg. Our message is that: to refer to the ground-state structure of $^{32}$Mg as being part of an "island" is obscuring the discovery frontier of such structures, which must extend to higher excitation energies and broadly encompass nuclides in the region. This is a severe criticism of the misuse of language in science. A schematic view of the energies that contribute to intruder states is shown in Figure 46. A global view that recognizes the dominance of deformation in nuclear ground states is shown in Figure 47.

The challenge of the exploration of intruder states in nuclei is to arrive at the ability to predict their occurrence. With reference to Figure 46, there is a current interest in the so-called monopole energy contribution to the total energy that dictates the appearance of intruder states at low energy in various mass regions. This has received attention already a long time ago [166,167]; more recently, there has been attention from Heyde and collaborators [168], Zuker and collaborators [169], and a review by Otsuka et al. [91]. The theoretical formalism is not a critical concern; but identifying an empirical basis for fixing the relative magnitudes of the energy contributions shown in Figure 46 needs in-depth consideration. The problem is identifying manifestations of monopole energy effects that are free of correlations from pairing and from deformation. These correlations already feature in our chosen subject: they lie at the heart of emergent structures in nuclei, whether involving intruder states or not. Some mass regions of critical concern are addressed below in this Section and in Section 9.

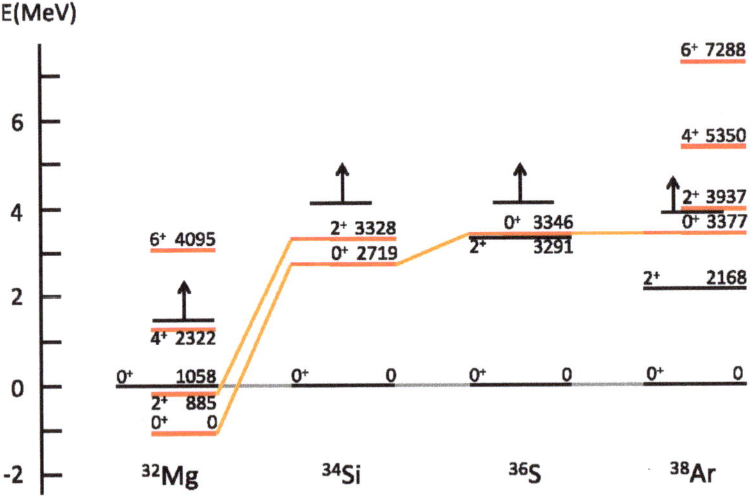


**Figure 44.** Shape coexistence and intruder states in the $N = 20$ isotones. The $0_2^+$ state identifications are made in: $^{32}$Mg [164] and $^{34}$Si [165]. The intruder states can be understood in an exactly parallel manner to the situation in the Sn isotopes. Thus, here, the 2p-2h configurations involve neutron pairs interacting with protons. The excitation pattern reflects proton subshell structure ($2s_{1/2}$, $1d_{3/2}$) as these orbitals are filled: this is beyond the present level of discussion. Note that the deformed bands in $^{32}$Mg and $^{38}$Ar possess nearly identical energy spacing. Taken from [8].

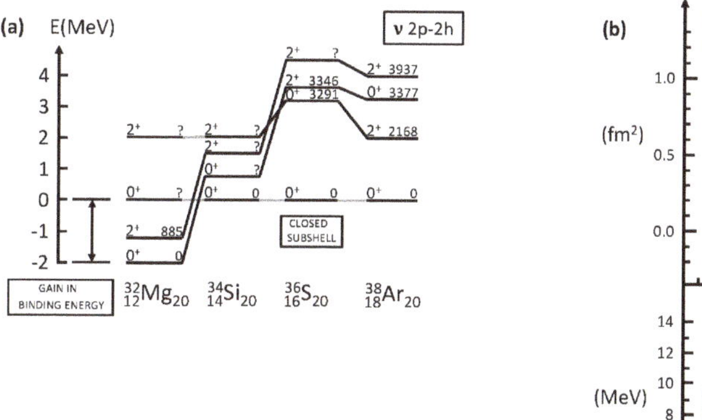


**Figure 45.** (**a**) Excited state systematics in the even-mass $N = 20$ isotones. The low-lying $2_1^+$ state in $^{32}$Mg is interpreted as resulting from a ground state intruder configuration. The ground state of $^{32}$Mg should have an anomalously larger mean-square radius. The ground-state binding energy of $^{32}$Mg has been reported variously as anomalous and normal. (**b**) Two-neutron separation energy, $S_{2n}$, and isotope shift, $\delta\langle r^2\rangle$, systematics for the neutron-rich Na isotopes [161,162]. The discontinuity at $N = 20$ indicates an increased ground-state mean-square charge radius and increased binding energy. Reproduced from [170].

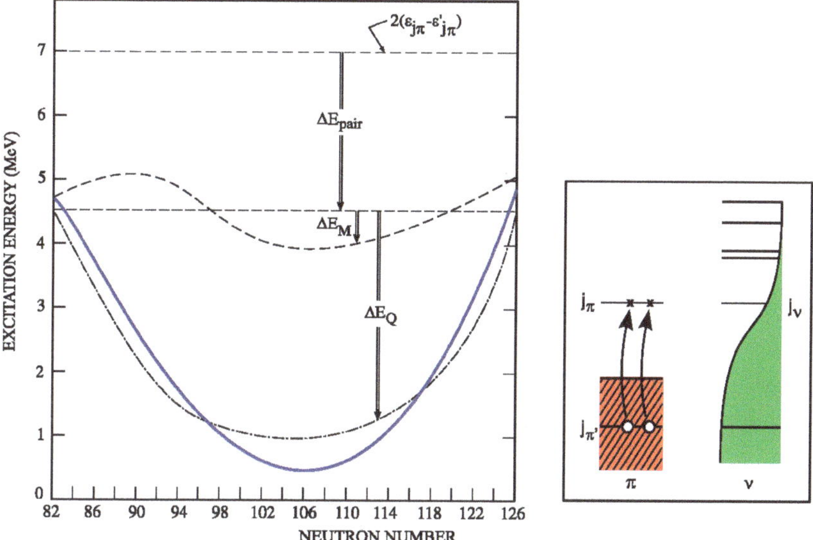


**Figure 46.** The different energy terms contributing to the energy of the lowest proton 2p-2h $0^+$ intruder state for heavy nuclei. On the right-hand side, a schematic view of the excitation is given. On the left-hand side, the unperturbed energy, the pairing energy, the monopole energy shift, and the quadrupole energy gain are presented, albeit in a schematic way. Reprinted with permission from [41]. Copyright (2011) by the American Physical Society.

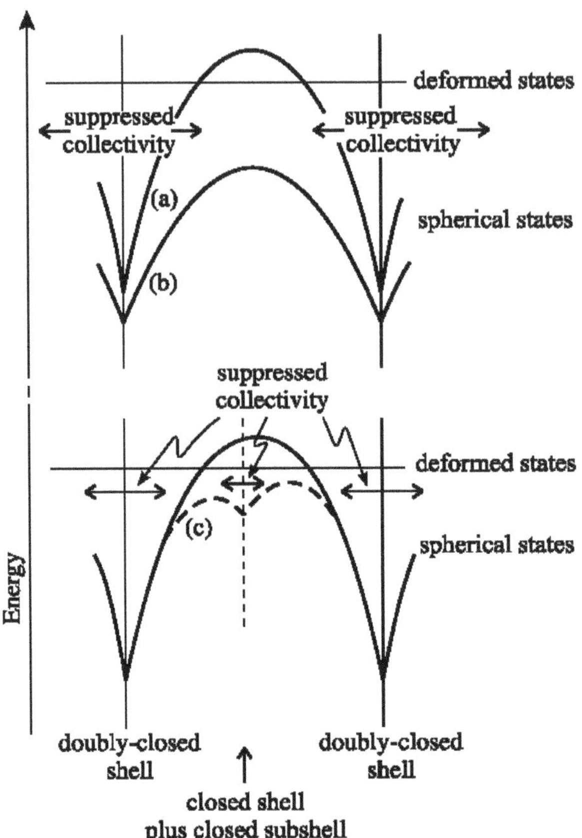


**Figure 47.** A schematic view of the intruder-state "parabolas", shown to dramatize the way that shells and subshells suppress the emergence of low-energy collectivity in nuclei. (**a**) The situation where deformed structures intrude to become the ground state at the middle of a singly closed shell, e.g., $^{32}$Mg. (**b**) The situation where the ground states for a sequence of singly closed shell nuclei remain spherical, but deformed structures form excited intruder bands, e.g., the Sn and Pb isotopes. (**c**) The situation where a subshell may suppress intrusion of a deformed structure from becoming the ground state or a low-lying excited band, e.g., $N = 50, 82$, cf. Figures 52–54. Reprinted with permission from [41]. Copyright (2011) by the American Physical Society.

The perspective that is presented in Figure 47 appears useful. This view "inverts" the parabolic energy perspective that can be applied to intruder states and, recognizing that most nuclei are deformed, and that shape coexistence probably occurs in all nuclei, expresses the occurrence of spherical shapes in nuclei as intruding to low energies only at and near closed shells. The competition between the controlling energies that lie behind this view devolve onto the configuration interaction problem that is foundational to the nuclear many-body problem.

### *7.2. Shell Model States*

The best view that one possesses of shell model states is of excited states in isotopic and isotonic sequences adjacent to closed shells. Examples are shown in Figures 17, 28 and 29, namely, the low-energy systematics of the odd-mass Tl, In and Sb isotopes, respectively. These views are the best because they are not dominated by pairing correlations with respect

to the unpaired nucleon: it is a single nucleon outside of a singly closed shell. Furthermore, the singly closed shell cores are dominated by spherical, seniority type excitations (when intruder states do not appear at low energy). Thus, one expects that the degrees of freedom of these closed-shell plus or minus one nucleon nuclei are dominated by independent-particle degrees of freedom. If there are any monopole energy shifts, i.e., changes in energy of shell model states across an isotopic sequence, they will be easy to see and easy to interpret. This was the universally held view until the observations on Coulomb excitation of $^{129}$Sb summarized in Figure 27, and the implications of these data when compared to the normal weak-coupling model case represented by $^{115}$In in Figure 26. The message from the data in Figures 27 and 26 is that, while a nucleon in a unique-parity configuration exhibits weak-coupling $E2$ strength, i.e., the summed strength in the odd-mass nucleus equals the singly closed-shell core strength, manifested in $B(E2; 0_1^+ \rightarrow 2_1^+)$, cf. Figure 26; when $j$ mixing occurs, the $E2$ strength may exceed the weak-coupling value, cf. Figure 27. Thus, $B(E2)$ data such as those in Figures 26 and 27 become a key focal point for exploring the emergence of collectivity in nuclei—$j$ mixing must be quantified. In turn, the issue of $j$ mixing is critical for assessing monople energy shifts in nuclei: any use of data in nuclei must first be assessed for $j$ mixing before single-$j$ energy shifts can be extracted.

The focus here on the role of $j$-mixing differs from the emphasis of the discussion of the increased $E2$ strength in $^{129}$Sb by Gray et al. [64], where the discussion in terms of shell model calculations identified that the collectivity of the neutron core was not increased by the addition of the extra proton, but rather the increased $E2$ strength arose primarily from the proton–neutron term, thus pointing to overall coherent contributions to the $E2$ strength. The role of $j$ mixing is not immediately evident in this approach.

To explore the role of $j$-mixing explicitly, schematic particle–vibration model calculations were performed for $^{115}$In and $^{129}$Sb. A code developed by one of the authors (AES) was employed (see Refs. [171–173]). With $^{115}$In modeled as a $1g_{9/2}$ proton hole coupled to the ground and first-excited state of $^{116}$Sn, the sum rule was confirmed for the weak-coupling case; however, it was observed that a shortfall in the summed $E2$ strength occurs when the particle–vibration coupling becomes finite. Turning to $^{129}$Sb, the low excitation states were described by allowing the odd proton to occupy the $\pi 1g_{7/2}$ and $\pi 2d_{5/2}$ orbits, coupled to the $^{128}$Sn core. This is a minimal model to describe the low-excitation positive-parity states shown in Figure 27. The sum rule was confirmed for the weak-coupling limit, and a short-fall in $E2$ excitation strength from the ground state was again observed when the particle–vibration coupling became finite. However, when the proton was also allowed to occupy the $\pi 2d_{3/2}$ orbit as well as $\pi 1g_{7/2}$ and $\pi 2d_{5/2}$, the summed $E2$ strength in $^{129}$Sb exceeded that of the $^{128}$Sn core.

The above calculations demonstrate the role of $j$-mixing in the enhanced $E2$ strength observed in $^{129}$Sb compared to $^{128}$Sn, perhaps in a more transparent way than the large-basis shell model calculations. No tension between the shell model and these schematic particle-vibration model calculations is seen. The important concept is that $j$ mixing means that the sphericity of the mean field has been broken. This is a fundamental point, based on symmetry, behind the emergence of nuclear collectivity. Some additional observations are made on $j$ mixing in the discussion that follows in this Section.

Let us take the above points and consider the systematic features of the odd-mass copper isotopes, shown in Figure 48. A natural first look at $j$ mixing is to assemble information for single-nucleon transfer reactions. As already noted, spectroscopic factors must be handled with caution. However, far more directly, fragmentation of $j$ strength is often observed, as shown in Figure 49 for the copper isotopes. One makes the following observation: $j$ is a quantum number characteristic of a spherical mean-field with spin–orbit coupling; if $j$ strength is fragmented, the mean field is not spherical.

A particular feature of note in the odd-mass copper isotopes is the sudden change in the relative energies of the lowest states with spin–parity $1/2^-$, $3/2^-$, and $5/2^-$ above $N = 40$, cf. Figure 48. These abrupt changes have been interpreted as a major illustration of monople energy shifts [91], based on the assumption that the observed states are the

$2p_{1/2}$, $2p_{3/2}$, and $1f_{5/2}$ shell model states. However, $E2$ transition strengths in $^{69,71,73}$Cu, compared to the closed shell cores, $^{68,70,72}$Ni, cf. Figure 33 reveal that $E2$ transition strength exceeds that in the cores, as seen also in $^{129}$Sb, cf. $^{128}$Sn (Figure 27). This collectivity is independent of intruder state structures, which are identified in Figure 48. However, the current status of monopole energy shifts requires detailed spectroscopy before it can be discussed quantitatively. Indeed, a recent paper [174] appears to give it dominant status in its role behind the appearance of intruder states at low energy. We would counsel greater caution, and consideration of the structures, interactions and energy dependencies of multiparticle-multihole intruder configurations as shown in Figure 40, in the pursuit of interpretations of the energies of intruder states.

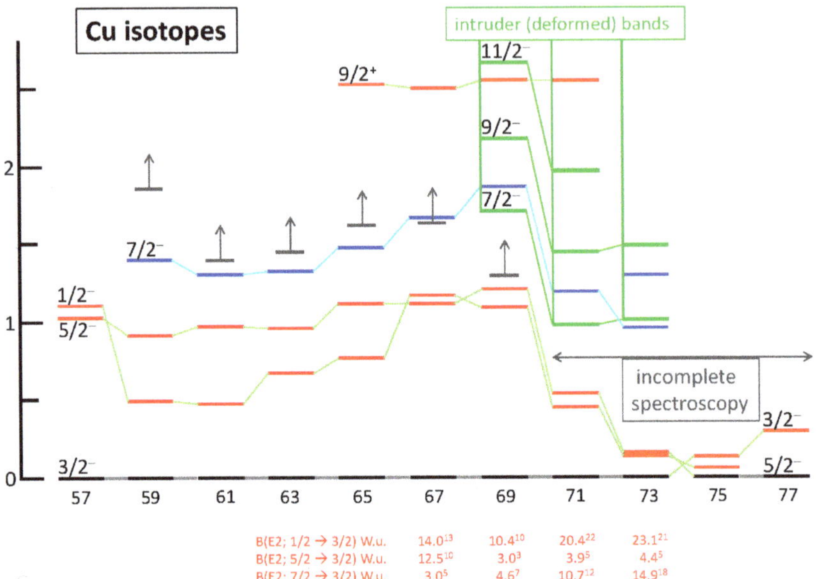


**Figure 48.** The lowest energy states in the odd-mass Cu isotopes. The naïve interpretation is that these states are simply the manifestation of the expected $fpg$ shell model states: $2p_{3/2}$, $2p_{1/2}$, $1f_{5/2}$, $1f_{7/2}$ and $1g_{9/2}$. The existence of rotational band patterns in $^{69,71,73}$Cu supports the $7/2^-$ states as the Nilsson state $7/2^-[303]$. The situation with respect to the other states remains confused: transfer reaction data (shown in Figure 49) reveal fragmentation of single-particle $j$ strength. The experimental situation in $^{71-77}$Cu is very incomplete and any interpretation is premature, except—see comments in the text. Horizontal bars with vertical arrows indicate excitation energies above which states are omitted. The $B(E2)$ data of Stefanescu et al. [175] for $^{67,69,71,73}$Cu can be compared with the Ni cores, cf. Figure 33: an investigation into collective enhancement in the odd-Cu isotopes relative to the corresponding Ni core nuclei appears to be in order.

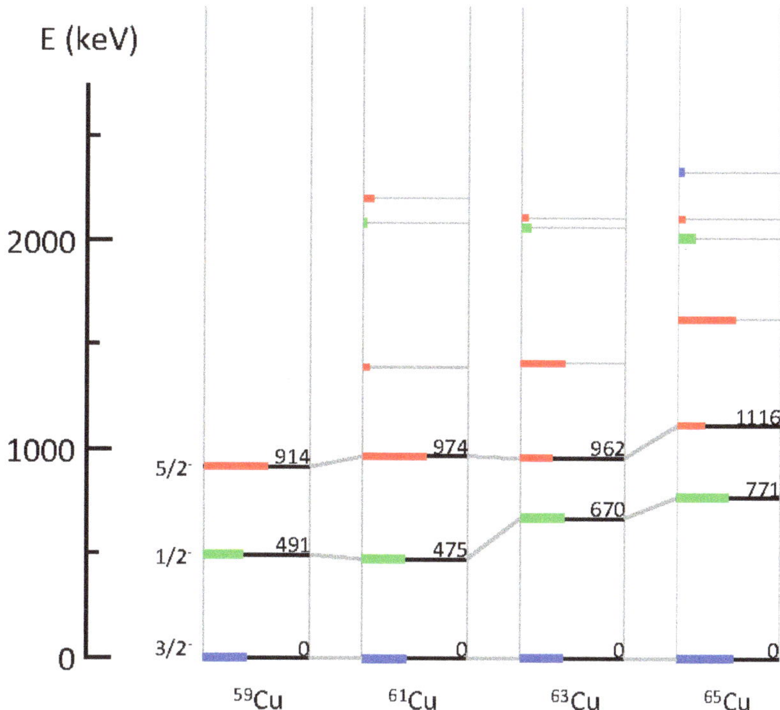


**Figure 49.** The three lowest states in $^{59-65}$Cu and states for which there is significant population in the one-proton transfer reaction ($^3$He,d). The lengths of the coloured bars are directly proportional to the strength of population of the states in the transfer reaction. The ground-states can be naïvely interpreted as the shell model configuration $2p_{3/2}$; similarly, the first excited states would appear to be the configuration $2p_{1/2}$. However, the assumption that these are spherical shell model states breaks down for the second excited state where the one-proton transfer strength is severely fragmented. The colour coding indicates the $\ell$ values of the transferred nucleon — blue ($\ell = 1$ with $j = \ell + 1/2$), green ($\ell = 1$ with $j = \ell - 1/2$), red ($\ell = 3$ with $j = \ell - 1/2$). The direct interpretation is that the $\ell$ quantum number is not a good quantum number in these nuclei, i.e.,they are deformed. Further discussion of such issues must await a more advanced level of treatment. The data are taken from [176]. Reproduced from [8].

Whereas the issue of monopole shifts in subshell energies remains open, no systematic study has been made; the idea has only been applied selectively where there is unfortunately a lack of detailed spectroscopic information [91]. However, detailed information exists, for example in the odd-mass Sc ($Z = 21$) isotopes, as shown in Figures 50 and 51. The parabolic pattern in these figures points to a dominance of deformation-producing forces controlling intruder state energies. Intruder states are strongly deformed structures with both large correlations that originate in their multi-shell structure and in their pairing structure. It would be interesting to make a thorough study of such structures across all nuclei to clarify the role of monopole energy shifts as a factor underlying intruder states and shape coexistence.

Deformation in nuclei immediately adjacent to closed shells has become a recent focus in the odd-mass F ($Z = 9$) isotopes [177]. The data are consistent with deformed ground states. This appears to lie outside of any shell model expectations. Indeed, the surprise that the double-closed shell nucleus $^{28}$O does not have a bound ground state, but its neighbour, $^{29}$F does, may be because the double-closed shell of $^{28}$O does not favour ground-state deformation, but $^{29}$F can deform in its ground state. This would appear to be a simple

explanation of the surprise that $^{28}$O is unbound (but, to our knowledge, has never been pointed out).

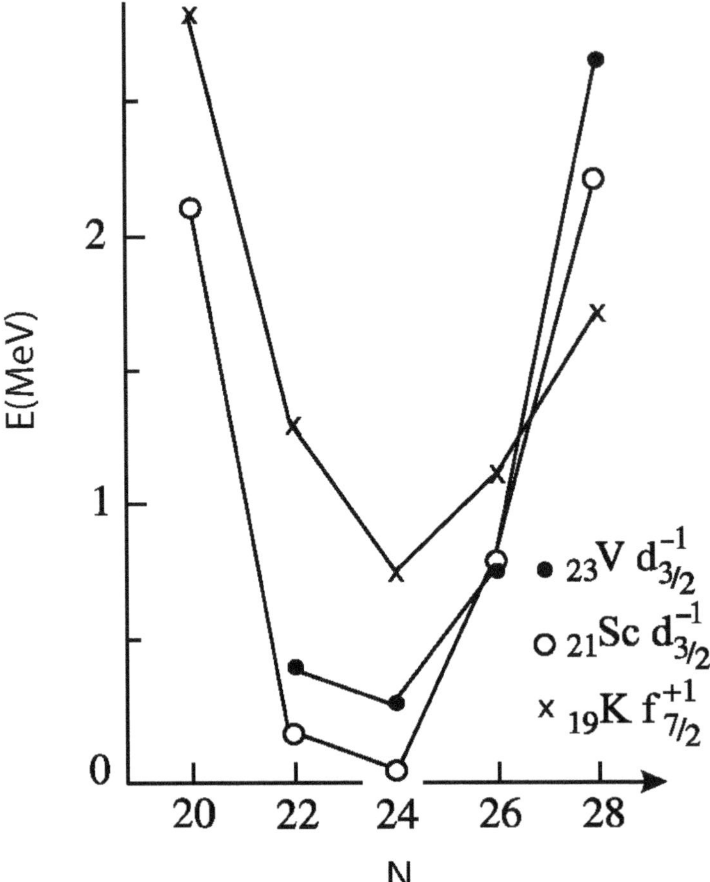


**Figure 50.** Intruder states in K, Sc, V isotopes. These states are the heads of bands, which are consistent with $K$ quantum numbers equal to the band head spins. $B(E2)$ data for $^{45}$Sc are shown in Figure 51. The pattern matches the parabolic trend shown schematically in Figure 46 and supports the dominant role of a quadrupole interaction between protons and neutrons. Note that $^{45}$Sc is almost an "island of inversion", if one ignores the complete range of occurrence of the structure across the entire shell. The figure is adapted from [41].

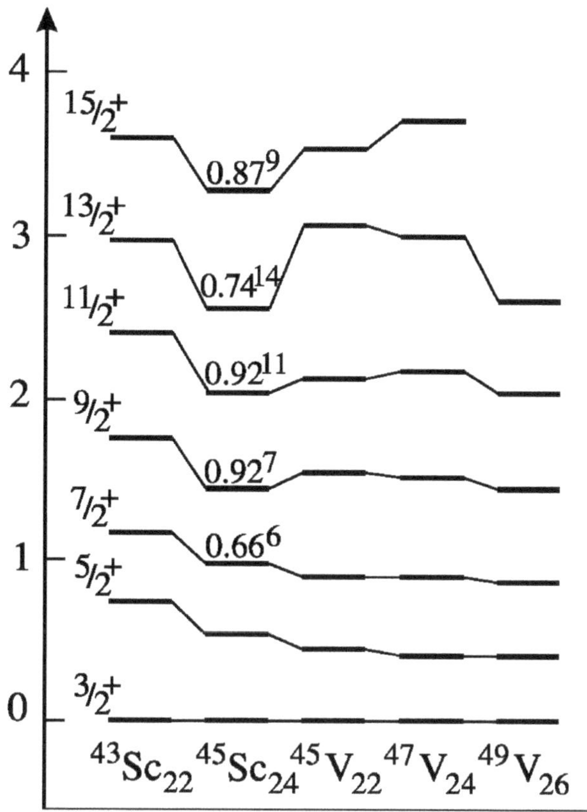


**Figure 51.** Bands built on the intruder states (cf. Figure 50) in the odd-mass Sc and V isotopes. The numbers given for $^{45}$Sc are magnitudes of the intrinsic quadrupole moments, $Q_0$ ($e\cdot$ b) deduced from $B(E2)$ measurements [178]. Adapted from [41].

At present, information on odd-mass nuclei adjacent to $N = 50$ and $N = 82$ remains very limited. Intruder states are observed in the $N = 49$ and $N = 81$ isotones as shown in Figures 52 and 53, respectively. These manifestations are not at the mid-shell points. Possibly, the proton structures, i.e., a subshell gap and/or proximity to a $j = 1/2$ subshell, at $Z = 40$ and $Z = 64$ have something to do with this. Excited $0^+$ states for the $N = 50$ and $N = 82$ isotones are shown in Figure 54. At present, the reason for the dissimilarity between $N = 50, 82$ and $Z = 50, 82$ remains an open question. Whether or not there are low-lying excited $0^+$ states in, e.g., $^{82}$Ge and $^{150}$Er, would be worth exploring. The situation at $N = 48$, i.e., in $^{80}$Ge, is of two contradicting reports [179,180] and a very recent result that casts further doubt on the existence of a low-energy excited $0^+$ state in $^{80}$Ge [181]; unlike at $Z = 48$ (the Cd isotopes) where low energy deformed excited $0^+$ states are well established (see, e.g., [41,116]).

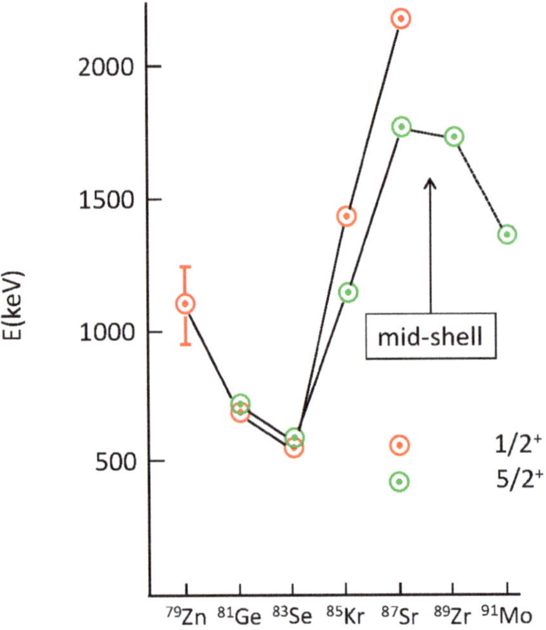


**Figure 52.** Intruder states in the $N = 49$ isotones. The state in $^{79}$Zn is from Orlandi et al. [182] and Yang et al. [183]. The configuration involved may be a prolate deformed structure built on the $1/2^+[431]$ Nilsson state. Note that these structures are nearly identical to the intruder state structures in the odd-In ($Z = 49$) isotopes, some details of which are noted in Section 4. Other data are taken from ENSDF [22]. For comments on the energy maximum at the mid-shell point, see Figure 54 caption.

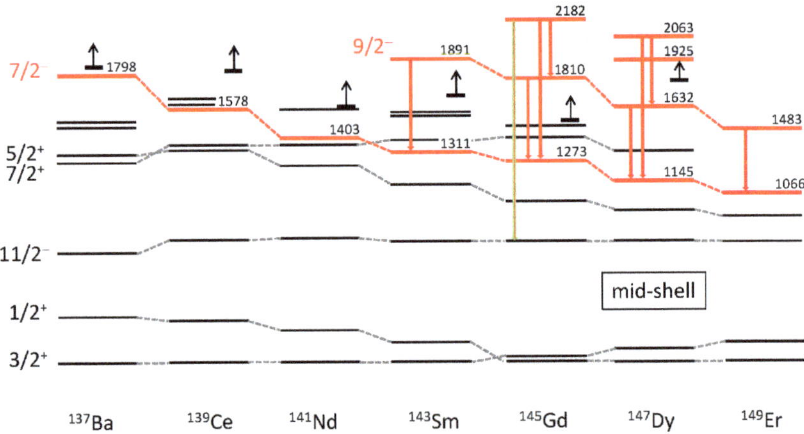


**Figure 53.** Intruder states in the $N = 81$ isotones (shown in red with all observed decay branches). Horizontal bars with vertical arrows indicate excitation energies above which states are omitted. The mid-shell point is indicated. The configuration involved may be an oblate deformed structure built on the $7/2^-[503]$ Nilsson state. The data are taken from ENSDF [22].

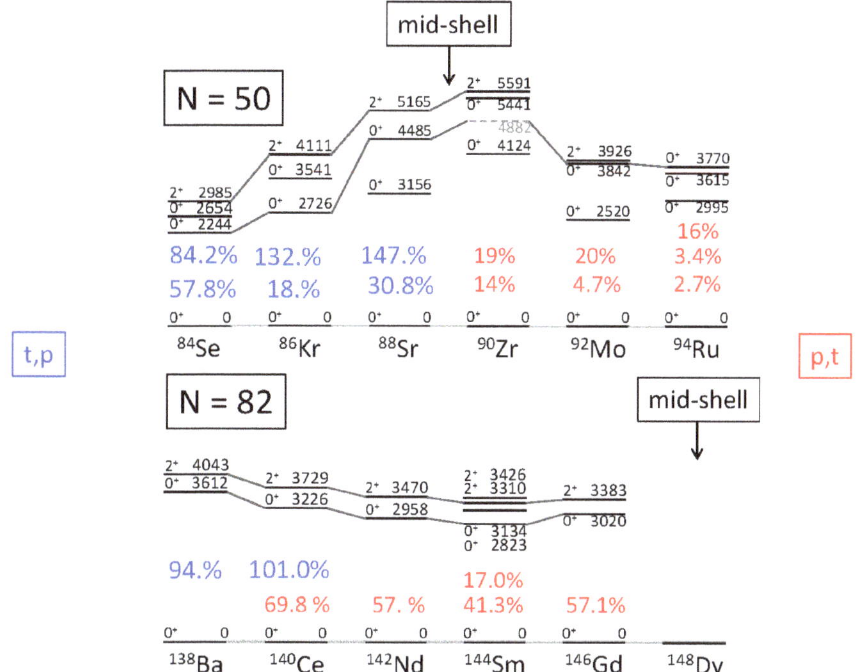


**Figure 54.** Excited $0^+$ and $2^+$ states in the $N = 50$ and $N = 82$ isotones which are candidate states for intruder configurations. They possess $\nu(2p - 2h)$ character as determined by two-neutron transfer studies [184–188]. Population of these states as percentages relative to the ground states are given as: blue for (t,p) and red for (p,t) reactions, respectively. These numbers are taken from [188–193]; other data are taken from ENSDF [22]. The mid-shell points are indicated. Possibly, the local high energy in $^{90}$Zr is due to a weak energy gap at $Z = 40$. See remarks on $^{82}$Ge in the text.

## 8. Survival of Seniority Structures Away from Closed Shells

The picture of the intrusion of deformed structures into the domain of spherical structures is summarized in the foregoing, but what about the survival of seniority structures away from closed shells? This is an issue with only a few circumstantial focal points; it has never been subjected to systematic study, to our knowledge.

A leading illustration of the survival of seniority away from closed shells is shown in Figure 55 for the even-even $N = 80$ isotones. The dominance of a neutron $1h_{11/2}^{-2}$ broken pair is manifested at $J = 10$. Furthermore, as $Z = 64$ is approached, $J = 10$ states involving a proton $1h_{11/2}$ broken pair appear. Magnetic moment data strongly complement this observation. More specifically, the $g$ factors of the $10_1^+$ states in $^{138}$Ce and $^{140}$Nd, $g = -0.176(10)$ and $g = -0.192(12)$, respectively, indicate their $\nu 1h_{11/2}^{-2}$ structure. For $^{144}$Gd, however, $g(10_1^+) = +1.276(14)$ [194] indicates the $\pi 1h_{11/2}^2$ configuration. But how far from closed shells does this broken-pair structure dominate $J = 10$ states, notably yrast states? A similar view is provided for the $J = 6$ state, due to the proton $1g_{7/2}^2$ broken pair in the tellurium isotopes in Figure 56. These and other issues are discussed in this Section.

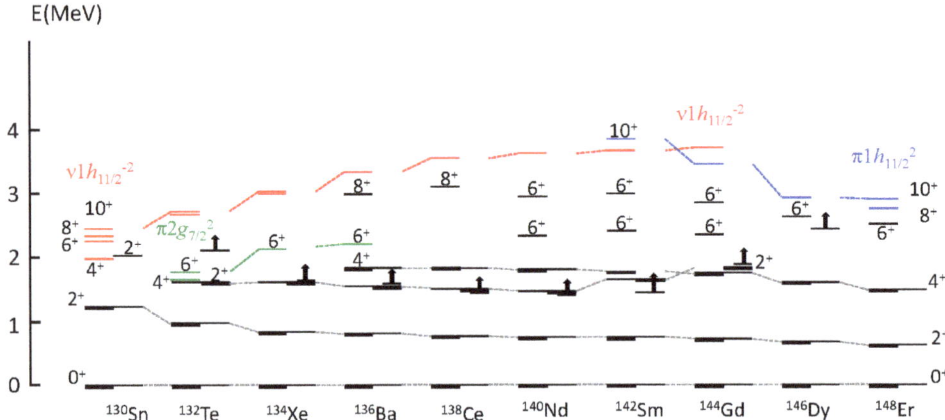


**Figure 55.** Low-energy systematics of the positive-parity states in the $N = 80$ isotones. The high-spin states are discussed in the text. Vertical arrows indicate energies above which other positive-parity states are observed. First-excited $0^+$ states are observed at (keV): $^{134}$Xe (1636), $^{136}$Ba (1579), $^{138}$Ce (1466), $^{140}$Nd (1413), $^{142}$Sm (1451), $^{144}$Gd (1887). The data are taken from ENSDF [22].

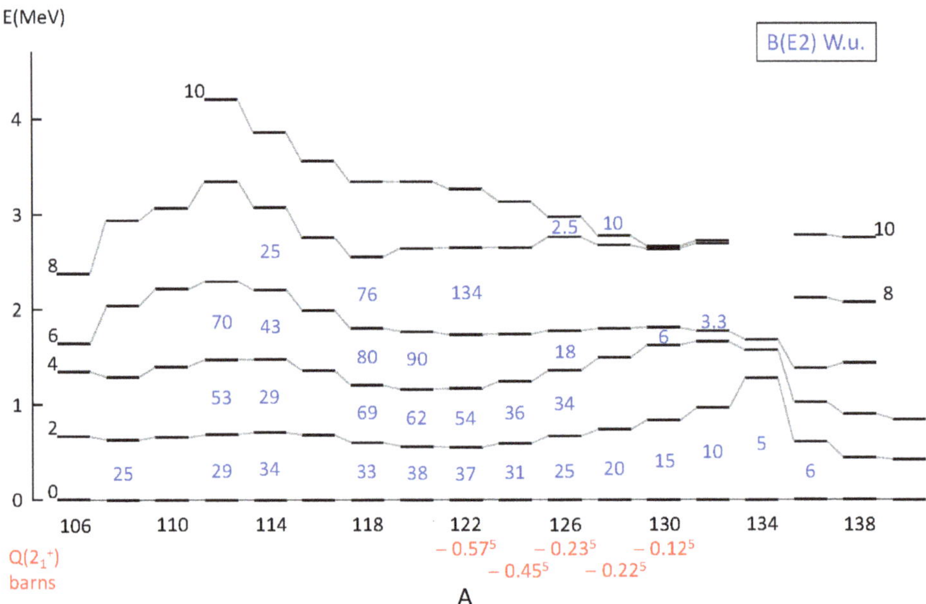


**Figure 56.** Yrast systematics in the even-mass Te ($Z = 52$) isotopes. Transition $B(E2)$ values in W.u., where measured, are shown in blue between levels. Quadrupole moments of $2_1^+$ states, where measured, are shown below the isotope mass numbers. Additional details are given in the text. Data from [195–199] and ENSDF [22]. Figure is from [8].

High-$J$ broken-pair states appear in localized regions across the entire mass surface. In spherical nuclei, they are manifested as seniority isomers; in deformed nuclei, they are manifested as $K$ isomers. The topic of $K$ isomerism is a time-honoured branch of nuclear structure study with comprehensive reviews [200–203]. The situation for transitional nuclei is poorly characterized. Two factors determine the excitation energies of high-

spin broken-pair states: pairing energy and rotational energy. Pairing contributions to broken-pair excitation energies are well understood and are well characterized. Rotational energy contributions to broken-pair excitation energies are epitomized by Figure 25. This aspect of nuclear structure is generally described as "rotational-alignment" effects: there is an enormous literature addressing this topic using the so-called cranked shell model. This model approximates the effects depicted in Figure 25 by "cranking" a deformed mean-field about a fixed axis at right angles to the symmetry axis of the deformed mean field. It has been extended to "tilting" the axis about which cranking occurs [204]. The cranked shell model has completely dominated the study of high-spin states in nuclei. Our concern here is with low-medium spin states in nuclei. Note that, at high spin, an axis of directional quantization approaches a semi-classical description in that the cone of uncertainty becomes narrow; thus, cranking about a fixed axis improves asymptotically with increasing total spin.

To move forward on the topic of the breaking of pairs away from closed shells, it is important to recognize that the prototype signature is properties of the $2_1^+$ states in nuclei, which manifestly involve breaking pairs. The leading question is: Which broken-pair configurations underlie a given $2_1^+$ state? While important insights can be gained through large-basis shell model calculations in the valence shell, the full answer must extend far beyond the valence shell, as manifested in the need for effective charges to describe $B(E2;2_1^+ \to 0_1^+)$ values (see Table 2). By looking at systematics of $B(E2)$ values near closed shells, one expects to learn something about this fundamental aspect of the emergence of collectivity in nuclei. A natural first step is to look at even-even nuclei with one valence proton pair and one valence neutron pair, particles or holes, as will now be discussed.

It turns out that $^{132}$Te is one of the more accessible nuclei for a detailed study of what might be termed "prototype emergence of quadrupole collectivity in nuclei". The region around $^{132}$Sn is attractive for this purpose because $^{132}$Sn is an $N > Z$ doubly magic core without low-excitation intruder states, and because detailed spectroscopic studies (including transfer reactions, $B(E2)$, and $g$-factor measurements) show it to be a "good" doubly magic core. However the challenge, which makes performing detailed spectroscopy difficult, is that $^{132}$Te is accessible to radioactive beams, by beta-minus decay and as a fission fragment—but not at stable-beam accelerators.

The current knowledge of excitations in $^{132}$Te is shown in Figure 57. The extent of detailed information is best described as "inadequate". For example, a naïve broken-pair view would predict two low-lying $2^+$ states, one due to a broken neutron (hole) pair, cf. $^{130}$Sn ($E(2_1^+) = 1221$ keV), the other due to a broken proton (particle) pair, cf. $^{134}$Te ($E(2_1^+) = 1279$ keV). Thus, (naïvely) there should be two excited $2^+$ states in $^{132}$Te at 1221 and 1279 keV. The lowest-lying $2^+$ states in $^{132}$Te are $2_1^+$ (974 keV), ($2^+$) (1665 keV), ($2^+$) (1778 keV), and then ($2^+$) states at 2249 and 2364 keV, where the parentheses indicate that the spin–parity assignment is tentative. To pursue the naïve view, the broken-pair configurations can be viewed as mixing and repelling, so that one resulting state appears pushed down by 1221 − 974 = 247 keV and the other state is at 1270 + 247 = 1517 keV, cf. (above) 1665 keV. This raises many questions, such as: What is the structure of the states at 1778, 2249 and 2364 keV? What are the detailed properties of these states? Are they $2^+$ states? What are their lifetimes, magnetic moments, and quadrupole moments? At present, all unanswered experimentally, except for some information on Coulomb excitation of the 1665 keV state. Note that the $J = 0$-coupled neutron-hole pair and the $J = 0$-coupled proton-particle pair also interact and cause an energy shift in the $0^+$ configuration that dominates the ground-state of $^{132}$Te; but $[(\pi^2)_J \otimes (\nu^{-2})_J]_0$ configurations can also be expected to contribute to the ground-state structure. Allowing for pair occupancies across the many shell-model subshells, there are many possibilities. In addition, note that there is an extensive literature that discusses the shell model configurations underlying the structure of $^{132}$Te [33,62,205–215].

To begin to answer some of the questions raised concerning the lowest few $2^+$ states in $^{132}$Te, one can note that there is a single low-excitation proton configuration that forms

a $2^+$ state, namely $\pi(1g_{7/2})^2_{2^+}$, with $g$ factor $g = 0.82$. In contrast, the neutron orbits $2d_{3/2}$, $3s_{1/2}$, and $1h_{11/2}$ are "almost degenerate", which means that low-excitation $2^+$ states can be formed by the two-neutron-hole configurations $\nu(2d_{3/2})^{-2}_{2^+}$, $\nu(1h_{11/2})^{-2}_{2^+}$, and $\nu(2d^{-1}_{3/2}3s^{-1}_{1/2})_{2^+}$, with $g$ factors 0.54, −0.24, and −0.27, respectively.

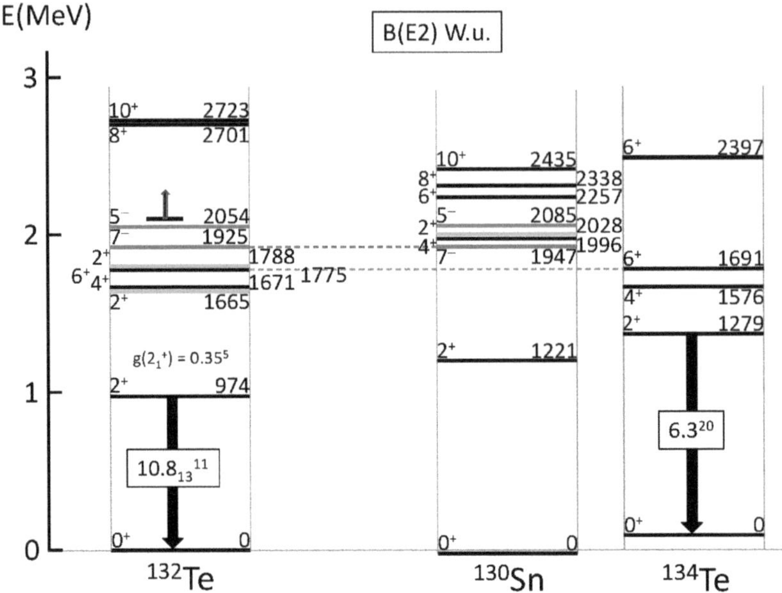


**Figure 57.** Excited states in $^{132}$Te cf. $^{130}$Sn and $^{134}$Te excited states. Naïvely there should be an appearance of both sets of states, corresponding to the independent "breaking" of the proton particle pair or the neutron hole pair. However, these broken-pair configurations can be expected to interact; thus, the $2^+_1$ in $^{132}$Te is lower in energy than that in $^{130}$Sn and $^{134}$Te. See the text for details. The data are taken from ENSDF [22].

Table 7 shows the results of shell model calculations for the lowest five $2^+$ states in $^{132}$Te. The calculations were performed with NuShellX [44] in the `jj55` basis space and with the `sn100` interactions; see Table 2 and [31,48,64,68] for additional details including the parameters of the effective $M1$ operator. Along with a comparison of the experimental and theoretical level energies, Table 7 lists the $g$ factors and the decomposition of the wavefunctions into the dominant proton and neutron components coupled to $0^+$ and $2^+$. Note that there is no relative phase information available in these structures. The mixing of the lowest two $2^+$ states discussed above in relation to Figure 57 is qualitatively consistent with the shell model calculations. The considerable variation in the calculated $g$ factors is an indication of the marked differences in the structures of these $2^+$ states. As collectivity emerges, the $g$ factors of all of the low-excitation states would be expected to approach the collective value, typically $g_{\text{coll}} \approx 0.7Z/A$. Such measurements are extremely challenging even for stable nuclides.

Given the complexity of the low-excitation states in $^{132}$Te due to the small energy spacing of the $2d_{3/2}$, $3s_{1/2}$, and $1h_{11/2}$ neutron hole orbits, one might consider nuclei like $^{136}$Te (approximately $\pi 1g_{7/2} \otimes \nu 2f_{7/2}$) and $^{212}$Po (approximately $\pi 1h_{9/2} \otimes \nu 2g_{9/2}$) as alternative "prototypes" to study the emergence of collectivity. Shell model calculations for these nuclei (see Table 2 for details of basis spaces and interactions) show that the configuration mixing in the lowest $2^+$ states of these nuclei is already considerable.

**Table 7.** Shell model calculations for the five lowest $2^+$ states in $^{132}$Te. The excitation energies and $g$ factors are shown, along with the structure of the state. The structure indicates how the angular momentum is apportioned between protons and neutrons. It is not the wavefunction. The weights indicated sum to unity when all contributions are included.

| $I_i^\pi$ | $E_x$ | | $g$ | Structure |
|---|---|---|---|---|
| | Exp | Theory | | |
| $2_1^+$ | 974 | 954 | 0.48 | $0.38\pi(0^+)\nu(2^+) + 0.51\pi(2^+)\nu(0^+) + 0.07\pi(2^+)\nu(2^+) + ...$ |
| $2_2^+$ | 1665 | 1645 | 0.37 | $0.49\pi(0^+)\nu(2^+) + 0.32\pi(2^+)\nu(0^+) + 0.02\pi(2^+)\nu(2^+) + ...$ |
| $2_3^+$ | 1788 | 1931 | 0.03 | $0.71\pi(0^+)\nu(2^+) + 0.11\pi(2^+)\nu(0^+) + 0.13\pi(2^+)\nu(2^+) + ...$ |
| $2_4^+$ | 2249 | 2258 | 0.30 | $0.83\pi(0^+)\nu(2^+) + 0.03\pi(2^+)\nu(0^+) + 0.07\pi(2^+)\nu(2^+) + ...$ |
| $2_5^+$ | 2364 | 2468 | 0.98 | $0.06\pi(0^+)\nu(2^+) + 0.77\pi(2^+)\nu(0^+) + 0.01\pi(2^+)\nu(2^+) + ...$ |

There are limited simple and accessible cases to study in detail the proton plus neutron broken-pair structures of $2^+$ states adjacent to a closed shell. Extending beyond this simplest case, the stable Te isotopes below $^{132}$Te provide the opportunity for detailed spectroscopy, including (n,n′ $\gamma$) studies [216], Coulomb excitation, and $g$-factor measurements [115], to track the emergence of collectivity as increasing numbers of neutron holes are added to the two protons outside the $Z = 50$ shell closure. The stable Xe isotopes, with four protons, are likewise accessible to detailed measurements [30,32,217–223]. In these iotopes, the cancellation of $E2$ strength for four protons in the $1g_{7/2}$ orbit (see Equation (1)) makes the observed $E2$ strengths in the Xe isotopes below $^{136}$Xe sensitive to the breakdown of the seniority structure and emerging collectivity.

Returning to the high-spin broken-pair states in this region, specifically the $J = 10$ broken-neutron-pair configurations and the $J = 10$ broken-proton-pair configurations, these do not mix strongly, as manifested in Figure 55, cf. $^{142}$Sm and $^{144}$Gd. This suggests that broken-pair high-$j$, high-spin configurations do not play a role in the emergence of collectivity. Figure 56 suggests survival of both the proton-broken-pair and the neutron-broken-pair structures, respectively for $J = 6$ and $J = 10$ in $^{126-132}$Te. The $g$ factor data, where available, support this suggestion. In the $N = 82$ case of $^{134}$Te, $g(6_1^+) = +0.847(25)$ [224], as expected for the $\pi 1g_{7/2}^2$ configuration. The $g$ factors of the $2_1^+$ [48] and $4_1^+$ [225] states in $^{134}$Te are consistent with $g(6_1^+)$, and hence the same configuration. In $^{132}$Te, with two neutron holes, $g(2_1^+) = +0.46(5)$ [206,210,226,227] is closer to the collective $g \approx Z/A \approx 0.39$, whereas $g(6_1^+) = +0.78(8)$ [228] remains consistent with that of the pure $\pi 1g_{7/2}^2$ configuration. Recent work at the Australian National University (ANU) tracks the persistence and eventual weakening of the proton-broken-pair structure in the $4^+$ states of $^{124,126,128,130}$Te [115].

Indeed, discontinuities in yrast state energies persist throughout the open-shell, $Z > 50$, $N < 82$ region and as an example yrast $\gamma$-ray energies, $E_\gamma$, versus the spins of the initial states, $I_i$, are shown for the Ba isotopes in Figure 58. It is important to note that $K$ isomerism can emerge in this region, as manifested in Figure 59, which shows the yrast sequences for the even-even $N = 74$ isotones. The band structures show that the deformation increases from $^{128}$Xe to $^{140}$Dy. An important issue in the emergence of collectivity in nuclei is: where and how is the validity of the $K$ quantum number established?

In principle, measurements of the magnetic dipole and electric quadrupole moments along the sequence of isotones could help answer this question. Unfortunately, the data are limited. The $g$ factors of the $8^-$ isomers in $^{128}$Xe and $^{130}$Ba have been measured to be $-0.036(9)$ [229] and $-0.0054(35)$ [230], respectively. The quadrupole moment of the isomer in $^{130}$Ba has also been measured to be $Q = +2.77(30)$b [230], which corresponds to a deformation of $\epsilon_2 \simeq 0.2$.

In $^{128}$Xe, the configuration of the isomer is assigned as $\nu(h_{11/2} \otimes g_{7/2})_{8^-}$. Evaluating the $g$ factor of this configuration with the spin $g$ factor, $g_s$, quenched from the free-neutron value by the standard factor of 0.7 gives $g(8^-) = -0.046$, consistent with

experiments. Empirical values for $g(1g_{7/2})$ and $g(1h_{11/2})$ from neighboring nuclei give $g(8^-) \approx -0.07$, somewhat larger than the experiment. For $^{130}$Ba, the isomer is assigned as $9/2^-[514] \otimes 7/2^-[404]$. The parentage of these Nilsson orbits is $\nu h_{11/2}$ and $\nu g_{7/2}$, i.e., as assigned to the isomer in $^{128}$Xe. Evaluating the $g$ factor of the $K$-isomer with standard Nilsson wavefunctions at $\epsilon_2 = 0.2$ and again quenching $g_s$ by the standard 0.7 factor, gives $g(8^-) = -0.003$, in excellent agreement with the experiment. This result is not sensitive to the deformation. Thus, the moment data suggest that the validity of the $K$ quantum number is established in $^{130}$Ba. It appears not to be established in $^{128}$Xe. Further insights could be gained from observation and characterization of bands built on the isomers.

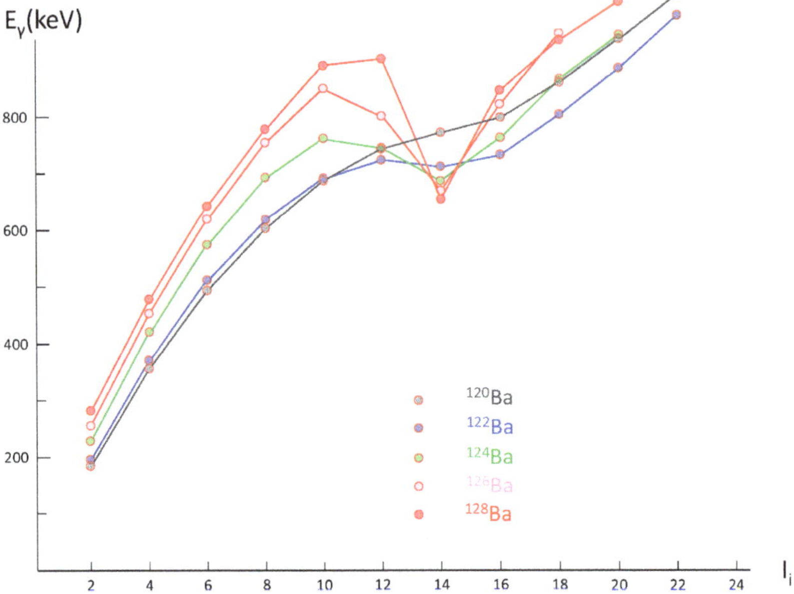


**Figure 58.** Yrast $E_\gamma$ vs. $I_i$ for $^{120-128}$Ba ($Z = 56$). Note the discontinuity above spin 10. The interpretation is that this is due to dominance of a broken neutron pair, $1h^2_{11/2}$. See discussion in the text. The data are taken from ENSDF [22].

Finally, with respect to the data shown in Figure 59, note that hindrance of the $E1$ isomeric transitions appears to increase with decreasing deformation: this appears counterintuitive. $E1$ transitions are an observable for which systematic features often remain elusive. In the normal valence space, they are forbidden. When looking at $E1$ strength, probably this involves the net result of many small contributions to the matrix element. Nevertheless, there is a visible systematic trend in Figure 59, which lacks an explanation.

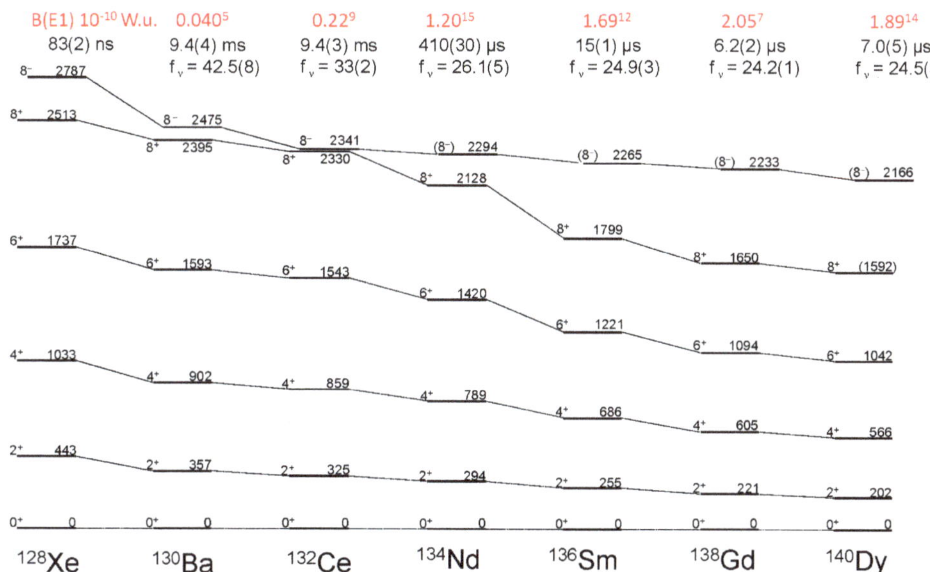


**Figure 59.** $K$ isomers in $N = 74$ isotones. The figure is based on a more limited view presented in Królas et al. [231]. Hindrance factors are $f_\nu^{-7} = B(E1)$ (in W.u.); the exponent is given by $\Delta K - \lambda$ where $\Delta K = 8$ and $\lambda = 1$, i.e., decay between the $K^\pi = 8^-$ isomer and $8^+$ state of the $K = 0$ ground-state band occurs by $E1$ multipole radiation. See text for details. The data are taken from ENSDF [22].

## 9. $A \sim 56$ and $N \sim 28$: New Regions of Shape Coexistence at Closed Shells

The observation of a deformed band in the double-closed shell nucleus $^{56}$Ni [232] suggested that an extension of shape coexistence from the $^{40}$Ca region to the $A \sim 56$ region seemed promising. This has not yet materialized. A leading factor is lack of stable targets for multi-nucleon transfer reactions. However, an initiative is underway, with a high-resolution internal-pair spectrometer, the *Super-e*, at the ANU to explore the occurrence of $E0$ transitions in this mass region [128,129,233–236]. However, there is evidence that shape coexistence is present in the $N = 28$ isotones $^{54}$Fe and $^{52}$Cr from transfer reactions: this is placed in a broad framework in Figure 60. Consistent with the transfer data, the ANU group observes $E0$ transitions in these $N = 28$ isotones.

The program of $E0$ decay studies at ANU has some surprises: while one expects a pattern of shape coexistence adjacent to $^{56}$Ni, associated with excited $0^+$ states, similar to that observed in nuclei adjacent to $^{40}$Ca, this is not apparent in the studies done so far. Evidence of $E0$ transitions other than for $Z = 28$ isotopes and $N = 28$ isotones is limited. However, note that the use of $E0$ transition strength as a spectroscopic fingerprint for shape coexistence depends on a combination of two factors: coexistence of two configurations with very different mean-square charge radii and strong mixing of these configurations [40].

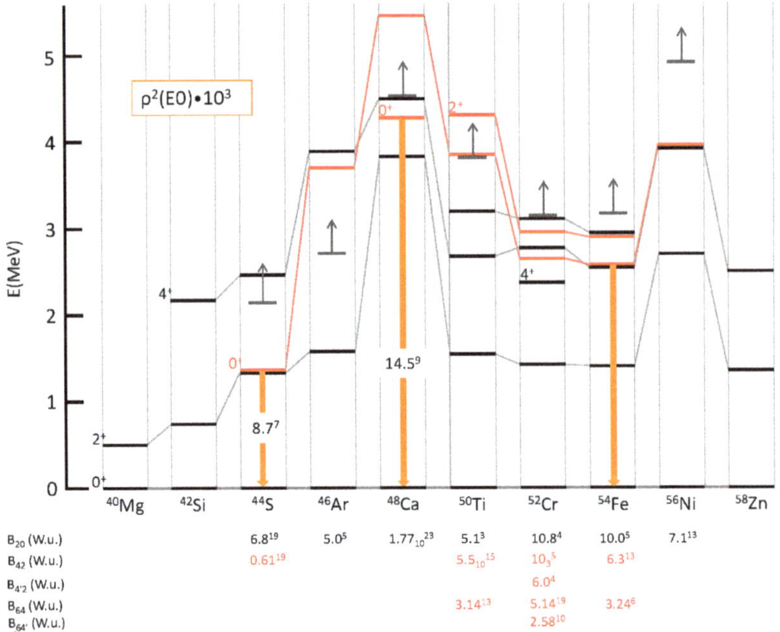


**Figure 60.** Systematics of the lowest positive-parity states in the $N = 28$ isotones. Candidate $\nu$ 2p-2h states are depicted in red. Electric monopole transitions from these states, where observed, are depicted as bold (orange) downwards-pointing arrows, the strengths are indicated where known. The 4282 keV $0^+$ state in $^{48}$Ca, a $\pi$ 2p - 2h state is also depicted in red. These assignments are based on two-neutron transfer reaction data [145,237–240]. The $^{58}$Ni(p,t)$^{56}$Ni reaction does not strongly populate $0^+$ states below 6.5 MeV [241]. Data for $^{46}$Ar are taken from [45,59,242]. The $B(E2; 2_1^+ \to 0_1^+)$ data are taken from [199]. Other data are taken from ENSDF [22], except the $B(E2)$ value for the decay of the first $4^+$ state in $^{52}$Cr, as quoted in ENSDF is in error; the value presented here is calculated from the half life quoted in ENSDF.

With reference to Figure 60, a search for deformed bands built on $0^+$ and $2^+$ states in $^{52}$Cr and $^{54}$Fe would be of great interest. Based on the energies of the first excited $2^+$ states, deformation appears to dominate the ground-state structure of $^{40}$Mg and $^{42}$Si (lifetime data would help to support this suggestion). Strengths of $E2$ transitions are shown for the transitions depopulating the $2_1^+$, $4_1^+$, and $6_1^+$ states where lifetime data are available. In $^{52}$Cr, two $4^+$ states are observed: the lowest is dominated by a seniority $v = 4$ $\pi 1f_{7/2}^4$ configuration and the upper is dominated by a seniority $v = 2$ $\pi 1f_{7/2}^4$ configuration. The weak $B_{42}$ value in $^{44}$S has been interpreted as due to $K$ isomerism, but band structure is not observed [243]; a seniority isomer resulting from a $\pi 1d_{5/2}$ broken pair is equally plausible.

The evidence for shape coexistence in nuclei with $N \sim 28$ and $Z < 20$ is highlighted in Figure 61, which summarizes the connection between intruder states in odd-mass nuclei and low-energy excited $0^+$ states in neighbouring even-even nuclei. The accumulation of the necessary data to establish shape coexistence in such neutron-rich nuclei is very demanding with respect to technique and accelerator running times.

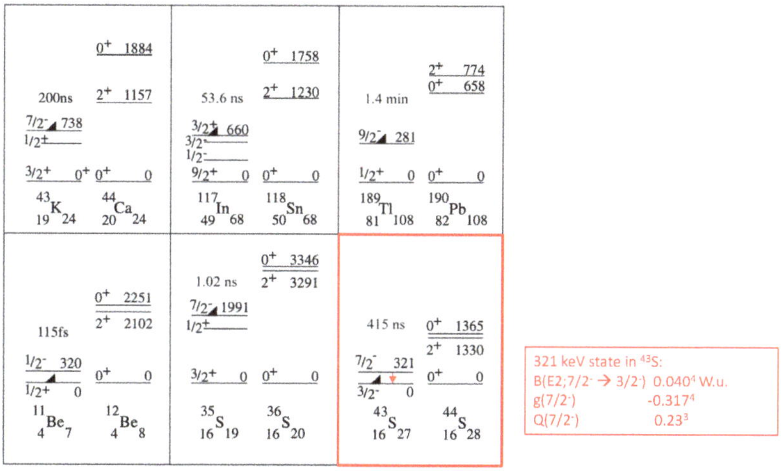

Figure 61. Selected pairs of nuclei showing a possible relationship between lp-2h intruder states (marked with solid triangles) and low lying excited $0^+$ states. The data are from [244–249] and ENSDF [22]. The figure is adapted from [41].

## 10. A Quantum Mechanical Perspective on Emergent Structures in the Nuclear Many-Body Problem

Nuclei are finite many-body quantum systems that self-organize to yield well-defined sizes and moments. The size of the nucleus determines the energy scale of quantization by virtue of the confinement (specific length) of nucleons (specific mass), scaled by $\hbar$. Defining the nucleon position and momentum observables, $x_j$ and $p_j$, with $j = 1, 2, 3$, and the nucleon mass $m$, this leads to the stationary states of any given nucleus via the definition of a Hamiltonian and the fundamental relationships

$$[x_j, p_j] = i\hbar, \tag{9}$$

and

$$p_j = m\dot{x}_j. \tag{10}$$

The consideration herein is limited to a discussion of model forms of independent-particle potentials and residual two-body interactions between the nucleons. At this point of inception, a representation of the problem for determining the eigenstates of the Hamiltonian must be chosen. This involves the use of symmetries of the Hamiltonian. If the nucleus has spherical symmetry, the handling of the independent-particle part of the Hamiltonian is greatly simplified by using the familiar representation that is a factorization into angular momentum and radial degrees of freedom. This extends to labeling states with angular momentum quantum numbers. The familiar radial confining potentials—the infinite square well and the harmonic oscillator—are solvable in closed form.

The factorization into radial and angular degrees of freedom equips us with the powerful algebra of angular momentum,

$$L_k \equiv x_i p_j - x_j p_i, \tag{11}$$

and

$$[L_i, L_j] = i\hbar\epsilon_{ijk} L_k, \tag{12}$$

where $\epsilon_{ijk}$ is the permutation symbol. The power of this algebra, the so(3) Lie algebra, is that it permits an enormous reduction of computational labor via the classification of states and operators as so(3) tensors, with their associated irreducible representations, Kronecker

products, and Wigner–Eckart theorem. Spins of nucleons are simply accommodated by extension of so(3) to its isomorphic algebra, su(2). The Hamiltonian becomes block diagonal in su(2) irreps, reducing computational labor; selection rules emerge; many transitions of interest appear in ratios that depend only on Clebsch–Gordan coefficients.

But, when one factorizes the shell model problem into angular and radial parts, and arrives at the so(3) algebra of angular momentum, one does not look any further for algebraic structures in the problem. However, there is another algebra "right under our noses": the radial degree of freedom possess an su(1,1) algebra. This is a so-called dual algebra for the shell model. Details are presented in pedagogical form in [250] and in a more advanced form in [6]. This is not found in any quantum mechanics textbook. It can be used to evaluate radial matrix elements in shell model computations, and this has recently been explored [251]. The algebra is defined by

$$T_1 \equiv \boldsymbol{r} \cdot \boldsymbol{r}, \quad T_2 \equiv \tfrac{1}{2}(\boldsymbol{r} \cdot \boldsymbol{p} + \boldsymbol{p} \cdot \boldsymbol{r}), \quad T_3 \equiv \boldsymbol{p} \cdot \boldsymbol{p}, \tag{13}$$

which, via linear combinations of the $\{T_i\}$ and scale factors, leads to the commutator brackets recognizable as su(1,1) (see [250]). Indeed, radial matrix elements possess simple relationships including "cancellations", which reflect properties of su(1,1) irreps and a su(1,1) Wigner–Eckart theorem.

Thus, what other algebraic structures can one expect in nuclei that emerge from functions of $x_i$, $p_i$ and Equation (9)? The clue comes from the dominance of quadrupole deformation in nuclei. One can define "quadrupole" coordinates, $x_i x_j$: these are rank-2 symmetric Cartesian tensors and there are six of them—using $x_i = x$, $x_j = y$, $x_k = z$, they are $xx$, $xy$, $xz$, $yy$, $yz$, $zz$. From these, in a straightforward manner, combinations such as $xp_x$, etc., and $p_x p_x$, etc., are obtained, yielding a Lie algebra with 21 generators, called sp(3,R). Details are presented in pedagogical form in [5] and full details are presented in [6]. The Lie algebra possesses many useful subalgebras: so(3), su(3), and others which need not concern us here; however, note that the su(3) subalgebra is that of the historical Elliott model [160]. A characteristic of the majority of nuclei is that they possess a very large value for the leading sp(3,R) quantum number, $\mathcal{N}$—the total number of shell model oscillator quanta carried by the sum of all the nucleons—counting the number of oscillator quanta for each nucleon partitioned across the entire occupancy of the oscillator shells of the given nucleus. For example, for $^{168}$Er, $\mathcal{N} = 814$ [252]. This leads to contraction in nuclei dominated by the su(3) subalgebra, yielding a (near) rigid rotor with properties that closely match observations [253] with the use of effective charges $e_p = +e$, $e_n = 0$ [254]. Contraction is a process where a Lie product, e.g., Equation (12), approaches zero asymptotically as quantum numbers become very large: for a state with angular momentum $L = 100$ and projection $m_L = +100$, the cone of indeterminacy appears almost identical to a classical angular momentum vector with three sharp Cartesian components. The origin of the concept of contraction is in a paper by Inonu and Wigner [255]; and the process is often called Inonu–Wigner contraction.

Thus, how does the shell model stand in relationship to the foregoing categorization? The shell model utilizes the su(2) spin-angular momentum algebra and adopts a central potential, but one does not find use of the su(1,1) algebra. This leaves open the functional form of the central potential: an su(1,1) algebraic structure is only realized for four central potentials—the Coulomb potential, the harmonic oscillator potential and their less well-known modifications—through augmentation with a $1/r^2$ term—the Kratzer potential and the Davidson potential [256]. Furthermore, the shell model does not make use of the sp(3,R) algebra because of spin–orbit coupling. Such an interaction lies "outside" of the symplectic model and must be treated as a perturbation. While the shell structure of nuclei and the dependence of magic numbers on spin–orbit coupling appear to invalidate the symplectic model, $Q \cdot Q$ interactions shift shell structures by up to 100 MeV, as manifested in observed shape coexisting structure; thus, $L \cdot S$ and $Q \cdot Q$ interactions have their respective domains of influence in nuclear structure. Indeed, the dividing line of their influence epitomizes the primary focus of this contribution. Notably, where the $L \cdot S$ interaction

dominates, $J$ emerges as a good quantum number and pairing interactions result in the emergence of seniority structure and its underlying quasispin su(2) algebra. These few mathematical structures appear to cover all the structures manifested in nuclei, observed so far, and as summarized in this contribution.

The foregoing leaves open the answer to the question posed by the title of this paper. The shell model versus the symplectic model approaches, with their respective dominance by spin–orbit coupling versus quadrupole–quadrupole coupling, each go some way to describing the structure of transitional nuclei. A shell model description can be achieved by using effective interactions. However, it should be noted that it is beginning to emerge that the effective interactions used in ab initio shell model calculations appear to be dominated by just those components that are compatible with symplectic model structures [257–259].

## 11. Conclusions

The present exploration of the interface between shell model and collective nuclear structure, which we term "emergence of nuclear collectivity", raises many questions. From a summary of systematic features in data, this paper has focused on the effective charge problem, which reveals itself already in the reduced transition strengths between the first-excited state and the ground state, $B(E2; 2_1^+ \to 0_1^+)$, in nuclei possessing two valence nucleons coupled to a doubly closed shell. A notable puzzle is the neutron effective charge needed for $^{18}$O compared to the well-known value of $e_n \sim +0.5e$ in $^{17}$O. It would appear that applying state-of-the-art shell model calculations beyond these simple structures needs great caution; and claims of successful descriptions in such nuclei deserve skepticism. Let us note the issue of spectroscopic factors as deduced from proton knockout by quaiselastic electron scattering [81] (see also [260,261]). The occupancies of particle configurations above the shell closures in doubly closed shell nuclei, shown in Figure 62, indicate that one is likely never dealing with simple shell model configurations when confronting data.

We have suggested directions in which shell model states should be explored as one moves away from closed shells, in the guise of seniority isomers (which involve pairing correlations). We have suggested criteria for exploring the validity of the language of deformation (proton–neutron correlations) in describing weakly deformed nuclei. Notably, nuclei that are termed "transitional" are severely neglected in the spectroscopic data base: we have outlined focal points for experimental study. We concluded with a sketch of details that leads shell model philosophy into the symplectic shell model: in the framework of this model, specific multi-shell configurations are emerging as a major clue to what is going on in low-energy nuclear excitations, and towards which state-of-the-art shell model activity needs to move.

We close with the view: "Data will have the last word in this Shakespearian drama" and "All the [nuclear] World's a [data] stage, and all the protons and neutrons merely players." (Adapted from *As You Like It* by W. Shakespeare). The message is that *one needs precision spectroscopy across the mass surface*, as well as pushing to exotic nuclei towards the limits of nuclear stability.

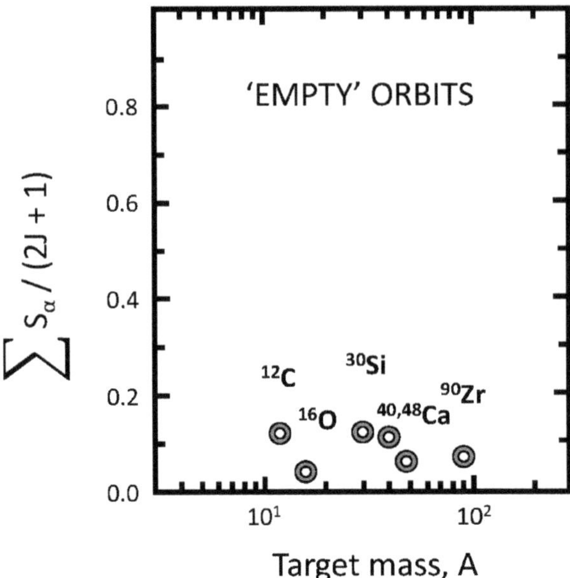


**Figure 62.** Quasiparticle strength for states just above the Fermi surface, observed in the reaction $(e,e'p)$ as a function of the target mass. All strengths are integrated to an excitation energy of about 20 MeV. Reprinted from [81], Copyright (1993), with permission from Elsevier. The language used in the original paper, from which this figure is taken, needs some clarification. 'Empty' orbits refers to shell model configurations above the shell closure, which are conventionally regarded as empty in doubly closed shell nuclei. However, in the $(e,e'p)$ studies, these configurations must have proton occupancy in the doubly closed shell target nuclei to explain the pattern of protons that are knocked out. Thus, one must conclude that the shells are not "closed".

An underlying theme that emerges in this look at nuclear structure is the role of algebraic structures in the quantum mechanics of the nuclear many-body problem. Two structures are widely manifested where many nucleon configurations are involved. In singly closed shell nuclei, the seniority coupling scheme dominates. This coupling scheme is explained by an $\mathtt{su(2)}$ algebra for correlated pairs in $j$-shell configurations. This stems from dominance of spin–orbit coupling imposed on a spherical mean-field independent-particle description. In open-shell nuclei, the Bohr unified model coupling scheme dominates. This can be traced to an $\mathtt{sp(3,R)}$ algebra with contraction on the very large quantum number values involved. Thus, we suggest that a way forward is to explore algebraic structures based on the shell model. This is being pursued, as noted, in the symmetry-adapted and symmetry-guided approaches [257–259], wherein effective charges are not needed.

**Author Contributions:** Conceptualization, J.L.W. and A.E.S.; methodology, J.L.W. and A.E.S.; validation, J.L.W. and A.E.S.; formal analysis, J.L.W. and A.E.S.; investigation, J.L.W. and A.E.S.; writing—original draft preparation, J.L.W.; writing—review and editing, J.L.W. and A.E.S.; visualization, J.L.W. and A.E.S.; funding acquisition, A.E.S. All authors have read and agreed to the published version of the manuscript.

**Funding:** This research was funded in part by the Australian Research Council Grant No. DP170101673.

**Data Availability Statement:** Data available in the references given.

**Conflicts of Interest:** The authors declare no conflict of interest. The funders had no role in the design of the study; in the collection, analyses, or interpretation of data; in the writing of the manuscript, or in the decision to publish the results.

## Abbreviations

The following abbreviations are used in this paper:

| | |
|---|---|
| ANU | The Australian National University |
| BCS | Bardeen–Cooper–Schrieffer theory of superconductivity |
| BMF | Beyond Mean Field |
| ENSDF | Evaluated Nuclear Structure Data File [22] |
| PTRM | Particle Triaxial Rotor Model [102] |
| RPA | Random Phase Approximation [72] |
| SM | Shell Model |

## References


1. Mayer, M.G. On Closed Shells in Nuclei. II. *Phys. Rev.* **1949**, *75*, 1969–1970. [CrossRef]
2. Haxel, O.; Jensen, J.H.D.; Suess, H.E. On the "magic numbers" in nuclear structure. *Phys. Rev.* **1949**, *75*, 1766–1766. [CrossRef]
3. Heyde, K.L. *The Nuclear Shell Model*; Springer: Berlin/Heidelberg, Germany, 1990. [CrossRef]
4. Bohr, A.; Mottelson, B.R. *Nuclear Structure*; World Scientific Publishing Company: Singapore, 1998. [CrossRef]
5. Heyde, K.; Wood, J.L. *Quantum Mechanics for Nuclear Structure*; IOP Publishing: Bristol, UK, 2020; Volume 2, pp. 2053–2563. [CrossRef]
6. Rowe, D.J.; Wood, J.L. *Fundamentals of Nuclear Models*; World Scientific: Singapore, 2010. [CrossRef]
7. Rosiak, D.; Seidlitz, M.; Reiter, P.; Naïdja, H.; Tsunoda, Y.; Togashi, T.; Nowacki, F.; Otsuka, T.; Colò, G.; Arnswald, K.; et al. Enhanced quadrupole and octupole strength in doubly magic $^{132}$Sn. *Phys. Rev. Lett.* **2018**, *121*, 252501. [CrossRef] [PubMed]
8. Jenkins, D.G.; Wood, J.L. *Nuclear Data. A Primer*; IOP Publishing: Bristol, UK, 2021. [CrossRef]
9. Mayer, M.G. Nuclear configurations in the spin–orbit coupling model. I. Empirical evidence. *Phys. Rev.* **1950**, *78*, 16–21. [CrossRef]
10. Mayer, M.G. Nuclear configurations in the spin–orbit coupling model. II. Theoretical considerations. *Phys. Rev.* **1950**, *78*, 22–23. [CrossRef]
11. Kerman, A. Pairing forces and nuclear collective motion. *Ann. Phys.* **1961**, *12*, 300–329. [CrossRef]
12. Racah, G. Theory of complex spectra. I. *Phys. Rev.* **1942**, *61*, 186–197. [CrossRef]
13. Racah, G. Theory of complex spectra. II. *Phys. Rev.* **1942**, *62*, 438–462. [CrossRef]
14. Racah, G. Theory of complex spectra. III. *Phys. Rev.* **1943**, *63*, 367–382. [CrossRef]
15. Flowers, B.H.; Peierls, R.E. Studies in *jj*-coupling. I. Classification of nuclear and atomic states. *Proc. R. Soc. Lond. Ser. Math. Phys. Sci.* **1952**, *212*, 248–263. [CrossRef]
16. Helmers, K. Symplectic invariants and Flowers' classification of shell model states. *Nucl. Phys.* **1961**, *23*, 594–611. [CrossRef]
17. Lawson, R.; Macfarlane, M. The quasi-spin formalism and the dependence of nuclear matrix elements on particle number. *Nucl. Phys.* **1965**, *66*, 80–96. [CrossRef]
18. Rowe, D.J.; Carvalho, M.J.; Repka, J. Dual pairing of symmetry and dynamical groups in physics. *Rev. Mod. Phys.* **2012**, *84*, 711–757. [CrossRef]
19. Anderson, P.W. Random-phase approximation in the theory of superconductivity. *Phys. Rev.* **1958**, *112*, 1900–1916. [CrossRef]
20. Cooper, L.N. Bound electron pairs in a degenerate Fermi gas. *Phys. Rev.* **1956**, *104*, 1189–1190. [CrossRef]
21. Jungclaus, A.; Cáceres, L.; Górska, M.; Pfützner, M.; Pietri, S.; Werner-Malento, E.; Grawe, H.; Langanke, K.; Martínez-Pinedo, G.; Nowacki, F.; et al. Observation of isomeric decays in the *r*-process waiting-point nucleus $^{130}$Cd$_{82}$. *Phys. Rev. Lett.* **2007**, *99*, 132501. [CrossRef]
22. Evaluated Nuclear Structure Data File. Available online: https://www.nndc.bnl.gov/ensdf/(accessed on 5 March 2022).
23. Kameda, D.; Kubo, T.; Ohnishi, T.; Kusaka, K.; Yoshida, A.; Yoshida, K.; Ohtake, M.; Fukuda, N.; Takeda, H.; Tanaka, K.; et al. Observation of new microsecond isomers among fission products from in-flight fission of 345 MeV/nucleon $^{238}$U. *Phys. Rev. C* **2012**, *86*, 014319. [CrossRef]
24. Palacz, M.; Nyberg, J.; Grawe, H.; Sieja, K.; de Angelis, G.; Bednarczyk, P.; Blazhev, A.; Curien, D.; Dombradi, Z.; Dorvaux, O.; et al. $N = 50$ core excited states studied in the $^{96}_{46}$Pd$_{50}$ nucleus. *Phys. Rev. C* **2012**, *86*, 014318. [CrossRef]
25. Watanabe, H.; Lorusso, G.; Nishimura, S.; Xu, Z.Y.; Sumikama, T.; Söderström, P.A.; Doornenbal, P.; Browne, F.; Gey, G.; Jung, H.S.; et al. Isomers in $^{128}$Pd and $^{126}$Pd: Evidence for a robust shell closure at the neutron magic number 82 in exotic palladium isotopes. *Phys. Rev. Lett.* **2013**, *111*, 152501. [CrossRef]
26. Garnsworthy, A.B.; Regan, P.H.; Pietri, S.; Sun, Y.; Xu, F.R.; Rudolph, D.; Górska, M.; Cáceres, L.; Podolyák, Z.; Steer, S.J.; et al. Isomeric states in neutron-deficient *A* 80–90 nuclei populated in the fragmentation of $^{107}$Ag. *Phys. Rev. C* **2009**, *80*, 064303. [CrossRef]
27. Park, J.; Krücken, R.; Lubos, D.; Gernhäuser, R.; Lewitowicz, M.; Nishimura, S.; Ahn, D.S.; Baba, H.; Blank, B.; Blazhev, A.; et al. Properties of $\gamma$-decaying isomers and isomeric ratios in the $^{100}$Sn region. *Phys. Rev. C* **2017**, *96*, 044311. [CrossRef]
28. McNeill, J.H.; Blomqvist, J.; Chishti, A.A.; Daly, P.J.; Gelletly, W.; Hotchkis, M.A.C.; Piiparinen, M.; Varley, B.J.; Woods, P.J. Exotic N=82 nuclei $^{153}$Lu and $^{154}$Hf and filling of the $\pi h_{11/2}$ subshell. *Phys. Rev. Lett.* **1989**, *63*, 860–863. [CrossRef] [PubMed]

29. Lozeva, R.L.; Simpson, G.S.; Grawe, H.; Neyens, G.; Atanasova, L.A.; Balabanski, D.L.; Bazzacco, D.; Becker, F.; Bednarczyk, P.; Benzoni, G.; et al. New sub-$\mu$s isomers in $^{125,127,129}$Sn and isomer systematics of $^{124-130}$Sn. *Phys. Rev. C* **2008**, *77*, 064313. [CrossRef]
30. Peters, E.E.; Van Isacker, P.; Chakraborty, A.; Crider, B.P.; Kumar, A.; Liu, S.H.; McEllistrem, M.T.; Mehl, C.V.; Prados-Estévez, F.M.; Ross, T.J.; et al. Seniority structure of $^{136}$Xe$_{82}$. *Phys. Rev. C* **2018**, *98*, 034302. [CrossRef]
31. Stuchbery, A.E. Gyromagnetic ratios of excited states and nuclear structure near $^{132}$Sn. *Aip Conf. Proc.* **2014**, *1625*, 52–58. [CrossRef]
32. Peters, E.E.; Stuchbery, A.E.; Chakraborty, A.; Crider, B.P.; Ashley, S.F.; Kumar, A.; McEllistrem, M.T.; Prados-Estévez, F.M.; Yates, S.W. Emerging collectivity from the nuclear structure of $^{132}$Xe: Inelastic neutron scattering studies and shell-model calculations. *Phys. Rev. C* **2019**, *99*, 064321. [CrossRef]
33. Brown, B.A.; Stone, N.J.; Stone, J.R.; Towner, I.S.; Hjorth-Jensen, M. Magnetic moments of the $2_1^+$ states around $^{132}$Sn. *Phys. Rev. C* **2005**, *71*, 044317. [CrossRef]
34. Teruya, E.; Yoshinaga, N.; Higashiyama, K.; Odahara, A. Shell-model calculations of nuclei around mass 130. *Phys. Rev. C* **2015**, *92*, 034320. [CrossRef]
35. Sonzogni, A. Nuclear data sheets for $A = 136$. *Nucl. Data Sheets* **2002**, *95*, 837–994. [CrossRef]
36. Bazzacco, D.; Brandolini, F.; Löwenich, K.; Pavan, P.; Rossi-Alvarez, C.; Maglione, E.; de Poli, M.; Haque, A. Transient-field g-factor measurement of the first $2^+$ states in the $N = 82$ nuclei $^{140}$Ce, $^{142}$Nd and $^{144}$Sm. *Nucl. Phys. A* **1991**, *533*, 541–552. [CrossRef]
37. Lawson, R.D. $(\pi h11/2)^n$ States expected in$^{150}_{68}$Er$_{82}$, $^{151}_{69}$Tm$_{82}$ and $^{152}_{70}$Yb$_{82}$. *Z. für Phys. Atoms Nucl.* **1981**, *303*, 51–61. [CrossRef]
38. Schiffer, J.P.; True, W.W. The effective interaction between nucleons deduced from nuclear spectra. *Rev. Mod. Phys.* **1976**, *48*, 191–217. [CrossRef]
39. Bron, J.; Hesselink, W.; Van Poelgeest, A.; Zalmstra, J.; Uitzinger, M.; Verheul, H.; Heyde, K.; Waroquier, M.; Vincx, H.; Van Isacker, P. Collective bands in even mass Sn isotopes. *Nucl. Phys. A* **1979**, *318*, 335–351. [CrossRef]
40. Wood, J.; Heyde, K.; Nazarewicz, W.; Huyse, M.; van Duppen, P. Coexistence in even-mass nuclei. *Phys. Rep.* **1992**, *215*, 101–201. [CrossRef]
41. Heyde, K.; Wood, J.L. Shape coexistence in atomic nuclei. *Rev. Mod. Phys.* **2011**, *83*, 1467–1521. [CrossRef]
42. Rowe, D.J.; Wood, J.L. A relationship between isobaric analog states and shape coexistence in nuclei. *J. Phys. Nucl. Part. Phys.* **2018**, *45*, 06LT01. [CrossRef]
43. Dove, J.; Kerns, B.; McClellan, R.E.; Miyasaka, S.; Morton, D.H.; Nagai, K.; Prasad, S.; Sanftl, F.; Scott, M.B.C.; Tadepalli, A.S.; et al. The asymmetry of antimatter in the proton. *Nature* **2021**, *590*, 561–565. [CrossRef]
44. The Shell-Model Code NuShellX@MSU. *Nucl. Data Sheets* **2014**, *120*, 115–118. [CrossRef]
45. Calinescu, S.; Cáceres, L.; Grévy, S.; Sorlin, O.; Dombrádi, Z.; Stanoiu, M.; Astabatyan, R.; Borcea, C.; Borcea, R.; Bowry, M.; et al. Coulomb excitation of $^{44}$Ca and $^{46}$Ar. *Phys. Rev. C* **2016**, *93*, 044333. [CrossRef]
46. Radford, D.; Baktash, C.; Barton, C.; Batchelder, J.; Beene, J.; Bingham, C.; Caprio, M.; Danchev, M.; Fuentes, B.; Galindo-Uribarri, A.; et al. Coulomb excitation and transfer reactions with rare neutron-rich isotopes. *Nucl. Phys. A* **2005**, *752*, 264–272. [CrossRef]
47. Radford, D.C.; Baktash, C.; Barton, C.J.; Batchelder, J.; Beene, J.R.; Bingham, C.R.; Caprio, M.A.; Danchev, M.; Fuentes, B.; Galindo-Uribarri, A.; et al. Coulomb excitation and transfer reactions with neutron-rich radioactive beams. *Eur. Phys. J. A* **2005**, *25*, 383–387. [CrossRef]
48. Stuchbery, A.E.; Allmond, J.M.; Galindo-Uribarri, A.; Padilla-Rodal, E.; Radford, D.C.; Stone, N.J.; Batchelder, J.C.; Beene, J.R.; Benczer-Koller, N.; Bingham, C.R.; et al. Electromagnetic properties of the $2_1^+$ state in $^{134}$Te: Influence of core excitation on single-particle orbits beyond $^{132}$Sn. *Phys. Rev. C* **2013**, *88*, 051304. [CrossRef]
49. Kocheva, D.; Rainovski, G.; Jolie, J.; Pietralla, N.; Blazhev, A.; Astier, A.; Altenkirch, R.; Ansari, S.; Braunroth, T.; Cortés, M.L.; et al. A revised $B(E2; 2_1^+ \rightarrow 0_1^+)$ value in the semi-magic nucleus $^{210}$Po. *Eur. Phys. J. A* **2017**, *53*, 175.[CrossRef]
50. Warburton, E.K.; Brown, B.A. Effective interactions for the $0p1s0d$ nuclear shell-model space. *Phys. Rev. C* **1992**, *46*, 923–944. [CrossRef]
51. Warburton, E.K. Second-forbidden unique $\beta$ decays of $^{10}$Be, $^{22}$Na, and $^{26}$Al. *Phys. Rev. C* **1992**, *45*, 463–466. [CrossRef]
52. Brown, B.A.; Richter, W.A. New "USD" Hamiltonians for the *sd* shell. *Phys. Rev. C* **2006**, *74*, 034315. [CrossRef]
53. Brown, B.; Sherr, R. Charge-dependent two-body interactions deduced from displacement energies in the $1f_{7/2}$ shell. *Nucl. Phys. A* **1979**, *322*, 61–91. [CrossRef]
54. Honma, M.; Otsuka, T.; Brown, B.A.; Mizusaki, T. New effective interaction for $pf$-shell nuclei and its implications for the stability of the $N = Z = 28$ closed core. *Phys. Rev. C* **2004**, *69*, 034335. [CrossRef]
55. Honma, M.; Otsuka, T.; Brown, B.A.; Mizusaki, T. Shell-model description of neutron-rich pf-shell nuclei with a new effective interaction GXPF 1. *Eur. Phys. J. A* **2005**, *25*, 499–502. [CrossRef]
56. Utsuno, Y.; Otsuka, T.; Brown, B.A.; Honma, M.; Mizusaki, T.; Shimizu, N. Shape transitions in exotic Si and S isotopes and tensor-force-driven Jahn-Teller effect. *Phys. Rev. C* **2012**, *86*, 051301. [CrossRef]

57. Horie, H.; Ogawa, K. Effective proton-neutron interaction and spectroscopy of the nuclei with N=29. *Prog. Theor. Phys.* **1971**, *46*, 439–461. [CrossRef]
58. Warburton, E.K.; Brown, B.A. Appraisal of the Kuo-Herling shell-model interaction and application to $A = 210 - 212$ nuclei. *Phys. Rev. C* **1991**, *43*, 602–617. [CrossRef] [PubMed]
59. Mengoni, D.; Valiente-Dobón, J.J.; Gadea, A.; Lunardi, S.; Lenzi, S.M.; Broda, R.; Dewald, A.; Pissulla, T.; Angus, L.J.; Aydin, S.; et al. Lifetime measurements of excited states in neutron-rich $^{44,46}$Ar populated via a multinucleon transfer reaction. *Phys. Rev. C* **2010**, *82*, 024308. [CrossRef]
60. Scheit, H.; Glasmacher, T.; Brown, B.A.; Brown, J.A.; Cottle, P.D.; Hansen, P.G.; Harkewicz, R.; Hellström, M.; Ibbotson, R.W.; Jewell, J.K.; et al. New Region of Deformation: The Neutron-rich sulfur isotopes. *Phys. Rev. Lett.* **1996**, *77*, 3967–3970. [CrossRef] [PubMed]
61. Gade, A.; Bazin, D.; Campbell, C.M.; Church, J.A.; Dinca, D.C.; Enders, J.; Glasmacher, T.; Hu, Z.; Kemper, K.W.; Mueller, W.F.; et al. Detailed experimental study on intermediate-energy Coulomb excitation of $^{46}$Ar. *Phys. Rev. C* **2003**, *68*, 014302. [CrossRef]
62. Wang, H.K.; Ghorui, S.K.; Chen, Z.Q.; Li, Z.H. Analysis of low-lying states, neutron-core excitations, and electromagnetic transitions in tellurium isotopes $^{130--134}$Te. *Phys. Rev. C* **2020**, *102*, 054316. [CrossRef]
63. Qi, C. Shell-model configuration-interaction description of quadrupole collectivity in Te isotopes. *Phys. Rev. C* **2016**, *94*, 034310. [CrossRef]
64. Gray, T.J.; Allmond, J.M.; Stuchbery, A.E.; Yu, C.H.; Baktash, C.; Gargano, A.; Galindo-Uribarri, A.; Radford, D.C.; Batchelder, J.C.; Beene, J.R.; et al. Early signal of emerging nuclear collectivity in neutron-rich $^{129}$Sb. *Phys. Rev. Lett.* **2020**, *124*, 032502. [CrossRef]
65. Allmond, J.M.; Stuchbery, A.E.; Galindo-Uribarri, A.; Padilla-Rodal, E.; Radford, D.C.; Batchelder, J.C.; Bingham, C.R.; Howard, M.E.; Liang, J.F.; Manning, B.; et al. Investigation into the semimagic nature of the tin isotopes through electromagnetic moments. *Phys. Rev. C* **2015**, *92*, 041303. [CrossRef]
66. Ellegaard, C.; Barnes, P.; Eisenstein, R.; Romberg, E.; Bhatia, T.; Canada, T. Inelastic scattering of deuterons, protons and tritons on $^{210}$Po. *Nucl. Phys. A* **1973**, *206*, 83–96. [CrossRef]
67. Allmond, J.M.; Stuchbery, A.E.; Baktash, C.; Gargano, A.; Galindo-Uribarri, A.; Radford, D.C.; Bingham, C.R.; Brown, B.A.; Coraggio, L.; Covello, A.; et al. Electromagnetic moments of radioactive $^{136}$Te and the emergence of collectivity 2p $\oplus$ 2n outside of double-magic $^{132}$Sn. *Phys. Rev. Lett.* **2017**, *118*, 092503. [CrossRef] [PubMed]
68. Stuchbery, A.E.; Allmond, J.M.; Danchev, M.; Baktash, C.; Bingham, C.R.; Galindo-Uribarri, A.; Radford, D.C.; Stone, N.J.; Yu, C.H. First-excited state $g$ factor of $^{136}$Te by the recoil in vacuum method. *Phys. Rev. C* **2017**, *96*, 014321. [CrossRef]
69. Schielke, S.; Hohn, D.; Speidel, K.H.; Kenn, O.; Leske, J.; Gemein, N.; Offer, M.; Gerber, J.; Maier-Komor, P.; Zell, O.; et al. Evidence for $^{40}$Ca core excitation from $g$ factor and $B(E2)$ measurements on the $2_1^+$ states of $^{42,44}$Ca. *Phys. Lett. B* **2003**, *571*, 29–35. [CrossRef]
70. Nuclear Electromagnetic Moments. Available online: https://www-nds.iaea.org/nuclearmoments/ (accessed on 5 March 2022).
71. East, M.C.; Stuchbery, A.E.; Chamoli, S.K.; Pinter, J.S.; Crawford, H.L.; Wilson, A.N.; Kibédi, T.; Mantica, P.F. Relative $g$-factor measurements in $^{54}$Fe, $^{56}$Fe, and $^{58}$Fe *Phys. Rev. C* **2009**, *79*, 024304. [CrossRef]
72. Ring, P.; Schuck, P. *The Nuclear Many-Body Problem*; Springer: Berlin/Heidelberg, Germany, 1980.
73. Hamamoto, I. Giant resonances and polarization effects. *Phys. Scr.* **1972**, *6*, 266–269. [CrossRef]
74. Horikawa, Y.; Hoshino, T.; Arima, A. Polarization effects for higher multipoles. *Phys. Lett. B* **1976**, *63*, 134–138. [CrossRef]
75. Brown, B.A.; Arima, A.; McGrory, J.B. E2 core-polarization charge for nuclei near $^{16}$O and $^{40}$Ca. *Nucl. Phys. A* **1977**, *277*, 77–108. [CrossRef]
76. Sagawa, H. Core polarization charge for $E6$ transition. *Phys. Rev. C* **1979**, *19*, 506–510. [CrossRef]
77. Duguet, T.; Hagen, G. Ab initio approach to effective single-particle energies in doubly closed shell nuclei. *Phys. Rev. C* **2012**, *85*, 034330. [CrossRef]
78. Duguet, T.; Hergert, H.; Holt, J.D.; Somà, V. Nonobservable nature of the nuclear shell structure: Meaning, illustrations, and consequences. *Phys. Rev. C* **2015**, *92*, 034313. [CrossRef]
79. Pandharipande, V.R.; Sick, I.; Huberts, P.K.D. Independent particle motion and correlations in fermion systems. *Rev. Mod. Phys.* **1997**, *69*, 981–991. [CrossRef]
80. Paschalis, S.; Petri, M.; Macchiavelli, A.; Hen, O.; Piasetzky, E. Nucleon-nucleon correlations and the single-particle strength in atomic nuclei. *Phys. Lett. B* **2020**, *800*, 135110. [CrossRef]
81. Lapikás, L. Quasi-elastic electron scattering off nuclei. *Nucl. Phys. A* **1993**, *553*, 297–308. [CrossRef]
82. Vanhalst, M. Quantifying Short-Range Correlations in Nuclei. Ph.D. Thesis, Ghent University, Ghent, Belgium, 2014. Available online: https://biblio.ugent.be/publication/5713663 (accessed on 5 March 2022).
83. Launey, K.D.; Dytrych, T.; Draayer, J.P. Symmetry-guided large-scale shell-model theory. *Prog. Part. Nucl. Phys.* **2016**, *89*, 101–136. [CrossRef]
84. Kay, B.P.; Schiffer, J.P.; Freeman, S.J. Quenching of cross sections in nucleon transfer reactions. *Phys. Rev. Lett.* **2013**, *111*, 042502. [CrossRef]
85. Müther, H.; Polls, A.; Dickhoff, W.H. Momentum and energy distributions of nucleons in finite nuclei due to short-range correlations. *Phys. Rev. C* **1995**, *51*, 3040–3051. [CrossRef]
86. Dickhoff, W.H. Determining and calculating spectroscopic factors from stable nuclei to the drip lines. *J. Phys. Nucl. Part. Phys.* **2010**, *37*, 064007. [CrossRef]

87. Baranger, M. A definition of the single-nucleon potential. *Nucl. Phys. A* **1970**, *149*, 225–240. [CrossRef]
88. Macfarlane, M.H.; French, J.B. Stripping reactions and the structure of light and intermediate nuclei. *Rev. Mod. Phys.* **1960**, *32*, 567–691. [CrossRef]
89. Kay, B.P.; Schiffer, J.P.; Freeman, S.J.; Tang, T.L.; Cropper, B.D.; Faestermann, T.; Hertenberger, R.; Keatings, J.M.; MacGregor, P.T.; Smith, J.F.; et al. Consistency of nucleon-transfer sum rules in well-deformed nuclei. *Phys. Rev. C* **2021**, *103*, 024319. [CrossRef]
90. Crawford, H.L.; Macchiavelli, A.O.; Fallon, P.; Albers, M.; Bader, V.M.; Bazin, D.; Campbell, C.M.; Clark, R.M.; Cromaz, M.; Dilling, J.; et al. Unexpected distribution of $\nu 1f_{7/2}$ strength in $^{49}$Ca. *Phys. Rev. C* **2017**, *95*, 064317. [CrossRef]
91. Otsuka, T.; Gade, A.; Sorlin, O.; Suzuki, T.; Utsuno, Y. Evolution of shell structure in exotic nuclei. *Rev. Mod. Phys.* **2020**, *92*, 015002. [CrossRef]
92. Atomic Mass Evaluation—AME2020, Atomic Mass Data Center. Available online: https://www-nds.iaea.org/amdc/ (accessed on 5 March 2022).
93. Newton, J.; Cirilov, S.; Stephens, F.; Diamond, R. Possible oblate shape of $9/2^-$ isomer in $^{199}$Tl. *Nucl. Phys. A* **1970**, *148*, 593–614. [CrossRef]
94. Heyde, K.; Van Isacker, P.; Waroquier, M.; Wood, J.; Meyer, R. Coexistence in odd-mass nuclei. *Phys. Rep.* **1983**, *102*, 291–393. [CrossRef]
95. Meyer-Ter-Vehn, J. Collective model description of transitional odd-A nuclei: (I). The triaxial-rotor-plus-particle model. *Nucl. Phys. A* **1975**, *249*, 111–140. [CrossRef]
96. Meyer-Ter-Vehn, J. Collective model description of transitional odd-A nuclei: (II). Comparison with unique parity states of nuclei in the A = 135 and A = 190 mass regions. *Nucl. Phys. A* **1975**, *249*, 141–165. [CrossRef]
97. Wiebicke, H.; Münchow, L. A weak coupling model for high-spin states. *Phys. Lett. B* **1974**, *50*, 429–432. [CrossRef]
98. Gizon, J.; Gizon, A.; Meyer-Ter-Vehn, J. Triaxial shape and onset of high-spin bands in $^{129}$Ba. *Nucl. Phys. A* **1977**, *277*, 464–476. [CrossRef]
99. Wiedenhöver, I.; Yan, J.; Neuneyer, U.; Wirowski, R.; von Brentano, P.; Gelberg, A.; Yoshida, N.; Otsuka, T. Non-yrast states in $^{125}$Xe. *Nucl. Phys. A* **1995**, *582*, 77–108. [CrossRef]
100. Modamio, V.; Jungclaus, A.; Algora, A.; Bazzacco, D.; Escrig, D.; Fraile, L.M.; Lenzi, S.; Marginean, N.; Martinez, T.; Napoli, D.R.; et al. New high-spin isomer and quasiparticle-vibration coupling in $^{187}$Ir. *Phys. Rev. C* **2010**, *81*, 054304. [CrossRef]
101. Guo, S.; Zhou, X.H.; Petrache, C.M.; Lawrie, E.A.; Mthembu, S.H.; Fang, Y.D.; Wu, H.Y.; Wang, H.L.; Meng, H.Y.; Li, G.S.; et al. Probing the nature of the conjectured low-spin wobbling bands in atomic nuclei. *Phys. Lett. B* **2022**, *828*, 137010. [CrossRef]
102. Ragnarsson, I.; Semmes, P.B. Description of nuclear moments and nuclear spectra in the particle-rotor model. *Hyperfine Interact.* **1988**, *43*, 423–440. [CrossRef]
103. Lieberz, D.; Gelberg, A.; Granderath, A.; von Brentano, P.; Ragnarsson, I.; Semmes, P. Triaxial rotor plus particle description of negative-parity states in $^{125}$Xe. *Nucl. Phys. A* **1991**, *529*, 1–19. [CrossRef]
104. Lieberz, D.; Gelberg, A.; von Brentano, P.; Ragnarsson, I.; Semmes, P. Signatures of $\gamma$ deformation in nuclei and an application to $^{125}$Xe. *Phys. Lett. B* **1992**, *282*, 7–12. [CrossRef]
105. Stephens, F.S.; Diamond, R.M.; Leigh, J.R.; Kammuri, T.; Nakai, K. Decoupled yrast states in odd-mass nuclei. *Phys. Rev. Lett.* **1972**, *29*, 438–441. [CrossRef]
106. Bäcklin, A.; Fogelberg, B.; Malmskog, S. Possible deformed states in $^{115}$In and $^{117}$In. *Nucl. Phys. A* **1967**, *96*, 539–560. [CrossRef]
107. Tuttle, W.K.; Stelson, P.H.; Robinson, R.L.; Milner, W.T.; McGowan, F.K.; Raman, S.; Dagenhart, W.K. Coulomb excitation of $^{113,115}$In. *Phys. Rev. C* **1976**, *13*, 1036–1048. [CrossRef]
108. Gray, T.J.; Stuchbery, A.E.; Fuderer, L.A.; Allmond, J.M. *E*2 collectivity in shell-model calculations for odd-mass nuclei near $^{132}$Sn. *EPJ Web Conf.* **2020**, *232*, 04007. [CrossRef]
109. Eberz, J.; Dinger, U.; Huber, G.; Lochmann, H.; Menges, R.; Neugart, R.; Kirchner, R.; Klepper, O.; Kühl, T.; Marx, D.; et al. Spins, moments and mean square charge radii of $^{104-127}$In determined by laser spectroscopy. *Nucl. Phys. A* **1987**, *464*, 9–28. [CrossRef]
110. Brown, B.A. New Skyrme interaction for normal and exotic nuclei. *Phys. Rev. C* **1998**, *58*, 220–231. [CrossRef]
111. DiJulio, D.D.; Cederkall, J.; Fahlander, C.; Ekström, A.; Hjorth-Jensen, M.; Albers, M.; Bildstein, V.; Blazhev, A.; Darby, I.; Davinson, T.; et al. Coulomb excitation of $^{107}$In. *Phys. Rev. C* **2013**, *87*, 017301. [CrossRef]
112. Vaquero, V.; Jungclaus, A.; Aumann, T.; Tscheuschner, J.; Litvinova, E.V.; Tostevin, J.A.; Baba, H.; Ahn, D.S.; Avigo, R.; Boretzky, K.; et al. Fragmentation of single-particle strength around the doubly magic nucleus $^{132}$Sn and the position of the $0f_{5/2}$ proton-hole state in $^{131}$In. *Phys. Rev. Lett.* **2020**, *124*, 022501. [CrossRef] [PubMed]
113. Kumar, K. Intrinsic quadrupole moments and shapes of nuclear ground states and excited states. *Phys. Rev. Lett.* **1972**, *28*, 249–253. [CrossRef]
114. Cline, D. Nuclear shapes studied by Coulomb excitation. *Annu. Rev. Nucl. Part. Sci.* **1986**, *36*, 683–716. [CrossRef]
115. Coombes, B.J. Emergence of Nuclear Collectivity in Cd and Te Isotopes from *M*1 and *E*2 Observables. Ph.D. Thesis, The Australian National University, Canberra, Australia, 2022.
116. Garrett, P.E.; Wood, J.L.; Yates, S.W. Critical insights into nuclear collectivity from complementary nuclear spectroscopic methods. *Phys. Scr.* **2018**, *93*, 063001. [CrossRef]

117. Leoni, S.; Fornal, B.; Mărginean, N.; Sferrazza, M.; Tsunoda, Y.; Otsuka, T.; Bocchi, G.; Crespi, F.C.L.; Bracco, A.; Aydin, S.; et al. Multifaceted Quadruplet of Low-Lying Spin-Zero States in $^{66}$Ni: Emergence of shape isomerism in light nuclei. *Phys. Rev. Lett.* **2017**, *118*, 162502. [CrossRef]
118. Flavigny, F.; Pauwels, D.; Radulov, D.; Darby, I.J.; De Witte, H.; Diriken, J.; Fedorov, D.V.; Fedosseev, V.N.; Fraile, L.M.; Huyse, M.; et al. Characterization of the low-lying $0^+$ and $2^+$ states in $^{68}$Ni via $\beta$ decay of the low-spin $^{68}$Co isomer. *Phys. Rev. C* **2015**, *91*, 034310. [CrossRef]
119. Ishii, T.; Asai, M.; Makishima, A.; Hossain, I.; Ogawa, M.; Hasegawa, J.; Matsuda, M.; Ichikawa, S. Core-excited states in the doubly magic $^{68}Ni$ and its neighbor $^{69}Cu$. *Phys. Rev. Lett.* **2000**, *84*, 39–42. [CrossRef]
120. Elman, B.; Gade, A.; Janssens, R.V.F.; Ayangeakaa, A.D.; Bazin, D.; Belarge, J.; Bender, P.C.; Brown, B.A.; Campbell, C.M.; Carpenter, M.P.; et al. Probing the role of proton cross-shell excitations in $^{70}$Ni using nucleon knockout reactions. *Phys. Rev. C* **2019**, *100*, 034317. [CrossRef]
121. Prokop, C.J.; Crider, B.P.; Liddick, S.N.; Ayangeakaa, A.D.; Carpenter, M.P.; Carroll, J.J.; Chen, J.; Chiara, C.J.; David, H.M.; Dombos, A.C.; et al. New low-energy $0^+$ state and shape coexistence in $^{70}$Ni. *Phys. Rev. C* **2015**, *92*, 061302. [CrossRef]
122. Chiara, C.J.; Weisshaar, D.; Janssens, R.V.F.; Tsunoda, Y.; Otsuka, T.; Harker, J.L.; Walters, W.B.; Recchia, F.; Albers, M.; Alcorta, M.; et al. Identification of deformed intruder states in semi-magic $^{70}$Ni. *Phys. Rev. C* **2015**, *91*, 044309. [CrossRef]
123. Morales, A.I.; Benzoni, G.; Watanabe, H.; Nishimura, S.; Browne, F.; Daido, R.; Doornenbal, P.; Fang, Y.; Lorusso, G.; Patel, Z.; et al. Low-lying excitations in $^{72}$Ni. *Phys. Rev. C* **2016**, *93*, 034328. [CrossRef]
124. Morales, A.; Benzoni, G.; Watanabe, H.; de Angelis, G.; Nishimura, S.; Coraggio, L.; Gargano, A.; Itaco, N.; Otsuka, T.; Tsunoda, Y.; et al. Is seniority a partial dynamic symmetry in the first $\nu g_{9/2}$ shell? *Phys. Lett. B* **2018**, *781*, 706–712. [CrossRef]
125. Mazzocchi, C.; Grzywacz, R.; Batchelder, J.; Bingham, C.; Fong, D.; Hamilton, J.; Hwang, J.; Karny, M.; Krolas, W.; Liddick, S.; et al. Low energy structure of even–even Ni isotopes close to $^{78}$Ni. *Phys. Lett. B* **2005**, *622*, 45–54. [CrossRef]
126. Kolos, K.; Miller, D.; Grzywacz, R.; Iwasaki, H.; Al-Shudifat, M.; Bazin, D.; Bingham, C.R.; Braunroth, T.; Cerizza, G.; Gade, A.; et al. Direct lifetime measurements of the excited states in $^{72}$Ni. *Phys. Rev. Lett.* **2016**, *116*, 122502. [CrossRef]
127. Gottardo, A.; de Angelis, G.; Doornenbal, P.; Coraggio, L.; Gargano, A.; Itaco, N.; Kaneko, K.; Van Isacker, P.; Furumoto, T.; Benzoni, G.; et al. Transition strengths in the neutron-rich $^{73,74,75}$Ni isotopes. *Phys. Rev. C* **2020**, *102*, 014323. [CrossRef]
128. Evitts, L.; Garnsworthy, A.; Kibédi, T.; Smallcombe, J.; Reed, M.; Brown, B.; Stuchbery, A.; Lane, G.; Eriksen, T.; Akber, A.; et al. Identification of significant $E0$ strength in the $2^+_2 \rightarrow 2^+_1$ transitions of $^{58,60,62}$Ni. *Phys. Lett. B* **2018**, *779*, 396–401. [CrossRef]
129. Evitts, L.J.; Garnsworthy, A.B.; Kibédi, T.; Smallcombe, J.; Reed, M.W.; Stuchbery, A.E.; Lane, G.J.; Eriksen, T.K.; Akber, A.; Alshahrani, B.; et al. $E0$ transition strength in stable Ni isotopes. *Phys. Rev. C* **2019**, *99*, 024306. [CrossRef]
130. Passoja, A.; Julin, R.; Kantele, J.; Luontama, M. High-resolution study of $E0$ internal pair decay of excited $0^+$ states in $^{58,60,62}$Ni. *Nucl. Phys. A* **1981**, *363*, 399–412. [CrossRef]
131. Evers, D.; Assmann, W.; Rudolph, K.; Skorka, S.; Sperr, P. The ($^3$He, n) reaction on even (f, p) shell nuclei at 18 and 21 MeV (I). *Nucl. Phys. A* **1972**, *198*, 268–288. [CrossRef]
132. Stein, N.; Sunier, J.W.; Woods, C.W. Correspondence between $\alpha$-transfer and two-proton and two-neutron transfer reactions to the nickel isotopes. *Phys. Rev. Lett.* **1977**, *38*, 587–591. [CrossRef]
133. Hanson, D.; Stein, N.; Sunier, J.; Woods, C.; Hansen, O. High resolution study of the reactions $^{54,56,58}$Fe($^{16}$O, $^{12}$C)$^{58,60,62}$Ni and a comparison with ($^6$Li, d) $\alpha$-transfer spectroscopy. *Nucl. Phys. A* **1979**, *321*, 471–489. [CrossRef]
134. Hanson, D.; Stein, N.; Woods, C.; Sunier, J.; Hansen, O.; Nilsson, B. The $^{58}$Fe($^{16}$O, $^{14}$C)$^{60}$Ni reaction at 50 MeV. *Nucl. Phys. A* **1980**, *336*, 290–298. [CrossRef]
135. Amusa, A.; Lawson, R.D. High spin states in $^{94}$Ru and $^{95}$Rh. *Z. Phys. A* **1982**, *307*, 333–337. [CrossRef]
136. Mach, H.; Korgul, A.; Górska, M.; Grawe, H.; Matea, I.; Stănoiu, M.; Fraile, L.M.; Penionzkevich, Y.E.; Santos, F.D.O.; Verney, D.; et al. Ultrafast-timing lifetime measurements in $^{94}$Ru and $^{96}$Pd: Breakdown of the seniority scheme in $N = 50$ isotones. *Phys. Rev. C* **2017**, *95*, 014313. [CrossRef]
137. Tarasov, O.B.; Ahn, D.S.; Bazin, D.; Fukuda, N.; Gade, A.; Hausmann, M.; Inabe, N.; Ishikawa, S.; Iwasa, N.; Kawata, K.; et al. Discovery of $^{60}$Ca and implications for the stability of $^{70}$Ca. *Phys. Rev. Lett.* **2018**, *121*, 022501. [CrossRef] [PubMed]
138. Pore, J.L.; Andreoiu, C.; Smith, J.K.; MacLean, A.D.; Chester, A.; Holt, J.D.; Ball, G.C.; Bender, P.C.; Bildstein, V.; Braid, R.; et al. Detailed spectroscopy of $^{46}$Ca: A study of the $\beta^-$ decay of $^{46}$K. *Phys. Rev. C* **2019**, *100*, 054327. [CrossRef]
139. Ash, J.; Iwasaki, H.; Mijatović, T.; Budner, T.; Elder, R.; Elman, B.; Friedman, M.; Gade, A.; Grinder, M.; Henderson, J.; et al. Cross-shell excitations in $^{46}$Ca studied with fusion reactions induced by a reaccelerated rare isotope beam. *Phys. Rev. C* **2021**, *103*, L051302. [CrossRef]
140. Lach, M.; Styczeń, J.; Meczyński, W.; Bednarczyk, P.; Bracco, A.; Grebosz, J.; Maj, A.; Merdinger, J.C.; Schulz, N.; Smith, M.B.; et al. In-beam $\gamma$-ray spectroscopy of $^{42}$Ca. *Eur. Phys. J. A* **2003**, *16*, 309–311. [CrossRef]
141. Bjerregaard, J.H.; Hansen, O. Violation of seniority in the reaction $^{43}$Ca$(d, p)^{44}$Ca. *Phys. Rev.* **1967**, *155*, 1229–1237. [CrossRef]
142. Lach, M.; Bednarczyk, P.; Bracco, A.; Grębosz, J.; Kadłuczka, M.; Kintz, N.; Maj, A.; Merdinger, J.C.; Męczyński, W.; Pedroza, J.L.; et al. High-spin states in $^{44}$Ca. *Eur. Phys. J. A* **2001**, *12*, 381–382. [CrossRef]
143. Fortune, H.T.; Al-Jadir, M.N.I.; Betts, R.R.; Bishop, J.N.; Middieton, R. $\alpha$ spectroscopic factors in $^{40}$Ca. *Phys. Rev. C* **1979**, *19*, 756–764. [CrossRef]
144. Middleton, R.; Garrett, J.; Fortune, H. Search for multiparticle-multihole states of $^{40}$Ca with the $^{32}$S($^{12}$C,$\alpha$) reaction. *Phys. Lett. B* **1972**, *39*, 339–342. [CrossRef]

145. Bjerregaard, J.; Hansen, O.; Nathan, O.; Chapman, R.; Hinds, S.; Middleton, R. The (t, p) reaction with the even isotopes of Ca. *Nucl. Phys. A* **1967**, *103*, 33–70. [CrossRef]
146. Petersen, J.; Parkinson, W. The $^{40}$Ar($\tau$, n)42Ca reaction. *Phys. Lett. B* **1974**, *49*, 425–427. [CrossRef]
147. Fortune, H.; Betts, R.; Bishop, J.; AL-Jadir, M.; Middleton, R. Location of $0^+$ 4p-2h and 6p-4h configurations in $^{42}$Ca. *Nucl. Phys. A* **1978**, *294*, 208–212. [CrossRef]
148. Fortune, H.; Vermeulen, J.; Saha, A.; Drentje, A.; Put, L.; de Ruyter van steveninck, R.; van Hienen, J. Configuration of 3.59 MeV $0^+$ state in $^{44}$Ca. *Phys. Lett. B* **1978**, *79*, 205–208. [CrossRef]
149. Peng, J.C.; Stein, N.; Sunier, J.W.; Drake, D.M.; Moses, J.D.; Cizewski, J.A.; Tesmer, J.R. Study of the Reactions $^{46,48}$Ti($^{14}$C, $^{16}$O)$^{44,46}$Ca and $^{50,52}$Cr($^{14}$C, $^{16}$O)$^{48,50}$Ti at 51 MeV. *Phys. Rev. Lett.* **1979**, *43*, 675–678. [CrossRef]
150. Rowe, D.J.; Thiamova, G.; Wood, J.L. Implications of deformation and shape coexistence for the nuclear shell model. *Phys. Rev. Lett.* **2006**, *97*, 202501. [CrossRef]
151. Thiamova, G.; Rowe, D.; Wood, J. Coupled-SU(3) models of rotational states in nuclei. *Nucl. Phys. A* **2006**, *780*, 112–129. [CrossRef]
152. Rowe, D.J. The fundamental role of symmetry in nuclear models. *Aip Conf. Proc.* **2013**, *1541*, 104–136. [CrossRef]
153. Zheng, D.C.; Berdichevsky, D.; Zamick, L. Near degeneracies of the intrinsic state energies of many-particle, many-hole deformed configurations. *Phys. Rev. Lett.* **1988**, *60*, 2262–2265. [CrossRef] [PubMed]
154. Zuker, A.P.; Poves, A.; Nowacki, F.; Lenzi, S.M. Nilsson-SU3 self-consistency in heavy $N = Z$ nuclei. *Phys. Rev. C* **2015**, *92*, 024320. [CrossRef]
155. Hadyńska-Klęk, K.; Napiorkowski, P.J.; Zielińska, M.; Srebrny, J.; Maj, A.; Azaiez, F.; Valiente Dobón, J.J.; Kici ńska Habior, M.; Nowacki, F.; Naïdja, H.; et al. Superdeformed and triaxial states in $^{42}$Ca. *Phys. Rev. Lett.* **2016**, *117*, 062501. [CrossRef]
156. Caurier, E.; Menéndez, J.; Nowacki, F.; Poves, A. Coexistence of spherical states with deformed and superdeformed bands in doubly magic $^{40}$Ca: A shell-model challenge. *Phys. Rev. C* **2007**, *75*, 054317. [CrossRef]
157. Ellegaard, C.; Lien, J.; Nathan, O.; Sletten, G.; Ingebretsen, F.; Osnes, E.; Tjøm, P.; Hansen, O.; Stock, R. The $(1f_{7/2})^2$ multiplet in $^{42}$Ca. *Phys. Lett. B* **1972**, *40*, 641–644. [CrossRef]
158. Ideguchi, E.; Kibédi, T.; Dowie, J.T.H.; Hoang, T.H.; Kumar Raju, M.; Aoi, N.; Mitchell, A.J.; Stuchbery, A.E.; Shimizu, N.; Utsuno, Y.; et al. Electric monopole transition from the superdeformed band in $^{40}$Ca *Phys. Rev. Lett.* **2022**, *128*, 252501. [CrossRef]
159. Mahzoon, M.H.; Charity, R.J.; Dickhoff, W.H.; Dussan, H.; Waldecker, S.J. Forging the link between nuclear reactions and nuclear structure. *Phys. Rev. Lett.* **2014**, *112*, 162503. [CrossRef]
160. Elliott, J.P. Collective motion in the nuclear shell model. I. Classification schemes for states of mixed configurations. *Proc. R. Soc. Lond. Ser. Math. Phys. Sci.* **1958**, *245*, 128–145. [CrossRef]
161. Thibault, C.; Klapisch, R.; Rigaud, C.; Poskanzer, A.M.; Prieels, R.; Lessard, L.; Reisdorf, W. Direct measurement of the masses of $^{11}$Li and $^{26-32}$Na with an online mass spectrometer. *Phys. Rev. C* **1975**, *12*, 644–657. [CrossRef]
162. Huber, G.; Touchard, F.; Büttgenbach, S.; Thibault, C.; Klapisch, R.; Duong, H.T.; Liberman, S.; Pinard, J.; Vialle, J.L.; Juncar, P.; et al. Spins, magnetic moments, and isotope shifts of $^{21-31}$Na by high resolution laser spectroscopy of the atomic $D_1$ line. *Phys. Rev. C* **1978**, *18*, 2342–2354. [CrossRef]
163. Détraz, C.; Guillemaud, D.; Huber, G.; Klapisch, R.; Langevin, M.; Naulin, F.; Thibault, C.; Carraz, L.C.; Touchard, F. Beta decay of $^{27-32}$Na and their descendants. *Phys. Rev. C* **1979**, *19*, 164–176. [CrossRef]
164. Wimmer, K.; Kröll, T.; Krücken, R.; Bildstein, V.; Gernhäuser, R.; Bastin, B.; Bree, N.; Diriken, J.; Van Duppen, P.; Huyse, M.; et al. Discovery of the shape coexisting $0^+$ state in $^{32}$Mg by a two neutron transfer reaction. *Phys. Rev. Lett.* **2010**, *105*, 252501. [CrossRef] [PubMed]
165. Rotaru, F.; Negoita, F.; Grévy, S.; Mrazek, J.; Lukyanov, S.; Nowacki, F.; Poves, A.; Sorlin, O.; Borcea, C.; Borcea, R.; et al. Unveiling the Intruder Deformed $0_2^+$ State in $^{34}$Si. *Phys. Rev. Lett.* **2012**, *109*, 092503. [CrossRef] [PubMed]
166. Sorensen, R.A. Shifts of the spherical single-particle levels. *Nucl. Phys. A* **1984**, *420*, 221–236. [CrossRef]
167. Goodman, A.L. The $h_{9/2}$ " "intruder" state in odd mass Au and Tl isotopes. *Nucl. Phys. A* **1977**, *287*, 1–12. [CrossRef]
168. Smirnova, N.A.; Heyde, K.; Bally, B.; Nowacki, F.; Sieja, K. Nuclear shell evolution and in-medium $NN$ interaction. *Phys. Rev. C* **2012**, *86*, 034314. [CrossRef]
169. Duflo, J.; Zuker, A.P. The nuclear monopole Hamiltonian. *Phys. Rev. C* **1999**, *59*, R2347–R2350. [CrossRef]
170. Wood, J. New developments in nuclear shape coexistence. In *Proceedings of the International Conference on Nuclear Structure Through Static and Dynamic*; Moments; Bolotin, H., Ed.; Conference Proceedings Press: Melbourne, Australia, 1987; Volume II, p. 209.
171. Stuchbery, A.E.; Chamoli, S.K.; Kibédi, T. Particle-rotor versus particle-vibration features in $g$ factors of $^{111}$Cd and $^{113}$Cd. *Phys. Rev. C* **2016**, *93*, 031302. [CrossRef]
172. Coombes, B.J.; Stuchbery, A.E.; Blazhev, A.; Grawe, H.; Reed, M.W.; Akber, A.; Dowie, J.T.H.; Gerathy, M.S.M.; Gray, T.J.; Kibédi, T.; et al. Spectroscopy and excited-state $g$ factors in weakly collective $^{111}$Cd: Confronting collective and microscopic models. *Phys. Rev. C* **2019**, *100*, 024322. [CrossRef]
173. Tee, B.P.E.; Stuchbery, A.E.; Vos, M.; Dowie, J.T.H.; Lee, B.Q.; Alotiby, M.; Greguric, I.; Kibédi, T. High-resolution conversion electron spectroscopy of the $^{125}$I electron-capture decay. *Phys. Rev. C* **2019**, *100*, 034313. [CrossRef]
174. Mărginean, N.; Little, D.; Tsunoda, Y.; Leoni, S.; Janssens, R.V.F.; Fornal, B.; Otsuka, T.; Michelagnoli, C.; Stan, L.; Crespi, F.C.L.; et al. Shape coexistence at zero spin in $^{64}$Ni driven by the monopole tensor interaction. *Phys. Rev. Lett.* **2020**, *125*, 102502. [CrossRef] [PubMed]

175. Stefanescu, I.; Georgiev, G.; Balabanski, D.L.; Blasi, N.; Blazhev, A.; Bree, N.; Cederkäll, J.; Cocolios, T.E.; Davinson, T.; Diriken, J.; et al. Interplay between single-particle and collective effects in the odd-$A$ Cu isotopes beyond $N = 40$. *Phys. Rev. Lett.* **2008**, *100*, 112502. [CrossRef] [PubMed]
176. Schiffer, J.P.; Hoffman, C.R.; Kay, B.P.; Clark, J.A.; Deibel, C.M.; Freeman, S.J.; Honma, M.; Howard, A.M.; Mitchell, A.J.; Otsuka, T.; et al. Valence nucleon populations in the Ni isotopes. *Phys. Rev. C* **2013**, *87*, 034306. [CrossRef]
177. Macchiavelli, A.O.; Clark, R.M.; Crawford, H.L.; Fallon, P.; Lee, I.Y.; Morse, C.; Campbell, C.M.; Cromaz, M.; Santamaria, C. Core of $^{25}$F in the rotational model. *Phys. Rev. C* **2020**, *102*, 041301. [CrossRef]
178. Bednarczyk, P.; Styczeń, J.; Broda, R.; Lach, M.; Męczyński, W.; Bazzacco, D.; Brandolini, F.; de Angelis, G.; Lunardi, S.; Müller, L.; et al. High spin structure study of the light Odd-$A$ $f_{7/2}$ nuclei: $^{45}$Sc, $^{45}$Ti and $^{43}$Ca. *Eur. Phys. J. A* **1998**, *2*, 157–171. [CrossRef]
179. Gottardo, A.; Verney, D.; Delafosse, C.; Ibrahim, F.; Roussière, B.; Sotty, C.; Roccia, S.; Andreoiu, C.; Costache, C.; Delattre, M.C.; et al. First evidence of shape coexistence in the $^{78}$Ni region: Intruder $0_2^+$ state in $^{80}$Ge. *Phys. Rev. Lett.* **2016**, *116*, 182501. [CrossRef]
180. Garcia, F.H.; Andreoiu, C.; Ball, G.C.; Bell, A.; Garnsworthy, A.B.; Nowacki, F.; Petrache, C.M.; Poves, A.; Whitmore, K.; Ali, F.A.; et al. Absence of Low-Energy Shape Coexistence in $^{80}$Ge: The nonobservation of a proposed excited $0_2^+$ level at 639 keV. *Phys. Rev. Lett.* **2020**, *125*, 172501. [CrossRef]
181. Sekal, S.; Fraile, L.M.; Lică, R.; Borge, M.J.G.; Walters, W.B.; Aprahamian, A.; Benchouk, C.; Bernards, C.; Briz, J.A.; Bucher, B.; et al. Low-spin states in $^{80}$Ge populated in the $\beta$ decay of the $^{80}$Ga $3^-$ isomer. *Phys. Rev. C* **2021**, *104*, 024317. [CrossRef]
182. Orlandi, R.; Mücher, D.; Raabe, R.; Jungclaus, A.; Pain, S.; Bildstein, V.; Chapman, R.; de Angelis, G.; Johansen, J.; Van Duppen, P.; et al. Single-neutron orbits near $^{78}$Ni: Spectroscopy of the $N = 49$ isotope $^{79}$Zn. *Phys. Lett. B* **2015**, *740*, 298–302. [CrossRef]
183. Yang, X.F.; Wraith, C.; Xie, L.; Babcock, C.; Billowes, J.; Bissell, M.L.; Blaum, K.; Cheal, B.; Flanagan, K.T.; Garcia Ruiz, R.F.; et al. Isomer Shift and Magnetic Moment of the Long-Lived $1/2^+$ Isomer in $^{79}_{30}$Zn$_{49}$: Signature of Shape Coexistence near $^{78}$Ni. *Phys. Rev. Lett.* **2016**, *116*, 182502. [CrossRef] [PubMed]
184. Mullins, S.M.; Watson, D.L.; Fortune, H.T. $^{82}$Se(t,p)$^{84}$Se reaction at 17 MeV. *Phys. Rev. C* **1988**, *37*, 587–594. [CrossRef] [PubMed]
185. Flynn, E.R.; Sherman, J.D.; Stein, N.; Olsen, D.K.; Riley, P.J. $^{84}$Kr$(t,p)^{86}$Kr and $^{86}$Kr$(t,p)^{88}$Kr reactions. *Phys. Rev. C* **1976**, *13*, 568–577. [CrossRef]
186. Ragaini, R.C.; Knight, J.D.; Leland, W.T. Levels of $^{88}$Sr from the $^{86}$Sr$(t,p)^{88}$Sr reaction. *Phys. Rev. C* **1970**, *2*, 1020–1036. [CrossRef]
187. Flynn, E.; Cizewski, J.; Brown, R.E.; Sunier, J. 136,138Ba(t, p) and the systematics of neutron pairing vibrations at N = 82. *Phys. Lett. B* **1981**, *98*, 166–168. [CrossRef]
188. Mulligan, T.J.; Flynn, E.R.; Hansen, O.; Casten, R.F.; Sheline, R.K. $(t,p)$ and $(p,t)$ Reactions on even Ce isotopes. *Phys. Rev. C* **1972**, *6*, 1802–1814. [CrossRef]
189. Ball, J.B.; Auble, R.L.; Roos, P.G. Study of the zirconium isotopes with the $(p,t)$ reaction. *Phys. Rev. C* **1971**, *4*, 196–214. [CrossRef]
190. Ball, J.B.; Larsen, J.S. Systematics of $L = 0$ transitions observed in the $(p,t)$ reaction of nuclei near $N = 50$. *Phys. Rev. Lett.* **1972**, *29*, 1014–1017. [CrossRef]
191. Ball, J.; Fulmer, C.; Larsen, J.; Sletten, G. Energy levels of $^{94}$Ru observed with the $^{96}$Ru(p, t) reaction. *Nucl. Phys. A* **1973**, *207*, 425–432. [CrossRef]
192. Ball, J.; Auble, R.; Rapaport, J.; Fulmer, C. Levels in $^{140,142,144}$Nd and a search for pairing vibrations at $N = 82$ with the (p, t) reaction. *Phys. Lett. B* **1969**, *30*, 533–535. [CrossRef]
193. Flynn, E.R.; van der Plicht, J.; Wilhelmy, J.B.; Mann, L.G.; Struble, G.L.; Lanier, R.G. $^{146}$Gd and $^{144}$Sm excited by the (p,t) reaction on radioactive targets. *Phys. Rev. C* **1983**, *28*, 97–104. [CrossRef]
194. Häusser, O.; Taras, P.; Trautmann, W.; Ward, D.; Alexander, T.K.; Andrews, H.R.; Haas, B.; Horn, D. $g$ factors of high-spin yrast traps in $^{146,147}$Gd. *Phys. Rev. Lett.* **1979**, *42*, 1451–1454. [CrossRef]
195. Möller, O.; Warr, N.; Jolie, J.; Dewald, A.; Fitzler, A.; Linnemann, A.; Zell, K.O.; Garrett, P.E.; Yates, S.W. E2 transition probabilities in $^{114}$Te: A conundrum. *Phys. Rev. C* **2005**, *71*, 064324. [CrossRef]
196. Mihai, C.; Pasternak, A.A.; Filipescu, D.; Ivaşcu, M.; Bucurescu, D.; Căta Danil, G.; Căta Danil, I.; Deleanu, D.; Ghiţă, D.; Glodariu, T.; et al. Side feeding patterns and nuclear lifetime determinations by the Doppler shift attenuation method in $(\alpha, n\gamma)$ reactions. *Phys. Rev. C* **2010**, *81*, 034314. [CrossRef]
197. Mihai, C.; Pasternak, A.A.; Pascu, S.; Filipescu, D.; Ivaşcu, M.; Bucurescu, D.; Căta Danil, G.; Căta Danil, I.; Deleanu, D.; Ghiţă, D.G.; et al. Lifetime measurements by the Doppler-shift attenuation method in the $^{115}$Sn$(\alpha,n\gamma)^{118}$Te reaction. *Phys. Rev. C* **2011**, *83*, 054310. [CrossRef]
198. Saxena, M.; Kumar, R.; Jhingan, A.; Mandal, S.; Stolarz, A.; Banerjee, A.; Bhowmik, R.K.; Dutt, S.; Kaur, J.; Kumar, V.; et al. Rotational behavior of $^{120,122,124}$Te. *Phys. Rev. C* **2014**, *90*, 024316. [CrossRef]
199. Pritychenko, B.; Birch, M.; Singh, B.; Horoi, M. Tables of $E2$ transition probabilities from the first $2^+$ states in even–even nuclei. *At. Data Nucl. Data Tables* **2016**, *107*, 1–139. [CrossRef]
200. Walker, P.; Dracoulis, G. Energy traps in atomic nuclei. *Nature* **1999**, *399*, 35–40. [CrossRef]
201. Dracoulis, G.D. Isomers, nuclear structure and spectroscopy. *Phys. Scr.* **2013**, *T152*, 014015. [CrossRef]
202. Dracoulis, G.D.; Walker, P.M.; Kondev, F.G. Review of metastable states in heavy nuclei. *Rep. Prog. Phys.* **2016**, *79*, 076301. [CrossRef]
203. Jain, A.K.; Maheshwari, B.; Goel, A. *Nuclear Isomers. A Primer*; Springer International Publishing: Berlin/Heidelberg, Germany, 2021. [CrossRef]

204. Frauendorf, S. Spontaneous symmetry breaking in rotating nuclei. *Rev. Mod. Phys.* **2001**, *73*, 463–514. [CrossRef]
205. Radford, D.C.; Baktash, C.; Beene, J.R.; Fuentes, B.; Galindo-Uribarri, A.; Gross, C.J.; Hausladen, P.A.; Lewis, T.A.; Mueller, P.E.; Padilla, E.; et al. Coulomb excitation of radioactive $^{132,134,136}$Te beams and the low $B(E2)$ of $^{136}$Te. *Phys. Rev. Lett.* **2002**, *88*, 222501. [CrossRef] [PubMed]
206. Stone, N.J.; Stuchbery, A.E.; Danchev, M.; Pavan, J.; Timlin, C.L.; Baktash, C.; Barton, C.; Beene, J.; Benczer-Koller, N.; Bingham, C.R.; et al. First nuclear moment measurement with radioactive beams by the recoil-in-vacuum technique: The $g$ factor of the $2_1^+$ state in $^{132}$Te. *Phys. Rev. Lett.* **2005**, *94*, 192501. [CrossRef]
207. Biswas, S.; Palit, R.; Navin, A.; Rejmund, M.; Bisoi, A.; Sarkar, M.S.; Sarkar, S.; Bhattacharyya, S.; Biswas, D.C.; Caamaño, M.; et al. Structure of $^{132}_{52}$Te$_{80}$: The two-particle and two-hole spectrum of $^{132}_{50}$Sn$_{82}$. *Phys. Rev. C* **2016**, *93*, 034324. [CrossRef]
208. Hughes, R.O.; Zamfir, N.V.; Radford, D.C.; Gross, C.J.; Barton, C.J.; Baktash, C.; Caprio, M.A.; Casten, R.F.; Galindo-Uribarri, A.; Hausladen, P.A.; et al. $\gamma$-ray spectroscopy of $^{132}$Te through $\beta$ decay of a $^{132}$Sb radioactive beam. *Phys. Rev. C* **2005**, *71*, 044311. [CrossRef]
209. Severyukhin, A.P.; Arsenyev, N.N.; Pietralla, N.; Werner, V. Impact of variational space on $M1$ transitions between first and second quadrupole excitations in $^{132,134,136}$Te. *Phys. Rev. C* **2014**, *90*, 011306. [CrossRef]
210. Danchev, M.; Rainovski, G.; Pietralla, N.; Gargano, A.; Covello, A.; Baktash, C.; Beene, J.R.; Bingham, C.R.; Galindo-Uribarri, A.; Gladnishki, K.A.; et al. One-phonon isovector $2^+_{1,MS}$ state in the neutron-rich nucleus $^{132}$Te. *Phys. Rev. C* **2011**, *84*, 061306. [CrossRef]
211. Bao, M.; Jiang, H.; Zhao, Y.M.; Arima, A. Low-lying states of even-even $N = 80$ isotones within the nucleon-pair approximation. *Phys. Rev. C* **2020**, *101*, 014316. [CrossRef]
212. Terasaki, J.; Engel, J.; Nazarewicz, W.; Stoitsov, M. Anomalous behavior of $2^+$ excitations around $^{132}$Sn. *Phys. Rev. C* **2002**, *66*, 054313. [CrossRef]
213. Fleischer, P.; Klüpfel, P.; Reinhard, P.G.; Maruhn, J.A. Skyrme energy functional and low lying $2^+$ states in Sn, Cd, and Te isotopes. *Phys. Rev. C* **2004**, *70*, 054321. [CrossRef]
214. Sau, J.; Heyde, K.; Chéry, R. Shell-model description of the nucleus $^{132}$Te. *Phys. Rev. C* **1980**, *21*, 405–418. [CrossRef]
215. Gargano, A.; Coraggio, L.; Covello, A.; Itaco, N. Realistic shell-model calculations and exotic nuclei. *J. Phys. Conf. Ser.* **2014**, *527*, 012004. [CrossRef]
216. Hicks, S.F.; Stuchbery, A.E.; Churchill, T.H.; Bandyopadhyay, D.; Champine, B.R.; Coombes, B.J.; Davoren, C.M.; Ellis, J.C.; Faulkner, W.M.; Lesher, S.R.; et al. Nuclear structure of $^{130}$Te from inelastic neutron scattering and shell model analysis. *Phys. Rev. C* 2022, *in press*.
217. Jakob, G.; Benczer-Koller, N.; Kumbartzki, G.; Holden, J.; Mertzimekis, T.J.; Speidel, K.H.; Ernst, R.; Stuchbery, A.E.; Pakou, A.; Maier-Komor, P.; et al. Evidence for proton excitations in $^{130,132,134,136}$Xe isotopes from measurements of $g$ factors of $2_1^+$ and $4_1^+$ states. *Phys. Rev. C* **2002**, *65*, 024316. [CrossRef]
218. Coquard, L.; Pietralla, N.; Ahn, T.; Rainovski, G.; Bettermann, L.; Carpenter, M.P.; Janssens, R.V.F.; Leske, J.; Lister, C.J.; Möller, O.; et al. Robust test of E(5) symmetry in $^{128}$Xe. *Phys. Rev. C* **2009**, *80*, 061304. [CrossRef]
219. Coquard, L.; Pietralla, N.; Rainovski, G.; Ahn, T.; Bettermann, L.; Carpenter, M.P.; Janssens, R.V.F.; Leske, J.; Lister, C.J.; Möller, O.; et al. Evolution of the mixed-symmetry $2^+_{1,ms}$ quadrupole-phonon excitation from spherical to $\gamma$-soft Xe nuclei. *Phys. Rev. C* **2010**, *82*, 024317. [CrossRef]
220. Rainovski, G.; Pietralla, N.; Ahn, T.; Coquard, L.; Lister, C.; Janssens, R.; Carpenter, M.; Zhu, S.; Bettermann, L.; Jolie, J.; et al. How close to the O(6) symmetry is the nucleus $^{124}$Xe? *Phys. Lett. B* **2010**, *683*, 11–16. [CrossRef]
221. Coquard, L.; Rainovski, G.; Pietralla, N.; Ahn, T.; Bettermann, L.; Carpenter, M.P.; Janssens, R.V.F.; Leske, J.; Lister, C.J.; Möller, O.; et al. O(6)-symmetry breaking in the $\gamma$-soft nucleus $^{126}$Xe and its evolution in the light stable xenon isotopes. *Phys. Rev. C* **2011**, *83*, 044318. [CrossRef]
222. Radich, A.J.; Garrett, P.E.; Allmond, J.M.; Andreoiu, C.; Ball, G.C.; Bianco, L.; Bildstein, V.; Chagnon-Lessard, S.; Cross, D.S.; Demand, G.A.; et al. Ground-state and pairing-vibrational bands with equal quadrupole collectivity in $^{124}$Xe. *Phys. Rev. C* **2015**, *91*, 044320. [CrossRef]
223. Morrison, L.; Hadyńska-Klęk, K.; Podolyák, Z.; Doherty, D.T.; Gaffney, L.P.; Kaya, L.; Próchniak, L.; Samorajczyk-Pyśk, J.; Srebrny, J.; Berry, T.; et al. Quadrupole deformation of $^{130}$Xe measured in a Coulomb-excitation experiment. *Phys. Rev. C* **2020**, *102*, 054304. [CrossRef]
224. Wolf, A.; Cheifetz, E. Magnetic moment of the $6^+$ isomeric state of $^{134}$Te. *Phys. Rev. Lett.* **1976**, *36*, 1072–1074. [CrossRef]
225. Goodin, C.; Stone, N.J.; Ramayya, A.V.; Daniel, A.V.; Stone, J.R.; Hamilton, J.H.; Li, K.; Hwang, J.K.; Luo, Y.X.; Rasmussen, J.O.; et al. $g$ factors, spin–parity assignments, and multipole mixing ratios of excited states in $N = 82$ isotones $^{134}$Te, $^{135}$I. *Phys. Rev. C* **2008**, *78*, 044331. [CrossRef]
226. Stuchbery, A.E.; Stone, N.J. Recoil in vacuum for Te ions: Calibration, models, and applications to radioactive-beam $g$-factor measurements. *Phys. Rev. C* **2007**, *76*, 034307. [CrossRef]
227. Benczer-Koller, N.; Kumbartzki, G.; Gürdal, G.; Gross, C.; Stuchbery, A.; Krieger, B.; Hatarik, R.; O'Malley, P.; Pain, S.; Segen, L.; et al. Measurement of $g$ factors of excited states in radioactive beams by the transient field technique: $^{132}$Te. *Phys. Lett. B* **2008**, *664*, 241–245. [CrossRef]
228. Fogelberg, B.; Stone, C.; Gill, R.; Mach, H.; Warner, D.; Aprahamian, A.; Rehfield, D. $g$-Factor of the $6_1^+$ state in $^{132}$Te. *Nucl. Phys. A* **1986**, *451*, 104–112. [CrossRef]

229. Lönnroth, T.; Vajda, S.; Kistner, O.C.; Rafailovich, M.H. The $g$-factors of isomeric states in $^{127,128}$Xe. *Z. für Phys. Atoms Nucl.* **1984**, *317*, 215–223. [CrossRef]
230. Moore, R.; Bruce, A.; Dendooven, P.; Billowes, J.; Campbell, P.; Ezwam, A.; Flanagan, K.; Forest, D.; Huikari, J.; Jokinen, A.; et al. Character of an $8^-$ isomer of $^{130}$Ba. *Phys. Lett. B* **2002**, *547*, 200–204. [CrossRef]
231. Królas, W.; Grzywacz, R.; Rykaczewski, K.P.; Batchelder, J.C.; Bingham, C.R.; Gross, C.J.; Fong, D.; Hamilton, J.H.; Hartley, D.J.; Hwang, J.K.; et al. First observation of the drip line nucleus $^{140}$Dy : Identification of a $7\mu$s $K$ isomer populating the ground state band. *Phys. Rev. C* **2002**, *65*, 031303. [CrossRef]
232. Rudolph, D.; Baktash, C.; Brinkman, M.J.; Caurier, E.; Dean, D.J.; Devlin, M.; Dobaczewski, J.; Heenen, P.H.; Jin, H.Q.; LaFosse, D.R.; et al. Rotational bands in the doubly magic nucleus $^{56}Ni$. *Phys. Rev. Lett.* **1999**, *82*, 3763–3766. [CrossRef]
233. Eriksen, T.; Kibedi, T.; Reed, M.W.; deVries, M.; Stuchbery, A.E.; Akber, A.; Dowie, J.; Evitts, L.J.; Garnsworthy, A.B.; Gerathy, M.; et al. Systematic studies of $E0$ transitions in $^{54,56,58}$Fe. *Proc. Sci.* **2017**, *INPC2016*, 69. [CrossRef]
234. Dowie, J.T.H.; Kibédi, T.; Stuchbery, A.E.; Akber, A.; Avaa, A.; Bignell, L.J.; Chisapi, M.V.; Coombes, B.J.; Eriksen, T.K.; Gerathy, M.S.M.; et al. Evidence for shape coexistence in $^{52}$Cr through conversion-electron and pair-conversion spectroscopy. *EPJ Web Conf.* **2020**, *232*, 04004. [CrossRef]
235. Eriksen, T. Investigation of the Hoyle State in $^{12}$C and the Related Triple Alpha Reaction Rate. Ph.D. Thesis, Department of Nuclear Physics, Research School of Physics and Engineering, The Australian National University, Camberra, Australia, 2018. [CrossRef]
236. Dowie, J.T.H. An Investigation into Nuclear Shape Coexistence Through high Precision Conversion Electron and Electron-Positron Pair Spectroscopy. Ph.D. Thesis, The Australian National University, Camberra, Australia, 2021. [CrossRef]
237. Nowak, K.; Wimmer, K.; Hellgartner, S.; Mücher, D.; Bildstein, V.; Diriken, J.; Elseviers, J.; Gaffney, L.P.; Gernhäuser, R.; Iwanicki, J.; et al. Spectroscopy of $^{46}$Ar by the $(t,\, p)$ two-neutron transfer reaction. *Phys. Rev. C* **2016**, *93*, 044335. [CrossRef]
238. Hinds, S.; Middleton, R. A study of the (t,p) reactions leading to $^{48}$Ti and $^{50}$Ti. *Nucl. Phys. A* **1967**, *92*, 422–432. [CrossRef]
239. Casten, R.F.; Flynn, E.R.; Hansen, O.; Mulligan, T.J. Strong $L = 0$ $(t, p)$ Transitions in the even isotopes of Ti, Cr, and Fe. *Phys. Rev. C* **1971**, *4*, 130–137. [CrossRef]
240. Suehiro, T.; Kokame, J.; Ishizaki, Y.; Ogata, H.; Sugiyama, Y.; Saji, Y.; Nonaka, I.; Itonaga, K. Core-excited states of $^{54}$Fe from the $^{56}$Fe(p, t)$^{54}$Fe reaction at 52 MeV. *Nucl. Phys. A* **1974**, *220*, 461–476. [CrossRef]
241. Nann, H.; Benenson, W. Levels of $^{56}$Ni. *Phys. Rev. C* **1974**, *10*, 1880–1888. [CrossRef]
242. Dombrádi, Z.; Sohler, D.; Sorlin, O.; Azaiez, F.; Nowacki, F.; Stanoiu, M.; Penionzhkevich, Y.E.; Timár, J.; Amorini, F.; Baiborodin, D.; et al. Search for particle–hole excitations across the $N = 28$ shell gap in $^{45,46}$Ar nuclei. *Nucl. Phys. A* **2003**, *727*, 195–206. [CrossRef]
243. Parker, J.J.; Wiedenhöver, I.; Cottle, P.D.; Baker, J.; McPherson, D.; Riley, M.A.; Santiago-Gonzalez, D.; Volya, A.; Bader, V.M.; Baugher, T.; et al. Isomeric character of the lowest observed $4^+$ state in $^{44}$S. *Phys. Rev. Lett.* **2017**, *118*, 052501. [CrossRef]
244. Coenen, E.; Deneffe, K.; Huyse, M.; Duppen, P.V.; Wood, J.L. $\alpha$ decay of neutron-deficient odd Bi nuclei: Shell-model intruder states in Tl and Bi isotopes. *Phys. Rev. Lett.* **1985**, *54*, 1783–1786. [CrossRef]
245. Ajzenberg-Selove, F. Energy levels of light nuclei $A = 11$–$12$. *Nucl. Phys. A* **1990**, *506*, 1–158. [CrossRef]
246. Endt, P. Supplement to energy levels of $A = 21$–$44$ nuclei (VII). *Nucl. Phys. A* **1998**, *633*, 1–220. [CrossRef]
247. Grévy, S.; Negoita, F.; Stefan, I.; Achouri, N.L.; Angélique, J.C.; Bastin, B.; Borcea, R.; Buta, A.; Daugas, J.M.; De Oliveira, F.; et al. Observation of the $0_2^+$ state in $^{44}$S. *Eur. Phys. J. A* **2005**, *25*, 111–113. [CrossRef]
248. Shimoura, S.; Ota, S.; Demichi, K.; Aoi, N.; Baba, H.; Elekes, Z.; Fukuchi, T.; Gomi, T.; Hasegawa, K.; Ideguchi, E.; et al. Lifetime of the isomeric $0_2^+$ state in $^{12}$Be. *Phys. Lett. B* **2007**, *654*, 87–91. [CrossRef]
249. Gaudefroy, L.; Daugas, J.M.; Hass, M.; Grévy, S.; Stodel, C.; Thomas, J.C.; Perrot, L.; Girod, M.; Rossé, B.; Angélique, J.C.; et al. Shell Erosion and Shape Coexistence in $^{43}_{16}S_{27}$. *Phys. Rev. Lett.* **2009**, *102*, 092501. [CrossRef]
250. Heyde, K.; Wood, J.L. *Quantum Mechanics for Nuclear Structure*; IOP Publishing: Bristol, UK, 2019; Volume 1, pp. 2053–2563. [CrossRef]
251. McCoy, A.E.; Caprio, M.A. Algebraic evaluation of matrix elements in the Laguerre function basis. *J. Math. Phys.* **2016**, *57*, 021708. [CrossRef]
252. Rowe, D.J.; McCoy, A.E.; Caprio, M.A. The many-nucleon theory of nuclear collective structure and its macroscopic limits: An algebraic perspective. *Phys. Scr.* **2016**, *91*, 033003. [CrossRef]
253. Jarrio, M.; Wood, J.; Rowe, D. The SU(3) structure of rotational states in heavy deformed nuclei. *Nucl. Phys. A* **1991**, *528*, 409–435. [CrossRef]
254. Bahri, C.; Rowe, D. SU(3) quasi-dynamical symmetry as an organizational mechanism for generating nuclear rotational motions. *Nucl. Phys. A* **2000**, *662*, 125–147. [CrossRef]
255. Inonu, E.; Wigner, E.P. On the contraction of groups and their representations. *Proc. Natl. Acad. Sci. USA* **1953**, *39*, 510–524. [CrossRef]
256. Cooke, T.H.; Wood, J.L. An algebraic method for solving central force problems. *Am. J. Phys.* **2002**, *70*, 945–950. [CrossRef]
257. Dytrych, T.; Launey, K.D.; Draayer, J.P.; Rowe, D.J.; Wood, J.L.; Rosensteel, G.; Bahri, C.; Langr, D.; Baker, R.B. Physics of nuclei: Key role of an emergent symmetry. *Phys. Rev. Lett.* **2020**, *124*, 042501. [CrossRef] [PubMed]
258. McCoy, A.E.; Caprio, M.A.; Dytrych, T.C.V.; Fasano, P.J. Emergent Sp$(3,\mathbb{R})$ dynamical symmetry in the nuclear many-body system from an ab initio description. *Phys. Rev. Lett.* **2020**, *125*, 102505. [CrossRef] [PubMed]
259. Sargsyan, G.H.; Launey, K.D.; Baker, R.B.; Dytrych, T.; Draayer, J.P. SU(3)-guided realistic nucleon–nucleon interactions for large-scale calculations. *Phys. Rev. C* **2021**, *103*, 044305. [CrossRef]

260. Kramer, G.; Blok, H.; Van Den Brand, J.; Bulten, H.; Ent, R.; Jans, E.; Lanen, J.; Lapikás, L.; Nann, H.; Quint, E.; et al. Proton ground-state correlations in $^{40}$Ca studied with the reaction $^{40}$Ca(e, e′p)$^{39}$K. *Phys. Lett. B* **1989**, *227*, 199–203. [CrossRef]
261. Crespo, R.; Arriaga, A.; Wiringa, R.; Cravo, E.; Mecca, A.; Deltuva, A. Many-body effects in (p,pN) reactions within a unified approach. *Phys. Lett. B* **2020**, *803*, 135355. [CrossRef]

physics

MDPI

*Review*

# Emerging Concepts in Nuclear Structure Based on the Shell Model

**Takaharu Otsuka** [1,2,3]

1 RIKEN Nishina Center, 2-1 Hirosawa, Wako, Saitama 351-0198, Japan; otsuka@phys.s.u-tokyo.ac.jp
2 Department of Physics, The University of Tokyo, 7-3-1 Hongo, Bunkyo, Tokyo 113-0033, Japan
3 Advanced Science Research Center, Japan Atomic Energy Agency, Tokai, Ibaraki 319-1195, Japan

**Abstract:** Some emerging concepts of nuclear structure are overviewed. (i) Background: the many-body quantum structure of atomic nucleus, a complex system comprising protons and neutrons (called nucleons collectively), has been studied largely based on the idea of the quantum liquid (à la Landau), where nucleons are quasiparticles moving in a (mean) potential well, with weak "residual" interactions between nucleons. The potential is rigid in general, although it can be anisotropic. While this view was a good starting point, it is time to look into kaleidoscopic aspects of the nuclear structure brought in by underlying dynamics and nuclear forces. (ii) Methods: exotic features as well as classical issues are investigated from fresh viewpoints based on the shell model and nucleon–nucleon interactions. The 70-year progress of the shell–model approach, including effective nucleon–nucleon interactions, enables us to do this. (iii) Results: we go beyond the picture of the solid potential well by activating the monopole interactions of the nuclear forces. This produces notable consequences in key features such as the shell/magic structure, the shape deformation, the dripline, etc. These consequences are understood with emerging concepts such as shell evolution (including type-II), T-plot, self-organization (for collective bands), triaxial-shape dominance, new dripline mechanism, etc. The resulting predictions and analyses agree with experiment. (iv) Conclusion: atomic nuclei are surprisingly richer objects than initially thought.

**Keywords:** nuclear structure; shell model; exotic nuclei; shell evolution; type-II shell evolution; nuclear shape; self-organization; dripline; monopole interaction; monopole-quadrupole interplay

**Citation:** Otsuka, T. Emerging Concepts in Nuclear Structure Based on the Shell Model. *Physics* **2022**, *4*, 258–285. https://doi.org/10.3390/physics4010018

Received: 5 January 2022
Accepted: 8 February 2022
Published: 22 February 2022

**Publisher's Note:** MDPI stays neutral with regard to jurisdictional claims in published maps and institutional affiliations.

**Copyright:** © 2022 by the authors. Licensee MDPI, Basel, Switzerland. This article is an open access article distributed under the terms and conditions of the Creative Commons Attribution (CC BY) license (https://creativecommons.org/licenses/by/4.0/).

## 1. Introduction

The atomic nucleus is in a unique position in physics in that it is an isolated object but comprises many quantum ingredients. Some emerging concepts for the structure of atomic nuclei are overviewed in this paper, focusing on the works in which the author was involved. Obviously, those concepts have been found or clarified thanks to the great progress of nuclear-structure physics over 70 years, including the shell model.

In fact, the understanding of nuclear structure is based, to a great extent, on the shell model, which was introduced by Mayer [1] and Jensen [2] in 1949. Since then, the shell model has been developed significantly in many ways: an initial phase as many-body physics was presented, for instance, by Talmi in [3], in contrast to Mayer-Jensen's independent-particle model. The subsequent developments are reviewed, for instance by Caurier et al. in [4] up to 2005, and in this volume up to date. I would like to sketch emerging concepts of nuclear structure based on recent shell–model studies involving the author, as many other studies are to be presented in other papers of the same volume.

The atomic nucleus comprises $Z$ protons and $N$ neutrons. Their sum is called the mass number $A = Z + N$. Among atomic nuclei, stable nuclei are characterized by their infinite or practically infinite life times and are characterized by rather balanced $Z$ to $N$ ratios, with $N/Z$ ranging from about 1 up to about 1.5. There are about 300 nuclear species of this category. Other nuclei are called exotic (or unstable) nuclei. The total number

of them is unknown but seems to be between 7000 and 10,000, providing a huge show window of various features as well as the paths of nucleosynthesis in the cosmos (see, for instance, [5–7]). The exotic nuclei decay, by $\beta$ (i.e., weak) processes, to other nuclei where $Z$ and $N$ are better balanced, as the $\beta$ decay alters a neutron to a proton or vice versa. This decay occurs successively, until the process terminates at a stable nucleus. Thus, only stable nuclei exist on earth, while exotic nuclei do not, being exotic literally.

Some of the emerging concepts were conceived in the study of exotic nuclei, particularly by looking at the shell structure and magic numbers of them. The obtained concepts were found later not to be limited to exotic nuclei. In this way, after the initial trigger by exotic nuclei, the overall picture of the nuclear shell structure has been renewed, and Section 2 of this paper is devoted to a sketch of it with two major keywords, the *monopole interaction* and the *shell evolution*.

We then focus on the deformation of the nuclear surface. The surface deformation from the sphere has been a very important subject since the 1950s, as initiated by Rainwater [8] and by Bohr and Mottelson in [9–13]. In particular, the shape coexistence phenomenon is discussed as the crossroad between the shell evolution and the deformation, leading to the concept of *type-II shell evolution*. Although I do not discuss extensively the methodology of the shell model calculation in this paper because of the length limitation, the *T-plot* of the Monte Carlo Shell Model (MCSM) is mentioned as an essential theoretical tool for many physics cases of this paper. These are the main subjects of Section 3.

The in-depth clarification of the collective band is connected to the fundamental question on the relation between the single-particle degrees of freedom and the collective motion of nucleons. These two must be connected through nuclear forces. This question has not been clarified enough as also addressed by G. E. Brown [14]. I shall focus, in Section 4, on how this question may be understood more deeply, by introducing the *self-organization* aspect of the collective bands and by raising the importance of the triaxiality of nuclear shapes including the ground states.

The interplay between the monopole interaction and the quadrupole deformation is shown to be a major mechanism of the determination of the neutron driplines. This approach explains neutron driplines observed recently. We are led to two dripline mechanisms: the traditional one with the single-particle origin and the present one. The monopole–quadrupole interplay responsible for this new dripline mechanism is explained in detail in Section 5. As an alternative case, spherical isotopes, such as Ca, Ni, Sn, and Pb, are predicted to exhibit a different pattern.

The intention of this paper is to show the major flow of basic ideas and related results without going into details. I hope that the reader can grasp this flow and could become interested in watching further developments. The past 70 years are really great for the shell model, but the coming years look equally or even more brilliant. I apologize for not covering many of the major developments in the last 70 years, as such coverage is not possible within this paper, but the other contributions of this volume are expected to help.

## 2. Shell Evolution Due to Monopole Interaction

### 2.1. *Mayer–Jensen's Shell Model and Observed Magic Numbers*

Mayer [1] and Jensen [2] proposed, in 1949, the model of the shell structure and magic numbers of atomic nuclei. This model provided major guides for a deeper and wider understanding of the structure of atomic nuclei. While this is a similar situation to electrons in atoms, there are some differences. Figure 1 depicts the basic idea and consequences of the Mayer–Jensen's scheme. We start with the nuclear matter composed of protons and neutrons. This matter shows an almost constant density of nucleons (collective name of protons and neutrons) inside the surface, which is a sphere as a natural assumption (see Figure 1a). Because of the short-range character of nuclear forces, this constant density results in a mean potential with a constant depth inside the surface, as shown in Figure 1b. Let us assume that the density distribution is isotropic, producing an isotropic mean potential. Figure 1b also suggests that the Harmonic Oscillator (HO) potential is a good

approximation to this mean potential as long as the mean potential shows negative values as a function of $r$, the radius from the center of the nucleus. We then switch from the mean potential to the HO potential, which is analytically more tractable. Thus, the HO potential can be introduced from the constant density (sometimes referred to as "density saturation") and the short-range attraction due to nuclear forces.

The eigenstates of the HO potential are single-particle states shown in the far-left column of Figure 1c with associated magic numbers and HO quanta, N. These HO magic numbers do not change by adding the minor correction of the $\ell^2$ term, the scalar product of the orbital angular momentum $\vec{l}$ (see the second column from left in Figure 1c; for details see [12]).

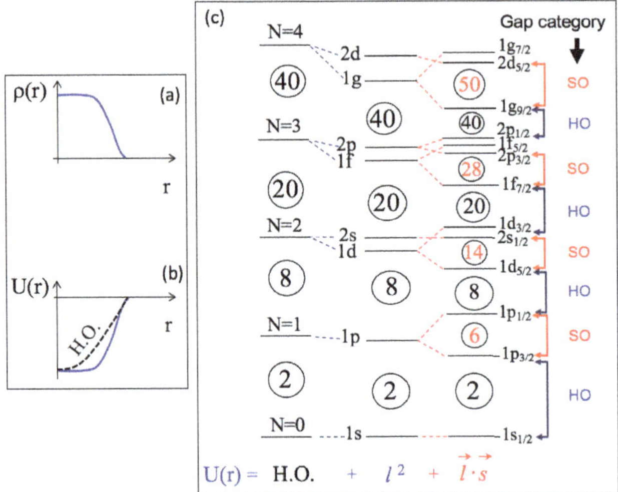


**Figure 1.** Schematic illustration of (**a**) density distribution of nucleons in atomic nuclei, (**b**) a mean potential (solid line) produced by nucleons in atomic nuclei and an approximation by a Harmonic Oscillator (HO) potential (dashed line). The abscissa, $r$, implies the radius from the center of the nucleus. (**c**) The shell structure produced with resulting magic numbers in circles. Left column: only the HO potential is taken with HO quanta shown as N = 0, N = 1, … (N here does not mean the neutron number, $N$.) Middle column: the $\ell^2$ term is aded to the HO potential, where the magic gaps are shown in circles. The single-particle orbits are labeled in the standard way to the left. Right column: the spin-orbit term, $(\vec{l} \cdot \vec{s})$, is included further, and magic gaps emerging from this term are shown in red. The single-particle orbits are labeled to the right, including $\vec{j} = \vec{l} + \vec{s}$. The magic gaps are classified as "HO" and "SO" for the HO potential and spin-orbit origins, respectively. Taken from Figure 2 of [15], which was based on [16].

The crucial factor introduced by Mayer and Jensen was the spin-orbit (SO) term, $(\vec{l} \cdot \vec{s})$, the effect of which is shown in the third column from the left in Figure 1c. The two orbits with the same orbital angular momentum, $\ell$, and the same HO quanta are denoted as,

$$j_> = \ell + 1/2 \text{ and } j_< = \ell - 1/2, \tag{1}$$

where $1/2$ is due to the spin, $s = 1/2$. The notation of $j_>$ and $j_<$ is used frequently in this paper. The spin-orbit term,

$$v_{\rm ls} = f(\vec{l} \cdot \vec{s}), \tag{2}$$

is added to the HO + $\ell^2$ potential, where $f$ is the strength parameter. With $f < 0$ as is the case for nuclear forces, the $j_>$ state is lowered in energy, whereas the $j_<$ state is raised. The value of $f$ is known empirically to be about $-20A^{-2/3}$ MeV (see Equation (2-132) of [12]).

The final pattern of the single-particle energies (SPE) is shown schematically in Figure 1c. The single-particle states are labeled in the standard way up to their $j$ values, and both HO and spin-orbit magic gaps are indicated in black and red, respectively. The magic numbers have been considered to be $Z, N = 2, 8, 20, 28, 50, 82$, and 126, because the effect of the spin-orbit term becomes stronger as $j$ becomes larger. In fact, the magic numbers 28, 50, 82, and 126 are all due to this effect. Instead, the HO magic numbers beyond 20 were considered to be absent or show only minor effects. We shall look back on them, from modern views of the nuclear structure covering stable and exotic nuclei.

We now investigate to what extent magic gaps in Figure 1c have been observed. Figure 2 displays the observed excitation energies of the first $2^+$ states of even-even nuclei as a function of $N$, where even-even stands for even-$Z$-even-$N$. These excitation energies tend to be high at the magic numbers, because excitations across the relevant magic gap are needed. The conventional magic numbers of Mayer and Jensen, $N = 2, 8, 20, 28, \ldots 126$ are expected to arise, and we indeed see sharp spikes at these magic numbers in Figure 2a where the excitation energies are shown for stable and long-lived (i.e., meta stable) nuclei. Figure 2b includes all measured first $2^+$ excitation energies as of 2016. In addition to the spikes in Figure 2a, one sees some new ones. One of them is at $N = 40$, which corresponds to $^{68}Ni_{40}$, representing a HO magic gap at $N = 40$. There are three others corresponding to the nuclei, $^{24}O_{16}$, $^{52}Ca_{32}$, and $^{54}Ca_{34}$, as marked in red. The $2^+$ excitation energies of these nuclei are about a factor of two higher than the overall trend, suggesting that $N = 16, 32$ and 34 can be magic numbers, although none of them is present in Figure 1c.

These new possible magic numbers are consequences of what are missing in the argument for deriving magic gaps in Figure 1c. We now turn to follow some passages along which this subject has been studied.

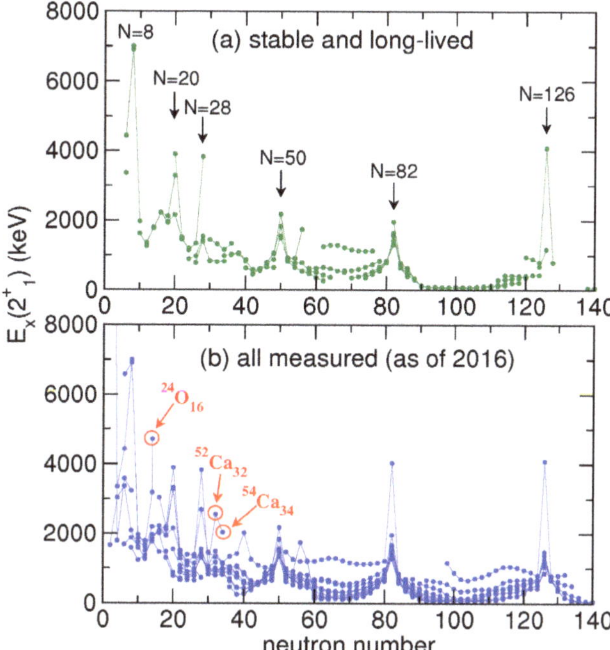


**Figure 2.** Systematics of the first $2^+$ excitation energies ($E_x(2_1^+)$), for (**a**) stable and long-lived nuclei and (**b**) all nuclei measured up to 2016, as functions of the neutron number. Peaks in (**a**) are labelled by the neutron number ($N$), while the names of the nuclei are displayed for some new points in (**b**). Taken from Figure 4 of [15].

### 2.2. Monopole Interaction

The change from the Mayer–Jensen scheme is discussed from the viewpoint of the nucleon-nucleon ($NN$) interaction. The Hamiltonian is written as,

$$\hat{H} = \hat{H}_0 + \hat{V}\,, \tag{3}$$

where $\hat{H}_0$ denotes the one-body term given by

$$\hat{H}_0 = \Sigma_j\, \epsilon^p_{0;j}\, \hat{n}^p_j + \Sigma_j\, \epsilon^n_{0;j}\, \hat{n}^n_j\,, \tag{4}$$

and $\hat{V}$ stands for the $NN$ interaction. Here, $\hat{n}^{p,n}_j$ means the proton- or neutron-number operator for the orbit $j$, and $\epsilon^{p,n}_{0;j}$ implies proton or neutron SPE of the orbit $j$. This SPE is composed of the kinetic energy of the orbit $j$ and the binding energy on the orbit $j$ generated by all nucleons in the inert core. We note that the interaction $\hat{V}$ in Equation (3) can be any interaction between two nucleons in the following discussions but actually refers to effective $NN$ interactions between valence (i.e., active) nucleons.

The interaction $\hat{V}$ can be decomposed, in general, into the two components: monopole and multipole interactions [17], irrespectively of its origin, derivation, or parameters. The monopole interaction, denoted as $\hat{V}^{\rm mono}$, is expressed in terms of the monopole matrix element, which is defined for single-particle orbits $j$ and $j'$ as,

$$V^{\rm mono}(j,j') = \frac{\Sigma_{(m,m')}\,\langle j,m\,;\,j',m'|\hat{V}|j,m\,;\,j',m'\rangle}{\Sigma_{(m,m')}\,1}\,, \tag{5}$$

where $m$ and $m'$ are magnetic substates of $j$ and $j'$, respectively, and the summation over $m, m'$ is taken for all ordered pairs allowed by the Pauli principle. The monopole matrix element represents, as displayed schematically in Figure 3, an orientation average for two nucleons in the orbits $j$ and $j'$. See [15] for more detailed descriptions.

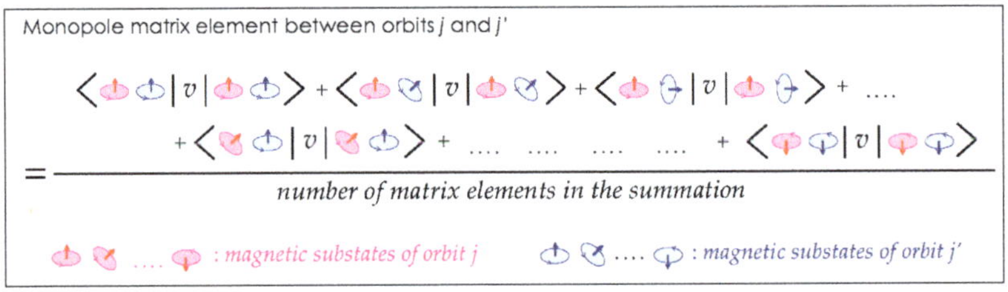

Monopole matrix element between orbits $j$ and $j'$

= (sum of matrix elements $\langle \cdots |v| \cdots \rangle$) / *number of matrix elements in the summation*

: *magnetic substates of orbit j*    : *magnetic substates of orbit j'*

**Figure 3.** Schematic illustration of the monopole matrix element for a two-body interaction $v$. See text for details. Taken from Figure 7 of [15].

The monopole interaction between two neutrons is then given as

$$\hat{V}^{\rm mono}_{nn} = \Sigma_j\, V^{\rm mono}_{nn}(j,j)\,\frac{1}{2}\,\hat{n}^n_j\,(\hat{n}^n_j - 1) + \Sigma_{j<j'}\, V^{\rm mono}_{nn}(j,j')\,\hat{n}^n_j\,\hat{n}^n_{j'}\,. \tag{6}$$

The monopole interaction between two protons is given similarly. The monopole interaction between a proton and a neutron can be given as

$$\begin{aligned}\hat{V}^{\rm mono}_{pn} &= \Sigma_{j\neq j'}\,\frac{1}{2}\Big\{V^{\rm mono}_{T=0}(j,j') + V^{\rm mono}_{T=1}(j,j')\Big\}\,\hat{n}^p_j\,\hat{n}^n_{j'} \\ &+ \Sigma_j\,\frac{1}{2}\Big\{V^{\rm mono}_{T=0}(j,j)\frac{2j+2}{2j+1} + V^{\rm mono}_{T=1}(j,j)\frac{2j}{2j+1}\Big\}\,\hat{n}^p_j\hat{n}^n_j\,,\end{aligned} \tag{7}$$

where $V^{\rm mono}_{T=0,1}(j,j')$ stands for the monopole matrix element for the isospin $T$ = 0 or 1 channel, respectively, defined by Equation (5) including isospin-symmetry effects (see

Sec. III A of Ref. [15] for details). Note that $V^{\rm mono}_{T=1}(j, j')$ implies $V^{\rm mono}_{nn,pp}(j, j')$. The second term on the right-hand-side (r.h.s.) of Equation (7) is slightly different from the first term on the r.h.s. of Equation (7) due to the special isospin property for the cases of $j = j'$. Obviously, $\hat{V}^{\rm mono}_{pn}$ can be rewritten as

$$\hat{V}^{\rm mono}_{pn} = \Sigma_{j,j'} \tilde{V}^{\rm mono}_{pn}(j, j')\, \hat{n}^p_j\, \hat{n}^n_{j'}, \tag{8}$$

with $\tilde{V}^{\rm mono}_{pn}(j, j')$ defined so as to reproduce Equation (7).

The functional forms in Equations (6) and (8) appear to be in accordance with the intuition from the averaging over all orientations: no dependencies on angular properties (e.g., coupled $J$ values) between the two interacting nucleons and the sole dependence on the number of particles in those orbits.

The (total) monopole interaction is written as

$$\hat{V}^{\rm mono} = \hat{V}^{\rm mono}_{pp} + \hat{V}^{\rm mono}_{nn} + \hat{V}^{\rm mono}_{pn}, \tag{9}$$

and the monopole Hamiltonian is defined as,

$$\hat{H}^{\rm mono} = \hat{H}_0 + \hat{V}^{\rm mono} = \Sigma_j\, \epsilon^p_{0;j}\, \hat{n}^p_j + \Sigma_j\, \epsilon^n_{0;j}\, \hat{n}^n_j + \hat{V}^{\rm mono}. \tag{10}$$

The multipole interaction is introduced as

$$\hat{V}^{\rm multi} = \hat{V} - \hat{V}^{\rm mono}, \tag{11}$$

and the (total) Hamiltonian is written as $\hat{H} = \hat{H}^{\rm mono} + \hat{V}^{\rm multi}$. The multipole interaction becomes crucial in many aspects of nuclear structure, for instance, the shape deformation, as touched upon in later sections of this article. The monopole interaction has been studied over decades with many works, for example, [17–20] (see [15] for more details).

We define the effective SPE (ESPE) of the proton (neutron) orbit $j$, denoted by $\hat{\epsilon}^p_j$ ($\hat{\epsilon}^n_j$), as the change of the monopole Hamiltonian, $\hat{H}^{\rm mono}$ in Equation (10), due to the addition of one proton (neutron) into the orbit $j$. This change is nothing but the difference, when $n^{p,n}_j$ is replaced by $n^{p,n}_j$+1. For instance, the first term on the r.h.s. of Equation (10) contributes to $\hat{\epsilon}^p_j$ by a constant, $\epsilon^p_{0;j}$. As another example, the r.h.s. of Equation (8) contributes by $\Sigma_{j'}\, \tilde{V}^{\rm mono}_{pn}(j,j')\{(\hat{n}^p_j + 1)\, \hat{n}^n_{j'} - \hat{n}^p_j\, \hat{n}^n_{j'}\} = \Sigma_{j'}\, \tilde{V}^{\rm mono}_{pn}(j,j')\hat{n}^n_{j'}$. Combining all terms, the ESPE of the proton orbit $j$ is given as,

$$\hat{\epsilon}^p_j = \epsilon^p_{0;j} + \Sigma_{j'}\, V^{\rm mono}_{pp}(j, j')\, \hat{n}^p_{j'} + \Sigma_{j'}\, \tilde{V}^{\rm mono}_{pn}(j, j')\, \hat{n}^n_{j'}. \tag{12}$$

The second and third terms on the r.h.s. are obviously contributions from valence protons and neutrons, respectively. The neutron ESPE is expressed similarly as

$$\hat{\epsilon}^n_j = \epsilon^n_{0;j} + \Sigma_{j'}\, V^{\rm mono}_{nn}(j, j')\, \hat{n}^n_{j'} + \Sigma_{j'}\, \tilde{V}^{\rm mono}_{pn}(j', j)\, \hat{n}^p_{j'}. \tag{13}$$

In many practical cases, an appropriate expectation value of the ESPE operator is also called the ESPE with an implicit reference to some state characterizing the structure, e.g., the ground state.

The ESPE as an expectation value is often discussed in terms of the difference between two states, e.g., $\Psi$ and $\Psi'$. The states $\Psi$ and $\Psi'$ may belong to the same nucleus or to two different nuclei. We here show the formulas for this difference. First we introduce the symbol $\Delta\mathcal{O}$ for an operator $\hat{\mathcal{O}}$ implying the difference, $\langle\Psi\,|\hat{\mathcal{O}}|\,\Psi\rangle - \langle\Psi'\,|\hat{\mathcal{O}}|\,\Psi'\rangle$. Such differences of the ESPE values are expressed as,

$$\Delta\epsilon^p_j = \Sigma_{j'}\, V^{\rm mono}_{pp}(j,j')\, \Delta n^p_{j'} + \Sigma_{j'}\, \tilde{V}^{\rm mono}_{pn}(j, j')\, \Delta n^n_{j'}, \tag{14}$$

and
$$\Delta\epsilon_j^n = \Sigma_{j'} V_{nn}^{\rm mono}(j,j')\, \Delta n_{j'}^n + \Sigma_{j'} \tilde{V}_{pn}^{\rm mono}(j',j)\, \Delta n_{j'}^p . \quad (15)$$

If $\Psi'$ is a doubly closed shell and $\Psi$ is an eigenstate with some valence protons and neutrons on top of this closed shell, these quantities stand for the evolution of ESPEs as functions of $Z$ and $N$. One can thus see various physics cases represented by $\Psi$ and $\Psi'$. Such ESPEs can provide picturesque prospects and great help in intuitive understanding without resorting to complicated numerical calculations. The notion of the ESPE has been well utilized, for instance, in empirical studies in [6,21], in certain ways related to the present article.

The interaction $\hat{V}$ can be decomposed into several parts according to some classifications. The discussions in this subsection can then be applied to each part separately: the monopole interaction of a particular part of $\hat{V}$ can be extracted, and its resulting ESPEs can be evaluated. Examples are presented in the subsequent subsections.

We note that the definition of the ESPE can have certain variants with similar consequences, for instance, the combination of $n_j^{p,n} - 1/2$ and $n_j^{p,n} + 1/2$ instead of $n_j^{p,n}$ and $n_j^{p,n} + 1$. Appendix A shows a note on the relation to Baranger's ESPE.

### 2.3. Central, Two-Body Spin-Orbit and Tensor Parts of the NN Interaction

With these formulations, we can discuss a variety of subjects ranging from the shell structure, to the collective bands, and to the driplines. Let us start with the shell structure. While the discussions in Section 2.1 are based on basic nuclear properties, some aspects are missing. One of them is the orbital dependencies of the monopole matrix element. This dependence generally appears but shows up more crucially in certain cases.

As we shall see, some parts of the $NN$ interaction, $\hat{V}$, show characteristic and substantial orbital dependencies. Such parts can be specified in terms of their spin properties, as the $NN$ interaction involves a spin operator, an axial vector $\vec{\sigma}$ of nucleon. We first take the part where no spin operator is included or spin operators are coupled to scalar terms, like $(\vec{\sigma}_1 \cdot \vec{\sigma}_2)$ with $\vec{\sigma}_{1,2}$ denoting the spin operator of the nucleon 1 or 2, and $(\,\cdot\,)$ being a scalar product. This part is called the *central force*, and its effects are discussed in Section 2.4. In the second part, spin operators are coupled to axial vectors. Such axial vectors must be coupled with other axial vectors such as the orbital angular momentum. The *two-body spin-orbit force* belongs to this case, and its effects are discussed in Section 2.8, while the effects remain quite modest except for special orbital combinations. As presented in Section 2.5, significant contributions arise from the *tensor force*, where spin operators are coupled to a (rank-2) tensor, $[\vec{\sigma}_1 \times \vec{\sigma}_2]^{(2)}$, where the last superscript means rank 2. This is a very complicated coupling, and this term must be coupled, in the interaction, with another (rank-2) tensor of the coordinates, in order to form a scalar. Similar terms appear in the electromagnetic interaction, but their effects are minor. The tensor force is, however, crucial in the nuclear case, because the pion exchange process produces it as its primary source. Section 2.5 presents monopole properties of the lowest-order contribution of the tensor force, while higher-order contributions are largely included in the central force of the effective $NN$ interaction mentioned above.

### 2.4. Monopole Interaction of the Central Force

We now discuss the monopole interaction of the central-force component of $NN$ interactions. Because the $NN$ interaction is characterized by intermediate-range ($\sim$1 fm) attraction after modifications or renormalizations, the monopole matrix elements gain large magnitudes with a negative sign (i.e., attractive), if radial wave functions of the single-particle orbits, $j$ and $j'$ in Equation (5), are similar to each other. This similarity is visible, if these orbits are spin-orbit partners ($j = j_>$ and $j' = j_<$) with the identical radial wave functions (see Equation (1)), for instance $1f_{7/2}$ and $1f_{5/2}$. Another example is the coupling between unique-parity orbits, such as $1g_{9/2}$ and $1h_{11/2}$, for which the radial wave functions are similar because of no radial node. These types of strong correlations were

pointed out by Federman and Pittel in [22], where the total effect of the $^3S_1$ channel of the $NN$ interaction was discussed without the reference to the monopole interaction.

### *2.5. Monopole Interaction of the Tensor Force*

Another important source of the monopole interaction with strong orbital dependences is the tensor force. The tensor force produces very unique effects on the ESPE. This is shown in Figure 4: the intuitive argument in [15,23] proves that the monopole interaction of the tensor force is attractive between a nucleon in an orbit $j_<$ and another nucleon in an orbit $j'_>$, whereas it becomes repulsive for combinations, $(j_>, j'_>)$ or $(j_<, j'_<)$. The magnitude of such monopole interaction varies also. For example, it is strong in magnitude between spin-orbit partners or between unique-parity orbits, etc. [15].

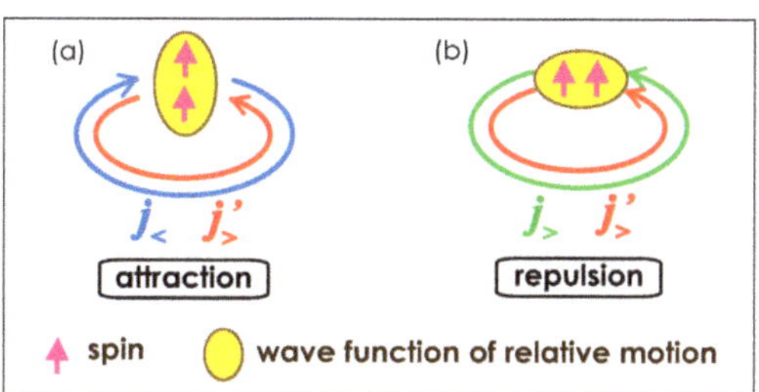


**Figure 4.** Monopole interaction of the tensor force for (**a**) between the orbits $j_<$ and $j'_>$, and (**b**) between the orbits $j_>$ and $j'_>$. See Equation (1) for the definitions of the orbits $j_>$ and $j_<$. See text for more details. Taken from [23].

The ESPE is shifted in very specific ways as exemplified in Figure 5b: if neutrons occupy a $j'_>$ orbit, the ESPE of the proton orbit $j_>$ is raised, whereas that of the proton orbit $j_<$ is lowered. This is nothing but a reduction in a proton spin-orbit splitting due to a specific neutron configuration. The amount of the shift is proportional to the number of neutrons in this configuration, as shown in Equation (14) and in Figure 5c. Other cases follow the same rule shown in Figure 4. These general features have been pointed out in [23] with an analytic formula and an intuitive description of its origin.

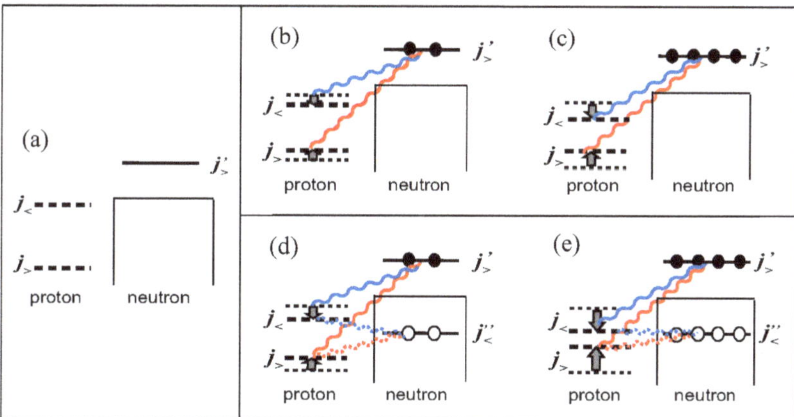


**Figure 5.** Schematic picture of the effective single-paticle energy (ESPE) change (i.e., shell evolution) due to the monopole interaction of the tensor force. (**a**) Single-particle energies (SPE) with no neutrons in the orbit $j'_>$. (**b**) The shifts in the proton ESPEs due to two (valence) neutrons in the orbit $j'_>$. (**c**) Same as (**b**) except for four neutrons. (**d**,**e**) Type-II shell evolution due to neutron particle–hole excitations. See text for more details. Taken from [24].

*2.6. Monopole-Interaction Effects from the Central and Tensor Forces Combined*

The combined effects of the central and tensor forces were discussed in [25] in terms of realistic shell–model interactions, USD [26], and GXPF1A [27]. These interactions were obtained in two steps: the starting point was given by microscopic G-matrix $NN$ interactions proposed initially by Kuo and Brown [28,29], and as the second step, certain phenomenological improvements were made by the fit to large numbers of experimental energy levels. It is mentioned that some main features, for instance, the tensor-force component, remain unchanged by this fit [25]. Many other valuable shell–model interactions, for instance, KB3 [17], Kuo-Herling [30], sn100pn [31], and LNPS [32] interactions, have been constructed from the G-matrix interactions sometimes with refinements like monopole adjustments. It should be noticed that these shell–model interactions are derived microscopically to a large extent and that they should be distinguished from purely phenomenological interactions in earlier times, e.g., [33]. The M3Y interaction [34] is related to the G-matrix, too. We appreciate the original contribution of the G-matrix approach to the effective $NN$ interaction [28,29].

The $V_{MU}$ interaction was then introduced as a general and simple shell–model $NN$ interaction. Its central part consists of Gaussian interactions with spin/isospin dependencies, and their strength parameters are determined so as to simulate the overall features of the monopole matrix elements of the central part of USD [26] and GXPF1A [27] interactions. Its tensor part is taken from the standard $\pi$- and $\rho$-meson exchange potentials [23,35,36]. Thus, the $V_{MU}$ interaction is defined as a function of the relative distance of two nucleons with spin/isospin dependences, which enables us to use it in a variety of regions of the nuclear chart, as we shall see. A wide model space, typically a HO shell or more, is required in order to obtain reasonable results, though.

Figure 6 depicts some examples: Figure 6a displays the transition from a standard (à la Mayer–Jensen) $N$ = 20 magic gap to an exotic $N$ = 16 magic gap by plotting $\langle \hat{\epsilon}^n_j \rangle$ within the filling scheme (see Equation (13)), as $Z$ decreases from 20 to 8. The tensor monopole interaction between the proton $d_{5/2}$ and the neutron $d_{3/2}$ orbits plays an important role. The small $N$ = 20 magic gap for $Z$ = 8–12 is consistent with the island of inversion picture (see reviews, e.g., [4,15]). Figure 6b depicts the inversion between the proton $f_{5/2}$ and $p_{3/2}$ orbits as $N$ increases in Ni isotopes, by showing $\langle \hat{\epsilon}^p_j \rangle$ (see Equation (12)). The figure exhibits exotically ordered single-particle orbits for $N > 44$. The tensor monopole interactions between the proton $f_{7/2,5/2}$ and the neutron $g_{9/2}$ orbits produce crucial effects. Figure 6c shows significant changes in the neutron single-particle levels from $^{90}$Zr to $^{100}$Sn, in terms of $\langle \hat{\epsilon}^n_j \rangle$. Without the tensor force, the approximate degeneracy of $g_{7/2}$ and $d_{5/2}$ orbits in $^{100}$Sn does not show up.

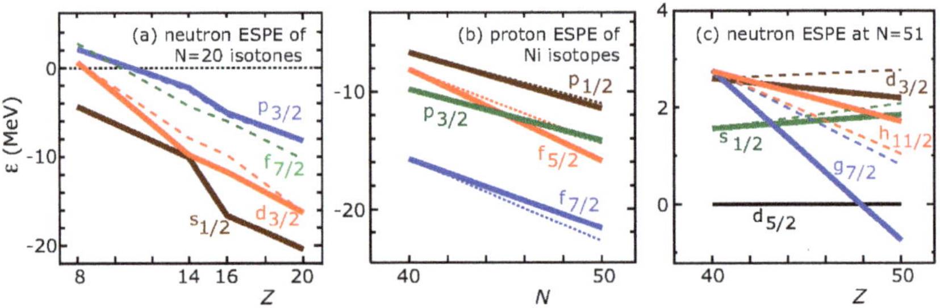


**Figure 6.** ESPEs calculated by the $V_{MU}$ interaction. The dashed lines are obtained by the central force only, while the solid lines include both the central-force and the tensor-force contributions. See text for more details. Taken from [25].

These changes in the shell structure as a function of Z and/or $N$ were collectively called *shell evolution* in [23]. The splitting between proton $g_{7/2}$ and $h_{11/2}$ in Sb isotopes shows a substantial widening as $N$ increases from 64 to 82 as pointed out by Schiffer et al. [37],

which was one of the first experimental supports to the shell evolution partly because this was not explained otherwise. Note that while the origin of the shell evolution can be any part of the $NN$ interaction, its appearance is exemplified graphically in Figure 5a–c for the tensor force. The shell-evolution trend depicted in Figure 6 appears to be consistent with experiment [15,25,38–42]. The monopole properties discussed in this subsection are consistent with the results shown by Smirnova et al. [43] obtained through the spin-tensor decomposition (see e.g., [15] for some account) for the "well-fitted realistic interaction for the *sdpf* shell–model space" [43].

### 2.7. *N = 34 New Magic Number as a Consequence of the Shell Evolution*

Among various cases of shell evolution, a notable impact was made by predicting a new magic number $N = 34$. Figure 7 displays the shell evolution of some neutron orbits from Ni back to Ca isotopes, as $Z$ decreases from 28 to 20. The $1f_{5/2}$ orbit is between the $2p_{3/2}$ and $2p_{1/2}$ in Mayer–Jensen's shell model (see Figure 1). By loosing eight protons lying in the $1f_{7/2}$ orbit of Ni isotopes (blue circles in Figure 7 left), this canonical shell structure is destroyed as the $1f_{5/2}$ orbit moves up above the $2p_{1/2}$ orbit. This movement of $1f_{5/2}$ orbit creates the $N = 32$ gap as a byproduct [44]. The energy shift in the $1f_{5/2}$ orbit is due to the central and tensor forces by almost equal amounts. We mention that the $N = 34$ magic gap would not appear, if the Mayer–Jensen scheme holds, as expected, in Ni isotopes but this shift did not occur. The appearance of the $N = 34$ magic number was predicted as a result of a spin–isospin interaction in [45]. However, 12 years were required [46] until the experimental verification became feasible [47] (see Figure 7 right). The measured $2^+$ energy levels are included in Figure 2b. More details are presented in [15]. Further evidences have been obtained recently by different experimental probes as reported in [48,49].

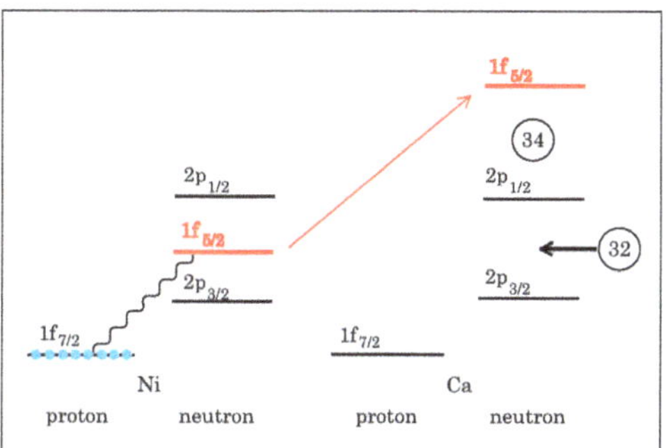


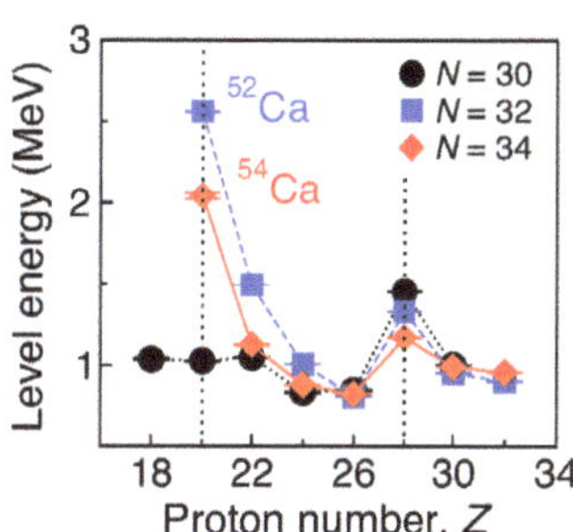


**Figure 7.** **Left**: Schematic illustration of the shell evolution from Ni back to Ca for neutron orbits. Blue circles denote protons. The wavy line is the interaction between the proton $1f_{7/2}$ orbit and the neutron $1f_{5/2}$ orbit. The numbers in circles indicate magic numbers. Taken from Figure 3 of [24]. **Right**: Observed excitation energies of the $2_1^+$ states. Taken from Figure 2c of [47].

### 2.8. *Monopole Interaction of the Two-Body Spin-Orbit Force*

It is a natural question what effect can be expected from the two-body spin-orbit force of the $NN$ interaction. This force can be well described by the M3Y interaction, and the monopole effects of the two-body spin-orbit force were described in detail in [15], particularly in its supplementary document. Although the monopole effects of this force contributes to the spin-orbit splitting [15], the effect is much weaker than the tensor force in most cases, as also discussed in the article by Utsuno in this volume.

An interesting case is found in the coupling between an $s$ orbit and $p_{3/2,1/2}$ orbits. There is no monopole effect from the tensor force, if an $s$ orbit is involved. Instead, the $s$-$p$ coupling due to the two-body spin-orbit force can be exceptionally strong as intuitively stressed in [15]. Figure 8 shows that the possible significant change in the neutron $2p_{3/2}$-

$2p_{1/2}$ gap between $^{35}$Si and $^{37}$S is explained to a good extent by the shell evolution due to the two-body spin-orbit force.

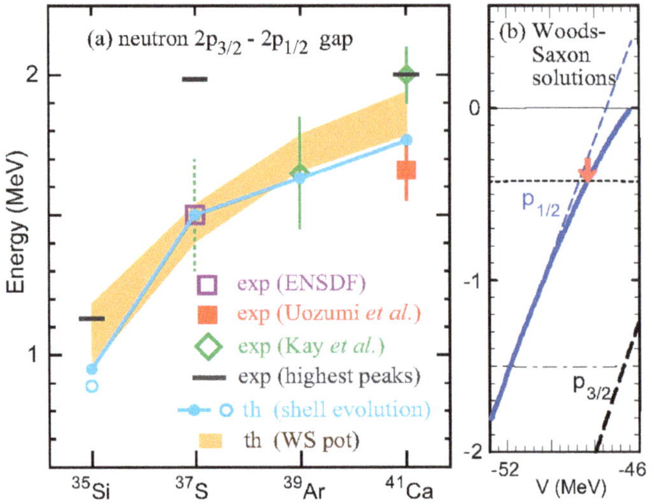

**Figure 8.** (**a**) Neutron $2p_{3/2}$-$2p_{1/2}$ splitting for N = 21 isotones. The symbols are the centroids for $^{37}$S [42], $^{39}$Ar [50], and $^{41}$Ca [50,51]. The horizontal bars are the energy differences between relevant highest peaks [50,52]. Shell evolution predictions are shown by blue closed symbols and the solid line connecting them. The loose binding effect for $^{35}$Si is included in the open circle. The calculation with Woods–Saxon potential with parameters adjusted are shown by the yellowish shaded area [50]. (**b**) Neutron $2p_{1/2}$ single-particle energy (blue solid line) by a Woods–Saxon potential [12] for varying depth parameter, V. The linear dependence of the deeply bound region is linearly extrapolated (blue dashed line) and is compared to the curved dependence that results from the proximity of the continuum. The dashed line is for the $2p_{3/2}$ orbit, and the loose-binding contribution to the present splitting appears to be 0.06 MeV against 1.5 MeV splitting itself. Taken from Figure 8 of Supplementary Material of [15].

### 2.9. Monopole Interaction from the Three-Nucleon Force

The three-nucleon force (3NF) is currently of intense interest (see, for instance, a review [53]). Among various aspects, we showed [54] the characteristic feature of the monopole interaction of the effective *NN* interaction derived from the Fujita-Miyazawa 3NF [55]. Figure 9a displays the effect of the Δ excitation in nucleon–nucleon interaction. The Δ-hole excitation from the inert core changes the SPE of the orbit $j$ as shown in Figure 9b, where $m$ is one of the magnetic substates of the orbit $j$, and $m'$ means any state. This diagram renormalizes the SPE, and observed SPE should include this contribution. If there is a valence nucleon in the state $m'$ as in Figure 9c, the process in Figure 9b is Pauli-forbidden. However, in the shell–model and other nuclear-structure calculations, the SPE containing the effect of Figure 9b is used. One has to somehow incorporate the Pauli effect of Figure 9c, and a solution is the introduction of the process in Figure 9d. In this process, the state $m'$ doubly appears in the intermediate state, but one can evaluate the Pauli effect by including Figure 9b,d consistently. This is a usual mathematical trick and enables us to correctly treat the Pauli principle within the simple framework. Figure 9d is equivalent to Figure 9e, which is nothing but the Fujita–Miyazawa 3NF, where the state $m'$ appears in double. Similar treatment is carried out in the chiral Effective Field Theory (EFT) framework. Figure 9f corresponds to Figure 9e, but the violation of the Pauli principle is slightly hidden, because of a vertex in the middle (depicted by a square) instead of the Δ-hole excitation.

In this argument, the 3NF produces a repulsive monopole $NN$ interaction in the valence space, after the summation over the hole states of the inert core (see Figure 9 bottom right), which corresponds to the normal ordering in other works.

The plot Figure 9 top right indicates an example of the repulsive effect on the ground-state energy of oxygen isotopes, locating the oxygen dripline at the right place or solving the oxygen anomaly [54]. This is rather strong repulsive monopole interaction, which is a consequence of the inert core. This means that the present case is irrelevant to the no-core shell model or other many-body approaches without the inert core (e.g., Green's Function Monte Carlo calculation [56]). This feature has caused some confusions in the past, but the difference is clear. The present repulsive monopole effect is much stronger than the other effects of the 3NF [57], and the latter will be better clarified by further developments of the chiral EFT for 3NF in the future. I note that the repulsive $T = 1$ $NN$ effect was empirically noticed by Talmi in the 1960s [3].

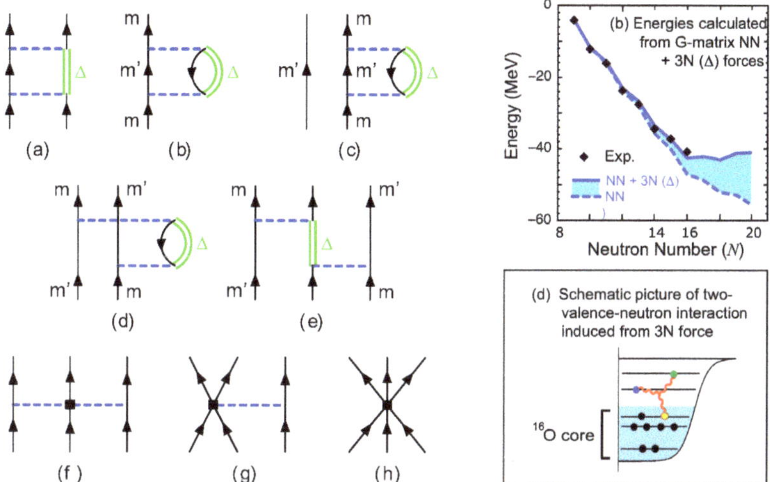


**Figure 9.** Schematic illustration of the three-nucleon force (3NF). **Left**: The diagrams (**a–e**) show how Δ-hole excitation effects are incorporated in accordance with Pauli principles, with the final form shown in (**e**), as described in the text. The diagrams in (**f–h**) represent three contributions from 3NF obtained in the chiral Effective Field Theory. **Top right:** the ground-state energy of oxygen isotopes, calculated with and without the 3NF and observed experimentally. **Bottom right**: the intuitive explanation of the diagrams in (**d,e**) of the left panel with the $^{16}$O inert core. Based on Figures 3 and 4 of [54].

### 2.10. Short Summary of This Section

The shell evolution phenomena are seen in many isotopic and isotonic chains and sometimes result in the formation of new magic gaps or the vanishing of old ones. Figure 2b displays the emergence of such new magic numbers $N = 16$, 32, and 34, whereas the lowering of some $2^+$ levels can mean the weakening of some magic numbers. More changes may appear in the future studies. Thus, the characteristic monopole features of the central, tensor, two-body $LS$, and 3NF-based $NN$ interactions and the resulting shell evolution are among the emerging concepts of the nuclear structure. Interestingly, these findings are neither isolated nor limited to particular aspects but are related to other aspects of the nuclear structure. We now move on to such a case.

## 3. Type-II Shell Evolution and Shape Coexistence

### 3.1. Type-II Shell Evolution

The shell evolution shown in Figure 5b,c are due to the addition of two or four neutrons into the orbit $j'_>$, respectively. Instead of adding, one can put neutrons into the orbit $j'_>$ by taking the neutrons from some orbits below $j'_>$, or equivalently by creating holes there, as shown in Figure 5d. If such a lower orbit happens to be the $j''_<$ orbit as in Figure 5d, its monopole matrix elements show just the opposite trends compared to the $j'_>$ orbit. However, because holes are created in $j''_<$, the sign of the monopole-interaction effect is reversed, and the final effect has the same sign as the monopole effect form the orbit $j'_>$ (see Figure 5d). Thus, the particle–hole (ph) excitation of the two neutrons Figure 5d reduces the proton $j_>$-$j_<$ splitting even more than in Figure 5b. This reduction becomes stronger with the ph excitations of four neutrons, as depicted in panel Figure 5e. Such strong reduction in the spin-orbit splitting produces interesting consequences beyond shell-structure changes. This type of the shell-structure change within the same nucleus is called *type-II shell evolution*.

### 3.2. A Doubly-Closed Nucleus $^{68}$Ni

The Type-II shell evolution was first discussed in [58] for $^{68}$Ni as an example. Figure 10 shows its theoretical and experimental energy levels. The theoretical results were obtained for the A3DA-m interaction by the MCSM [59–62], which is a powerful methodology for the shell model calculation but is not discussed in this article due to the length limitation.

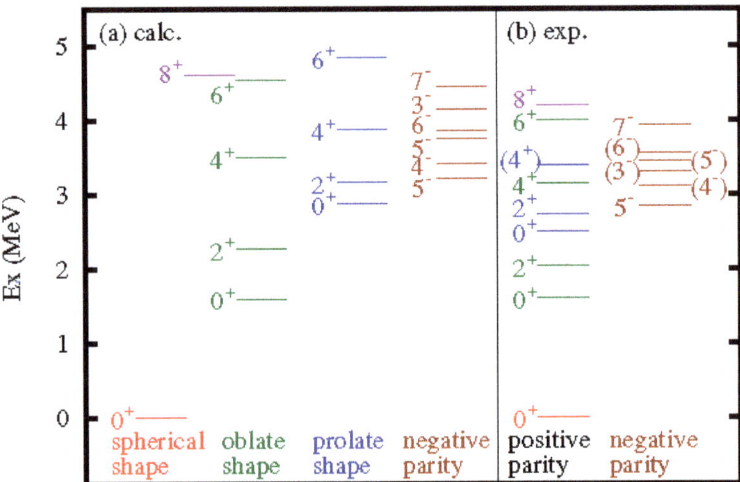


**Figure 10.** Level scheme of $^{68}$Ni. Taken from Figure 2 of [58].

Because $Z = 28$ is an SO magic number and $N = 40$ is an HO magic number (see Figure 1), the ground state of $^{68}$Ni is primarily a doubly closed shell. Indeed, in the theoretical ground state, the occupation of the neutron $g_{9/2}$ orbit is negligibly small. In contrast, the $0^+_3$ state located at the excitation energy, Ex$\sim$3 MeV, is the band head of a rotational band of an ellipsoidal shape, and its neutron $g_{9/2}$ occupation number is as large as $\sim$4. The mechanism shown in Figure 5e is then switched on, reducing the proton $f_{5/2}$–$f_{7/2}$ splitting. A reduced splitting facilitates more configuration mixing between these two orbits, which can produce notable effects on the quadrupole deformation as stated below.

### 3.3. Coexistence between Spherical and Deformed Shapes

We here quickly overview the quadrupole deformation or the shape deformation from a sphere to an ellipsoid [13]. The quadrupole deformation is driven by the quadrupole interaction, a part of the multipole interaction in Equation (11). The quadrupole interaction is a somewhat vague idea because of a certain mathematical complication, but its main effects can be simulated by the (scalar) coupling of the quadrupole moment operators. If the quadrupole moments are larger, i.e., a stronger quadrupole deformation occurs, the nucleus gains more binding energy from the quadrupole interaction. This is a very general phenomenon, and because of this the ground states of many nuclei are deformed, although $^{68}$Ni is not among them.

The energy of $^{68}$Ni (intrinsic state) is graphically illustrated in Figure 11 left for various ellipsoidal shapes, *spherical*, *prolate*, *oblate* and in between (called *triaxial*). The energy is calculated by the constraint Hartree–Fock (CHF) calculation with the same shell–model Hamiltonian as in Figure 10. The imposed constraints are given by the quadrupole moments in the intrinsic (body-fixed) frame, represented usually by $Q_0$ and $Q_2$ [13]. This plot is usually called the Potential Energy Surface (PES). The minimum energy occurs at the spherical shape (red sphere), with $Q_0 = Q_2 = 0$. The constraints are changed to a more prolate deformed ellipsoid (blue object) along the upper-right axis ("prolate deformation" in the figure), where $Q_0$ increases but $Q_2 = 0$. (Between two axes in Figure 11, $Q_2 \neq 0$. We come back to this point below.) The energy relative to the minimum energy climbs up by 6 MeV first. This is because protons and neutrons must be excited across the magic gaps from the doubly closed shell in order to create states of deformed shapes (see Figure 1). The energy then starts to come down, as the quadrupole moments increase, thanks to the quadrupole interaction. It is lowered by 3 MeV from the local peak to the local minimum. Beyond the local-minimum area, the effect of the quadrupole interaction is saturated, and it cannot compete with the energy needed for exciting more protons and neutrons across the gaps required by the constraints. This energy variation appears as the basin in the three-dimensional PES. This is the usual explanation of the local deformed minimum. The appearance of two (or more) different shapes with a rather small energy difference is one of the phenomena frequently seen and is called the shape coexistence [63]. The quadrupole interaction is undoubtedly among the essential factors of the shape coexistence. However, this may not be a full story.

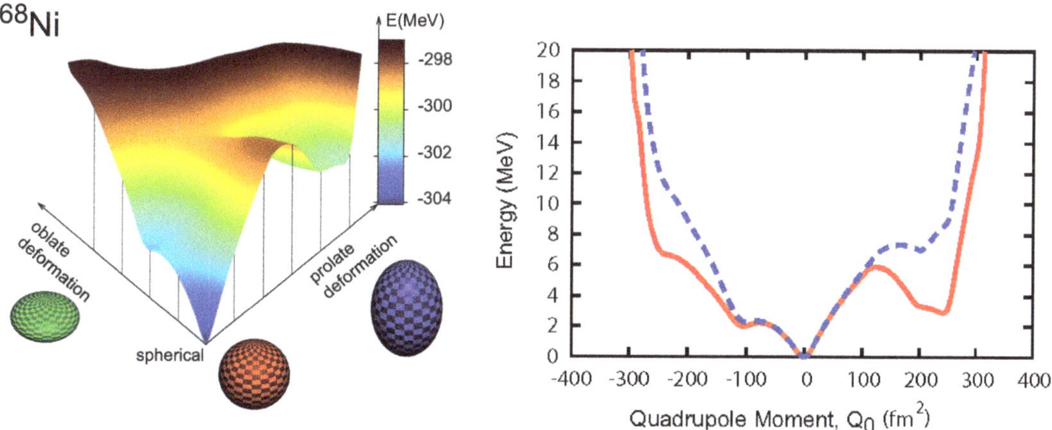


**Figure 11. Left**: Potential energy surface (PES) of $^{68}$Ni. Taken from Figure 5 of [24]. **Right**: PES of $^{68}$Ni with axially symmetric shapes. The solid line shows the PES of the full Hamiltonian, whereas the dashed line is the PES with practically no tensor-force contribution. Taken from Figure 6 of [24].

Figure 11 right exhibits the same energy along the axis lines of Figure 11 left, where $Q_0$ is varied from $-400$ fm$^2$ to 400 fm$^2$ while $Q_2 = 0$ is kept. The positive (intrinsic) quadrupole moments ($Q_0 > 0$) imply prolate shapes (blue object in Figure 11 left), whereas the negative ones imply ($Q_0 < 0$) oblate shapes (green object). The red solid line shows the CHF results of the full Hamiltonian, whereas for the dashed line, the tensor monopole interactions between the neutron $g_{9/2}$ orbit and the proton $f_{5/2,7/2}$ orbits are practically removed. This removal means no effects depicted in Figure 5d,e. The dashed line displays a less-pronounced prolate local minimum at weaker deformation with much higher excitation energy. The significant difference between the solid and dashed lines suggests that the monopole effects are crucial to lower this local minimum and stabilize it. We now discuss the mechanism for this difference. With the tensor monopole interaction, once sufficient neutrons are in $g_{9/2}$, the proton $f_{5/2}$–$f_{7/2}$ splitting is reduced, and this reduced splitting facilitates the mixing between these two orbits driven by the quadrupole interaction. The resulting deformation is stronger compared to no tensor-force case. In parallel to this, the tensor monopole interaction involving the neutron $g_{9/2}$ orbit produces extra binding energy, if more protons are in $f_{5/2}$ and less are in $f_{7/2}$. This extra binding energy lowers the deformed states, otherwise they are high in energy because of the energy cost for promoting neutrons from the $pf$ shell to $g_{9/2}$. Thus, a strong interplay emerges between the monopole interaction and the quadrupole interaction, and type-II shell evolution materializes this interplay in the present case. It enhances the deformation and lowers the energy of deformed states. Without this interplay, as indicated by blue dashed line in Figure 11 right, the rotational band corresponding to the local minimum is pushed up by 4 MeV and may be dissolved into the sea of many other states. It is obvious that this interplay mechanism works self-consistently.

### 3.4. T-Plot Analysis

The *T-plot* was introduced in the same Ref. [58], in order to clarify what shapes are more relevant to individual eigenstates of the shell–model calculation. Let us take an example. Figure 12 [64] depicts the PES of $^{66}$Ni with the same Hamiltonian as in Figure 10. The small circles on the PES are the T-plot. The T-plot is obtained from MCSM eigenstate. We therefore briefly explain the MCSM eigenstate. An MCSM eigenstate, $\Psi$, is written, with the ortho-normalization, as

$$\Psi = \sum_k f_k \hat{\mathcal{P}}_{J^\pi} \phi_k , \tag{16}$$

where $f_k$ denotes amplitude; $\hat{\mathcal{P}}_{J^\pi}$ means the projection operator on to the spin/parity $J^\pi$ (this part is more complicated in practice); and $\phi_k$ stands for a Slater determinant called ($k$-th) MCSM basis vector: $\phi_k = \Pi_i c_i^{(k)\dagger} |0\rangle$. Here, $|0\rangle$ is the inert core (closed shell); $c_i^{(k)\dagger}$ refers to a superposition of usual single-particle states,

$$c_i^{(k)\dagger} = \sum_n D_{i,n}^{(k)} a_n^\dagger , \tag{17}$$

with $a_n^\dagger$ being the creation operator of a usual single-particle state, for instance, that of the HO potential, and $D_{i,n}^{(k)}$ denoting a matrix element. By choosing an optimum matrix $D^{(k)}$, we can select $\phi_k$ so that such $\phi_k$ better contributes to the lowering of the corresponding energy eigenvalue. Thus, the determination of $D^{(k)}$ is the core of the MCSM calculation. The index $k$ runs up to 50–100 but sometimes to 300 at maximum. These are much smaller than the dimension of the many-body Hilbert space.

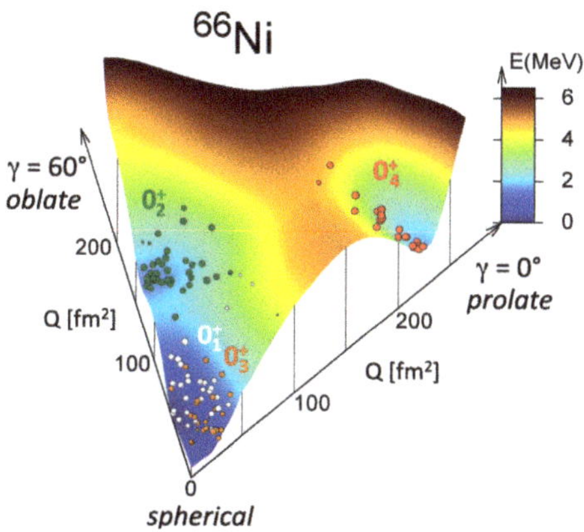


**Figure 12.** PES and T-plot for $^{66}$Ni. Taken from Figure 1 of [64].

Each $\phi_k$ has intrinsic quadrupole moments ($\langle\phi_k|\hat{Q}_0|\phi_k\rangle$ and $\langle\phi_k|\hat{Q}_2|\phi_k\rangle$), where $\hat{Q}_{0,2}$ imply the operators for $Q_{0,2}$ mentioned above. The T-plot circle for $\phi_k$ is placed according to those values on the PES with its area proportional to the overlap probability with the corresponding eigenstate, i.e., $\Psi$ in Equation (16). Such T-plot circles are shown in Figure 12. The white circles represent the MCSM basis vectors for the ground state, while the red circles indicate the MCSM basis vectors for the $0_4^+$ state, which is strongly deformed. Although there is no local minimum for oblate shape, the $0_2^+$ state is shown to be moderately oblate deformed. The T-plot can thus give partial labeling to fully correlated eigenstates for mean values as well as fluctuations with respect to their quadrupole shapes. The advantages of mean-field approaches are now nicely incorporated into the shell model.

### 3.5. *Short Summary of This Section*

Type-II shell evolution occurs in various cases, especially in a number of shape coexistence cases, providing deformed states with stronger deformation, lower excitation energies, and more stabilities. It is an appearance of the monopole–quadrupole interplay and plays crucial roles in various phenomena including the first-order quantum phase transition (Zr isotopes [65–67]), the second-order quantum phase transition (Sn isotopes [68]), the multiple even-odd quantum phase transitions (Hg isotopes [69]), as well as the raising of the intruder band due to the suppression of the type-II shell evolution (lighter Ni isotopes [64,70]). As the involvement of the monopole interaction in this manner had not been recognized, type-II shell evolution appears to be among the emerging concepts of nuclear structure. The type-II shell evolution has been clarified by the T-plot in many cases. Including other contributions, the T-plot is undoubtedly one of the emerging concepts of nuclear structure, apart from its impact on the computational methodology.

## 4. Self-Organization and Collective Bands in Heavy Nuclei

We now proceed to more general cases of the monopole–quadrupole interplay. This interplay leads to unexpected consequences in the underlying mechanism of collective bands of heavy nuclei [71], beyond the standard textbooks.

The MCSM has become powerful enough [62] to reproduce collective bands of heavy nuclei such as $^{154}$Sm and $^{166}$Er, with one and half HO major shells [71]. We sketch the new findings by using the results of such most-advanced MCSM calculations.

### 4.1. *Shape Coexistence in $^{154}$Sm*

Figure 13 shows low-lying energy levels of $^{154}$Sm. The present MCSM calculation can describe the four low-lying bands including the negative-parity one. The agreement between the experimental levels in Figure 13a and the theoretical levels in Figure 13b is

rather good. Although the importance of the quadrupole interaction is evident for the formation of deformed rotational bands, one can investigate to what extent the monopole interaction is involved. The monopole interaction here was obtained from the shell–model interactions, comprising the central, tensor, and other components.

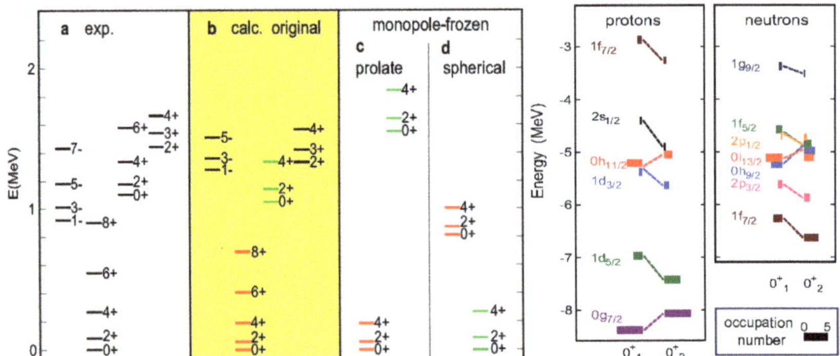


**Figure 13.** **Left**: (**a**) Experimental energy levels [42], (**b**) calculated original energy levels and (**c**,**d**) monopole-frozen energy levels of $^{154}$Sm. **Right**: ESPE (vertical position) and occupation number (horizontal width). Taken from Figures 2 and 3 of [71].

The monopole interaction is an operator, but we "freeze" it now: its ESPE expectation values $\langle \hat{\epsilon}_j^{p,n} \rangle$ are calculated for the state to be specified, and the obtained values are adopted as the SPEs, $\epsilon_{0;j}^{p,n}$ in Equation (4), with the monopole interaction removed. We then perform the shell-model calculation and draw the PES. This toy game is called the "monopole-frozen" analysis [71], as the monopole properties are included only through the specified state. Figure 13c exhibits the energy levels obtained by the monopole-frozen analysis referring to the ground state. The band built on the $0_2^+$ state (often called the $\beta$ band) is lifted up by 0.5 MeV (~50% of the original excitation energy), suggesting that the active monopole interaction produces a substantial lowering of this state. Figure 13d shows the monopole-frozen analysis referring to the spherical HF state: the ground state is no longer prolate, but triaxial, with the wave function close to the $0_2^+$ state of the original Hamiltonian. Thus, the crucial effect of the monopole interaction is verified.

Figure 13 right shows the actual values of $\langle \hat{\epsilon}_j^{p,n} \rangle$ for the $0_1^+$ and $0_2^+$ states. This figure demonstrates the significant differences between two sets of the ESPE values. The occupation numbers are also different: there are more half-filled orbits for the $0_2^+$ state, which is indicative of its triaxial nature. The smaller occupation numbers of unique-parity orbits are also consistent with the tendency away from the prolate shape.

We now introduce the deformation parameters $\beta_2$ and $\gamma$ [13], and their meanings are sketched in Figure 14a. The parameter $\beta_2$ represents the magnitude of the ellipsoidal deformation from sphere. The ellipsoid has three axes: the longest, middle, and shortest. The parameter $\gamma$ is an angle between 0° and 60° and represents mutual relations among the lengths of these axes: $\gamma$ = 0° means that the middle and shortest axes have the same length (prolate); $\gamma$ = 60° implies that the longest and the middle ones have the same length (oblate); and $\gamma$ values in between stand for intermediate situations, called triaxial. Figures 11 and 12 include them. The $\beta_2$ and $\gamma$ parameters can be obtained, in some approximation, from intrinsic quadrupole moments through the formulas [72],

$$\beta_2 = \sqrt{5/16\pi}\,\{(e + e'_p + e'_n)/e\}\,(4\pi/3R_0^2A^{5/3})\sqrt{(Q_0)^2 + 2(Q_2)^2}, \tag{18}$$

and

$$\gamma = \arctan(\sqrt{2}Q_2/Q_0), \tag{19}$$

where $e$ is the unit charge; $e'_p$ ($e'_n$) denotes proton (neutron) effective charge induced by in-medium (or core-polarization) effects; and $R_0$ stands for the radius parameter of the droplet model (spherical background) (see [73] for some detailed explanation). The relations in Equations (18) and (19) worked very well in many works, for instance [64,69–71].

Figure 14c,d shows the T-plot for the original interaction, where the PES is shown by using $\beta_2$ and $\gamma$ as coordinates (see Figure 14a). Figure 14e,f depicts the T-plot for the monopole-frozen interaction obtained with the spherical HF state. The T-plot patterns are consistent with the above features suggested by the shell–model diagonalization. The cut of the PES shown in Figure 14b suggests that the local minimum is raised by the monopole-frozen process referring to the ground state.

Figure 14c,d depict a valley of the PES with a local minimum around $\gamma = 15^\circ$. Similar valleys are seen in the PES obtained by the mean-field calculations [74,75], implying that this valley likely has a common origin. On the other hand, one can state that the present monopole effect results in not only the valley but also the local minimum, and the latter plays essential roles in the formation and stability of the side bands. It is of interest to refine the monopole interaction in mean-field models.

Regarding the $\beta$ vibration picture of the $0_2^+$ state, the present view is opposed to such a conventional view. The triaxial deformation is shared by the members not only of the $0_2^+$ band but also of the $2_3^+$ band (usually called $\gamma$ band), as can be verified by their T-plots. Namely, the $0_2^+$ state is the "ground" state of the triaxial states to which both the $0_2^+$ and $2_3^+$ bands belong. In short, this is a shape coexistence between the prolate and triaxial shapes assisted by the interplay between the monopole interaction and the quadrupole deformation. It is noted that the $\beta$ vibration picture of the $0_2^+$ states has been investigated from experimental viewpoints [76,77].

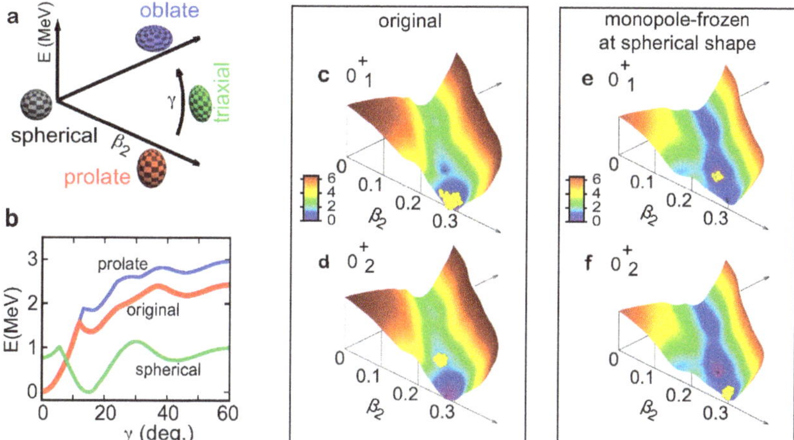


**Figure 14.** Properties of the $0^+_{1,2}$ states of $^{154}$Sm. (**a**) Deformation parameters and shapes. (**b**) Lowest values of PES for a given $\gamma$ value for the original case (red) as well as for the prolate (blue) and spherical (green) monopole-frozen cases. (**c–f**) Three-dimensional T-plot in the original and spherical monopole-frozen cases. Based on Figure 3 of [71].

### 4.2. Collective Bands and $\gamma$ Vibration in $^{166}$Er

The features of the collective motion in $^{166}$Er have been studied by the MCSM similarly well (see Figure 15a). Among rotational nuclei, $^{166}$Er is characterized by particularly low-lying $2_2^+$ state and the $\gamma$ band built on it. Aage Bohr stressed that this $2_2^+$ state was a $\gamma$ vibration from the prolate ground state [9–11,13]. The relatively strong $2_2^+ \to 0_1^+$ E2 transition (B(E2)$\sim$5 W.u., see Figure 15a, was ascribed to the annihilation of one $\gamma$ phonon in the $2_2^+$ state. This was one of the major points of the Nobel lecture by Aage Bohr and has been a common sense as stated in many textbooks of nuclear physics. We now challenge

this traditional belief, by utilizing the recent MCSM calculation. It is reminded that no firm experimental evidence to uniquely pin down the $\gamma$-vibration nature of $^{166}$Er has been reported and also that in a systematic calculation of many heavy nuclei [78], the excitation energies of the $2_2^+$ states in the $\gamma$ band appeared to be about twice higher than the observed values, despite much better description of those of the $2_1^+$ state in the ground band.

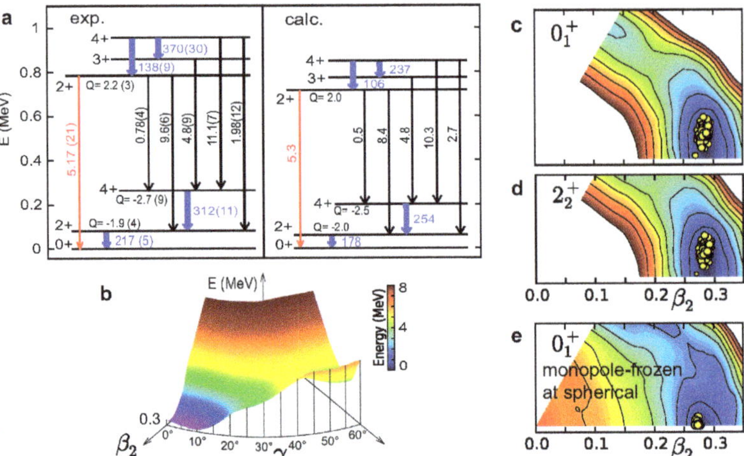

**Figure 15.** Experimental and calculated properties of the lowest states of $^{166}$Er. (**a**) Energy levels and electromagnetic transitions (W.u.) [42] as well as spectroscopic electric quadrupole moments (eb) [79]. (**b**) Three-dimensional PES and its cut surface for $\beta_2$ = 0.3. (**c**–**e**) T-plots for the $0_1^+$ and $2_2^+$ states and for the monopole-frozen $0_1^+$ state at spherical shape. Based on Figure 4 of [71].

Figure 15b shows the calculated PES, which shows the minimum not at $\gamma = 0°$ but around $\gamma = 9°$ (see also [80]). The T-plot is shown for the $0_1^+$ and $2_2^+$ states in Figure 15c and Figure 15d, respectively. The patterns of the T-plot circles are nearly identical between these two panels. This is consistent with a (rigid) triaxial interpretation, and indeed E2 transition strengths follow the predictions of the Davydov triaxial model [81,82] with $\gamma = 9°$. Certainly, a pure rigid triaxiality is not the correct picture, and there are quantum fluctuations, as evident from Figure 15c,d [80]. After all, the displacement from the $\gamma = 0°$ is obvious. The triaxiality of $^{166}$Er is also suggested by the triaxial projected shell model, although the rigid-triaxiality is not an outcome but an assumption [83,84].

The experimentally known $J^\pi = 4^+$ state around 2 MeV excitation energy provides a long-standing puzzle [85,86]: the observed relatively strong E2 transition from this state to the $2_2^+$ state looks like a sign that the $2_2^+$ state and this $J^\pi = 4^+$ states are the single- and double-phonon states in the $\gamma$ vibration picture (à la A. Bohr [9,10]), respectively, but the excitation energy of this $J^\pi = 4^+$ state is too high for a double-phonon excitation. The present calculation, on the other hand, reproduces both the excitation energy and the E2 transition strength, and this $J^\pi = 4^+$ state appears as the $K^\pi = 4^+$ member of the triaxial states including the $0_1^+$ and $2_{1,2}^+$ states (see Figure 15c,d) [71,80]. Thus, the triaxiality is shown to be one of the key aspects for understanding/predicting the shapes of heavy nuclei.

The monopole-frozen analysis referring to the spherical CHF state shows that the ground state moves to $\gamma = 0°$, confirming the important role of the monopole interaction activated. The triaxial ground states are now shown to appear in a large number of nuclei in the nuclear chart, besides the known triaxial domain [87].

### 4.3. A Historical Touch and a Short Summary of This Section

The collective bands in heavy nuclei have traditionally been understood in terms of the ground band with axially symmetric prolate shape and the side bands with the $\beta$ or $\gamma$ vibrational excitations from the ground state. This picture is consistent with the Nilsson model [88] and was confirmed by the Pairing + Quadrupole-Quadrupole (P+QQ) model [89,90], where the monopole interaction is not included, however. It has been shown in this section that the monopole interaction is crucial also for the collective bands in heavy nuclei. We just note that in lighter nuclei, the situation can be different mainly because of small model spaces comprising single or a few active orbits, where the rotational motion has been nicely described by symmetry-based approaches, e.g., SU(3) model of Elliott for the *sd* shell [91,92], and by realistic calculations, e.g., on $^{48}$Cr [93].

Regarding heavy nuclei, for individual rotational bands, the monopole interaction contributes differently, and the intrinsic structure is determined not only by the quadrupole interaction but also by the monopole interaction, as verified by the monopole-frozen analyses. Thus, the monopole–quadrupole interplay arises. The monopole interaction does not directly drive the deformation but optimizes the ESPEs so that more binding energy is gained. This gain is state-dependent and even can alter the ordering of bands as mentioned above. The present monopole–quadrupole interplay can be described also from the viewpoint of the *self-organization* [71]: the nucleus is changed from a disorder (original SPEs) to an order (ESPEs tailored to the shape of interest) by activating the monopole interaction. As this occurs "purposely" towards certain shapes with positive feedback, particularly between the monopole and quadrupole effects, the whole picture fits well the (quantal) self-organization [71]. The self-organization for collective bands is among the emerging concepts of nuclear structure, showing novel consequences. For example, the dominant fraction of the ground states of heavy nuclei are expected to show triaxial shapes, as another emerging concept of nuclear structure, in contrast to the traditional view of the prolate shape dominance in those states.

Appendix B presents a possible extension or generalization of the current idea to "many-ingredient" systems outside nuclear physics.

## 5. Dripline Mechanism

### 5.1. Traditional View

Figure 16a shows the left-lower part of the nuclear chart (Segrè chart) for $Z \leq 16$. The black squares represent stable nuclei while the orange ones exotic nuclei (see Section 1). An isotopic chain is a horizontal belt, and its neutron-rich end is called neutron dripline. The location of the dripline in the nuclear chart implies the extent of the isotopes and is of fundamental importance to nuclear science. The experimental determination of the dripline is a very difficult task. Very recently, as shown by red squares in Figure 16a, the driplines of F and Ne isotopes and its candidate of Na isotope were reported [94].

The traditional view of the dripline is shown in Figure 16b: all bound single-particle orbits are occupied, and the next neutron goes away. It is an open question whether this view is valid for all nuclei or not. We look into this question now [57].

The structure of neutron-rich exotic isotopes of F, Ne, Na, and Mg can be well described by the shell–model calculation with the full $sd$+$pf$ shells and the EEdf1 interaction [95]. This interaction was derived from the chiral EFT interaction of Machleidt and Entem [96], first processed by the $V_{\text{low-}k}$ method [97,98] and then processed by the EKK (Extended Krenciglowa-Kuo) method [99–101]. The $V_{\text{low-}k}$ method is used to transform the nuclear forces in the free space into a tractable form for further treatments. The $V_{\text{low-}k}$ method has been adopted for the derivation of other modern shell–model interactions, for instance, the one by Coraggio et al. for Sn and Cr-Fe regions [102,103].

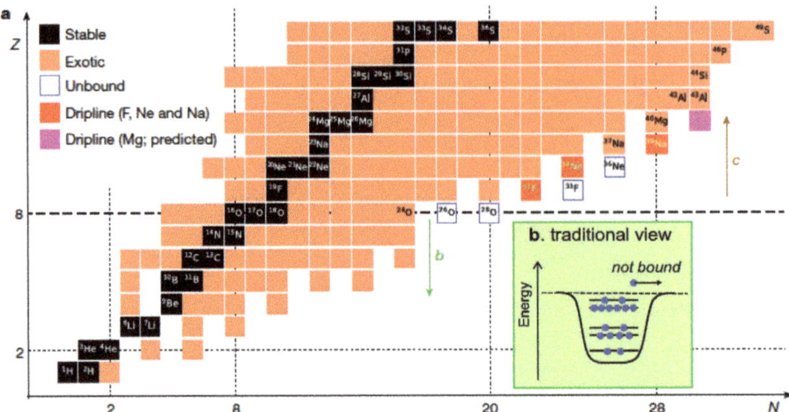


**Figure 16.** (**a**) Left-lower part of the nuclear chart with stable (black square), exotic (orange) and (confirmed) unbound (blank) nuclei as well as dripline nuclei (red, and purple). (**b**) Schematic illustration of the traditional view of the dripline. Based on Figure 2 of [57].

The present work is unique in the usage of the EKK method, which enlarges the scope of the approaches based on the many-body perturbation theory (MBPT) [29]. The MBPT produced the G-matrix interactions in its early formulations [28], from which many useful shell-model interactions have been constructed (see Section 2.6). However, the resulting G-matrix interaction shows a limitation that if two major shells are merged, the results may diverge [101]. As the gap between two shells often vanishes or becomes smaller in exotic nuclei, this difficulty can be fatal there, although it is irrelevant to one-major-shell calculations. The EKK method nicely avoids this difficulty besides other merits.

Here, I present a very quick sketch of the formal aspect of the EKK method focusing on the logical flow based on Refs. [99–101] particularly the last one. This paragraph is not so relevant for understanding later parts of the article and can be skipped. In this paragraph, the symbol ^ for operators is omitted for clarity. The EKK method starts from the separation of the Hamiltonian $H$ with a parameter $\xi$ as

$$H = \begin{pmatrix} \xi & 0 \\ 0 & QH_0Q \end{pmatrix} + \begin{pmatrix} P(H-\xi)P & PVQ \\ QVP & QVQ \end{pmatrix}, \tag{20}$$

where $P$ stands for the projection onto the Hilbert space explicitly treated (called $P$ space usually), and $Q = 1 - P$. From this equation, we obtain the effective Hamiltonian for the $P$ space at the $n$-th stage of the successive process,

$$\tilde{H}_{\text{eff}}^{(n)} = \tilde{H}_{\text{BH}}(\xi) + \sum_{k=1}^{\infty} Q_k(\xi)\{\tilde{H}_{\text{eff}}^{(n-1)}\}^k, \tag{21}$$

where $\tilde{O}$ means $O - \xi$ for any operator $O$, e.g., $\tilde{H}_{\text{BH}}(\xi) = H_{\text{BH}}(\xi) - \xi$. Here, the Bloch–Horowitz Hamiltonian is written as,

$$H_{\text{BH}}(\xi) = PHP + PVQ\frac{1}{\xi - QHQ}QVP, \tag{22}$$

where the second term on the r.h.s. is called the $Q$-box. The quantity $Q_k$ in Equation (21) represents its $k$-th derivative with respect to $\xi$. Provided that $\tilde{H}_{\text{eff}}^{(n)} \approx \tilde{H}_{\text{eff}}^{(n-1)}$ is achieved, we can regard and use them as the effective Hamiltonian, $\tilde{H}_{\text{eff}}$. The effective interaction, like the EEdf1 interaction, is obtained as $V_{\text{eff}} = H_{\text{eff}} - PH_0P$ with $H_0$ being the unperturbed Hamiltonian (usually the SPEs). The solution of the given many-body problem remains (almost) unchanged within a certain range of $\xi$. In fact, the $\xi$ parameter can be interpreted

as the origin point of a Tayler expansion in a generalized sense. The divergence due to the energy denominator does not occur if the adopted $\xi$ values are far from the poles causing the divergence. I would like to stress that by construction, this effective Hamiltonian produces the exact solutions, once the convergence is achieved. This sketch is expected to depict that the EKK method is an expansion but not a perturbation one. This can be exemplified by the feature that the final result is independent of the $\xi$ parameter, in contrast to the perturbation expansion.

The EEdf1 interaction has thus been derived in an ab initio way by the $V_{\text{low-}k}$ and EKK methods from the chiral EFT interaction of Machleidt and Entem [96]. Some effects of 3NF are included in terms of the effective $NN$ interaction by averaging over the hole states in the inert core, of which the monopole part is discussed in Section 2.9. While the Fujita–Miyazawa 3NF was used so far, other 3NF can be taken [57]. The EEdf1 interaction describes the properties of the ground and low-lying states of F, Ne, Na, and Mg isotopes quite well [57,95].

### 5.2. Monopole–Quadrupole Interplay for the Driplines

Figure 17 shows the ground-state energies of F, Ne, Na, and Mg isotopes as functions of the neutron number $N$. These energies are decomposed into several pieces according to their origins: SPE (on top of the $^{16}$O inert core), monopole, pairing, and rest terms. The Coulomb contribution is ignored in the following discussion, because it is of virtually no relevance. Here, the multipole interaction is divided into the pairing and rest terms. The pairing is the BCS-type pairing interaction acting on two neutrons coupled to $J^{\pi} = 0^{+}$ and on two protons coupled to $J^{\pi} = 0^{+}$. The rest term means the multipole interaction subtracted by the pairing term. Although the rest term contains many different pieces, its major effects in the present discussion is simulated by the quadrupole interaction. This is the reason why the rest term is associated with "(quadrupole etc)" in the figure.

The lower edges of the red areas exhibit the ground-state energies as functions of the neutron number $N$, while only even $N$ values are taken. These values show a good agreement with measured values shown by black dots. As long as the ground-state energy becomes lower as $N$ increases, the isotope gains more binding energy by having more neutrons, and the isotope chain is stretched. However, if the ground-state energy is not lowered, there is no gain in the binding energy by having these extra neutrons; these extra neutrons are emitted, and the neutron dripline implies the nucleus with the lowest ground-state energy. The driplines obtained by the present calculation are shown by red arrows for each isotopic chain, reproducing experimental driplines for F, Ne, and Na isotopes [94].

We focus on the lower edge of the green areas in Figure 17. This represents the monopole contributions comprising the SPE and the monopole interaction. For Ne, Na, and Mg isotopes, this edge is lowered almost linearly as $N$ increases from $N = 16$ to each dripline. We then fit the edge with pink dashed, purple dotted, and black solid lines for Ne, Na, and Mg isotopes, respectively. The lines of Ne and Na isotopes are copied to the panel for Mg, with their positions adjusted at a certain $N$. It is evident that the lines become steeper almost linearly as $Z$ increases. This edge is almost flat for F isotopes for $N \geq 16$, and this feature is discussed below.

Figure 17 indicates that the effect of the pairing term shows small variations. In contrast, the rest term changes more, which is largely due to the quadrupole interaction. Figure 18a schematically indicates the variation of the effect of the quadrupole interaction: The effect is small at the far-left position with a spherical shape. As some neutrons are added, the shape is deformed, and the ground-state energy is lowered due to the quadrupole interaction. This trend continues but becomes its maximum at a certain value of $N$ (red object in the figure). However, the dripline is not determined just by this maximum point.

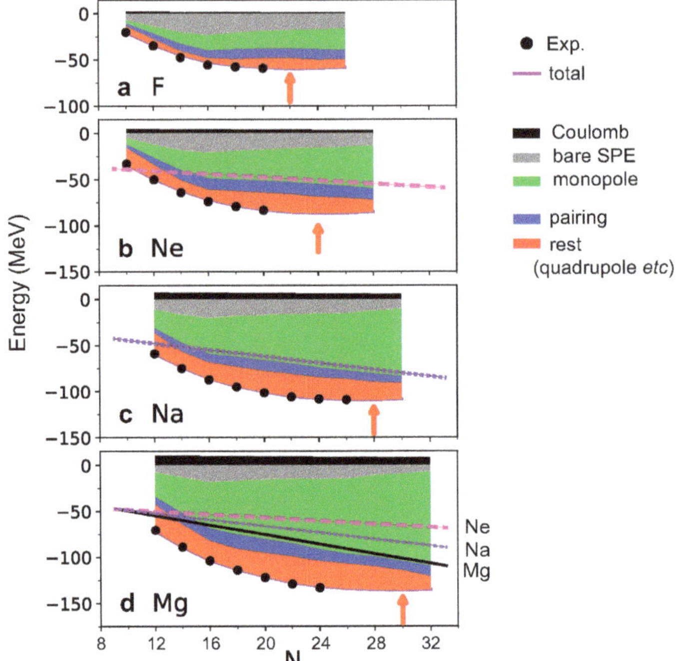


**Figure 17.** Ground-state energies of even-N isotopes of (**a**) F, (**b**) Ne, (**c**) Na and (**d**) Mg, relative to the $^{16}$O value. Colored segments exhibit decompositions into various effects from the monopole (green), pairing (blue) and rest (such as quadrupole) (red) components of the effective nucleon–nucleon interaction as well as those from Coulomb interaction (black) and single-particle energies (bare SPE; grey). The monopole effect grows steadily as a function of $N$ in all cases, as highlighted by straight lines: dashed (Ne), dotted (Na) and solid (Mg). The experimental values are indicated by black circles [42]. The theoretical driplines indicated by red arrows. Modified from Figure 4 of [57].

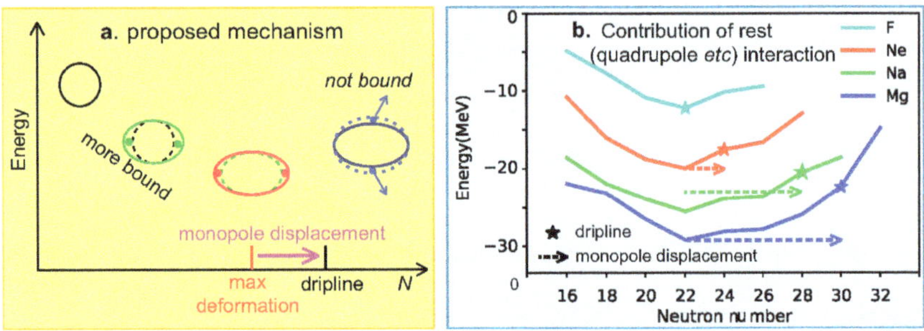


**Figure 18.** (**a**) Presently proposed mechanism based on shape evolution and the resulting change in the ground-state energy. (**b**) The rest-term contribution to the ground-state energies for F, Ne, Na, and Mg isotopes. Dashed arrows indicate the monopole displacement. See text for more details. Modified from Figures 2 and 6 of [57].

Figure 18b depicts the actual effect of the rest term. It follows the trend illustrated in Figure 18a, with the maximum effect at $N = 22$ in all four chains. However, the driplines are different among these four. This is due to the monopole interaction. Let me explain it by taking the Mg isotopes as an example. The black straight line of the monopole effect in Figure 17d depicts about 3 MeV lowering per additional neutron, implying about 6 MeV

for an additional two neutrons. After $N = 22$, the rest effect loses its magnitude. If the loss is less than the monopole gain ($\sim$6 MeV), this loss is compensated by the monopole effect. However, the loss becomes larger for $N$ larger, and at a certain point, the loss exceeds the monopole compensation. The dripline thus arises with the "monopole displacement" from $N = 22$ to $N = 30$ as shown in Figure 18b (and also in Figure 18a schematically).

The monopole effect depends directly on the number of protons, as visualized by three straight lines in Figure 17. Consequently, the monopole displacement is $\Delta N = 2$ (6) for Ne (Na) isotopes. For F isotopes, the monopole effect is negligibly small for $N \geq 16$, and the dripline is located at the maximum rest (quadrupole etc.) effect.

### 5.3. Stability of Spherical Isotopes and the Monopole-Quadrupole Interplay

An immediate lemma of the present dripline mechanism is that the driplines of spherical nuclei, such as Ca, Sn, and Pb isotopes, can be further away from the stability line than other elements. One can assume a basically constant pairing contribution and a minor rest-term contribution. These two are thus irrelevant to the driplines of these isotopes. The remaining monopole effect gradually changes, pushing the driplines away.

### 5.4. A Short Summary of This Section

The present new dripline mechanism [57] involves the monopole–quadrupole interplay and is one of the emerging concepts. It definitely differs from the traditional mechanism of the single-particle origin, where a neutron halo arises at extremes [104,105]. In the new mechanism, the coupling to continuum may be visible if the monopole effect vanishes like heavy F isotopes [57]. As $Z$ changes, two dripline mechanisms may appear alternatively, but the present one may be more relevant to heavier nuclei where the deformation develops more. Finally, I would like to point out that the Bethe–Weizäcker mass formula does not include a deformation energy term, at least, explicitly .

## 6. Prospect

As this article is a kind of summary, I am afraid that a summary section may be redundant. I state some prospects. First of all, ab initio no-core Monte Carlo shell–model calculations became feasible recently up to $^{12}$C and beyond [106], and as an example, we can look into $\alpha$ clustering in light nuclei, e.g., the Hoyle state, with correlations produced by nuclear forces [107]. This direction will produce a major outcome from the shell model. This includes clarifications of $\alpha$ decay, $\alpha$ knockout, etc. Another major frontier is the quest for fission dynamics and superheavy elements, with (almost) full inclusion of the correlations due to nuclear forces.

Although more computer power and further advancements in computational methodology are needed also, the perspectives of the shell model look unlimited, to me. *May the (nuclear) force be with you.*

**Funding:** This work was supported in part by MEXT as "Program for Promoting Researches on the Super computer Fugaku" (Simulation for basic science: from fundamental laws of particles to creation of nuclei) and by JICFuS. This work was supported by JSPS KAKENHI Grant Numbers JP19H05145, JP21H00117.

**Data Availability Statement:** Not applicable.

**Acknowledgments:** The author is grateful to A. Gargano and S.M. Lenzi for their interests and for the invitation to this valuable program. He acknowledges T. Abe, Y. Akaishi, B.A. Brown, P. Van Duppen, B. Fornal, R. Fujimoto, A. Gade, H. Grawe, R.V.F. Janssens, M. H.-Jensen, K. Heyde, J. Holt, M. Honma, M. Huyse, Y. Ichikawa, S. Leoni, T. Mizusaki, N. Pietralla, P. Ring, E. Sahin, J. P. Schiffer, A. Schwenk, N. Shimizu, O. Sorlin, Y. Sun, T. Suzuki, K. Takayanagi, T. Togashi, K. Tsukiyama, N. Tsunoda, Y. Tsunoda, Y. Utsuno, S. Yoshida, and H. Ueno for their valuable direct collaborations towards the works presented in this article and thanks many others for their useful comments and help. The comment by Y. Tsunoda on Appendix A is appreciated.

**Conflicts of Interest:** The author declares no conflict of interest.

## Appendix A. Note on the Relation between the Present ESPE and the Baranger's ESPE

This is a short note on the the relation between the present ESPE and Baranger's ESPE [19] discussed in [15]. A possible problem was pointed out by Y. Tsunoda. Although the relevant arguments and results in [15] are basically correct, the following term is found to be added to Equation (43) of [15]: $-1/(2j+1)V^m(j,j)\langle 0|\hat{n}_j|0\rangle$, where $j$ includes the index, proton, or neutron. So, this is the contribution from the interaction between a neutron orbit $j$ and the same neutron orbit $j$ (or between protons similarly), of which the monopole interaction is known to be weak. In addition, the factor $1/(2j+1)$ reduces this quantity. Because of all these factors combined, the correction is quite minor. This correction does not change the basic equivalence relation between the two schemes.

## Appendix B. Self-Organization and Its Extension to Other "Many-Body" Systems

We here discuss briefly how the present self-organization mechanism may be applied to other systems comprising many constituents, including human societies. One of the essential points is two interactions with different characters: one drives the system into specific modes, as denoted by the mode-driving force. The mode here generally refers to a collective phenomenon involving many constituents, like the shape of an atomic nucleus. A certain resistance usually exists against the mode development. The other interaction is to control the resistance, called the resistance-control force. The monopole interaction in this work is an example. The resistance-control force does not create any mode, being neutral. However, it can change the disorder in the original environment (=original SPE in this work) to the order where the resistance is weakened for certain modes (ESPE tailored to the shape). This order thus gives extra stability to the system, to varying degrees depending on the modes. Thus, the resistance-control force can be a crucial factor in determining which mode gains the maximum stability (i.e., binding energy). Obviously, in many systems, only the maximum-stability mode matters, which may not be the one most favored by the driving force. If this general idea can be applied to various problems, including social/economical issues, it is of great interest. While the mode varies over different systems, the mode-driving force may be visible. The resistance-control force, however, may not be so, because it exhibits less characteristics (like the monopole interaction in atomic nuclei). Studies in this direction can be of interest. What are the resistance and its control force in human societies?

## References


1. Goeppert Mayer, M. On closed shells in nuclei. II. *Phys. Rev.* **1949**, *75*, 1969. [CrossRef]
2. Haxel, O.; Jensen, J.H.D.; Suess, H.E. On the "magic numbers" in nuclear structure. *Phys. Rev.* **1949**, *75*, 1766. [CrossRef]
3. Talmi, I. Effective Interactions and Coupling Schemes in Nuclei. *Rev. Mod. Phys.* **1962**, *34*, 704–722. [CrossRef]
4. Caurier, E.; Martínez-Pinedo, G.; Nowacki, F.; Poves, A.; Zuker, A.P. The shell model as a unified view of nuclear structure. *Rev. Mod. Phys.* **2005**, *77*, 427–488. [CrossRef]
5. Gade, A.; Glasmacher, T. In-beam nuclear spectroscopy of bound states with fast exotic ion beams. *Prog. Part. Nucl. Phys.* **2008**, *60*, 161–224. [CrossRef]
6. Sorlin, O.; Porquet, M.-G. Nuclear magic numbers: New features far from stability. *Prog. Part. Nucl. Phys.* **2008**, *61*, 602–673. [CrossRef]
7. Nakamura, T.; Sakurai, H.; Watanabe, H. Exotic nuclei explored at in-flight separators. *Prog. Part. Nucl. Phys.* **2017**, *97*, 53–122. [CrossRef]
8. Rainwater, J. Nuclear energy level argument for a spheroidal nuclear model. *Phys. Rev.* **1950**, *79*, 432. [CrossRef]
9. Bohr, A. The coupling of nuclear surface oscillations to the motion of individual nucleons. *Dan. Mat. Fys. Medd.* **1952**, *26*, 14.
10. Bohr, A. Rotational motion in nuclei. In *Nobel Lectures, Physics 1971–1980*; Lundqvist, S., Ed.; World Scientific: Singapore, 1992; pp. 213–232. Available online: https://www.nobelprize.org/prizes/physics/1975/bohr/facts/ (accessed on 1 February 2022).
11. Bohr, A.; Mottelson, B.R. Collective and individual-particle aspects of nuclear structure. *Dan. Mat. Fys. Medd.* **1953**, *27*, 16.
12. Bohr, A.; Mottelson, B.R. *Nuclear Structure I*; Benjamin: New York, NY, USA, 1969.
13. Bohr, A.; Mottelson, B.R. *Nuclear Structure II*; Benjamin: New York, NY, USA, 1975.
14. Schäfer, T. Fermi liquid theory: A brief survey in memory of Gerald E. Brown, *Nucl. Phys. A* **2014**, *928*, 180–189. [CrossRef]
15. Otsuka, T.; Gade, A.; Sorlin, O.; Suzuki, T.; Utsuno, Y. Evolution of shell structure in exotic nuclei. *Rev. Mod. Phys.* **2020**, *92*, 015002. [CrossRef]
16. Ragnarsson, I.; Nilsson, S. *Shapes and Shells in Nuclear Structure*; Cambridge University Press: Cambridge, UK, 1995.

17. Poves, A.; Zuker, A. Theoretical spectroscopy and the fp shell. *Phys. Rep.* **1981**, *70*, 235–314. [CrossRef]
18. Bansal, R.K.; French, J.B. Even-parity-hole states in f7/2-shell nuclei. *Phys. Lett.* **1964**, *11*, 145–148. [CrossRef]
19. Baranger, M. A definition of the single-nucleon potential. *Nucl. Phys. A* **1970**, *149*, 225. [CrossRef]
20. Storm, M.; Watt, A.; Whitehead, R. Crossing of single-particle energy levels resulting from neutron excess in the sd shell. *J. Phys. G* **1983**, *9*, L165–L168. [CrossRef]
21. Grawe, H. Shell model from a practitioner's point of view. In *The Euroschool Lectures on Physics with Exotic Beams*; Al-Khalili, J., Roeckl, E., Eds.; Springer, Berlin/Heidelberg, Germany, 2004; Volume I, pp. 33–75.
22. Federman, P.; Pittel, S. Towards a unified microscopic description of nuclear deformation. *Phys. Lett. B* **1977**, *69*, 385–388. [CrossRef]
23. Otsuka, T.; Suzuki, T.; Fujimoto, R.; Grawe, H.; Akaishi, Y. Evolution of the nuclear shells due to the tensor force. *Phys. Rev. Lett.* **2005**, *95*, 232502. [CrossRef]
24. Otsuka, T.; Tsunoda, Y. The role of shell evolution in shape coexistence. *J. Phys. G* **2016**, *43*, 024009. [CrossRef]
25. Otsuka, T.; Suzuki, T.; Honma, M.; Utsuno, Y.; Tsunoda, N.; Tsukiyama, K.; Hjorth-Jensen, M. Novel features of nuclear forces and shell evolution in exotic nuclei. *Phys. Rev. Lett.* **2010**, *104*, 012501. [CrossRef]
26. Brown, B.A.; Wildenthal, B.H. Status of the nuclear shell model. *Annu. Rev. Nucl. Part. Sci.* **1988**, **38**, 29–66. [CrossRef]
27. Honma, M.; Otsuka, T.; Brown, B.A.; Mizusaki, T. Effective interaction for pf-shell nuclei. *Phys. Rev. C* **2004**, *65*, 061301. [CrossRef]
28. Kuo, T.T.S.; Brown, G.E. Structure of finite nuclei and the free nucleon-nucleon interaction An application to $^{18}$O and $^{18}$F. *Nucl. Phys.* **1966**, *85*, 40. [CrossRef]
29. Hjorth-Jensen, M.; Kuo, T.T.S.; Osnes, E. Realistic effective interactions for nuclear systems. *Phys. Rep.* **1995**, *261*, 125–270. [CrossRef]
30. Brown, B.A. Double-octupole states in $^{208}$Pb. *Phys. Rev. Lett.* **2000**, *85*, 5300. [CrossRef]
31. Brown, B.A.; Stone, N.J.; Stone, J.R.; Towner, I.S.; Hjorth-Jensen, M. Magnetic moments of the $2_1^+$ states around $^{132}$Sn. *Phys. Rev. C* **2005**, *71*, 044317. [CrossRef]
32. Lenzi, S.M.; Nowacki, F.; Poves, A.; Sieja, K. Island of inversion around $^{64}$Cr. *Phys. Rev. C* **2010**, *82*, 054301. [CrossRef]
33. Cohen, S.; Kurath, D. Effective interactions for the 1p shell. *Nucl. Phys.* **1965**, *73*, 1. [CrossRef]
34. Bertsch, G.; Borysowicz, J.; McManus, H.; Love, W.G. Interactions for inelstic scattering derived from realistic potentials. *Nucl. Phys. A* **1977**, *284*, 399–419. [CrossRef]
35. Osterfeld, F. Nuclear spin and isospin excitations. *Rev. Mod. Phys.* **1992**, *64*, 491–557. [CrossRef]
36. Bäckman, S.-O.; Brown, G.E.; Niskanen, J.A. The nucleon-nucleon interaction and the nuclear many-body problem. *Phys. Rep.* **1985**, *124*, 1–68. [CrossRef]
37. Schiffer, J.P.; Freeman, S.J.; Caggiano, J.A.; Deibel, C.; Heinz, A.; Jiang, C.L. Is the nuclear spin-orbit interaction changing with neutron excess? *Phys. Rev. Lett.* **2004**, *92*, 162501. [CrossRef]
38. Sahin, E.; Garrote, F.B.; Tsunoda, Y.; Otsuka, T.; De Angelis, G.; Görgen, A. Shell evolution towards $^{78}$Ni: Low-lying states in $^{77}$Cu. *Phys. Rev. Lett.* **2017**, *118*, 242502. [CrossRef] [PubMed]
39. Ichikawa, Y.; Nishibata, H.; Tsunoda, Y.; Takamine, A.; Imamura, K.; Fujita, T. Interplay between nuclear shell evolution and shape deformation revealed by the magnetic moment of $^{75}$Cu. *Nat. Phys.* **2019**, *15*, 321–325. [CrossRef]
40. Liddick, S. N.; Grzywacz, R.; Mazzocchi, C.; Page, R.D.; Rykaczewski, K.P.; Batchelder, J.C. Discovery of $^{109}$Xe $^{105}$Te: Superallowed decay doubly magic $^{100}$Sn. *Phys. Rev. Lett.* **2006**, *97*, 082501. [CrossRef]
41. Seweryniak, D.; Carpenter, M.P.; Gros, S.; Hecht, A.A.; Hoteling, N.; Janssens, R.V.F. Single-neutron states $^{101}$Sn. *Phys. Rev. Lett.* **2007**, *99*, 022504. [CrossRef] [PubMed]
42. Evaluated Nuclear Structure Data File. Available online: http://www.nndc.bnl.gov/ensdf/ (accessed on 1 February 2022).
43. Smirnova, N.A.; Bally, B.; Heyde, K.; Nowacki, F.; Sieja, K. Shell evolution and nuclear forces. *Phys. Lett. B* **2010**, *686*, 109–113. [CrossRef]
44. Huck, A.; Klotz, G.; Knipper, A.; Miehé, C.; Richard-Serre, C.; Walter, G.; Poves, A.; Ravn, H.L.; Marguier, G. Beta decay of the new isotopes $^{52}$K, $^{52}$Ca, and $^{52}$Sc; a test of the shell model far from stability. *Phys. Rev. C* **1985**, *31*, 2226. [CrossRef]
45. Otsuka, T.; Fujimoto, R.; Utsuno, Y.; Brown, B.A.; Honma, M.; Mizusaki, T. Magic numbers in exotic nuclei and spin-isospin properties of the *NN* interaction. *Phys. Rev. Lett.* **2001**, *87*, 082502. [CrossRef]
46. Janssens, R.V.F. Elusive magic numbers. *Nature* **2005**, *435*, 897–898. [CrossRef]
47. Steppenbeck, D.; Takeuchi, S.; Aoi, N.; Doornenbal, P.; Matsushita, M.; Wang, H. Evidence for a new nuclear 'magic number'from the level structure of $^{54}$Ca. *Nature* **2013**, *502*, 207–210. [CrossRef] [PubMed]
48. Michimasa, S.; Kobayashi, M.; Kiyokawa, Y.; Ota, S.; Ahn, D.S.; Baba, H. Magic nature of neutrons in 54Ca: First mass measurements of $^{55-57}$Ca. *Phys. Rev. Lett.* **2018**, *121*, 022506. [CrossRef] [PubMed]
49. Chen, S.; Lee, J.; Doornenbal, P.; Obertelli, A.; Barbieri, C.; Chazono, Y. Quasifree neutron knockout from $^{54}$Ca corroborates arising $N = 34$ neutron magic number. *Phys. Rev. Lett.* **2019**, *123*, 142501. [CrossRef] [PubMed]
50. Kay, B.P.; Hoffman, C.R.; Macchiavelli, A.O. Effect of Weak binding on the apparent spin-orbit splitting in nuclei. *Phys. Rev. Lett.* **2017**, *119*, 182502. [CrossRef]
51. Uozumi, Y.; Kikuzawa, N.; Sakae, T.; Matoba, M.; Kinoshita, K.; Sajima, S. Shell-Model Study of $^{40}$Ca with the 56-MeV $(\vec{d}, P)$ reaction. *Phys. Rev. C* **1994**, *50*, 263–274. [CrossRef]

52. Burgunder, G.; Sorlin, O.; Nowacki, F.; Giron, S.; Hammache, F.; Moukaddam, M. Experimental study of the two-Body spin-orbit force in nuclei. *Phys. Rev. Lett.* **2014**, *112*, 042502. [CrossRef]
53. Hammer, H.-W.; Nogga, A.; Schwenk, A. Colloquium: Three-body forces: Cold atoms nuclei. *Rev. Mod. Phys.* **2013**, *85*, 197. [CrossRef]
54. Otsuka, T.; Suzuki, T.; Holt, J.D.; Schwenk, A.; Akaishi, Y. Three-body forces and the limit of oxygen isotopes. *Phys. Rev. Lett.* **2010**, *105*, 032501. [CrossRef]
55. Fujita, J.; Miyazawa, H. Pion theory of three-body forces. *Prog. Theor. Phys.* **1957**, *17*, 360. [CrossRef]
56. Carlson, J.; Golfi, S.; Pederiva, F.; Pieper, S.C.; Schiavilla, R.; Schmidt, K.E.; Wiringa, R.B. Quantum Monte Carlo methods for nuclear physics. *Rev. Mod. Phys.* **2015**, *87*, 1067–1118. [CrossRef]
57. Tsunoda, N.; Otsuka, T.; Takayanagi, K.; Shimizu, N.; Suzuki, T.; Utsuno, Y.; Yoshida, S.; Ueno, H. The impact of nuclear shape on the emergence of the neutron dripline. *Nature* **2020**, *587*, 66–71. [CrossRef] [PubMed]
58. Tsunoda, Y.; Otsuka, T.; Shimizu, N.; Honma, M.; Utsuno, Y. Novel shape evolution in exotic Ni isotopes and configuration-dependent shell structure. *Phys. Rev. C* **2014**, *89*, 031301(R). [CrossRef]
59. Honma, M.; Mizusaki, T.; Otsuka, T. Diagonalization of Hamiltonians for many-body systems by auxiliary field quantum Monte Carlo technique. *Phys. Rev. Lett.* **1995**, *75*, 1284. [CrossRef] [PubMed]
60. Otsuka, T.; Mizusaki, T.; Honma, M. Structure of the $N = Z = 28$ closed shell studied by Monte Carlo shell model calculation. *Phys. Rev. Lett.* **1998**, *81*, 1588–1591. [CrossRef]
61. Otsuka, T.; Honma, M.; Mizusaki, T.; Shimizu, N.; Utsuno, Y. Monte Carlo shell model for atomic nuclei. *Prog. Part. Nucl. Phys.* **2001**, *47*, 319–400. [CrossRef]
62. Shimizu, N.; Abe, T.; Tsunoda, Y.; Utsuno, Y.; Yoshida, T.; Mizusaki, T.; Honma, M.; Otsuka, T. New-generation Monte Carlo shell model for the K computer era. *Prog. Theor. Exp. Phys.* **2012**, *2012*, 01A205. [CrossRef]
63. Heyde, K.; Wood, J.L. Shape coexistence in atomic nuclei. *Rev. Mod. Phys.* **2011**, *83*, 1467–1521. [CrossRef]
64. Leoni, S.; Fornal, B.; Mărginean, N.; Sferrazza, M.; Tsunoda, Y.; Otsuka, T. Multifaceted quadruplet of low-lying spin-zero states in $^{66}$Ni: Emergence of shape isomerism in light nuclei. *Phys. Rev. Lett.* **2017**, *118*, 162502. [CrossRef] [PubMed]
65. Togashi, T.; Tsunoda, Y.; Otsuka, T.; Shimizu, N. Quantum Phase Transition in the Shape of Zr isotopes. *Phys. Rev. Lett.* **2016**, **117**, 172502. [CrossRef]
66. Kremer, C.; Aslanidou, S.; Bassauer, S.; Hilcker, M.; Krugmann, A.; von Neumann-Cosel, P. First measurement of collectivity of coexisting shapes based on type II shell evolution: The case of case $^{96}$Zr. *Phys. Rev. Lett.* **2016**, *117*, 172503. [CrossRef]
67. Singh, P.; Korten, W.; Hagen, T.W.; Görgen, A.; Grente, L.; Salsac, M.D. Evidence for coexisting shapes through lifetime measurements in $^{98}$Zr. *Phys. Rev. Lett.* **2016**, *121*, 192501. [CrossRef] [PubMed]
68. Togashi, T.; Tsunoda, Y.; Otsuka, T.; Shimizu, N.; Honma, M. Novel Shape evolution in Sn isotopes from magic numbers 50 to 82. *Phys. Rev. Lett.* **2018**, *121*, 062501. [CrossRef]
69. Marsh, B.A.; Goodacre, T.; Sels, S.; Tsunoda, Y.; Andel, B.; Andreyev, A.N. Characterization of the shape-staggering effect in mercury nuclei. *Nat. Phys.* **2018**, *14*, 1163–1167. [CrossRef]
70. Mărginean, S.; Little, D.; Yusuke, T.; Leoni, S.; Janssens, R.V.F.; Fornal, B.; Takaharu, O.; Michelagnoli, C.; Stan, L.; Luigi, C.F.C.; et al. Shape coexistence at zero spin in $^{64}$Ni driven by the monopole tensor interaction. *Phys. Rev. Lett.* **2020**, *125*, 102502. [CrossRef] [PubMed]
71. Otsuka, T.; Tsunoda, Y.; Abe, T.; Shimizu, N.; Van Duppen, P. Underlying Structure of collective bands and self-organization in quantum systems. *Phys. Rev. Lett.* **2019**, *123*, 222502. [CrossRef]
72. Utsuno, Y.; Shimizu, N.; Otsuka, T.; Yoshida, T.; Tsunoda, Y. Nature of Isomerism in exotic Sulfur isotopes. *Phys. Rev. Lett.* **2015**, *114*, 032501. [CrossRef]
73. Otsuka, T.; Shimizu, N.; Tsunoda, Y. Moments and radii of exotic Na and Mg isotopes. *Phys. Rev. C* **2022**, *105*, 014319. [CrossRef]
74. Robeldo, L.M.; Rodríguez-Guzmán, R.R.; Sarriguren, P. Evolution of nuclear shapes in medium mass isotopes from a microscopic perspective. *Phys. Rev. C* **2008**, *78*, 034314. [CrossRef]
75. Li, Z.P.; Nikšić, T.; Vretenar, D.; Meng, J.; Lalazissis, G.A.; Ring, P. Microscopic analysis of nuclear quantum phase transitions in the $N \approx 90$ region. *Phys. Rev. C* **2009**, *79*, 054301. [CrossRef]
76. Garrett, P. E. Characterization of the $\beta$ vibration and $0_2^+$ states in deformed nuclei. *J. Phys. G* **2001**, *27*, R1. [CrossRef]
77. Sharpey-Schafer, J.F.; Bark, R.A.; Bvumbi, S.P.; Dinoko, T.R.S.; Majola, S.N.T. "Stiff" deformed nuclei, configuration dependent pairing and the $\beta$ and $\gamma$ degrees of freedom. *Eur. Phys. J. A* **2019**, *55*, 15. [CrossRef]
78. Delaroche, J.-P.; Girod, M.; Libert, J.; Goutte, H.; Hilaire, S.; Péru, S.; Pillet, N.; Bertsch, G.F. Structure of even-even nuclei using a mapped collective Hamiltonian and the D1S Gogny interaction. *Phys. Rev. C* **2010**, *81*, 104303. [CrossRef]
79. Stone, N.J. Table of nuclear electric quadrupole moments. *At. Data Nucl. Data Tables* **2016**, *11–112*, 1–28. [CrossRef]
80. Tsunoda, Y.; Otsuka, T. Triaxial rigidity of $^{166}$Er and its Bohr-model realization. *Phys. Rev. C* **2021**, *103*, L021303. [CrossRef]
81. Davydov, A.S.; Filippov, G.F. Rotational states in even atomic nuclei. *Nucl. Phys.* **1958**, *8*, 237–249. [CrossRef]
82. Davydov, A.S.; Rostovsky, V.S. Relative transition probabilities between rotational levels of non-axial nuclei. *Nucl. Phys.* **1959**, *12*, 58–68. [CrossRef]
83. Sun, Y.; Hara, K.; Sheikh, J.A.; Hirsch, J.G.; Velázquez, V.; Guidry, M. Multiphonon $\gamma$-vibrational bands and the triaxial projected shell model. *Phys. Rev. C* **2000**, *61*, 064323. [CrossRef]

84. Boutachkov, P.; Aprahamian, A.; Sun, Y.; Sheikh, J.A.; Frauendorf, S. In-band and inter-band B(E2) values within the Triaxial Projected Shell Model. *Eur. Phys. J. A* **2002**, *15*, 455. [CrossRef]
85. Fahlander, C.; Axelsson, A.; Heinebrodt, M.; Hartlein, T.; Schwalm, D. Two-phonon $\gamma$-vibrational states in $^{166}$Er. *Phys. Lett. B* **1996**, *388*, 475. [CrossRef]
86. Garrett, P.E.; Kadi, M.; Li, M.; McGrath, C.A.; Sorokin, V.; Yeh, M.; Yates, S.W. $K^{\pi} = 0^{+}$ and $4^{+}$ two-phonon $\gamma$-vibrational states in $^{166}$Er. *Phys. Rev. Lett.* **1997**, *78*, 4545. [CrossRef]
87. Hayashi, A.; Hara, K.; Ring, P. Existence of triaxial shapes in transitional nuclei. *Phys. Rev. Lett.* **1984**, *53*, 337–340. [CrossRef]
88. Nilsson, S.G. Binding states of individual nucleons in strongly deformed nuclei. *Dan. Mat. Fys. Medd.* **1955**, *29*, 16.
89. Kumar, K.; Baranger, M. Nuclear deformations in the pairing-plus-quadrupole model (III). Static nuclear shapes in the rare-earth region. *Nucl. Phys. A* **1968**, *110*, 529–554. [CrossRef]
90. Bes, D.R.; Sorensen, R.A. The Pairing-Plus-Quadrupole Model. In *Advances in Nuclear Physics;* Baranger, M., Vogt, E., Eds.; Plenum Press, New York, NY, USA, 1969. [CrossRef]
91. Elliott, J.P. Collective motion in the nuclear shell model I. Classification schemes for states of mixed configurations. *Proc. R. Soc. Lond. Ser. A* **1958**, *245*, 128–145.
92. Elliott, J.P. Collective motion in the nuclear shell model II. The introduction of intrinsic wave-functions. *Proc. R. Soc. Lond. Ser. A* **1958**, *245*, 562–581.
93. Caurier, E.; Zuker, A.; Poves, A.; Martiínez-Pinedo, G. Full *Pf* Shell Model Study $A$ = 48 Nuclei. *Phys. Rev. C* **1994**, *50*, 225–236. [CrossRef]
94. Ahn, D.S.; Fukuda, N.; Geissel, H.; Inabe, N.; Iwasa, N.; Kubo, T. Location of the neutron dripline at fluorine and neon. *Phys. Rev. Lett.* **2019**, *123*, 212501. [CrossRef]
95. Tsunoda, N.; Otsuka, T.; Shimizu, N.; Hjorth-Jensen, M.; Takayanagi, K.; Suzuki, T. Exotic neutron-rich medium-mass nuclei with realistic nuclear forces. *Phys. Rev. C* **2017**, *95*, 021304(R). [CrossRef]
96. Machleidt, R.; Entem, D.R. Chiral effective field theory and nuclear forces. *Phys. Rep.* **2011**, *503*, 1–75. [CrossRef]
97. Bogner, S.; Kuo, T.T.S.; Coraggio, L.; Covello, A.; Itaco, N. Low momentum nucleon-nucleon potential and shell model effective interactions. *Phys. Rev. C* **2002**, *65*, 051301. [CrossRef]
98. Nogga, A.; Bogner, S.K.; Schwenk, A. Low-momentum interaction in few-nucleon systems. *Phys. Rev. C* **2004**, *70*, 061002. [CrossRef]
99. Takayanagi, K. Effective interaction in non-degenerate model space. *Nucl. Phys. A* **2011**, *852*, 61–81. [CrossRef]
100. Takayanagi, K. Effective Hamiltonian in the extended Krenciglowa-Kuo method. *Nucl. Phys. A* **2011**, *864*, 91–112. [CrossRef]
101. Tsunoda, N.; Takayanagi, K.; Hjorth-Jensen, M.; Otsuka, T. Multi-shell effective interactions. *Phys. Rev. C* **2014**, *89*, 024313. [CrossRef]
102. Coraggio, L.; Covello, A.; Gargano, A.; Itaco, N. Similarity of nuclear structure in the $^{132}$Sn and $^{208}$Pb regions: Proton-neutron multiplets. *Phys. Rev. C* **2009**, *80*, 021305. [CrossRef]
103. Arnswald, K.; Braunroth, T.; Seidlitz, M.; Coraggio, L.; Reiter, P.; Birkenbach, B. Enhanced collectivity along the $N$=$Z$ line: Lifetime measurements in $^{44}$Ti, $^{48}$Cr, and $^{52}$Fe. *Phys. Lett. B* **2017**, *772*, 599–606. [CrossRef]
104. Tanihata, I.; Hamagaki, H.; Hashimoto, O.; Shida, Y.; Yoshikawa, N.; Sugimoto, K. Measurements of interaction cross sections and nuclear radii in the light p-shell region. *Phys. Rev. Lett.* **1985**, *55*, 2676–2679. [CrossRef]
105. Hansen, P.G.; Jonson, B. The neutron halo of extremely neutron-rich nuclei. *Europhys. Lett.* **1987**, *4*, 409–414. [CrossRef]
106. Abe, T.; Maris, P.; Otsuka, T.; Shimizu, N.; Utsuno, Y.; Vary, J.P. Ground-state properties of light 4*n* self-conjugate nuclei in ab initio no-core Monte Carlo shell model calculations with nonlocal *NN* interactions. *Phys. Rev. C* **2021**, *104*, 054315. [CrossRef]
107. Otsuka, T.; Abe, T.; Yoshida, T.; Tsunoda, Y.; Shimizu, N.; Itagaki, N.; Utsuno, Y.; Vary, J.; Maris, P.; Ueno, H. This Material Belongs to the Field of Nuclear Physics. 2022, *Unpublished work.*

 *physics* 

*Review*

# The Nuclear Shell Model towards the Drip Lines

**B. Alex Brown**

Department of Physics and Astronomy, and the Facility for Rare Isotope Beams, Michigan State University, East Lansing, MI 48824-1321, USA; brown@frib.msu.edu

**Abstract:** Applications of configuration-mixing methods for nuclei near the proton and neutron drip lines are discussed. A short review of magic numbers is presented. Prospects for advances in the regions of four new "outposts" are highlighted: $^{28}$O, $^{42}$Si, $^{60}$Ca and $^{78}$Ni. Topics include shell gaps, single-particle properties, islands of inversion, collectivity, neutron decay, neutron halos, two-proton decay, effective charge, and quenching in knockout reactions.

**Keywords:** nuclear shell model; configuration-interaction method; magic numbers; proton drip line; neutron drip line; proton decay; neutron decay; collectivity; islands of inverson; effective charge

**Citation:** Brown, B.A. The Nuclear Shell Model towards the Drip Lines. *Physics* **2022**, *4*, 525–547. https://doi.org/10.3390/physics4020035

Received: 7 March 2022
Accepted: 14 April 2022
Published: 12 May 2022

**Publisher's Note:** MDPI stays neutral with regard to jurisdictional claims in published maps and institutional affiliations.

**Copyright:** © 2022 by the authors. Licensee MDPI, Basel, Switzerland. This article is an open access article distributed under the terms and conditions of the Creative Commons Attribution (CC BY) license (https://creativecommons.org/licenses/by/4.0/).

## 1. Introduction

The starting point for the nuclear shell model is the establishment of model spaces that allow for tractable configuration-interaction (CI) calculations from which we are able to understand and predict the properties of low-lying states [1–5]. This choice is based on the observation that a few even–even nuclei can be interpreted in terms of having magic numbers for $Z$ (atomic number) or $N$ (nucleon number) and doubly-magic numbers for a given $(Z, N)$. These magic numbers can be inferred from experimental excitation energies of $2^+$ states shown for the low end of the nuclear chart in Figure 1. Magic numbers are those values of $Z$ or $N$ for nuclei that have a relatively high $2^+$ energy within a series of isotopes or isotones.

Another measure of magic numbers is given by the double difference in the binding energy, BE, defined by

$$D(q) = (-1)^q [2\text{BE}(q) - \text{BE}(q+1) - \text{BE}(q-1)] \tag{1}$$

for isotopes ($q = N$ with $Z$ held fixed) or isotones ($q = Z$ with $N$ held fixed) can also be used to measure shell gaps [6]. An example for the neutron-rich calcium isotopes is shown in Figure 2 (the dashed line extrapolation to $N = 40$ is discussed below.) The value of $D(N)$ at these magic numbers gives the effective shell gap. In between the magic numbers, $D(N)$ gives the pairing energy [6]. The excitation energies of the $2^+$ states at $N = 28$, 32 and 34, also shown in Figure 2, are close to the $D(N)$ values at these magic numbers. The neutron gaps at $N = 32$ and 34 are weaker than the gap at $N = 28$, but they are strong enough to allow the configurations to be dominated by the orbitals, shown in Figure 2.

In the simplest model, the magic number is associated with a ground state that has a closed-shell configuration for the given value of $Z$ or $N$. The following is from footnote 9 in [7]. It was Eugene Paul Wigner who coined the term "magic number". Steven A. Moszkowski, who was a student of Maria Goeppert-Mayer, in a talk presented at the American Physical Society meeting in Indianapolis, 4 May 1996 said: "Wigner believed in the liquid drop model, but he recognized, from the work of Maria Mayer, the very strong evidence for the closed shells. It seemed a little like magic to him, and that is how the words 'Magic Numbers' were coined". The discovery of "magic numbers" lead M. Goeppert-Mayer, and independently J. Hans D. Jensen in Germany, one year later, in 1949, to the construction of the shell model with strong spin–orbit coupling, and to the Nobel Prize they shared with Wigner in 1963.

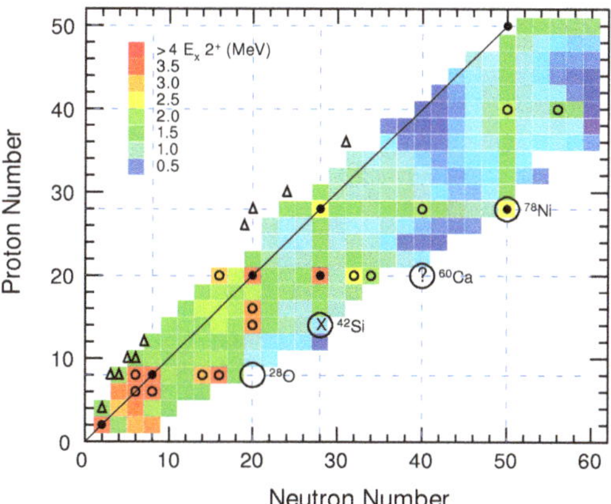


**Figure 1.** Lower mass region of the nuclear chart. The colors indicate the energy of the first $2^+$ state. In addition to the data from [8], recent data for $^{40}$Mn [9], $^{62}$Ti [10], $^{66}$Cr [11] and $^{70,72}$Fe [11] are added. The filled black circles show the doubly-magic nuclei associated with the most robust pairs of magic numbers 8, 20, 28 and 50. The small open circles show the doubly-magic nuclei associated with less robust magic numbers 6, 14, 16, 32, 34, and 40. The large open circles indicate the nuclei near the neutron drip lines that are the focus of this paper. The triangles are those nuclei observed to decay by two protons in the ground state. The cross indicates no magic number for protons or neutrons, and the question mark indicates that the doubly-magic status is not known.

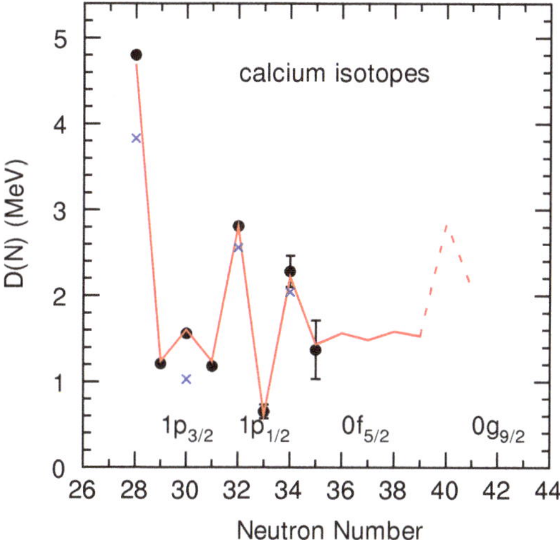


**Figure 2.** $D(N)$ as given by Equation (1). The black dots with error bars are the experimental data. The blue crosses are the excitation energies of the $2_1^+$ states. The orbitals that are being filled are shown. The red line is the results from the universal $fp$ calcium (UFP-CA) Hamiltonian [12]. The dashed line is the extrapolation based on the universal nuclear energy density functional (version zero) (UNEDF0) binding energies. for $^{60,61,62}$Ca [13].

The nuclei marked with closed circles in Figure 1 are commonly used to define the boundaries of CI model spaces. Those indicated by small open circles are usually contained within a larger CI model spaces. Historically, the size of the assumed model space has

depended on the computational capabilities. At the very beginning in the 1960s, they were the $0p$ model space bounded by $^{4}$He and $^{16}$O, and the $0f_{7/2}$ model space bounded by $^{40}$Ca and $^{56}$Ni.

For heavy nuclei, doubly-magic nuclei are associated with the shell gaps at 28, 50, 82 and 126. These gaps are created by the spin–orbit splitting of the high $\ell$ orbitals, which lowers the the $j = \ell + 1/2$ single-particle energies for $\ell = 3$ (28), $\ell = 4$ (50), $\ell = 5$ (82) and $\ell = 6$ (126). Since the two $j$ values for a given high $\ell$ value are split, 28, 50, 82 and 126 will be referred to as $jj$ magic numbers. The nuclei with $jj$ magic numbers for both protons and neutrons will be called double-$jj$ closed-shell nuclei. These are shown by the red circles in Figure 3: $^{208}$Pb, $^{132}$Sn, $^{100}$Sn, $^{78}$Ni and $^{56}$Ni. The open red circle for $^{100}$Sn indicates that it is expected to be double-$jj$ magic [14], but it has not yet been confirmed experimentally. The continuation of the double-$jj$ sequence with $\ell = 2$ (14) and $\ell = 1$ (6) is shown by the open blue circles for $^{42}$Si, $^{28}$Si, $^{18}$C and $^{12}$C on the lower left-hand side of Figure 3. As discussed below, the calculations for these nuclei show rotational bands with positive quadrupole moments indicative of an oblate intrinsic shape.

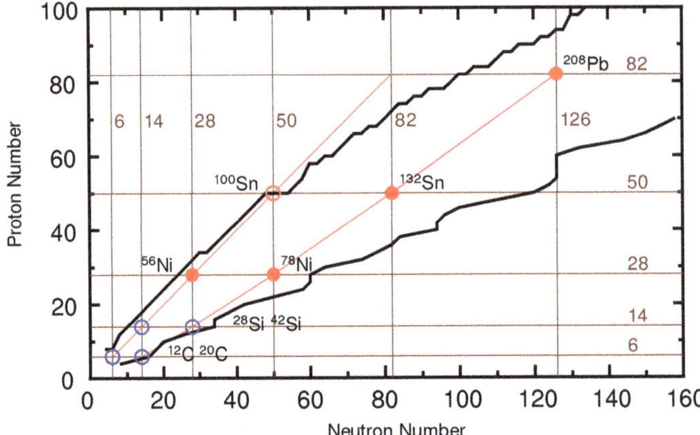


**Figure 3.** The nuclear chart showing the $jj$ magic numbers (see text for $jj$ definition). The black lines show where the two-proton (upper) and two-neutron (lower) separation energies obtained with the universal nuclear energy density functional (version one) (UNEDF1) [13] cross 1 MeV. The filled red circles show the locations of double-$jj$ magic nuclei established from experiment. The open red circle for $^{100}$Sn indicates a probably double-$jj$ magic nucleus that has not been confirmed by experiment. The blue circles in the bottom left-hand side are nuclei in the double-$jj$ magic number sequence that are oblate deformed.

In light nuclei, magic numbers 2, 8, 20 and 40 are associated with the filling of a major harmonic-oscillator shell with $N_o = (2n_r + \ell)$ ($n_r$ is the radial quantum number), where both members of the spin–orbit pair $j = \ell \pm 1/2$ are filled. Since one can recouple the two orbitals with the same $\ell$ value to total angular momentum $L$ and total spin $S$, 2, 8, 20 and 40 will be referred to as $LS$ magic numbers.

The $LS$ magic numbers for isotopes and isotones are shown by the thin brown lines in Figure 4. There are only three known double-$LS$ magic nuclei, $^{4}$He, $^{16}$O and $^{40}$Ca shown by the filled red circles in Figure 4. The next one in the sequence would be $^{80}$Zr, but in this case the $Z = N = 40$ gap is too small due to the lowering of the $0g_{9/2}$ single-particle energy from the spin–orbit splitting. As will be discussed below, $^{60}$Ca (the red open circle with a question mark) could be a "fourth" double-$LS$ magic nucleus. There are regions where the $LS$ magic numbers for isotopes or isotones dissappear as shown by the blue lines in Figure 4. These will be referred to as "islands of inversion" [15].

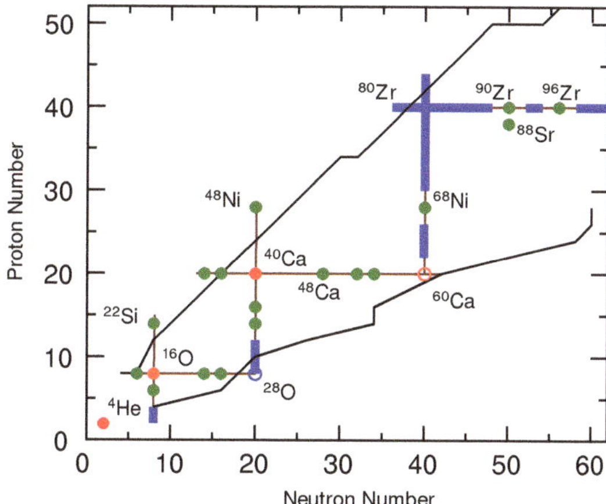


**Figure 4.** Lower mass region of the nuclear chart showing the *LS* magic numbers, 2, 8, 20 and 40 (see text for *LS* definitioon). The black lines show where the two-proton (upper) and two-neutron (lower) separation energies obtained with the UNEDF1 [13] functional cross 1 MeV. The filled red circles show the double-*LS* magic nuclei $^{4}$He, $^{16}$O and $^{40}$Ca. The open red circle for $^{60}$Ca indicates a possible doubly-magic nucleus that has not been confirmed by experiment. The green circles are doubly-magic nuclei associated with the *j*-orbital fillings. The blue lines indicate isotopes or isotones where the *LS* magic number is observed to be broken.

The nuclei with green circles in Figure 4 also have doubly-magic properties. The pattern is that when one type of nucleon (proton or neutron) has an *LS* magic number, then the other one has a magic number for the filling of each *j* orbital. These are 6 ($0p_{3/2}$), 8 ($0p_{1/2}$), 14 ($0d_{5/2}$), 16 ($1s_{1/2}$), 20 ($0d_{3/2}$), 28 ($0f_{7/2}$), 32 ($1p_{3/2}$), 34 ($1p_{1/2}$), 40 ($0f_{5/2}$), 50 ($0g_{9/2}$) and 56 ($1d_{5/2}$).

The only addition to the *jj* and *LS* closed-shell systematics discussed above is for $^{88}$Sr shown in Figure 4, where there is an energy gap between the proton $1p_{1/2}$ and $1p_{3/2}$,$0f_{5/2}$ states. In early calculations, $^{88}$Sr was used as the closed shell for the $1p_{1/2}$,$0g_{9/2}$ model space [16], but more recently the four-orbit model space of $0f_{5/2}$,$1p_{3/2}$,$1p_{1/2}$,$0g_{9/2}$ has been used for the $N = 50$ isotones [17,18].

For a given shell gap, the *LS* magic numbers are more robust than those for *jj*. The reason is that deformation for *jj* magic numbers starts with a one-particle one-hole ($1p$–$1h$) excitation of a nucleon in the $j = \ell + 1/2$ orbital to the other members of the same oscillator shell, $N_o = (2n_r + \ell)$. Since $1p$–$1h$ excitations across *LS* closed shell gaps change parity, ground-state deformation for *LS* magic numbers must come from $np$–$nh$ ($n \geq 2$) excitations across the *LS* closed shells as in the region of $^{32}$Mg [15].

Let us discuss here results, obtained with Hamiltonians. based on data-driven improvements to the two-body matrix elements, provided by ab initio methods. The ab initio methods are based on two-nucleon (NN) and three-nucleon (NNN) interactions obtained by model-dependent fits to nucleon-nucleon phase shifts and properties of nuclei with $A = 2$ to 4. For a given model space, these are renormalized for short-range correlations and for the truncations into the chosen model space to provide a set of two-body matrix elements (TBME) for nuclei near a chosen doubly-closed shell. From this starting point, one attempts to make minimal changes to the Hamiltonian to improve the agreement with energy data for a selected set of nuclei and states within the model space. A convenient way to do this is by using singular value decomposition (SVD) [19]. In many cases, one adjusts specific TBME or combinations of TBME. The most important are the monopole, pairing and quadrupole components. An important part of the universal Hamiltonian is in

the evolution of the effective single-particle energies (ESPE) as one changes the number of protons and neutrons. Starting with a closed shell with a given set of single-particle energies, the ESPE as a function of $Z$ and $N$ are determined by the monopole average parts of the TBME [5].

These methods provide "universal" Hamiltonians in the sense that a single set of single-particle energies and two-body matrix elements are applied to all nuclei in the model space, perhaps allowing for some smooth mass dependence. This has turned out to be a practical and useful approximation. As the ab initio, starting points are improved, these "universal" Hamiltonians were replaced by Hamiltonians for a more restricted set of nuclei, or even for individual nuclei as has been done in the valence-space in-medium similarity renormalization group (VS-IMSRG) method [4,20].

The empirical modifications to the effective Hamiltonian account for deficiencies in the more ab initio methods. Most ab initio calculations are carried out in a harmonic-oscillator basis due to its convenient analytical properties. Near the neutron drip lines, the radial wavefunctions become more extended, the single-particle energy spectrum becomes more compressed, and the continuum becomes explicitly more important. To take this into account, the ab initio methods require a very large harmonic-oscillator basis.

Due to the continuum, nuclei near the neutron drip line present a substantial theoretical challenge [2,21]. Methods have been developed that take the continuum into account explicitly. The density matrix renormalization group (DMRG) method [22,23] makes use of a single-particle potential together with a simplified interaction based on halo effective field theory [24,25]. In the Gamow shell model (GSM) [26–28], the many-body basis is constructed from a single-particle Berggren ensemble [29,30]. The DMRG and GSM methods rely on use of simplified two-body interactions with adjustable parameters. There is also the shell model embedded in the continuum formalism that can make use of the universal interactions [31]. Recent progress in the GSM method is presented in [32].

Ground-state nuclear halos are a unique feature of nuclei near the neutron drip line [33]. This is due to the loose binding of low-$\ell$ orbitals with extended radial wavefunctions. The most famous case is that for $^{11}$Li which was observed to have a rapid rise in the nuclear matter radius compared to the trends up to $^{9}$Li [34]. The wavefunction of $^{11}$Li is dominated by a pair of neutrons in the $1s_{1/2}$ orbital. As discussed below, halos in the region of $^{30}$Ne and $^{42}$Si are dominated by the $1p_{3/2}$ orbital. Proton halos are not so extreme due to the Coulomb barrier. The excited $1/2^{+}$ $(1s_{1/2})$ state of $^{17}$F is a good example of an excited-state halo as determined indirectly from its large Thomas–Ehrman energy shift of 0.87 MeV $^{17}$O to 0.49 MeV in $^{17}$F.

States above the (proton/neutron) separation energy have (proton/neutron) decay widths. In the conventional CI approach, one calculates states whose energy is taken to be the centroid energy of the decaying state. The decay width is calculated using the approximation $\Gamma = C^2S\,\Gamma_{sp}(Q)$, where $C^2S$ is the spectroscopic factor and $\Gamma_{sp}$ is the single-particle neutron decay width calculated with a a decay energy, $Q$, value taken from the shell-model centroid or the experimental centroid if known. The explicit addition of the continuum shifts down the energy relative to its CI energy [31]. Further, the continuum (finite-well potential) is responsible for the Thomas–Ehrman shift for states in proton-rich nuclei compared to those in the neutron-rich mirror nuclei [19].

In this review, I concentrate on four regions of neutron-rich "outposts" whose understanding are most important for future developments. These are shown in Figure 1: $^{28}$O, $^{42}$Si, $^{60}$Ca and $^{78}$Ni. $^{42}$Si is labeled by "$\times$" since it does not have a magic number for protons or neutrons. $^{78}$Ni is labeled by a filled circle since it is now known to be doubly magic [35]. $^{60}$Ca is known to be inside the neutron drip line [36], but its mass and excited states have not yet been measured.

Nuclei that are observed to decay by two protons are shown by the triangles in Figure 1. The two-proton ground-state decays for $^{45}$Fe, $^{48}$Ni, $^{54}$Zn and $^{67}$Kr have half-lives on the order of ms and compete with the $\beta$ decay of those nuclei. An experimental and theoretical summary of the results for those nuclei together with that of $^{19}$Mg has been

given in [37]. There is qualitative agreement between experiment and theory. In order to become more quantitative, the experimental errors in the partial half-lives need to be improved. Theoretical models need to be improved to incorporate three-body decay dynamics (presently based on single-orbit configurations) with the many-body CI calculations for the two-nucleon decay amplitudes. The correlations for two-nucleon transfer amplitudes via (t,p) or ($^3$He,n) are largely determined by the $(S, T) = (0, 1)$ structure of the triton or $^3$He, whereas two-nucleon decay is determined by the decay through the Coulomb and angular-momentum barriers that are dominated by the low-$\ell$ components. For the lightest nuclei, multi-proton emissions (shown in Figure 1 of [38]) are observed as broad resonances.

Knockout reactions are used to produce nuclei further from stability. The cross sections for these reactions can be compared to theoretical models in terms of the cross-section ratio $R_s = \sigma_{\text{exp}}/\sigma_{\text{th}}$; see [39] for a recent summary. It is observed for nuclei far from stability where $\Delta S = | S_{1p} - S_{1n} |$ is large ($S_1$ is the one nucleon separation energy) that $R_s$ is near unity when the knocked out nucleon is loosely bound but drops to approximately 0.3 for deeply bound nucleons. This has been attributed to the short- and long-ranged correlations that depletes the occupation of deeply-bound states [40]. The short-ranged correlations are connected to the high-momemtum tail observed in observed in high-energy electron scattering experiments [41]. The long-ranged correlations come fron particle-core coupling and pairing correlations beyond that included within the valence space. Another reason may be the approximations made in the sudden approximation for the dynamics used for the reaction [39]. In the analysis of [40], the $R_s$ factor for loosely-bound nucleons that comes mainly from the long-ranged correlations is expected to be 0.6–0.7 rather than unity. The analysis of $(p, 2p)$ experiments [42] find $R_s$ values that depend less on the proton separation energy going from 0.6 to 0.7.

The $\sigma_{\text{th}}$ depends on the CI calculations for the spectroscopic factors. An approximation that is made in CI calculations is that only the change in configurations for the knocked out nucleon contributes to the spectroscopic factor. The radial wavefunctions for all other nucleons in the parent and daughter nuclei are assumed to be the same. However, consider, as an example, the knockout of a deeply bound proton from $^{30}$Ne to $^{29}$F. The size of the neutrons orbtials in $^{30}$Ne and $^{29}$F are changing due to the proximity to the continuum, and the overlap of the spectator neutrons in the nuclei with the atomic mass numbers $A$ and $A - 1$ will be reduced from unity. This effect should be contained in ab initio and continuum models [43,44], but an understanding within these models requires an explicit separation of the one-nucleon removal overlaps in terms of the removed nucleon within the basis states for $(A, Z)$ and the radial overlaps between the nuclei with $A$ and $A - 1$.

## 2. The Region of $^{28}$O

The oxygen isotopes provided the first complete testing ground for theory and experiment from the proton drip line to the neutron drip line [45]. The prediction by the "universal" *sd*-shell (USD) Hamiltonian [1,46], in the 1980s that $^{24}$O was a doubly-magic nucleus was later confirmed experimentally in 2009 [47–49].

For the one-neutron decay of $^{25}$O, the USD charge-dependent (USDC) Hamiltonian in the *sd* shell [19] gives $Q = 1.15(15)$ MeV, to be compared to the experimental value of $Q = 0.749(10)$ MeV [50]. The explicit addition of the continuum will lower the calculated energy [31]. The calculated value of the spectroscopic factor is $(25/24)^2\, \text{C}^2\text{S}(0d_{3/2}) = 1.01(1)$ (the error, shown in the parentheses for the value last decimal, comes from the comparison of the four *sd*-shell Hamiltonians developed in [19]). For the calculated decay width, one obtains $\Gamma = (25/24)^2\ \text{C}^2\text{S}\ \Gamma_{sp}(Q) = 75(1)$ keV. $\Gamma_{sp} = 74(1)$ keV is obtained using the experimental $Q$ value and a Woods–Saxon potential. The experimental neutron decay width is $\Gamma = 88(6)$ keV [50]. The theoretical error in the width is probably dominated by the uncertainty in the parameters of the Woods–Saxon potential.

The measured masses of the Na isotopes [51] found more binding near $N = 20$ than could be accounted for by the pure $\Delta = 0$ configurations; here, the notation $\Delta = n$ is used where $n$ is the number of neutrons excited from *sd* to $pf$. Hartree–Fock calculations [52]

showed that these mass anomalies were associated with a large prolate deformation, where the $2\Omega^{\pi}$ [N,$n_z$,$\Lambda$] = $1^-$ [3,3,0] and $3^-$ [3,2,1] Nilsson orbitals from the $fp$ shell cross the $1^+$ [2,0,0] and $3^+$ [2,0,2] orbitals from the *sd* shell near a value for the deformation paramater of $\beta = +0.3$. The anomaly was confirmed by $\Delta = 0$, CI calculations in [53,54], where in [53] it was called the "collapse of the conventional shell-model". CI calculations that included $\Delta = 2$ components [15,55] showed that nuclei in this region have ground-state wavefunctions dominated by the $\Delta = 2$ component. This is due to a weakened shell gap at $N = 20$ below $Z = 14$, pairing correlations in the $\Delta = 2$ configurations, and proton–neutron quadrupole correlations that give rise to the Nilsson orbital inversion. In [15], the region of nuclei below $^{34}$Si involved in this inversion was called the "island-of-inversion".

The Hamiltonian, used in [15], was appropriate for pure $\Delta = n$ configurations. This Hamiltonian was modified to account for more recent data related to the energies of $\Delta = 1$ and $\Delta = 2$ configurations resulting in the new Florida State University (FSU) Hamiltonian [56]. As examples of the type of predictions, results, obtained with the FSU Hamiltonian, are shown for $^{34}$Si in Figure 5, $^{32}$Mg in Figure 6, and $^{29}$F in Figure 7. All of these calculaitons were carried out with NuShellX [57] code and allowed only for neutron excitations from $1s$–$0d$ to $1p$–$0f$. Calculations in a full $n\hbar\omega$ basis ($\hbar$ being the reduced Planck constant) with $n > 0$ also require the addition of proton excitations from $0p$ to $1s$–$0d$ and proton exicitations from $1s$–$0d$ to $1p$–$0f$. In full $n\hbar\omega$ basis, the $1\hbar\omega$ spurious states can be removed with the Gloeckner-Lawson method [58]. Comparison to calculations in the full $n\hbar\omega$ basis with the Oxbash code [59] show that the energies are lowered relative to the $\Delta$ basis by up to approximately 200 keV. This shows the $\Delta = 1, 2$ proton and proton–neutron components are small compared to the $\Delta = 1, 2$ neutron components for the low-lying states in these neutron-rich nuclei. For nuclei with $N \approx Z$, removal of the spurious states in the $n\hbar\omega$ basis is important.

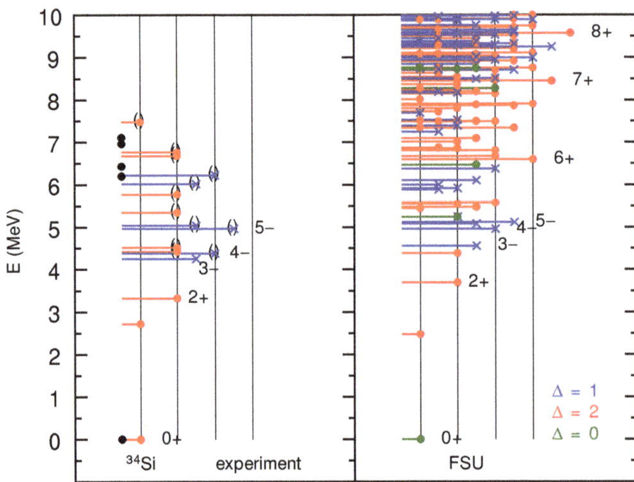


**Figure 5.** Spectrum of $^{34}$Si obtained with the Florida State University (FSU) Hamiltonian [56] compared to experiment. The length of the horizontal lines are proportional to the the angular momentum, $J$. The experimental parity is indicated by blue for negative parity and red for positive parity. Experimental spin-parity, $J^{\pi}$, values that are tentative are shown by "()", and those with multiple of no $J^{\pi}$ assignments are shown by the black points. The calculated results are obtained with the FSU Hamiltonian with pure $\Delta$ configurations. The parities are positive for $\Delta = 0$ (green) and $\Delta = 2$ (red) and negative for $\Delta = 1$ (blue).

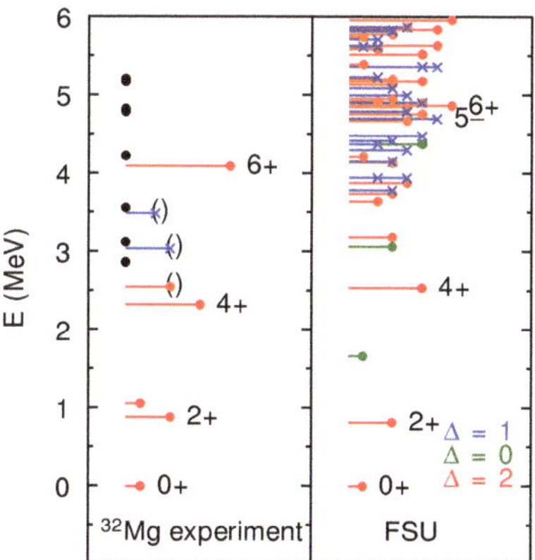


**Figure 6.** Spectrum of $^{32}$Mg obtained with the FSU Hamiltonian [56]. The results are obtained with pure $\Delta$ configurations. The spins are proportional to the length of the horizontal lines. The parities are positive for $\Delta = 0$ (green) and $\Delta = 2$ (red) and negative for $\Delta = 1$ (blue).

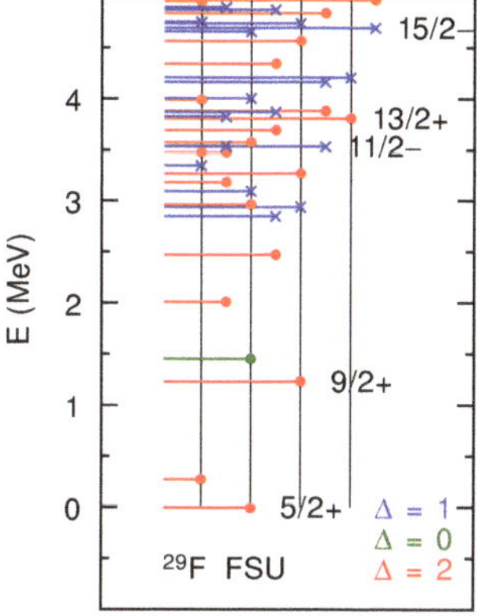


**Figure 7.** Spectrum of $^{29}$F obtained with the FSU Hamiltonian [56]. The results are obtained with pure $\Delta$ configurations. The spins are proportional to the length of the horizontal lines. The parities are positive for $\Delta = 0$ (green) and $\Delta = 2$ (red) and negative for $\Delta = 1$ (blue).

The barrier between the $\Delta = 0$ (spherical) and $\Delta = 2$ (deformed) configurations reduces the mixing between the lowest energy states of each configuration. When one combines the $\Delta = 0$ and $\Delta = 2$ configurations in CI calculations, the state that is dominated by $\Delta = 0$ is pushed down in energy by the mixing with many $\Delta = 2$ configurations mainly due to the increase in the pairing energy. If one were to start with the FSU Hamiltonian and add off-diagonal TBME of the type $< sd \mid V \mid fp >$, the components dominated by $\Delta = 0$ would be pushed down in energy due to this increase in pairing. However, this results in a double-counting since the *sd* part FSU interaction is already implicitly renormalized for the $fp$ admixtures. In addition, to achieve convergence in the mixed wavefunctions, one has to add $\Delta = 4$ and higher. This results in large matrix dimensions.

When one mixes the $\Delta$ components, one has to modify parts of the Hamiltonian that are diagonal in $\Delta$. This is sometimes performed by changing the pairing strength in the $J = 0$, $T = 1$ two-body matrix elements, so that the ground-state binding energies agree with experimental values. Hamiltonians that have been designed for mixed configurations are called SDPF-U-MIX [60] and SDPF-M [61,62]. Details about the modifications to SDPF-

U to obtain SDPF-U-MIX are given in the Appendix of [60]. In the remainder of this section, I discuss some examples, obtained with the FSU Hamiltonian with pure $\Delta$ configurations. This provides a starting point for more complete calculations with mixed $\Delta$ and those explicitly involving the continuum.

The $\Delta = 0$ (*sd*-shell) part of the FSU spectrum for $^{34}$Si (the green lines in Figure 5) has a simple interpretation. The ground state is dominated by the $(0d_{5/2})^6$ proton configuration. The 5.24 MeV $2^+$ and the 6.47 MeV $3^+$ states are dominated by the $(0d_{5/2})^5(1s_{1/2})^1$ proton configuration. In the two-proton transfer experiment from $^{36}$S [63], a $2^+$ state at 5.33 is observed that can be interpreted as two protons removed from $(0d_{5/2})^6(1s_{1/2})^2$ to make $(0d_{5/2})^5(1s_{1/2})^1$. The $(0d_{5/2})^4(1s_{1/2})^2$ 0+ state is predicted at 8.76 MeV. For the FSU Hamiltonian, all of these predictions are based on the USDB effective Hamiltonian [64]. The ESPE for the $0d_{5/2}$ and $1s_{1/2}$ proton states near $^{34}$Si are determined from the binding energies of $^{33}$Al, $^{34}$Si and $^{35}$P. Above 2.5 MeV the level density is dominated by the neutron $\Delta = 1$ and $\Delta = 2$ configurations. The $\Delta = 1$ states can be interpreted in terms of the low-lying $3/2^+$ and $1/2^+$ $1h$ states of $^{33}$Si coupled to the low-lying $7/2^-$ and $3/2^-$ $1p$ states of $^{35}$Si. The state with maximum spin-parity $J^\pi$ of $5^-$ predicted at 5.12 MeV can be compared to the proposed experimental $5^-$ state at 4.97 MeV [65]. The theoretical spectra from the mixed SDPF-U-Mix shown in [65] is similar to the FSU unmixed spectrum in Figure 5.

The FSU results for $^{32}$Mg are shown in Figure 6. Compared to $^{34}$Si, there is an inversion of the low-lying $\Delta = 0$ and $\Delta = 2$ configurations. For pure $\Delta$ configurations, the reduced electric-quadrupole transition strength $B(E2)$ for $2_1^+$ ($\Delta = 2$) to $0_2^+$ ($\Delta = 0$) is zero. Experimentally, $B(E2, 2_1^+ \rightarrow 0_2^+) = 48^{+75}_{-20}$ e$^2$ fm$^4$ compared to $B(E2, 2_1^+ \rightarrow 0_1^+) = 96(16)$ e$^2$ fm$^4$; see Table 1 in [66]. An improved half-life for the $0_2^+$ is important since it helps to determine the $\Delta$ mixing.

One of the key experiments for $^{32}$Mg is the two-neutron transfer from $^{30}$Mg (t,p), where the first two $0^+$ states were observed with approximately equal strength [67]. This observatiom has proven difficult to understand; see the references in [68]. Starting from a $\Delta = 0$ configuration for the $^{30}$Mg ground state, one can populate the $\Delta = 0, 0^+$ configuration in $^{32}$Mg by $(sd)^2$ transfer and the $\Delta = 2, 0^+$ configuration by $(fp)^2$ transfer. Macchiavelli et al. [68] analyzed the (t,p) cross sections by used centroid energies for the $\Delta$ = 0,2,4 configurations of 1.4, 0.2 and 0.0 MeV, respectively, obtained with the SDPF-U-MIX Hamiltonian [60]. This three-level model could account for the experimental observation with a ground state that is 4% $\Delta = 0$, 46% $\Delta = 2$ and 40% $\Delta = 4$ together with a ground-state wavefunction for $^{30}$Mg that has 97% $\Delta = 0$ and 3% $\Delta = 2$. In this three-level model for $^{32}$Mg, the main part of the $\Delta = 0$ configuration is in the $0_3^+$ state predicted to be near 2.2 MeV; see Table 1 in [68].

Two-proton knockout from $^{34}$Si provides more information. Starting with a pure $\Delta = 0$ configuration for the $^{34}$Si ground state, only $\Delta = 0, 0^+$ configurations in $^{32}$Mg can be made. In the two-proton knockout experiment of [69,70], strong $0^+$ strength is observed in the sum of the first two $0^+$ states; see Figure 9 in [70]. The strength to the $0_1^+$ and $0_2^+$ states cannot be separated due to the long lifetime of the $0_2^+$ state. Significant strength to $0^+$ states above 1.5 MeV was not observed, in contradiction to that predicted in the three-level model above [68] or the SDPF-M model. More needs to be done to understand the structure of $^{32}$Mg and how it connects to the experimental data discussed above.

Results from the FSU Hamiltonian provide an extrapolation down to $^{28}$O. $^{29}$F has been called a "lighthouse on the island-of-inversion" [71]. The FSU results for $^{29}$F are shown in Figure 7. The lowest state is $5/2^+$ with a $\Delta = 2$ configuration. The lowest $1/2^+, 3/2^+$, $7/2^+$ and $9/2^+$ $\Delta = 2$ states are dominated by the configuration with $0d_{5/2}$ coupled to the $\Delta = 2, 2^+$ state in $^{28}$O at 1.26 MeV. The $0d_{5/2}$ coupled to $2^+, 5/2^+$ configuration is spread over many higher $5/2^+$ states in $^{29}$F. The $\Delta = 3$ states for $^{29}$F start at 3.9 MeV. An excited state in $^{29}$F at 1.080(18) MeV [72] made from proton knockout from $^{30}$Ne was suggested to be $1/2^+$ on the basis comparisons to the SDPF-M calculations shown in [72].

With the FSU Hamiltonian, for $^{27}$F, the lowest $\Delta$ = 0, $5/2^{+}$ state is 1.9 MeV below the $\Delta$ = 2, $5/2^{+}$ state. The large FSU occupancy of 1.38 in $^{29}$F for the loosely bound $0p_{3/2}$ orbital may explain the observed neutron halo [73]. In particular, the two-neutron transfer amplitudes TNA[($0p_{3/2}$)]=0.62 for the $^{29}$F, $\Delta$ = 2, $5/2^{+}$ ground state going to the $^{27}$F, $\Delta$ = 0, $5/2^{+}$ ground state. Improved mass mesurements are needed for the neutron-rich fluorine and neon isotopes.

Results for these calculations depend on the ESPE extrapolation down to $^{28}$O contained in the FSU interaction. The ESPE for the neutron orbitals as a function of $Z$ obtained with the FSU Hamiltonian with ($\Delta$ = 0) are shown in Figure 8 (for $^{34}$Si I assume a $(0d_{5/2})^{6}$ configuration for the protons). These are compared with the results from the Skyrme-x energy density functional (Skx EDF) calculations [74].

For unbound states, the energies can be approximated by first increasing the EDF central potential to obtain a wavefunction bound by, for example, 0.2 MeV, and then taking the expectation value of the wavefunction value with original EDF Hamiltonian. This method provides a practical approximation to the centroid energy. Results for the unbound resonances could be calculated more exactly from neutron scattering on the EDF potential.

The results in Figure 8 show that the $N = 20$ shell gap decreases from approximately 7.0 MeV in $^{34}$Si to approximately 2.7 MeV in $^{28}$O. The major part of this decrease is due to the lowering energy for $1p_{3/2}$ relative to $0f_{7/2}$ as the states become more unbound. The energies for these two states cross at approximately $Z = 10$. Recent experimental information on the ESPE near $^{28}$Mg and their interpretation similar to those of Figure 8 with a Woods–Saxon potential is given in [75]. For the FSU Hamiltonian, the loose binding effects are implicitly built into the monopole components of the TBME from the SVD fit to data on the BE and excitations energies.

There is also an increase in the gap in $^{34}$Si due to the proton-neutron tensor interaction contribution to the spin–orbit splitting [5] that is built into the FSU Hamiltonian. The spin–orbit tensor interacton is zero in the double-$LS$ closed shell nuclei such as $^{28}$O and $^{40}$Ca. The tensor interaction is important for changing the effective single–particle energies as a function of proton and/or neutron number [5] or as a function of the state-dependent orbital occupancies [76].

The $fp$ ESPE obtained from the Skx EDF [74] from $^{30}$Ne to $^{78}$Ni are shown in Figure 9. The energies of $1p$ and $0f$ systematically shift due to the finite-well potential.

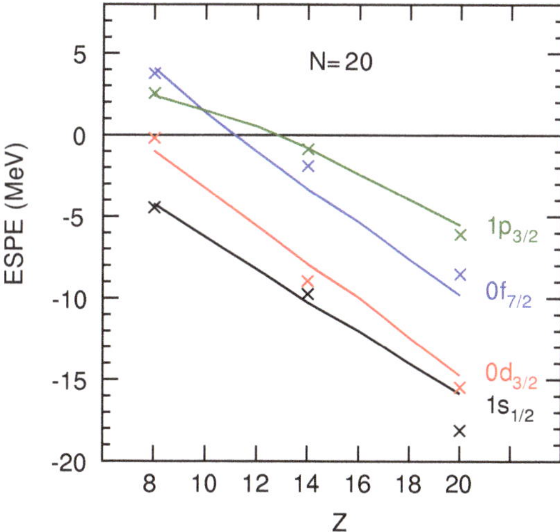


**Figure 8.** Effective single-particle energies (ESPE) for neutron orbitals as a function of proton number. The lines are from the Skyrme-x energy-density functional (Skx EDF) [74] calculations. The crosses are from the FSU [56] Hamiltonian calculations.

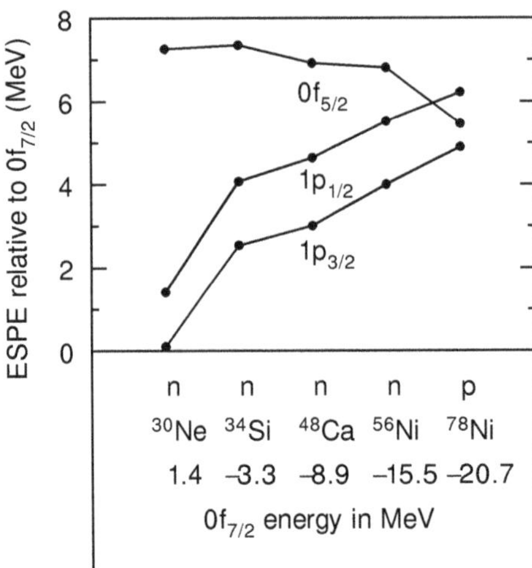


**Figure 9.** ESPE for the $fp$ proton (p) and neutron (n) orbitals obtained from the Skx EDF interaction [74] for a range of nuclei.

For nuclei near the neutron drip line, there are few bound states that can be studied by their gamma decay. States above the neutron separation energy neutron decay. These neutron decays can be complex both experimentally and theoretically. The neutron decay spectrum depends upon how the unbound states are populated. They are often made by proton and neutron knockout reactions. For one- and two-nucleon knockout, one can calculate spectroscopic factors that can be combined with a reaction model to find which states are most strongly populated. A recent example of this type of calculation was for two-proton knockout from $^{33}$Mg going to $^{31}$Ne [77]. One neutron decay can often go to excited states in the daughter [77]. Additionally, multi-neutron decay can occur. It is important to measure the neutrons in coincidence with the final nucleus and its gamma decays. On the theoretical side, one must use the calculated wavefunctions to obtain neutron decay spectra.

An example of multi-neutron decay is in the one-proton knockout from $^{25}$F to make $^{24}$O [78,79]. The calculated one-proton knockout spectroscopic factors showed that $0d_{5/2}$ knockout mainly leads to the ground state of $^{24}$O, and that $0p$ knockout leads to many negative-parity states above the neutron separation energy of $^{24}$O. These excited states multi-neutron decay to $^{21-23}$O [78]. However, in the (p,2p) reaction [79], it was suggested from the momentum-distribution of $^{23}$O that a low-lying positive-parity excited state in $^{24}$O above the neutron separation energy was strongly populated by $0d$ removal, in strong disagreement with the calculations of [78]. This experimental result should be confirmed.

The two-neutron decay of $^{26}$O has a remarkably small $Q$ value of 0.018(5) MeV [50]. The theoretical $Q$ value from USDC Hamiltonian [19] is 0.02(15) MeV. The decay width depends strongly on the $\ell$ for the $\ell^2$ two-nucleon decay amplitude. From Figure 2b of [80], pure $\ell^2$ two-nucleon decays widths with the experimental $Q$ value are approximately $10^{-4}$, $10^{-8}$ and $10^{-14}$ MeV for $\ell$ = 0, 1 and 2, respectively. The calculated TNA in the $sd$ model space with the USDC Hamiltonian are 0.99 for $(0d_{3/2})^2$ and 0.16 for $(1s_{1/2})^2$. Thus, $\Gamma = [\text{TNA}(1s_{1/2})^2]^2\,\Gamma_{sp}(Q) \approx 0.003$ keV. The $(1p_{3/2})^2$ TNA will be on the order of TMBE $< (0d_{3/2})^2 \mid V \mid (1p_{3/2})^2 > /2\Delta E$, where $\Delta E$ is the energy difference between the the $1p_{3/2}$ and $0d_{d3/2}$ states in $^{25}$O. With typical values of TMBE $< (0d_{3/2})^2 \mid V \mid (1p_{3/2})^2 >\approx 2$ MeV and $\Delta E \approx 2$ MeV [81] giving TNA = 0.5, the $(1p_{3/2})^2$ contribution to the two-neutron decay width will be small.

The nucleus $^{28}$O is unbound to four neutron decay. The theoretical understanding of this complex decay involves the four-body continuum [80]. These continuum calculations strongly depend upon the single-particle states involved; see Figure 2d in [80]. With the FSU Hamiltonian, the $\Delta = 2$ configuration for $^{28}$O lies 0.8 MeV below the $\Delta = 0$ (closed-shell) configuration due to the pairing correlations. The calculated four-neutron decay energy is 1.5 MeV. The energy should be lowered by an explicit treatment of the many-body continuum. Thus, the "island of inversion" may be a "peninsula of inversion" extending from $^{32}$Mg all the way to the neutron drip line; below, I discuss what may be the first true "island of inversion" between $^{60}$Ca and $^{78}$Ni. There are many paths for the four-neutron decay of $^{28}$O. For example, in the FSU $\Delta = 2$ model, it may proceed by a relatively fast $(1p_{3/2})^2$ decay to the $^{26}$O ground state followed by its decay to $^{24}$O.

## 3. The Region of $^{42}$Si

In this Section, results for two often used effective Hamiltonians for this model space, called SDPF-MU [82] and SDFP-U-SI [83], together with those based on the IM-SRG method [20] are compared. The MU and U-SI Hamiltonians are "universal" in the sense that a single Hamiltonian with a smooth mass-dependence is applied to a wide mass region. MU is used for all nuclei in this model space, while U-SI was designed for $Z \leq 14$ (the SDPF-U version was designed for $Z > 14$ [83]).

The $2^+$ energy in $^{42}$Si $[Z,N] = [14,28]$ (0.74 MeV) is low compared to those in $^{34}$Si [14,20] (3.33 MeV) and $^{48}$Ca [20,28] (3.83 MeV). $^{34}$Si and $^{48}$Ca are doubly magic due to the *LS* magic number 20. $^{28}$Si [14,14] has a known intrinsic oblate deformation [84].

The $2^+$ energy in $^{20}$C [6,14] (1.62 MeV) is low compared to those in $^{14}$C [6,8] (7.01 MeV) and $^{22}$O [8,14] (3.20 MeV). $^{14}$C and $^{22}$O are doubly magic due to the *LS* magic number 8. Hartree–Fock calculations [85] as well as CI calculations for the *Q* moment within the $p - sd$ model space [86] show that $^{12}$C and $^{20}$C have intrinsic oblate shapes.

The oblate shapes for $^{28}$Si and $^{42}$Si are shown by their *E*2 maps in Figures 10 and 11. The transition from spherical to oblate shapes for the *jj* doubly-magic numbers can be qualitatively understood in the Nilsson diagram as shown, for example, for $^{42}$Si in Figure 12. The highest filled Nilsson orbitals have rather flat energies between $\beta = 0$ and $\beta = -0.3$. The important aspect is the concave bend of the $2\Omega^{\pi}$ [N,n$_z$,$\Lambda$] = $1^+$ [2,2,0] proton and $1^-$ [3,3,0] neutron Nilsson orbitals for oblate shapes. For the heavier *jj* doubly-magic nuclei, $\ell$ increases and the $j = \ell + 1/2$ orbital decreases in energy, the bend will not be so large and the energy minima come closer to $\beta = 0$. This is illustrated in Figure 10. In Figure 10a and Figure 10c, the 0*d* spin–orbit gap is small enough to give an oblate rotational pattern. The oblate shape is manifest in the positive *Q* moments. In Figure 10a, the 0*d* spin–orbit gap is increased by 1 MeV and the rotational energy pattern is broken. The pattern in Figure 10a is similar to that obtained for $^{56}$Ni in the $fp$ model space as shown in Figure 13. An interesting feature for $^{56}$Ni is the relatively strong $0^+_2$ to $2^+_1$ $B(E2)$. I am not aware of a simple explanation for this.

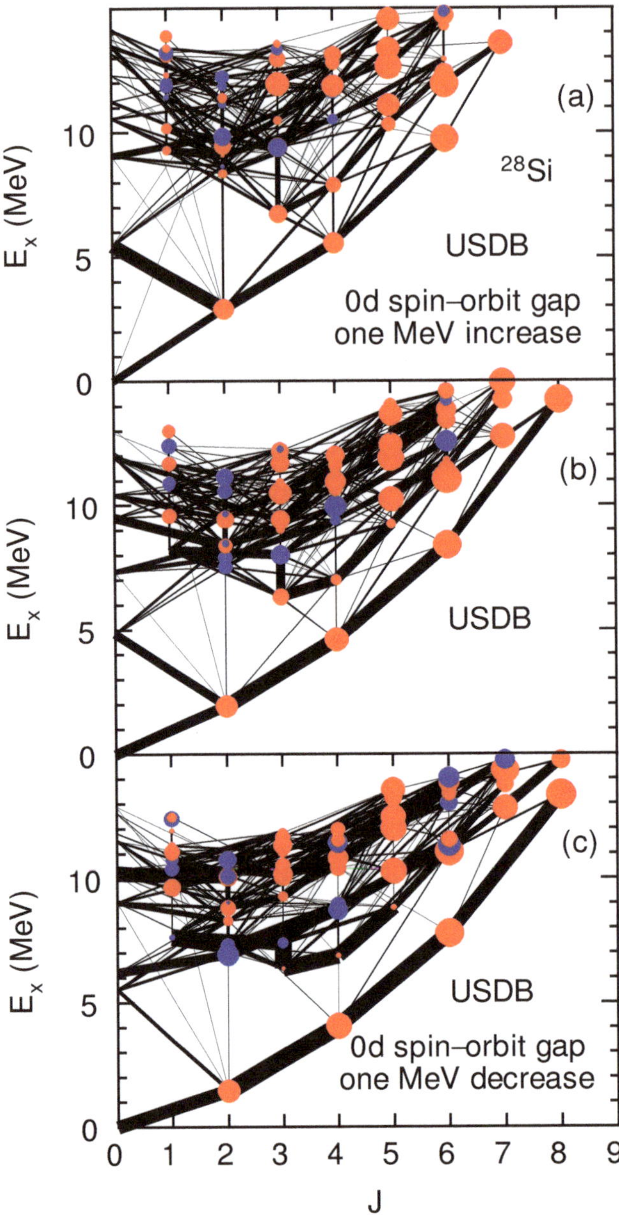


**Figure 10.** Electric quadrupole ($E2$) maps for $^{28}$Si. The results shown are based on the universal $sd$-shell version-B (USDB) Hamiltonian with (**a**) the $0d$ spin–orbit energy gap increased by 1 MeV, (**b**) in the $sd$ model space, and (**c**) with the $0d$ spin–orbit energy gap decreased by 1 MeV. For each $J$ value, ten states were calculated. The widths of the lines are proportional to the reduced electric quadrupole transition strength, $B(E2)$. Lines for $B(E2)$ less than 5% of the largest value are not shown. The radius of the circles are proportional to spectroscopic quadrupole moment, $Q_s$ (2). To set the scale, for panel (**b**) the $2_1^+$ to $0_1^+$ $B(E2) = 82$ e$^2$ fm$^4$ and $Q_s$ $(2_1^+) = +19$ e fm are used.

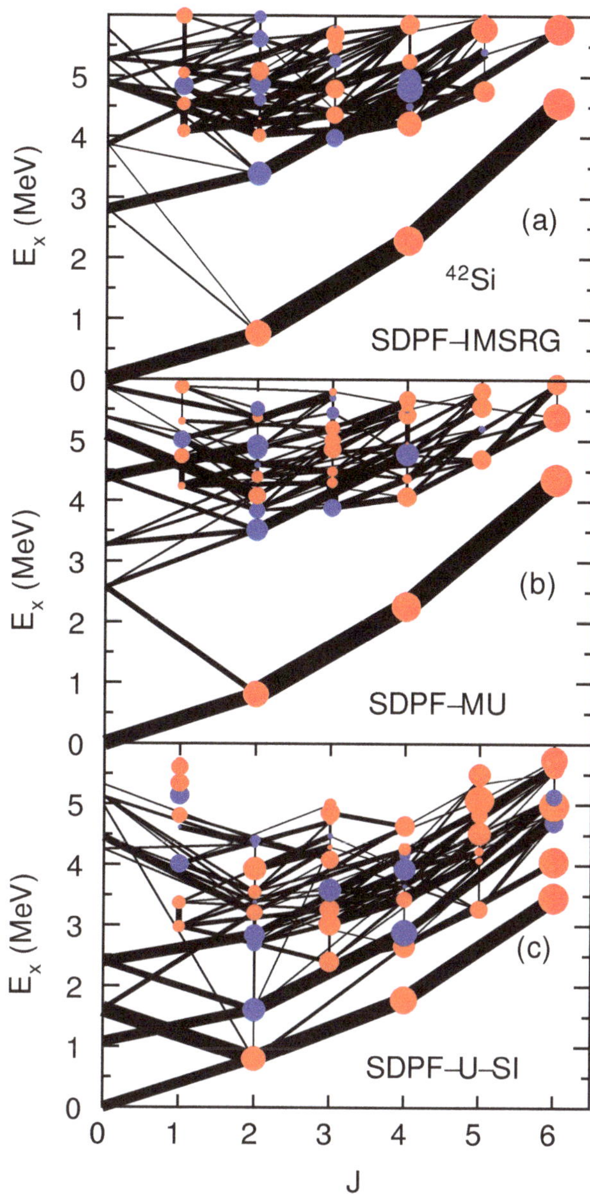


**Figure 11.** *E*2 maps for $^{42}$Si obtained with the SDPF-IMSRG [20] (**a**), SDPF-MU [82] (**b**), and SDPF-U-SI [83] (**c**) Hamiltonians. See Figure 10 and text for details.

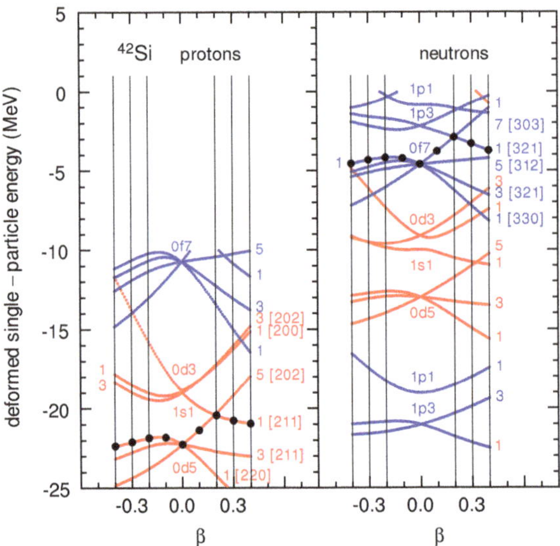


**Figure 12.** Nilsson diagram for $^{42}$Si. At the deformation parameter $\beta = 0$, the orbitals are labeled by $n_r, \ell, 2j$ (see text for definitions), and, at larger deformation, the orbitals are labeled by the Nilsson quantum numbers 2Ω [N,n$_z$,Λ]. The negative and positive parities is shown by the blue and red lines, respectively. The black dots show the highest Nilsson states occupied. This figure is made using the code WSBETA [87] with the potential choice ICHOIC = 3. The spin–orbit potential are reduced here for protons to make the spherical energies for the $0d_{3/2}$ and $1s_{1/2}$ orbitals approximately the same.

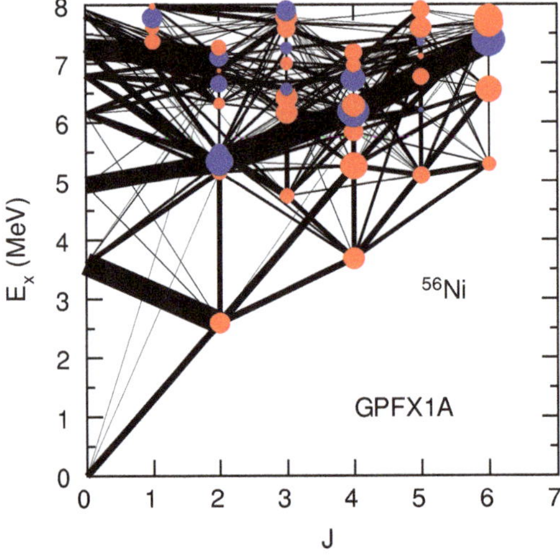


**Figure 13.** *E*2 map for $^{56}$Ni obtained with the GXFP1A Hamiltonian [88,89] in the full *fp* model space. See Figure 10 and text for details.

The oblate bands in $^{28}$Si and $^{42}$Si are linked to the $0d_{5/2}$ and $0f_{7/2}$ orbitals. For completeness, the *E*2 maps for $^{12}$C and $^{20}$C obtained with the WBP Hamiltonian [90] are shown in Figure 14. For these nuclei, the oblate ground-state bands are linked with the $0p_{3/2}$ and $0d_{5/2}$ orbitals.

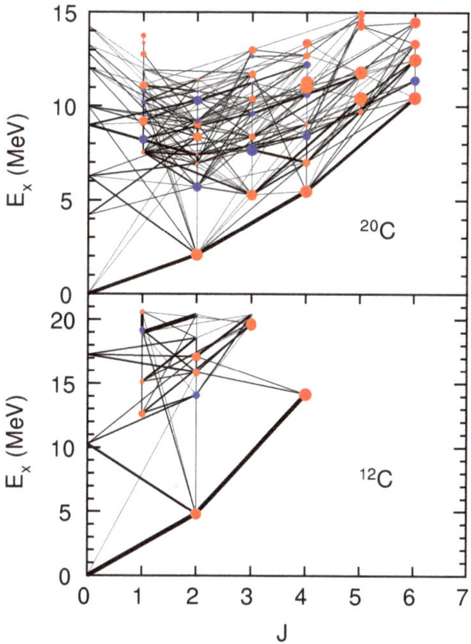


**Figure 14.** $E2$ maps for $^{12}$C (**bottom**) and $^{20}$C (**top**), obtained with the WBP Hamiltonian [90]. See Figure 10 and text for details.

For CI calculations, the $B(E2)$ values depend on the effective charge parameters $e_p$ and $e_n$. In the harmonic-oscillator basis, the $E2$ operator connects states within a major shell as well as those that change $N_o$ by two. The $E2$ strength function contains low-lying $\Delta N_o = 0$ strength as well "giant-quadrupole" strength near an energy of $2\hbar\omega$. The effective charges account for the renormalization of the proton and neutron components of the $E2$ matrix elements within the CI basis of a major shell due to admixtures of the $1p$–$1h$, $\Delta N_o = 2$ proton configurations. For the calculations, shown here, effective charges, which depend on the model space, are used. The effective charges are chosen to best reproduce observed $B(E2)$ values and quadrupole moments within that model space. These are the *sd* model space with $e_p = 0.45$ and $e_n = 0.36$ [91], the $fp$ model space with $e_p = e_n = 0.50$ [88] and the neutron-rich $sd - pf$ model space with $e_p = e_n = 0.35$ [82]. Since low-lying excitations in nuclei are mostly isoscalar, only $e_p + e_n$ is well determined. It takes special situations such as a comparison of B(E2) in mirror nuclei [92] to obtain the isovector combination $e_p - e_n$.

The isoscalar effective charge decreases for more neutron-rich nuclei (e.g., the drop from 0.5 in the $fp$ model space to 0.35 in the *sd* model space). This can be understood by the macroscopic model of Mottelson [93], by the microscopic Hartree–Fock calculations of Sagawa et al. [85], and by the microscopic models, discussed in [94,95]. Microscopic models also give an orbital dependence to the effective charge. A recent example of this is for the relatively small B(E2) value for the the $1/2^+$ to $5/2^+$ transition in $^{21}$O [96]. This transition is dominated by the $1s_{1/2}$–$0d_{5/2}$ $E2$ matrix element, and the relatively small neutron effective charge is due to the node in the $1s_{1/2}$ wavefunction.

The results for CI calculations for $^{42}$Si are shown in Figure 11 for three Hamiltonians. The IMSRG Hamiltonian is based on a VS-IMSRG calculation [20] similar to that used in [12].

The interpretation of the spectroscopic quadrupole moments, $Q_s$, shown in Figure 11 in terms of an intrinsic shape, $Q_o$, is given by the rotational formula [97],

$$Q_s = \frac{2K^2 - J(J+1)}{(J+1)(2J+3)} Q_o \, \text{e}, \tag{2}$$

with the Nilsson quantum number $K = 0$ for the ground-state bands in even–even nuclei. The MU [82] and IMSRG [20] calculations show an intrinsic oblate ground-state band, ($Q_s > 0$ and $Q_o < 0$), followed by a large energy gap to other more complex states. The U-SI Hamiltonian [83] also gives an oblate ground-state band, but there is also an intrinsic prolate band at relatively low energy. The presence of this low-lying prolate band dramatically increases the level density below 4 MeV [98,99].

The Nilsson diagram in Figure 12 shows a higher-energy prolate minimum related to a crossing of the $1^-$ [3,2,1] and $7^-$ [3,0,3] Nilsson orbitals near $\beta = +0.3$. At present, there is not enough experimental information to determine the energy of the prolate band in $^{42}$Si. The structure of $^{42}$Si is a touchstone for understanding all of the nuclei near the drip line in this mass region. More complete experimental results for the energy levels of $^{42}$Si are needed. The low-lying structure of $^{42}$Si depends on the details of the neutron ESPE that are affected by the continuum for the $0p$ orbitals. The deformed neutron ESPE need to be established by one-neutron transfer reactions on $^{42}$Si.

Deformation for $N = 28$ as a function of $Z$ is determined by how the proton Nilsson orbitals are filled in Figure 12. When six protons are added to make $^{48}$Ca with $Z = 20$, there is a sharp energy minimum for protons at $\beta = 0$, and thus $^{48}$Ca is doubly magic. For $^{44}$S, the protons have a intrinsic prolate minimum near $\beta = +0.2$ where the neutrons are near the crossing of the $2\Omega^{\pi} = 1^-$ and $7^-$ orbitals [100]. In $^{44}$S, a $K = 4^+$ isomer at 2.27 MeV coming from the two quasi-particle state made from these two neutron orbitals was observed [101]. In $^{43}$S rotational bands associated with these, two $\Omega$ states have been observed [102]. All of these features are reproduced by CI calculations based on the SDPF-MU [82] and SDPF-U [83] Hamiltonians. At higher excitation energy, the CI energy spectra are more complex than anything that could be easily understood by the collective model.

The $E2$ map obtained with the SDPF-MU Hamiltonian for $^{40}$Mg is shown in Figure 15. In this case, the ground-state band has an intrinsic prolate shape. In the nuclear chart, prolate shapes are most common [103], in contrast to the oblate shapes obtained for $jj$ magic numbers discussed above. The oblate shape for $^{40}$Mg can be understood in the Nilsson diagram of Figure 12. When two protons are removed, the energy minimum for protons shifts to positive $\beta$ in the $3^-$ [2,1,1] orbital. The experimental energy of the first $2^+$ is 500(14) keV [9] compared to the result of 718 keV obtained with the SDPF-MU Hamiltonian. Models that explicitly include the $\ell = 1$ levels in the continuum are needed.

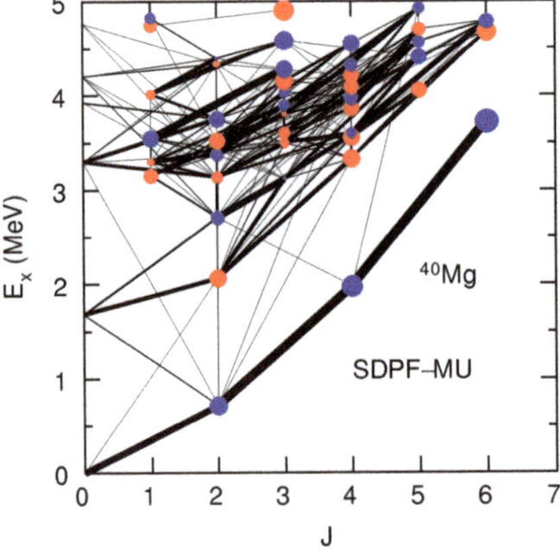


**Figure 15.** $E2$ map for $^{40}$Mg obtained with the SDFPF-MU Hamiltonian.

## 4. The Region of $^{60}$Ca

Many Hamiltonians have been developed for the calcium isotopes for the $fp$ model space. Near $^{42}$Ca, it is known that $\Delta = 2$ $sd$–$pf$ proton excitations are necessary for the low-lying intruder states and their mixing with the $fp$ configurations which greatly increase the B(E2) values compared to those obtained in the $fp$ model space [95]. In the doubly-magic nucleus $^{48}$Ca, the $sd - pf$ intruder states start with the $0^+$ state at 4.28 MeV [104]. The $^{48-55}$Ca nuclei exhibit low-lying spectra which are dominated by $fp$ configurations [12]. There are weak magic numbers at $N = 32$ and 34 as shown in Figure 2. The reason for the low value of the pairing for the $1p_{1/2}$ at $N = 33$ was discussed in [104].

The KB3G [105] and GXPF1A [88,89] Hamiltonians have provided predictions for the spectra in this region which have been a source of comparison for many experiments over the last 20 years. Both of these are "universal" Hamiltonians for the $pf$ model space. Recently, it has been shown that a data-driven Hamiltonian for the calcium isotopes improves the description of all of the known data [12]; this is called the UFP-CA (universal $fp$ for calcium) Hamiltonian. All of the known energy data for $N \geq 28$ can be described by an SVD-derived Hamiltonian that is close to the starting IMSRG Hamiltonian for $^{48}$Ca. UFP-CA is able to describe the energy data for $N \geq 28$ with an rms error of 120 keV. In particular, the calculated $D(N)$ values, shown by the red line in Figure 2, agree extremely well with the data (the black points).

The UFP-CA Hamiltonian does not explicitly involve the $2s$–$1d$–$0g$ orbitals, but the influence of these orbitals are present in their contributions to the renormalization into the $fp$ model space. This renormalization is contained microscopically in the IMSRG starting point, as well as empirically in the SVD fit.

The success of UFA-CA is similar to the success of the USD-type Hamiltonians in the $sd$ model space for all nuclei except those in the island of inversion. If the UFP-CA predictions for $^{55-59}$Ca turn out to be in relatively good agreement with experiment, the implication is that $^{60}$Ca will be a doubly-magic nucleus similar to that of $^{68}$Ni [12]. If that is the case, $^{60}$Ca will be the last doubly-magic nucleus to be discovered. In [12], EDF models are used to estimate the $0f_{5/2}$ $0g_{9/2}$ shell gap at $N = 40$ to be approximately 3.0 MeV. The implication of this for $D(N)$ is shown by the red dashed line in Figure 2. The $0g_{9/2}$ orbital will first appear as intruder states in the low-lying spectra of $^{55-60}$Ca. These nuclei can be reached by proton knockout on the scandium and titatium isotopes. The proton knockout will be dominated by $0f_{7/2}$ removal to the low-lying $fp$ neutron configurations. An example of this is the population of the ground state of $^{54}$Ca from $^{55}$Sc [106]. Protons will also be removed from the $1s_{1/2}$ and $0d_{3/2}$ orbitals to populate states at higher energy such as the negative parity state in $^{54}$Ca. These will mix with the $2s$–$1d$–$0g$ configurations and neutron decay to the lighter calcium isotopes. For example, in $^{57,59}$Ca, a $9/2^+$ ($0g_{9/2}$) state just above the neutron separation energy $S_n$ value would neutron decay to the ($0^+, 2^+, 4^+$) multiplet predicted in $^{56,58}$Ca; see Figure 1 in [12]. Calculations that include proton excited from $sd$ to $pf$ and neutrons excited from $pf$ to $sdg$ will be needed to understand the neutron dacays of these states.

The position of the $0g_{9/2}$ orbital is crucial for the structure of nuclei around $^{60}$Ca [107]. Lenzi et al. [108] have extrapolated the neutron effective single-particle energies from $Z = 28$ down to $Z = 20$ based on their LNPS Hamiltonian. Their $0f_{5/2}$-$0g_{9/2}$ ESPE gap for $^{60}$Ca is close to zero (see Figure 1 in [108]) and the structure of $^{60}$Ca is dominated by $\Delta = 4$ ($fp$ to $sdg$) configurations (see Table 1 in [108]). With LNPS, $^{60}$Ca is very different from $^{68}$Ni which is dominated by the closed $fp$-shell configuration ($\Delta = 0$). Below $^{68}$Ni, the nuclei $^{66-70}$Fe, [11] $^{64-66}$Cr [11] and $^{62}$Ti [10], have deformed spectra coming from $fp - sdf$ island of inversion for $N = 40$. Calculations with the LNPS Hamiltonian [108] show that these are all dominated by $\Delta = 4$. The $N = 40$ island of inversion is the topic of another contribution to this series of papers [109].

The existence of $^{60}$Ca, confirmed only recently, agrees with UFP-CA as well as with most of the other predictions [36]. It will be exciting to have more complete experimental data for nuclei around $^{60}$Ca from FRIB and other radioactive-beam facilities.

## 5. The Region of $^{78}$Ni

$^{78}$Ni has recently been established as a double-$jj$ magic nucleus from the relatively high energy of 2.6 MeV for the $2^+_1$ state [35]. More detailed magic properties can be obtained from the $D(N)$ and $D(Z)$, derived from new experiments on the masses around $^{78}$Ni. The ESPE can be established from the masses together with the low-lying spectra of $^{77}$Ni, $^{79}$Ni, $^{77}$Co and $^{79}$Cu. A proton knockout experiment from $^{80}$Zn has recently been used to establish excitation energies of low-lying states in $^{79}$Cu [35] In particular, the ground state and two lowest-lying states are likely associated with the triplet of states shown in Figure 9. In comparison with the extrapolations of CI calculations, shown in [35], the order is likely to be $0f_{5/2}$, $1p_{3/2}$ and $1p_{1/2}$. The single-particle nature of low-lying states around $^{78}$Ni will require one-nucleon transfer experiments.

The position of the proton $0g_{9/2}$ orbital above $^{78}$Ni is important for Gamow–Teller strength in the electron-capture rates for core-collapse supernovae similations [110,111]. The filling of the $0g_{9/2}$ orbital leads to $^{100}$Sn on the proton drip line. $^{100}$Sn has the largest calculated reduced Gamow-Teller transition probability, $B(GT)$, value (see Table A1 in [112]) due to nearly filled $0g_{9/2}$ orbital decaying into the nearly empty $0g_{7/2}$ orbital. The understanding of $^{100}$Sn [113] and other nuclei near the proton drip line in this mass region will be improved by radioactive-beam experiments.

As shown in Figure 4b of [35], large-scale CI calculations predict a deformed band with $\beta \approx +0.3$ at approximately 2.6 MeV. $^{56}$Ni is also spherical with a $2^+_1$ state observed at 2.7 MeV. For $^{56}$Ni, the deformed band is predicted to start at 5.0 MeV as shown in Figure 13. The relatively low-lying deformed band in $^{78}$Ni is predicted to lead to a "5th island-of-inversion" in $^{76}$Fe and other nuclei with $N = 50$ below $Z = 28$ [114].

## 6. Conclusions

I have discussed the new physics related to the properties of nuclei near the drip lines that will be studied by the next generation of rare-isotope beam experiments. In particular, I have focused on four "outposts" for the regions of $^{28}$O, $^{42}$Si, $^{60}$Ca and $^{78}$Ni, where new experiments will have the greatest impact on understanding the evolution of nuclear struture as one approaches the neutron drip line.

**Funding:** This research was funded by the National Science Foundation under Grant PHY-2110365.

**Acknowledgments:** I thank Ragnar Stroberg for providing the VS-IMSRG Hamiltonian for $^{42}$Si.

**Conflicts of Interest:** The author declares no conflict of interest.

## References

1. Brown, B.A.; Wildenthal, B.H. Status of the nuclear shell model. *Ann. Rev. Nucl. Part. Sci.* **1988**, *38*, 29–66. [CrossRef]
2. Brown, B.A. The nuclear shell model towards the drip lines. *Prog. Part. Nucl. Phys.* **2001**, *47*, 517–599. [CrossRef]
3. Caurier, E.; Martinez-Pinedo, G.; Nowacki, F.; Poves, A.; Zuker, A. The shell model as a unified view of nuclear structure. *Rev. Mod. Phys.* **2005**, *77*, 427–488. [CrossRef]
4. Stroberg, S.R.; Hergert, H.; Bogner, S.K.; Holt, J.D. Nonempirical interactions for the nuclear shell model: An update. *Ann. Rev. Nucl. Part. Sci.* **2019**, *69*, 307–362. [CrossRef]
5. Otsuka, T.; Gade, A.; Sorlin, O.; Suzuki, T.; Utsuno, Y. Evolution of shell structure in exotic nuclei. *Rev. Mod. Phys.* **2020**, *92*, 015002. [CrossRef]
6. Brown, B.A. Nuclear pairing gap: How low can it go? *Phys. Rev. Lett.* **2013**, *111*, 162502. [CrossRef]
7. Audi, G. The history of nuclidic masses and of their evaluation. *Int. J. Mass Spec.* **2006**, *251*, 85–94. [CrossRef]
8. Pritychenko, B.; Birch, M.; Singh, B.; Horoi, M. Tables of E2 transition probabilities from the first $2^+$ states in even-even nuclei. *Atom. Data Nucl. Data Tables* **2016**, *107*, 1–139. [CrossRef]
9. Crawford, H.L.; Fallon, P.; Macchiavelli, A.O.; Doornenbal, P.; Aoi, N.; Browne, F.; Campbell, C.M.; Chen, S.; Clark, R.M.; Cortes, M.L.; et al. First spectroscopy of the near drip-line nucleus $^{40}$Mg. *Phys. Rev. Lett.* **2019**, *122*, 052501. [CrossRef]
10. Cortes, M.L.; Rodriguez, W.; Doornenbal, P.; Obertelli, A.; Holt, J.D.; Lenzi, S.M.; Menendez, J.; Nowacki, F.; Ogata, K.; Poves, A.; et al. Shell evolution of $N = 40$ isotones towards $^{60}$Ca: First spectroscopy of $^{62}$Ti. *Phys. Lett. B* **2020**, *800*, 135071. [CrossRef]
11. Santamaria, C.; Louchart, C.; Obertelli, A.; Werner, V.; Doornenbal, P.; Nowacki, F.; Authelet, G.; Baba, H.; Calvet, D.; Chateau, F.; et al. Extension of the $N = 40$ island of inversion towards $N = 50$: Spectroscopy of $^{66}$Cr, $^{70}$Fe, $^{72}$Fe. *Phys. Rev. Lett.* **2015**, *115*, 192501. [CrossRef] [PubMed]

12. Magilligan, A.; Brown, B.A.; Stroberg, S.R. Data-driven configuration-interaction Hamiltonian extrapolation to $^{60}$Ca. *Phys. Rev. C* **2021**, *104*, L051302. [CrossRef]
13. Mass Explorer. Available online: http://massexplorer.frib.msu.edu/ (accessed on 10 April 2022).
14. Morris, T.D.; Simonis, J.; Stroberg, S.R.; Stumpf, C.; Hagen, G.; Holt, J.D.; Jansen, G.R.; Papenbrock, T.; Roth, R.; Schwenk, A. Structure of the lightest tin isotopes. *Phys. Rev. Lett.* **2018**, *120*, 152503. [CrossRef] [PubMed]
15. Warburton, E.K.; Becker, J.A.; Brown, B.A. Mass systematics for $A = 29$–44 nuclei: The deformed $A \sim 32$ region. *Phys. Rev. C* **1990**, *41*, 1147–1166. [CrossRef]
16. Serduke, F.J.D.; Lawson, R.D.; Gloeckner, D.H. Shell-model study of $N = 49$ isotones. *Nucl. Phys. A* **1976**, *256*, 45–86. [CrossRef]
17. Ji, X.D.; Wildenthal, B.H. Shell-model calculations for the energy-levels of the $N = 50$ isotones with $A = 80$–87. *Phys. Rev. C* **1989**, *40*, 389–398. [CrossRef]
18. Lisetskiy, A.; Brown, B.; Horoi, M.; Grawe, H. New $T = 1$ effective interactions for the $f_{5/2}p_{3/2}p1/2g_{9/2}$ model space: Implications for valence-mirror symmetry and seniority isomers. *Phys. Rev. C* **2004**, *70*, 044314. [CrossRef]
19. Magilligan, A.; Brown, B.A. New isospin-breaking "USD" Hamiltonians for the *sd* shell. *Phys. Rev. C* **2020**, *101*, 064312. [CrossRef]
20. Stroberg, S.R.; Holt, J.D.; Schwenk, A.; Simonis, J. Ab initio limits of atomic nuclei. *Phys. Rev. Lett.* **2021**, *126*, 022501. [CrossRef]
21. Forssén, C.; Hagen, G.; Hjorth-Jensen, M.; Nazarewicz, W.; Rotureau, J. Living on the edge of stability, the limits of the nuclear landscape. *Physica Scripta* **2013**, *2013*, 014022. [CrossRef]
22. Rotureau, J.; Michel, N.; Nazarewicz, W.; Płoszajczak, M.; Dukelsky, J. Density matrix renormalization group approach for many-body open quantum systems. *Phys. Rev. Lett.* **2006**, *97*, 110603. [CrossRef] [PubMed]
23. Rotureau, J.; Michel, N.; Nazarewicz, W.; Płoszajczak, M.; Dukelsky, J. Density matrix renormalization group approach to two-fluid open many-fermion systems. *Phys. Rev. C* **2009**, *79*, 014304. [CrossRef]
24. Bertulani, C.; Hammer, H.W.; Van Kolck, U. Effective field theory for halo nuclei: shallow p-wave states. *Nucl. Phys. A* **2002**, *712*, 37–58. [CrossRef]
25. Bedaque, P.; Hammer, H.W.; Van Kolck, U. Narrow resonances in effective field theory. *Phys. Lett. B* **2003**, *569*, 159–167. [CrossRef]
26. Id Betan, R.; Liotta, R.J.; Sandulescu, N.; Vertse, T. Two-particle resonant states in a many-body mean field. *Phys. Rev. Lett.* **2002**, *89*, 042501. [CrossRef]
27. Michel, N.; Nazarewicz, W.; Płoszajczak, M.; Bennaceur, K. Gamow shell model description of neutron-rich nuclei. *Phys. Rev. Lett.* **2002**, *89*, 042502. [CrossRef]
28. Michel, N.; Nazarewicz, W.; Płoszajczak, M.; Vertse, T. Shell model in the complex energy plane. *J. Phys. G: Nucl Part. Phys.* **2009**, *36*, 013101. [CrossRef]
29. Berggren, T.; Lind, P. Resonant state expansion of the resolvent. *Phys. Rev. C* **1993**, *47*, 768–778. [CrossRef]
30. Lind, P. Completeness relations and resonant state expansions. *Phys. Rev. C* **1993**, *47*, 1903–1920. [CrossRef]
31. Volya, A.; Zelevinsky, V. Discrete and continuum spectra in the unified shell model approach. *Phys. Rev. Lett.* **2005**, *94*, 052501. [CrossRef]
32. Li, J.; Ma, Y.; Michel, N.; Hu, B.; Sun, Z.; Zuo, W.; Xu, F. Recent progress in Gamow shell model calculations of drip line nuclei. *Physics* **2021**, *3*, 977–997. [CrossRef]
33. Tanihata, I.; Savajols, H.; Kanungo, R. Recent experimental progress in nuclear halo structure studies. *Prog. Part. Nucl. Phys.* **2013**, *68*, 215–313. [CrossRef]
34. Tanihata, I.; Hamagaki, H.; Hashimoto, O.; Shida, Y.; Yoshikawa, N.; Sugimoto, K.; Yamakawa, O.; Kobayashi, T.; Takahashi, N. Measurements of interaction cross-sections and nuclear radii in the light *p*-shell region. *Phys. Rev. Lett.* **1985**, *55*, 2676–2679. [CrossRef] [PubMed]
35. Olivier, L.; Franchoo, S.; Niikura, M.; Vajta, Z.; Sohler, D.; Doornenbal, P.; Obertelli, A.; Tsunoda, Y.; Otsuka, T.; Authelet, G.; et al. Persistence of the $Z = 28$ shell gap around $^{78}$Ni: First spectroscopy of $^{79}$Cu. *Phys. Rev. Lett.* **2017**, *119*, 192501. [CrossRef]
36. Tarasov, O.B.; Ahn, D.S.; Bazin, D.; Fukuda, N.; Gade, A.; Hausmann, M.; Inabe, N.; Ishikawa, S.; Iwasa, N.; Kawata, K.; et al. Discovery of $^{60}$Ca and implications for the stability of $^{70}$Ca. *Phys. Rev. Lett.* **2018**, *121*, 022501. [CrossRef]
37. Brown, B.A.; Blank, B.; Giovinazzo, J. Hybrid model for two-proton radioactivity. *Phys. Rev. C* **2019**, *100*, 054332. [CrossRef]
38. Jin, Y.; Niu, C.Y.; Brown, K.W.; Li, Z.H.; Hua, H.; Anthony, A.K.; Barney, J.; Charity, R.J.; Crosby, J.; Dell'Aquila, D.; et al. First observation of the four-proton unbound nucleus $^{18}$Mg. *Phys. Rev. Lett.* **2021**, *127*, 262502. [CrossRef]
39. Tostevin, J.A.; Gade, A. Updated systematics of intermediate-energy single-nucleon removal cross sections. *Phys. Rev. C* **2021**, *103*, 054610. [CrossRef]
40. Paschalis, S.; Petri, M.; Macchiavelli, A.O.; Hen, O.; Piasetzky, E. Nucleon-nucleon correlations and the single-particle strength in atomic nuclei. *Phys. Lett. B* **2020**, *800*, 135110. [CrossRef]
41. Hen, O.; Miller, G.A.; Piasetzky, E.; Weinstein, L.B. Nucleon-nucleon correlations, short-lived excitations, and the quarks within. *Rev. Mod. Phys.* **2017**, *89*, 045002. [CrossRef]
42. Panin, V.; Taylor, J.T.; Paschalis, S.; Wamers, F.; Aksyutina, Y.; Alvarez-Pol, H.; Aumann, T.; Bertulani, C.A.; Boretzky, K.; Caesar, C.; et al. Exclusive measurements of quasi-free proton scattering reactions in inverse and complete kinematics. *Phys. Lett. B* **2016**, *753*, 204–210. [CrossRef]
43. Jensen, O.; Hagen, G.; Hjorth-Jensen, M.; Brown, B.A.; Gade, A. Quenching of spectroscopic factors for proton removal in oxygen isotopes. *Phys. Rev. Lett.* **2011**, *107*, 032501. [CrossRef] [PubMed]

44. Wylie, J.; Okołowicz, J.; Nazarewicz, W.; Płoszajczak, M.; Wang, S.M.; Mao, X.; Michel, N. Spectroscopic factors in dripline nuclei. *Phys. Rev. C* **2021**, *104*, L061301. [CrossRef]
45. Brown, B.A. The oxygen isotopes. *Int. J. Mod. Phys. E* **2017**, *26*, 1740003. [CrossRef]
46. Wildenthal, B.H. Empirical strengths of spin operators in nuclei. *Prog. Part. Nucl. Phys.* **1984**, *11*, 5–51. [CrossRef]
47. Kanungo, R.; Nociforo, C.; Prochazka, A.; Aumann, T.; Boutin, D.; Cortina-Gil, D.; Davids, B.; Diakaki, M.; Farinon, F.; Geissel, H.; et al. One-neutron removal measurement reveals $^{24}$O as a new doubly magic nucleus. *Phys. Rev. Lett.* **2009**, *102*, 152501. [CrossRef] [PubMed]
48. Hoffman, C.R.; Baumann, T.; Bazin, D.; Brown, J.; Christian, G.; Denby, D.H.; DeYoung, P.A.; Finck, J.E.; Frank, N.; Hinnefeld, J.; et al. Evidence for a doubly magic $^{24}$O. *Phys. Lett. B* **2009**, *672*, 17–21. [CrossRef]
49. Janssens, R.V.F. Unexpected doubly magic nucleus. *Nature* **2009**, *459*, 1069–1070. [CrossRef]
50. Kondo, Y.; Nakamura, T.; Tanaka, R.; Minakata, R.; Ogoshi, S.; Orr, N.A.; Achouri, N.L.; Aumann, T.; Baba, H.; Delaunay, F.; et al. Nucleus $^{26}$O: A barely unbound system beyond the drip line. *Phys. Rev. Lett.* **2016**, *116*, 102503. [CrossRef]
51. Thibault, C.; Klapisch, R.; Rigaud, C.; Poskanzer, A.M.; Prieels, R.; Lessard, L.; Reisdorf, W. Direct measurement of the masses of $^{11}$Li and $^{26-32}$Na with an on-line mass spectrometer. *Phys. Rev. C* **1975**, *12*, 644–657. [CrossRef]
52. Campi, X.; Flocard, H.; Kerman, A.K.; Koonin, S. Shape transition in the neutron rich sodium isotopes. *Nucl. Phys. A* **1975**, *251*, 193–205. [CrossRef]
53. Wildenthal, B.H.; Chung, W. Collapse of the conventional shell-model ordering in the very-neutron-rich isotopes of Na and Mg. *Phys. Rev. C* **1980**, *22*, 2260–2262. [CrossRef]
54. Wildenthal, B.H.; Curtin, M.S.; Brown, B.A. Predicted features of the decay-decay of neutron-rich *sd*-shell nuclei. *Phys. Rev. C* **1983**, *28*, 1343–1366. [CrossRef]
55. Poves, A.; Retamosa, J. The onset of deformation at the $N = 20$ neutron shell closure far from stability. *Phys. Lett. B* **1987**, *184*, 311–315. [CrossRef]
56. Lubna, R.S.; Kravvaris, K.; Tabor, S.L.; Tripathi, V.; Volya, A.; Rubino, E.; Allmond, J.M.; Abromeit, B.; Baby, L.T.; Hensley, T.C. Structure of $^{38}$Cl and the quest for a comprehensive shell model interaction. *Phys. Rev. C* **2019**, *100*, 034308. [CrossRef]
57. Brown, B.A.; Rae, W.D.M. The Shell-model code NuShellX@MSU. *Nucl. Data Sheets* **2014**, *120*, 115–118. [CrossRef]
58. Gloeckner, D.H.; Lawson, R.D. Spurious center-of-mass motion. *Phys. Lett. B* **1974**, *53*, 313–318. [CrossRef]
59. Brown, B.A.; Etchegoyen, A.; Godwin, N.S.; Rae, W.D.M.; Richter, W.A.; Ormand, W.E.; Warburton, E.K.; Winfield, J.S.; Zhao, L.; Zimmerman, C.H. *Oxbash for Windows PC*; MSU-NSCL Report 1289; National Superconducting Cyclotron Laboratory MSU: East Lansing, MI, USA, 2004.
60. Caurier, E.; Nowacki, F.; Poves, A. Merging of the islands of inversion at $N = 20$ and $N = 28$. *Phys. Rev. C* **2014**, *90*, 014302. [CrossRef]
61. Utsuno, Y.; Otsuka, T.; Mizusaki, T.; Honma, M. Varying shell gap and deformation in $N \sim 20$ unstable nuclei studied by the Monte Carlo shell model. *Phys. Rev. C* **1999**, *60*, 054315. [CrossRef]
62. Otsuka, T.; Honma, M.; Mizusaki, T.; Shimizu, N.; Utsuno, Y. Monte Carlo shell model for atomic nuclei. *Prog. Part. Nucl. Phys.* **2001**, *47*, 319–400. [CrossRef]
63. Fifield, L.K.; Woods, C.L.; Bark, R.A.; Drumm, P.V.; Hotchkis, M.A.C. Masses and level schemes of $^{33}$Si and $^{34}$Si. *Nucl. Phys. A* **1985**, *440*, 531–542. [CrossRef]
64. Brown, B.A.; Richter, W.A. New "USD" Hamiltonians for the sd shell. *Phys. Rev. C* **2006**, *74*, 034315. [CrossRef]
65. Lica, R.; Rotaru, F.; Borge, M.J.G.; Grevy, S.; Negoita, F.; Poves, A.; Sorlin, O.; Andreyev, A.N.; Borcea, R.; Costache, C.; et al. Normal and intruder configurations in $^{34}$Si populated in the $\beta^-$ decay of $^{34}$Mg and $^{34}$Al. *Phys. Rev. C* **2019**, *100*, 034306. [CrossRef]
66. Elder, R.; Iwasaki, H.; Ash, J.; Bazin, D.; Bender, P.C.; Braunroth, T.; Brown, B.A.; Campbell, C.M.; Crawford, H.L.; Elman, B.; et al. Intruder dominance in the $0_2^+$ state of $^{32}$Mg studied with a novel technique for in-flight decays. *Phys. Rev. C* **2019**, *100*, 041301. [CrossRef]
67. Wimmer, K.; Kroell, T.; Kruecken, R.; Bildstein, V.; Gernhaeuser, R.; Bastin, B.; Bree, N.; Diriken, J.; Van Duppen, P.; Huyse, M.; et al. Discovery of the shape coexisting $0^+$ state in $^{34}$Mg by a two neutron transfer reaction. *Phys. Rev. Lett.* **2010**, *105*, 252501. [CrossRef]
68. Macchiavelli, A.O.; Crawford, H.L.; Campbell, C.M.; Clark, R.M.; Cromaz, M.; Fallon, P.; Jones, M.D.; Lee, I.Y.; Salathe, M.; Brown, B.A.; et al. The $^{30}$Mg$(t,p)^{32}$Mg "puzzle" reexamined. *Phys. Rev. C* **2016**, *94*, 051303. [CrossRef]
69. Kitamura, N.; Wimmer, K.; Poves, A.; Shimizu, N.; Tostevin, J.A.; Bader, V.M.; Bancroft, C.; Barofsky, D.; Baugher, T.; Bazin, D.; et al. Coexisting normal and intruder configurations in $^{32}$Mg. *Phys. Lett. B* **2021**, *822*, 136682. [CrossRef]
70. Kitamura, N.; Wimmer, K.; Miyagi, T.; Poves, A.; Shimizu, N.; Tostevin, J.A.; Bader, V.M.; Bancroft, C.; Barofsky, D.; Baugher, T.; et al. In-beam gamma-ray spectroscopy of $^{32}$Mg via direct reactions. *Phys. Rev. C* **2022**, *105*, 034318. [CrossRef]
71. Fortunato, L.; Casal, J.; Horiuchi, W.; Singh, J.; Vitturi, A. The $^{29}$F nucleus as a lighthouse on the coast of the island of inversion. *Nat. Comm. Phys.* **2020**, *3*, 132. [CrossRef]
72. Doornenbal, P.; Scheit, H.; Takeuchi, S.; Utsuno, Y.; Aoi, N.; Li, K.; Matsushita, M.; Steppenbeck, D.; Wang, H.; Baba, H.; et al. Low-Z shore of the "island of inversion" and the reduced neutron magicity toward $^{28}$O. *Phys. Rev. C* **2017**, *95*, 041301. [CrossRef]
73. Bagchi, S.; Kanungo, R.; Tanaka, Y.K.; Geissel, H.; Doornenbal, P.; Horiuchi, W.; Hagen, G.; Suzuki, T.; Tsunoda, N.; Ahn, D.S.; et al. Two-neutron halo is unveiled in $^{29}$F. *Phys. Rev. Lett.* **2020**, *124*, 222504. [CrossRef] [PubMed]

74. Brown, B. New Skyrme interaction for normal and exotic nuclei. *Phys. Rev. C* **1998**, *58*, 220–231. [CrossRef]
75. MacGregor, P.T.; Sharp, D.K.; Freeman, S.J.; Hoffman, C.R.; Kay, B.P.; Tang, T.L.; Gaffney, L.P.; Baader, E.F.; Borge, M.J.G.; Butler, P.A.; et al. Evolution of single-particle structure near the $N = 20$ island of inversion. *Phys. Rev. C* **2021**, *104*, L051301. [CrossRef]
76. Otsuka, T.; Tsunoda, Y. The role of shell evolution in shape coexistence. *J. Phys. G: Nucl Part. Phys.* **2016**, *43*, 024009. [CrossRef]
77. Chrisman, D.; Kuchera, A.N.; Baumann, T.; Blake, A.; Brown, B.A.; Brown, J.; Cochran, C.; DeYoung, P.A.; Finck, J.E.; Frank, N.; et al. Neutron-unbound states in $^{31}$Ne. *Phys. Rev. C* **2021**, *104*, 034313. [CrossRef]
78. Thoennessen, M.; Baumann, T.; Brown, B.A.; Enders, J.; Frank, N.; Hansen, R.G.; Heckman, P.; Luther, B.A.; Seitz, J.; Stolz, A.; et al. Single proton knock-out reactions from $^{24}$F, $^{25}$F, $^{26}$F. *Phys. Rev. C* **2003**, *68*, 044318. [CrossRef]
79. Tang, T.L.; Uesaka, T.; Kawase, S.; Beaumel, D.; Dozono, M.; Fujii, T.; Fukuda, N.; Fukunaga, T.; Galindo-Uribarri, A.; Hwang, S.H.; et al. How different is the core of $^{25}$F from $^{24}$O$_{g.s.}$ ? *Phys. Rev. Lett.* **2020**, *124*, 212502. [CrossRef]
80. Grigorenko, L.V.; Mukha, I.G.; Scheidenberger, C.; Zhukov, M.V. Two-neutron radioactivity and four-nucleon emission from exotic nuclei. *Phys. Rev. C* **2011**, *84*, 021303. [CrossRef]
81. Lepailleur, A.; Wimmer, K.; Mutschler, A.; Sorlin, O.; Thomas, J.C.; Bader, V.; Bancroft, C.; Barofsky, D.; Bastin, B.; Baugher, T.; et al. Spectroscopy of $^{28}$Na: Shell evolution toward the drip line. *Phys. Rev. C* **2015**, *92*, 054309. [CrossRef]
82. Utsuno, Y.; Otsuka, T.; Brown, B.A.; Honma, M.; Mizusaki, T.; Shimizu, N. Shape transitions in exotic Si and S isotopes and tensor-force-driven Jahn-Teller effect. *Phys. Rev. C* **2012**, *86*, 051301. [CrossRef]
83. Nowacki, F.; Poves, A. New effective interaction for $0\hbar\omega$ shell-model calculations in the $sd$–$pf$ valence space. *Phys. Rev. C* **2009**, *79*, 014310. [CrossRef]
84. Morris, L.; Jenkins, D.G.; Harakeh, M.N.; Isaak, J.; Kobayashi, N.; Tamii, A.; Adachi, S.; Adsley, P.; Aoi, N.; Bracco, A.; et al. Search for in-band transitions in the candidate superdeformed band in $^{28}$Si. *Phys. Rev. C* **2021**, *104*, 054323. [CrossRef]
85. Sagawa, H.; Zhou, X.; Zhang, X.Z. Deformations and electromagnetic moments in carbon and neon isotopes. *Phys. Rev. C* **2004**, *70*, 054316. [CrossRef]
86. Petri, M.; Fallon, P.; Macchiavelli, A.O.; Paschalis, S.; Starosta, K.; Baugher, T.; Bazin, D.; Cartegni, L.; Clark, R.M.; Crawford, H.L.; et al. Lifetime measurement of the $2_1^+$ state in $^{20}$C. *Phys. Rev. Lett.* **2011**, *107*, 102501. [CrossRef]
87. Cwiok, S.; Dudek, J.; Nazarewicz, W.; Skalski, J.; Werner, T. Single-particle energies, wave-functions, quadrupole-moments and g-factors in an axially deformed Woods-Saxson potential with applications to the 2-center-type nuclear problems. *Comp. Phys. Comm.* **1987**, *46*, 379–399. [CrossRef]
88. Honma, M.; Otsuka, T.; Brown, B.; Mizusaki, T. New effective interaction for pf-shell nuclei and its implications for the stability of the $N = Z = 28$ closed core. *Phys. Rev. C* **2004**, *69*, 034335. [CrossRef]
89. Honma, M.; Otsuka, T.; Brown, B.; Mizusaki, T. Shell-model description of neutron-rich pf-shell nuclei with a new effective interaction GXPF1. *Eur. Phys. J. A* **2005**, *25*, 499–502. [CrossRef]
90. Warburton, E.; Brown, B. Effective interactions for the 0p1s0d nuclear shell-model space. *Phys. Rev. C* **1992**, *46*, 923–944. [CrossRef]
91. Richter, W.A.; Mkhize, S.; Brown, B.A. *sd*-shell observables for the USDA and USDB Hamiltonians. *Phys. Rev. C* **2008**, *78*, 064302. [CrossRef]
92. du Rietz, R.; Ekman, J.; Rudolph, D.; Fahlander, C.; Dewald, A.; Moller, O.; Saha, B.; Axiotis, M.; Bentley, M.; Chandler, C.; et al. Effective charges in the $fp$ shell. *Phys. Rev. Lett.* **2004**, *93*, 222501. [CrossRef]
93. Bohr, A.; Mottelson, B.R. *Nuclear Structure*; W. A. Benjamin, Inc.: New York, NY, USA, 1975; Volume 2.
94. Brown, B.A.; Arima, A.; McGrory, J.B. E2 core-polarization charge for nuclei near $^{16}$O and $^{40}$Ca. *Nucl. Phys. A* **1977**, *277*, 77–108. [CrossRef]
95. Longfellow, B.; Weisshaar, D.; Gade, A.; Brown, B.A.; Bazin, D.; Brown, K.W.; Elman, B.; Pereira, J.; Rhodes, D.; Spieker, M. Quadrupole collectivity in the neutron-rich sulfur isotopes $^{38}$S, $^{40}$S, $^{42}$S, $^{44}$S. *Phys. Rev. C* **2021**, *103*, 054309. [CrossRef]
96. Heil, S.; Petri, M.; Vobig, K.; Bazin, D.; Belarge, J.; Bender, P.; Brown, B.A.; Elder, R.; Elman, B.; Gade, A.; et al. Electromagnetic properties of $^{21}$O for benchmarking nuclear Hamiltonians. *Phys. Lett. B* **2020**, *809*, 135678. [CrossRef]
97. Lobner, K.; Vetter, M.; Honig, V. Nuclear intrinsic quadrupole moments and deformation paramters. *Atom. Data Nucl. Data Tables* **1970**, *7*, 495–520. [CrossRef]
98. Tostevin, J.A.; Brown, B.A.; Simpson, E.C. Two-proton removal from $^{44}$S and the structure of $^{42}$Si. *Phys. Rev. C* **2013**, *87*, 027601. [CrossRef]
99. Gade, A.; Brown, B.A.; Tostevin, J.A.; Bazin, D.; Bender, P.C.; Campbell, C.M.; Crawford, H.L.; Elman, B.; Kemper, K.W.; Longfellow, B.; et al. Is the structure of $^{42}$Si understood? *Phys. Rev. Lett.* **2019**, *122*, 222501. [CrossRef]
100. Utsuno, Y.; Shimizu, N.; Otsuka, T.; Yoshida, T.; Tsunoda, Y. Nature of Isomerism in Exotic Sulfur Isotopes. *Phys. Rev. Lett.* **2015**, *114*. [CrossRef]
101. Santiago-Gonzalez, D.; Wiedenhoever, I.; Abramkina, V.; Avila, M.L.; Baugher, T.; Bazin, D.; Brown, B.A.; Cottle, P.D.; Gade, A.; Glasmacher, T.; et al. Triple configuration coexistence in $^{44}$S. *Phys. Rev. C* **2011**, *83*, 061305. [CrossRef]
102. Longfellow, B.; Weisshaar, D.; Gade, A.; Brown, B.A.; Bazin, D.; Brown, K.W.; Elman, B.; Pereira, J.; Rhodes, D.; Spieker, M. Shape changes in the $N = 28$ island of inversion: Collective structures built on configuration-coexistingsStates in $^{43}$S. *Phys. Rev. Lett.* **2020**, *125*, 232501. [CrossRef]

103. Sharon, Y.Y.; Benczer-Koller, N.; Kumbartzki, G.J.; Zamick, L.; Casten, R.F. Systematics of the ratio of electric quadrupole moments $Q(2_1^+)$ to the square root of the reduced transition probabilities $B(E2; 0_1^+ \to 2_1^+)$ in even-even nuclei. *Nucl. Phys. A* **2018**, *980*, 131–142. [CrossRef]
104. Brown, B.; Richter, W. Shell-model plus Hartree-Fock calculations for the neutron-rich Ca isotopes. *Phys. Rev. C* **1998**, *58*, 2099–2107. [CrossRef]
105. Poves, A.; Sanchez-Solano, J.; Caurier, E.; Nowacki, F. Shell model study of the isobaric chains $A = 50$, $A = 51$ and $A = 52$. *Nucl. Phys. A* **2001**, *694*, 157–198. [CrossRef]
106. Browne, F.; Chen, S.; Doornenbal, P.; Obertelli, A.; Ogata, K.; Utsuno, Y.; Yoshida, K.; Achouri, N.L.; Baba, H.; Calvet, D.; et al. Pairing forces govern population of doubly magic $^{54}$Ca from direct reactions. *Phys. Rev. Lett.* **2021**, *126*, 252501. [CrossRef] [PubMed]
107. Bhoy, B.; Srivastava, P.C.; Kaneko, K. Shell model results for $^{47-58}$Ca isotopes in the fp, $fpg_{9/2}$ and $fpg_{9/2}d_5/2$ model spaces. *J. Phys. G: Nucl. Part. Phys.* **2020**, *47*, 065105. [CrossRef]
108. Lenzi, S.M.; Nowacki, F.; Poves, A.; Sieja, K. Island of inversion around Cr-64. *Phys. Rev. C* **2010**, *82*, 054301. [CrossRef]
109. Gade, A. Reaching into the $N = 40$ Island of inversion with nucleon removal reactions. *Physics* **2021**, *3*, 1226–1236. [CrossRef]
110. Zamora, J.C.; Zegers, R.G.T.; Austin, S.M.; Bazin, D.; Brown, B.A.; Bender, P.C.; Crawford, H.L.; Engel, J.; Falduto, A.; Gade, A.; et al. Experimental constraint on stellar electron-capture rates from the $^{88}\mathrm{Sr}(t, {}^3\mathrm{He} + \gamma)\ {}^{88}\mathrm{Rb}$ reaction at 115 MeV/u. *Phys. Rev. C* **2019**, *100*, 032801(R). [CrossRef]
111. Titus, E.R.; Ney, E.M.; Zegers, R.G.T.; Bazin, D.; Belarge, J.; Bender, P.C.; Brown, B.A.; Campbell, C.M.; Elman, B.; Engel, J.; et al. Constraints for stellar electron-capture rates on $^{86}$Kr via the $^{86}\mathrm{Kr}(t, {}^3\mathrm{He} + \gamma)\ {}^{86}\mathrm{Br}$ reaction and the implications for core-collapse supernovae. *Phys. Rev. C* **2019**, *100*, 045805. [CrossRef]
112. Stroberg, S.R. Beta decay in medium-mass nuclei with the in-medium similarity renormalization group. *Particles* **2021**, *4*, 521–535. [CrossRef]
113. Lubos, D.; Park, J.; Faestermann, T.; Gernhaeuser, R.; Kruecken, R.; Lewitowicz, M.; Nishimura, S.; Sakurai, H.; Ahn, D.S.; Baba, H.; et al. Improved value for the Gamow-Teller strength of the $^{100}$Sn beta decay. *Phys. Rev. Lett.* **2019**, *122*, 222502. [CrossRef]
114. Nowacki, F.; Poves, A.; Caurier, E.; Bounthong, B. Shape coexistence in $^{78}$Ni as the portal to the fifth island of inversion. *Phys. Rev. Lett.* **2016**, *117*, 272501. [CrossRef] [PubMed]

*physics*

MDPI

*Review*

# Recent Progress in Gamow Shell Model Calculations of Drip Line Nuclei

**Jianguo Li [1], Yuanzhuo Ma [1], Nicolas Michel [2,3], Baishan Hu [1], Zhonghao Sun [1], Wei Zuo [2,3] and Furong Xu [1,*]**

1 State Key Laboratory of Nuclear Physics and Technology, School of Physics, Peking University, Beijing 100871, China; jianguo_li@pku.edu.cn (J.L.); yuanzhuoma@pku.edu.cn (Y.M.); hubsh@pku.edu.cn (B.H.); sunzhg07@tennessee.edu (Z.S.)
2 Institute of Modern Physics, Chinese Academy of Sciences, Lanzhou 730000, China; nicolas.michel@impcas.ac.cn (N.M.); zuowei@impcas.ac.cn (W.Z.)
3 School of Nuclear Science and Technology, University of Chinese Academy of Sciences, Beijing 100049, China
* Correspondence: frxu@pku.edu.cn

**Abstract:** The Gamow shell model (GSM) is a powerful method for the description of the exotic properties of drip line nuclei. Internucleon correlations are included via a configuration interaction framework. Continuum coupling is directly included at basis level by using the Berggren basis, in which, bound, resonance, and continuum single-particle states are treated on an equal footing in the complex momentum plane. Two different types of Gamow shell models have been developed: its first embodiment is that of the GSM defined with phenomenological nuclear interactions, whereas the GSM using realistic nuclear interactions, called the realistic Gamow shell model, was introduced later. The present review focuses on the recent applications of the GSM to drip line nuclei.

**Keywords:** Gamow shell model; realistic nuclear forces; phenomenological interactions; resonance; continuum; drip line nuclei

**Citation:** Li, J.; Ma, Y.; Michel, N.; Hu, B.; Sun, Z.; Zuo, W.; Xu, F. Recent Progress in Gamow Shell Model Calculations of Drip Line Nuclei. *Physics* **2021**, *3*, 977–997. https://doi.org/10.3390/physics3040062

Received: 17 August 2021
Accepted: 25 October 2021
Published: 8 November 2021

**Publisher's Note:** MDPI stays neutral with regard to jurisdictional claims in published maps and institutional affiliations.

**Copyright:** © 2021 by the authors. Licensee MDPI, Basel, Switzerland. This article is an open access article distributed under the terms and conditions of the Creative Commons Attribution (CC BY) license (https://creativecommons.org/licenses/by/4.0/).

## 1. Introduction

Exotic nuclei have been studied for many years using a new generation of accelerators, which are now able to reach nuclear drip lines [1–4]. Contrary to well-bound nuclei, which are closed quantum systems, drip line nuclei can be seen as open quantum systems, as they are either weakly bound or unbound with respect to the particle emission threshold [5]. Many interesting phenomena appear at drip lines, such as a halo structure [1,6,7] and particle emission in resonance states [4,8]. Continuum coupling plays an important role in these loosely bound and unbound nuclear systems [5]. The proper description of nuclei at drip lines is one of the main challenges of nuclear theory, which was mostly developed to account for the structure of well-bound nuclei [5,6].

A clear consequence of the strong intertwinings of the continuum degrees of freedom and internucleon correlations at drip lines consists of the odd–even staggering found in the helium chain [9,10]. Indeed, odd helium isotopes (except $^{3}$He) are all resonances and bear widths of several hundreds of keV [10–12]. Conversely, the even–even helium isotopes $^{4,6,8}$He are bound, with $^{6,8}$He both exhibiting halo properties [13–15]. To accurately reproduce nuclear halos, many-body wave functions in asymptotic regions must be treated properly, which demands to take into account continuum coupling [1,6,7,16–19]. Adding to that, these weakly bound and unbound drip line nuclei also provide good laboratories to understand the single-particle structure, continuum coupling, internucleon correlations, and nucleon-nucleon (NN) interactions, which are not well understood in these regions.

Most present nuclear models, such as the no-core shell model (NCSM) [20], self-consistent Green's function [21], coupled-cluster (CC) [22,23], in-medium similarity renormalization group (IM-SRG) [24], and standard shell model (SM) [25,26] have been developed for the study of well-bound nuclei, whereby continuum coupling is absent. Only few

models explicitly include continuum coupling. The main models including both internucleon correlations and continuum coupling in a unified picture are the no-core Gamow shell model (NCGSM) [27–29], the no-core shell model with continuum (NCSMC) [30], the complex CC [31,32], the Gamow IMSRG (Gamow-IMSRG) [33], continuum shell model (CSM) [34,35], and the Gamow shell model (GSM) [19,36–39], which are extensions of the NCSM, CC, IM-SRG, and SM, respectively. However, due to their huge model space dimensions, the NCGSM and NCSMC can only be used to describe light nuclei [27,28,30]. Furthermore, only nuclei in the vicinity of closed-shell nuclear systems can be investigated by the complex-CC and Gamow-IMSRG methods [31–33]. CSM [34,35] takes into account the continuum effect by projecting the model space onto the subspaces of bound and scattering states in a real-energy basis, in which, resonance states are not included. Within GSM, continuum coupling is treated at basis level by way of the Berggren basis [36–38]. The latter comprises bound, resonance, and continuum scattering states, with all of these states treated on an equal footing within the Berggren ensemble [40]. Internucleon correlations in GSM are induced by configuration mixing, similarly to conventional SM. GSM has been seen to successfully reproduce many situations of physical interest [5,38]; for example, the resonances of oxygen drip line nuclei [38,41,42] and the neutron halo structure of $^{31}$F [18].

The GSM was introduced in nuclear physics in 2002 [36,37], where only simple phenomenological nuclear potentials were used, while calculations were limited to only two valence neutrons outside of the inner core. After that, the GSM was extended to many valence particle systems, such as $^{8}$He [17] and *psd*-shell nuclei [43]. The realistic Gamow shell model was proposed in Refs. [44,45], with which, two- or three-particle systems could be investigated. An effective Hamiltonian based on realistic interactions was constructed by using the degenerate $\hat{Q}$-box approach; however, folded diagrams are neglected [45]. A folded diagram sums up the subset of diagrams to infinite order so as to include high-order effects. In 2017, we developed the realistic GSM method with the full $\hat{Q}$-box folded-diagram method using the nondegenerate Berggren basis [38]. We applied it to the case of the neutron-rich oxygen isotopes up to the neutron drip line. After that, many extensions of the realistic GSM were developed, such as performing the realistic GSM in the Gamow Hartree-Fock basis (GHF) [41].

In the present review, the framework of the two types of GSM (realistic GSM and phenomenological GSM) is first introduced in Section 2. Then, we review our recent applications of GSM, including the calculations of neutron-rich oxygen and fluorine isotopes [38,39,41,42,46], neutron-rich calcium isotopes [47], and proton decays in $^{16}$Ne and $^{18}$Mg [48]. Finally, a short summary of the review and the future challenges of the next GSM calculations are given.

## 2. Method

GSM is built within a configuration interaction framework based on the one-body Berggren basis [5,36–38]. The Berggren basis [40,49] is generated by a finite-range potential, which can be written as the solutions of the one-body Schröndinger equation in the complex momentum space, which reads

$$\frac{d^2u(k,r)}{dr^2} = \left[\frac{l(l+1)}{r^2} + \frac{2m}{\hbar^2}U(r) - k^2\right]u(k,r), \quad (1)$$

where $l$ is the orbital angular momentum of the nucleon motion, $m$ is the mass of the nucleon, $r$ stands for the radius, and $\hbar$ is the reduced Planck constant. The momentum $k$ and wave function $u(k,r)$ can be complex. $U(r)$ is the finite-range potential, which is, in practice, a Woods–Saxon (WS) [50] or GHF potential [33,44]. When considering protons, the Coulomb potential must be included in $U(r)$. Bound, resonance, and scattering states can then be generated. The eigenenergy of single-particle states in the above equation is complex in general, and reads $\tilde{e}_n = k^2/2m = e_n - i\gamma_n/2$, where $n$ denotes the state [40,49]. $e_n$ stands for the energy, whereas $\gamma_n$ represents the particle decay width, so that $\gamma_n = 0$ for bound states and $\gamma_n > 0$ for resonance states. A schematic Berggren basis set of states in the complex $k$-plane is illustrated in Figure 1. The wave function of a resonance state

is not square-integrable, as its exponential increase in modulus implies that the wave function of a resonance state cannot be normalized with conventional techniques [40,49]. Consequently, one has to rely on the complex scaling method, which has been seen to properly account for the normalization of resonance states [51].

The completeness relation borne by Berggren basis states [40,49] reads

$$\sum_n |n\rangle\langle n| + \int_{L_+} |k\rangle\langle k| dk = \mathbf{1}, \tag{2}$$

where $|n\rangle$ states are bound states and resonance states inside the $L_+$ contour of Figure 1. These states are called pole states, as they are the $S$-matrix poles of the finite-range potential. $|k\rangle$ states are scattering states and follow the $L_+$ contours in the complex $k$-plane, starting from $k = 0$ and going to $k \to +\infty$, as shown in Figure 1. Scattering states initially form a continuum. Hence, in order to be used in numerical applications, the scattering states along the $L_+$ contour must be discretized with a Gauss–Legendre quadrature [5,49]. It has been checked that 10–45 states per contour are necessary to have converged results [5,38]. Once discretized, the Berggren basis is, in effect, the same as that of the harmonic-oscillator (HO) states within the standard SM [5,49]. Concerning resonance states, only narrow resonance states contribute to the physical states, and thus are included in the real calculations, whereas broad resonance states are not included, as they lie below the $L_+$ contour.

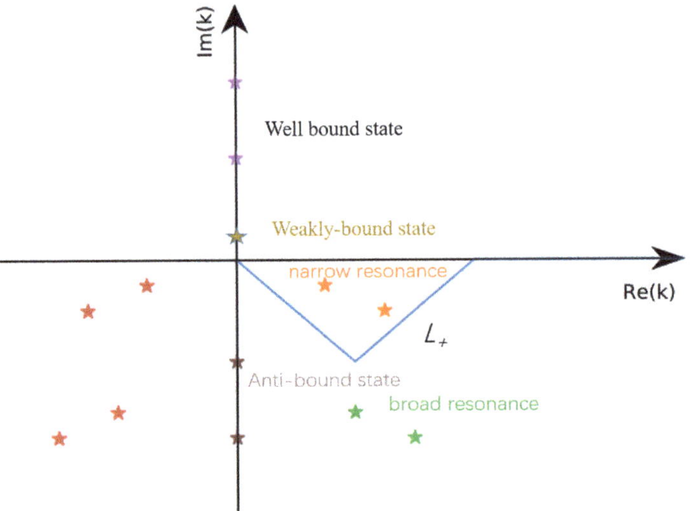


**Figure 1.** Depiction of the Berggren basis in the complex-momentum-$k$ plane for a fixed partial wave. Typical complex momenta of bound, narrow, and broad-resonance states, i.e., $S$-matrix pole, are provided. The $L_+$ contour of scattering states encompasses the $S$-matrix poles of interest.

In fact, the Berggren basis is the complex extension of the real-energy completeness relation of Newton [52], which consists of bound states and of a continuum of real-energy scattering states. Contrary to the Newton completeness relation [52], with which, only localized states can be expanded, the Berggren basis can expand unbound resonance states [40,49]. The many-body completeness relation is obtained by constructing Slater determinants from the one-body Berggren basis, which contains bound, resonance, and scattering states [5,49]. In the GSM, the Hamiltonian is represented by a complex symmetric matrix when using the one-body Berggren basis, which has to be diagonalized [5,49]. This process can be handled efficiently by using the complex symmetric extension of the Jacobi-Davidson method [49,53], where one can take advantage of the relatively small

coupling to continuum states in order to have a fast convergence of calculations [5]. The full configuration space is extremely large due to the many scattering states within the model space. In practical calculations, however, we often truncate basis model spaces so that only two particles can occupy scattering states. It has been checked that this is sufficient to obtain converged results for both the energy and decay width of many-body states [5,38,54].

In GSM calculations, an effective Hamiltonian must be constructed. There are two main methods to build the effective Hamiltonian in GSM calculations. One is to construct an effective Hamiltonian based on realistic nuclear force [38,39,41], and hence in the frame of the realistic GSM, whereas the other one consists of using an effective phenomenological nuclear potential [5,36,37,49], in which, the parameters of the potential are optimized to reproduce experimental data. In the following, we give details about these two versions of GSM.

### 2.1. Realistic Gamow Shell Model Calculations

Within realistic GSM, one starts from the intrinsic Hamiltonian of an $A$-body system, which reads

$$H = \sum_{i=1}^{A} \frac{\boldsymbol{p}_i^{\,2}}{2m} + \sum_{i<j}^{A} V_{\mathrm{NN}}^{ij} - \frac{\boldsymbol{P}^2}{2Am}, \tag{3}$$

where $\boldsymbol{p}_i$ is the nucleon momentum in laboratory frame, $\boldsymbol{P} = \sum_{i=1}^{A} \boldsymbol{p}_i$ is the center-of-mass (CoM) momentum of the system, and $V_{\mathrm{NN}}^{(ij)}$ is the realistic NN interaction, such as CD-Bonn [55] and $\mathrm{N^3LO}$ [56] interaction. In the above Hamiltonian, the CoM energy is removed. In order to construct the effective Hamiltonian to be used in GSM calculations, an auxiliary potential is usually introduced [38,57,58], so that the Hamiltonian can be rewritten as,

$$\begin{aligned} H &= \sum_{i=1}^{A} \left(\frac{\boldsymbol{p}_i^2}{2m} + U\right) + \sum_{i<j}^{A} \left(V_{\mathrm{NN}}^{(ij)} - U - \frac{\boldsymbol{p}_i^2}{2Am} - \frac{\boldsymbol{p}_i \cdot \boldsymbol{p}_j}{Am}\right) \\ &= H_0 + H_1, \end{aligned} \tag{4}$$

where $H_0 = \sum_{i=1}^{A} (\frac{\boldsymbol{p}_i^2}{2m} + U)$ has a one-body form, and $H_1 = \sum_{i<j}^{A} (V_{\mathrm{NN}}^{(ij)} - U - \frac{\boldsymbol{p}_i^2}{2Am} - \frac{\boldsymbol{p}_i \cdot \boldsymbol{p}_j}{Am})$ is the residual two-body interaction, including the correction issued from the CoM motion. For the auxiliary potential $U$, we usually take a WS finite-range potential. To speed up the convergence of many-body calculations, the bare force is often softened by a similarity renormalization group method [59], such as the similarity renormalization group (SRG) and $V_{\text{low-}k}$,in order to remove the strong short-range repulsive core.

The realistic NN interaction is firstly defined in a relative momentum space. However, the many-body problem is usually solved in the laboratory frame (with, e.g., the HO basis), so that a transformation from relative and CoM coordinates to the laboratory frame is necessary. This procedure can be conveniently carried out in the HO basis via Brody–Moshinsky brackets [60]. In the HO basis, the two-body completeness relation reads

$$\sum_{\alpha \le \beta} |\alpha\beta\rangle\langle\alpha\beta| = \mathbf{1}, \tag{5}$$

where $|\alpha\beta\rangle$ is the two-particle state of the HO basis. The two-body interaction in the HO basis is given by

$$V_{\mathrm{HO}} = \sum_{\alpha \le \beta}^{N_{\mathrm{shell}}} \sum_{\gamma \le \delta}^{N_{\mathrm{shell}}} |\alpha\beta\rangle\langle\alpha\beta|V_{\mathrm{low}-k}|\gamma\delta\rangle\langle\gamma\delta|, \tag{6}$$

where $N_{\mathrm{shell}} = 2n + l$, indicates that a finite summation is performed up to $N_{\mathrm{shell}}$. The GSM calculations are carried out in the Berggren basis, so that the transformation of the interaction matrix elements from the HO basis to the Berggren basis needs to be carried

out. This is achieved, in practice, by computing the overlaps between the Berggren and HO basis wave functions,

$$\langle ab|V|cd\rangle \approx \sum_{\alpha\leq\beta}^{N_{\text{shell}}}\sum_{\gamma\leq\delta}^{N_{\text{shell}}} \langle ab|\alpha\beta\rangle\langle\alpha\beta|V_{\text{low}-k}|\gamma\delta\rangle\langle\gamma\delta|cd\rangle, \quad (7)$$

where $|ab\rangle$ ($|cd\rangle$) is a two-particle state of the Berggren basis. For identical particles (proton–proton or neutron–neutron), the overlap of the two-body state reads

$$\langle ab|\alpha\beta\rangle = \frac{\langle a|\alpha\rangle\langle b|\beta\rangle - (-1)^{J-j_\alpha-j_\beta}\langle a|\beta\rangle\langle b|\alpha\rangle}{\sqrt{(1+\delta_{ab})(1+\delta_{\alpha\beta})}}, \quad (8)$$

where $J$ is the total angular momentum of the two-particle state, and $j$ is the angular momentum of a single-particle basis state. The $\langle a|\alpha\rangle$ ($\langle b|\beta\rangle$) is the overlap of the one-body basis state, and $\delta_{\alpha\beta}$ is the Kronecker delta. For the proton–neutron coupling, the overlap of the two-body state is simply given by

$$\langle ab|\alpha\beta\rangle = \langle a|\alpha\rangle\langle b|\beta\rangle. \quad (9)$$

The overlaps of one-body basis states are directly obtained from an integration in $r$-space

$$\langle a|\alpha\rangle = \int drr^2 u_a(r)R_\alpha(r)\delta_{l_a l_\alpha}\delta_{j_a j_\alpha}\delta_{t_a t_\alpha}, \quad (10)$$

where $u(r)$ and $R(r)$ are the radial parts of the single-particle Berggren and HO basis wave functions, with $l$, $j$, and $t$ being the orbital, total angular momentum, and isospin quantum number, respectively. The single-particle wave functions of resonance and continuum states are not localized and hence are not square-integrable. The transformation defined by Equation (7), in fact, utilizes the short-range nature of nuclear force. Indeed, the Gaussian fall-off of the HO wave function renders the overlaps integrable, even when one considers resonances or scattering states of complex energy. For long-range operators, such as the one-body kinetic energy and Coulomb potential, using Equation (7) is not suitable in practice. In this case, we use the exterior complex scaling technique [51] to treat the kinetic and Coulomb operator, i.e., terms proportional to $\boldsymbol{p}^2$ and $1/r$, respectively, with the Berggren basis.

The obtained interaction matrix elements in the Berggren basis are complex, and associated operators are non-Hermitian. The many-body perturbation theory (MBPT), named the full $\hat{Q}$-box folded-diagram method [61], is employed to construct the realistic complex effective Hamiltonian in the defined model space for GSM calculations. The complex-$k$ Berggren basis states in the model space are non-degenerate; therefore a non-degenerate $\hat{Q}$-box folded-diagram perturbation, i.e., the extended Kuo–Krenciglowa (EKK) method [62], has been used. For this, we first calculate the $\hat{Q}$-box using MBPT in the Berggren complex-$k$ basis,

$$\begin{aligned}\hat{Q}(E) &= PVP + PVQ\frac{Q}{E-QHQ}QVP \\ &= PVP + PVQ\frac{Q}{E-QH_0Q}QVP + \ldots,\end{aligned} \quad (11)$$

where $E$ is starting energy, $P$ and $Q$ represent the model space and the excluded space, respectively, with $P+Q=\mathbf{1}$. The $\hat{Q}$-box is composed of irreducible valence-linked diagrams [57,58], which can be built order-by-order. $V$ and $H$ are the two-body interaction and two-body Hamiltonian, respectively, and $H_0$ is the unperturbed one-body Hamiltonian. The derivatives of the $\hat{Q}$-box are defined as

$$\begin{aligned}\hat{Q}_s(E) &= \frac{1}{s!}\frac{d^s\hat{Q}(E)}{dE^s} \\ &= (-1)^s PVQ\frac{Q}{(E-QHQ)^{s+1}}QVP,\end{aligned} \tag{12}$$

where $s$ denotes the $s$-th derivative.

The effective Hamiltonian $H_{\text{eff}}$ can then be constructed in operator form [63], written as

$$\widetilde{H}_{\text{eff}} = \widetilde{H}_{BH}(E) + \sum_{k=1}^{\infty} \hat{Q}_k(E)\widetilde{H}_{\text{eff}}, \tag{13}$$

where $\widetilde{H}_{\text{eff}}$ stands for $\widetilde{H}_{\text{eff}} = H_{\text{eff}} - E$, and $\widetilde{H}_{BH}(E) = H_{BH}(E) - E$ is the Block–Horowitz Hamiltonian shifted by an energy $E$, with

$$\begin{aligned}H_{BH}(E) &= PH_0P + \hat{Q}(E) \\ &= PH_0P + PVP + PVQ\frac{Q}{E-QHQ}QVP.\end{aligned} \tag{14}$$

$\widetilde{H}_{\text{eff}}$ is obtained by performing iterations of Equation (13), which is equivalent to calculate folded-diagrams, where one considers high-order contributions by summing up the subsets of diagrams to finite order. When convergence is obtained, the effective Hamiltonian is given by $H_{\text{eff}} = \widetilde{H}_{\text{eff}} - E$, and the effective interaction reads $V_{\text{eff}} = H_{\text{eff}} - PH_0P$. The extended $\hat{Q}$-box folded-diagram calculations provide a useful approach for including effects from the continuum coupling and core polarization [57,58,61,62].

### 2.2. Gamow Shell Model with Phenomenological Nuclear Potential

Within the Gamow shell model with phenomenological nuclear potential, the nucleus is assumed to be a system of $N_v$ valence particles outside a frozen closed core, from which, core polarization is absent [5,36,37,49]. The GSM Hamiltonian, expressed with intrinsic nucleon-core cluster-orbital shell model coordinates [64], writes

$$H = \sum_{i=1}^{N_v}\left[\frac{\boldsymbol{p}_i^2}{2\mu_i} + U_{\text{core}}(i)\right] + \sum_{i<j=1}^{N_v}\left[V(i,j) + \frac{\boldsymbol{p}_i\boldsymbol{p}_j}{M_{\text{core}}}\right], \tag{15}$$

where $\boldsymbol{p}_i$ is the nucleon momentum in cluster-orbital shell model frame, $U_{\text{core}}$ is the single-particle nucleon-core potential, and $V$ is the phenomenological NN interaction between valence nucleons. $\mu_i$ and $M_{\text{core}}$ stand for the reduced mass of the nucleon and the mass of the core, respectively. The $\frac{\boldsymbol{p}_i\boldsymbol{p}_j}{M_{\text{core}}}$ term accounts for the two-body recoil term. As seen in Equation (15), the GSM has two components: the one-body part Hamiltonian $H_0 = \sum_{i=1}^{N_v}\left[\frac{\boldsymbol{p}_i^2}{2\mu_i} + U_{\text{core}}(i)\right]$ and the two-body part Hamiltonian $H_I = \sum_{i<j=1}^{N_v}\left[V(i,j) + \frac{\boldsymbol{p}_i\boldsymbol{p}_j}{M_{\text{core}}}\right]$. The core-valence particle potential $U_{\text{core}}$ is usually a WS potential, in which a spin-orbit term is included. The NN interaction $V$ takes the form of an effective phenomenological NN interaction, such as Minnesota [65], Furutani-Horiuchi-Tamagaki (FHT) [66,67], or effective field theory (EFT) [18,56] interactions. The parameters within the effective Hamiltonian in Equation (15), both one- and two-body interactions, need to be optimized to reproduce experimental data. For optimizations, the $\chi^2$ optimization method is employed, where one uses the Gauss–Newton algorithm augmented by the singular value decomposition technique to calculate the Jacobian pseudo-inverse [43].

## 3. Gamow Shell Model Calculations

### 3.1. Neutron-Rich Oxygen and Fluorine Isotopes

Neutron-rich oxygen isotopes form a particularly interesting chain for experimental and theoretical research. Firstly, the proton number $Z = 8$ shows magical properties for the neutron-rich oxygen isotopes, which provide a good laboratory to perform configuration interaction (shell-model) calculations [22,38,68,69]. Secondly, the nuclei $^{22}$O and $^{24}$O exhibit doubly magicity at the neutron number $N = 14$ and 16, respectively, [70–73]. Thirdly, experiments have shown that the $^{25}$O and $^{26}$O are unbound and decay by one- and two-neutron emission, respectively, [8,74]. Experimental studies suggest that $^{24}$O is the heaviest bound isotope of the oxygen chain [8,74]. However, the loosely unbound property of $^{26}$O, which is only $-18$ keV unbound [8], is a strong incentive to investigate the bound or unbound character of $^{28}$O, which should have a magicity of $N = 20$. Consequently, the neutron-rich oxygen isotopes provide an ideal laboratory to study many-body correlation, continuum coupling, and single-particle structure. By adding one valence proton to the neutron-rich oxygen isotopes, one obtains the fluorine isotopes at the neutron drip line, which can sustain six additional neutrons after $^{25}$F, hence, up to $^{31}$F, which is suspected to be at the neutron drip line of the fluorine chain [75]. This dramatic change is called an "oxygen anomaly". Moreover, many exotic properties develop at the neutron drip line for fluorine isotopes, such as halos in $^{29}$F [76] and $^{28}$F within the island of inversion around $N = 20$ [77], and thus fluorine isotopes provide a very interesting ground for theoretical studies.

#### 3.1.1. Realistic Gamow Shell Model Calculations

We have developed realistic GSM with the Berggren basis using a WS potential, while the realistic effective Hamiltonian is constructed within the model space using a nondegenerate $\hat{Q}$-box folded-diagram method [38]. We first employed it to investigate the neutron-rich oxygen isotopes up to and beyond the neutron drip line [38]. In our calculations, the realistic CD-Bonn potential [55] was used. To speed up the convergence of many-body calculations, the bare force is usually softened to remove the strong short-range repulsive core. The $V_{\text{low}-k}$ method [59] is used for that matter in Ref. [38].

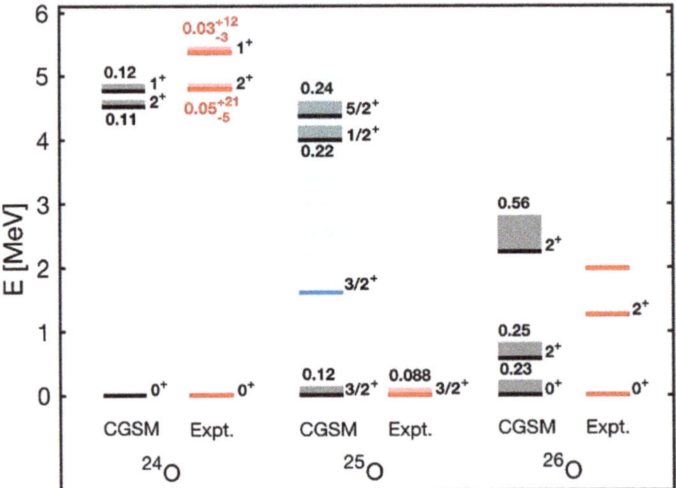


**Figure 2.** Calculated spectrum of $^{24,25,26}$O, compared with available experimental data [8,10,74]. The resonances are indicated by shades, and their widths (in MeV) are given by the number below or above the levels. The light blue shade indicates the $3/2^+$ many-body scattering states (with permissions from Ref. [38]). The "CGSM" stays for "core Gamow shell model".

Figure 2 shows the calculated low-lying states of $^{24-26}$O, along with experimental data [8,10,74]. Our realistic GSM calculations [38] reproduce the experimental excited-state spectrum well, including the observed resonance widths. The ground state energies and one-neutron separation energies $S_n$ of the neutron-rich oxygen isotopes are also calculated [38] (see Figure 3) and compared to the experimental data [8,10,74,78]. The WS parameters used, taken from Ref. [38], reproduce the experimental $1s_{1/2}$ and $0d_{3/2}$ single-particle energies well, including the decay width of the $0d_{3/2}$ state, but give the $0d_{5/2}$ energy as lower than the experimental data, at about 1.17 MeV [10]. The results presented in Figure 3 show that adopting the experimental $0d_{5/2}$ energy can dramatically improve calculations. Overbinding in the GSM calculations of oxygen isotopes after $^{24}$O is obtained in Ref. [38], which is caused by the absence of the three-nucleon force (3NF).

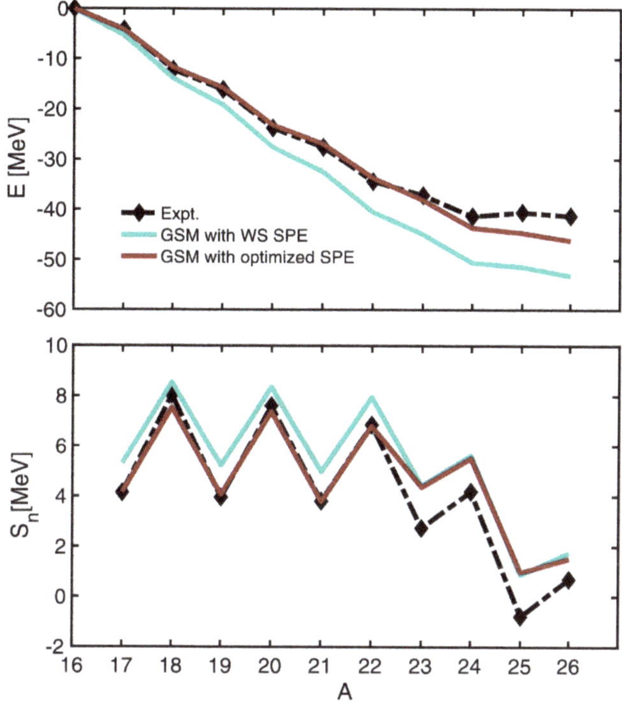


**Figure 3.** Calculated ground state energies of oxygen isotopes with respect to the $^{16}$O core (upper panel) and associated neutron separation energies $S_n$ (lower panel) as a function of atomic number compared with experimental data [8,74,78]. "GSM with WS SPE" indicates that the calculations were performed with Woods-Saxon (WS) single-particle energies (SPE), whereas "GSM with optimized SPE" means that the calculations were performed with the $0d_{5/2}$ SPE replaced by its experimental value (with permissions from Ref. [38]).

#### 3.1.2. *Ab-initio* Realistic GSM Calculations within GHF Basis

GSM is usually performed using a basis generated by a WS potential [5,19,36–39], whose parameters must be determined by fitting experimental single-particle energies and resonance widths. However, the single-particle energies and resonance widths in the multi-shell case are sometimes difficult to assess due to the lack of experimental data for that matter [10]. We then developed an ab initio realistic GSM approach by introducing the GHF basis as the Berggren basis [41]. The GHF basis is obtained by using the same interaction as the one used in the construction of the effective SM Hamiltonian [41], and thus there is no parameter introduced in the GHF Berggren basis. Starting from the chiral next-to-next-to-leading-order ($\text{NNLO}_{\text{opt}}$) force [79], we perform a nondegenerate $\hat{Q}$-box folded-diagram

calculation [38,62] in the GHF basis in order to construct a complex effective Hamiltonian. The energies and widths of single-particle orbitals can also be obtained self-consistently using the nondegenerate $\hat{S}$-box folded-diagram method [41]. The neutron-rich fluorine isotopes have been extended to the $pf$-shell, using a cross-shell effective Hamiltonian with the following model space : $\{1s_{1/2}, 0d_{5/2}, 0d_{3/2}\}$ for the valence proton, and $\{1s_{1/2}, 0d_{3/2} + d_{3/2}$ scattering states, $1p_{3/2} + p_{3/2}$ scattering states, $1p_{1/2} + p_{1/2}$ scattering states, $f_{7/2}$ scattering states$\}$ for valence neutrons. More details can be found in Ref. [41]. The constructed effective Hamiltonian was employed to study neutron-rich oxygen and fluorine drip line nuclei.

Figure 4 shows the calculated ground-state energies and neutron separation energies $S_n$ of oxygen and fluorine isotopic chains, as well as comparisons with experimental data [78] and other theoretical calculations [31,68,79–83]. The GSM calculations using a GHF basis and based on the NNLO$_{\text{opt}}$ [79] provide the correct location of the neutron drip line of oxygen isotopes and a good description of the unbound nuclei $^{25,26}$O, which lie beyond the neutron drip line (see the left panel of Figure 4). Note that, when using the standard SM calculations with the USDB interaction [68], conventional SM calculations with NN + 3NF [82], or valence–space IMSRG (VS-IMSRG) calculations with NN + 3NF [81], the resonance and continuum couplings are absent. Complex CC [31] and GSM calculations [80] are displayed in Figure 4 for comparison.

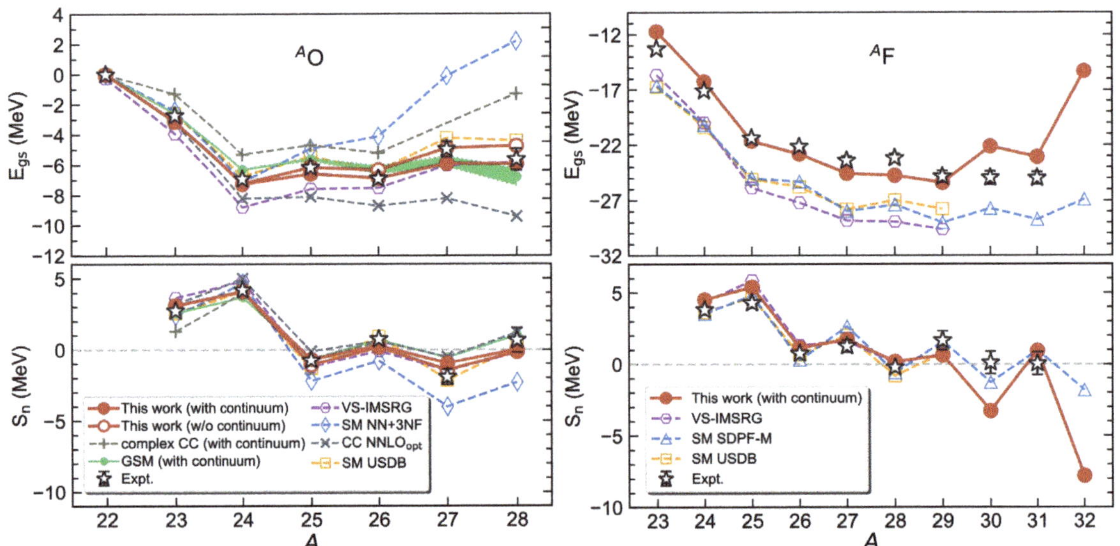


**Figure 4.** Calculated ground-state energies (**upper panel**) with respect to the $^{22}$O core and associated neutron separation energies $S_n$ (**lower panel**) for oxygen and fluorine isotopes, compared with experimental data [78] (the AME2016 extrapolated values are taken for $^{27,28}$O and $^{30,31}$F) and theoretical calculations from other groups: complex coupled-cluster (CC) with next-to-next-to-next-to-leading-order nucleon-nucleon CC with NNLO$_{\text{opt}}$ interaction [79], GSM [80], –space in-medium similarity renormalization group (VS-IMSRG) [81], SM with NN+3NF [82], SM with USDB [68], and SM with SDPF-M [83] (with permissions from Ref. [41]).

The results of fluorine isotopes are shown in the right panels of Figure 4. For comparison, standard SM calculations using USDB [68] and SDPF-M [83] effective interactions are also presented. All ground-state energies in Figure 4 are given with respect to the ground state of $^{22}$O. Experiments revealed that $^{31}$F is a neutron drip line nucleus [75]. Although our GSM calculations provide a lower energy of $^{31}$F compared to that of $^{30}$F, $^{31}$F is still unbound compared to $^{29}$F. However, our GSM calculations provide good descriptions of ground-state energies for $^{23-29}$F.

#### 3.1.3. GSM Calculations with Phenomenological Nuclear Potential

Many ab initio calculations, such as SM [82], VS-IMSRG [81], complex CC [31], and realistic GSM [38,39,41] calculations, have been employed for the description of neutron-rich oxygen and fluorine isotopes. However, these calculations bear a large theoretical uncertainty. Furthermore, results arising from ab initio calculations depend on the realistic nuclear forces used (a short summary of the VS-IMSRG calculations based on different chiral nuclear forces can be found in Ref. [80]). Moreover, continuum coupling is absent in the VS-IMSRG [81] and SM [82] calculations. Similar situations also occur for neutron-rich fluorine isotopes, where few calculations have been performed and most of the calculations are absent for the continuum coupling [83,84]. Based on these grounds, we performed the GSM calculations with a phenomenological nuclear interaction for neutron-rich oxygen and fluorine drip line nuclei.

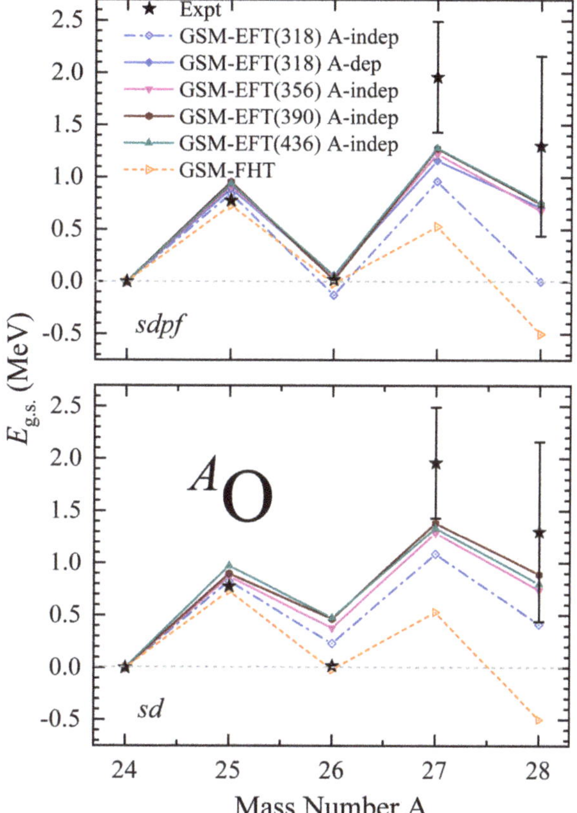


**Figure 5.** Energies of ground states in $^{24-28}$O, calculated by GSM within the *sdpf* ($s_{1/2}, p_{1/2,3/2}, d_{3/2,5/2}, f_{5/2,7/2}$ partial waves) (**upper**) and *sd* ($s_{1/2}, d_{3/2,5/2}$ partial waves) (**lower**) model spaces, using the effective field theory (EFT) EFT(318)(the value within braket stands momentum cut-offs), EFT(356), EFT(390), EFT(436), and Furutani–Horiuchi–Tamagaki (FHT) interactions, with A-independence (*A*-indep) or A-dependence (*A*-dep) (see details in Ref. [42]). Results are compared with the experimental data available, represented by a star. The data for $^{25,26}$O and $^{27,28}$O are taken from experiment (see Refs. [8,74]) and evaluations given in AME2016 [78] (with permissions from Ref. [42]).

For the considered neutron-rich oxygen isotopes, the closed-shell nucleus $^{22}$O is selected to be the inner core. The one-body potential is mimicked by a WS potential, whose

parameters are adjusted to reproduce the single-particle spectrum of $^{23}$O [10]. We use the pionless EFT interaction [85,86] as the two-body interaction. Owing to the few available data related to the oxygen drip line nuclei [10], only the leading order (LO) NN interaction of the EFT force is fitted to reproduce selected experimental data. The effect of the 3NF at LO is then effectively taken into account in the fitted parameters. Details can be found in Ref. [42]. We calculated the energies of the ground states of $^{24-28}$O with GSM within $sdpf$ ($s_{1/2}, p_{1/2,3/2}, d_{3/2,5/2}, f_{5/2,7/2}$ partial waves) and *sd* ($s_{1/2}, d_{3/2,5/2}$ partial waves) active spaces, using different EFT interactions (see details in Ref. [42]). The calculated ground-state energies of $^{24-28}$O are shown in Figure 5. The calculations within the *sd* space show that the $^{25-28}$O isotopes are unbound and that their binding energies are close to the experimental data [8,74,78] and to calculations performed within the $sdpf$ space. However, the calculations obtained in the *sd* space provide an unbound $^{26}$O ground state, by about 300 keV relative to the ground state of $^{24}$O, which is a little higher than its experimental value, which is about 20 keV unbound [8]. Though the energy difference obtained using the two different model spaces is small, the calculation performed within the $spdf$ space seems to be more reasonable. The GSM calculations performed within the $sdpf$ space provide good agreements of the $^{23-26}$O ground states with experimental data [8,74,78]; in particular, the two-neutron separation energy ($S_{2n}$) of $^{26}$O is about 20 keV [8]. The calculated ground state of $^{28}$O is unbound in all three cases and located about 700 keV above the ground state of $^{24}$O. The ground states of the $^{26,28}$O isotopes are unbound, but bear negligible widths. Together with the calculated one-body densities of the $^{26,28}$O isotopes in Ref. [42], the results suggest that the ground state of $^{28}$O exhibits four-neutron decay by way of $2n$-$2n$ emission via the $^{26}$O ground state, which is consistent with few-body [87] and the above ab initio GSM calculations [41].

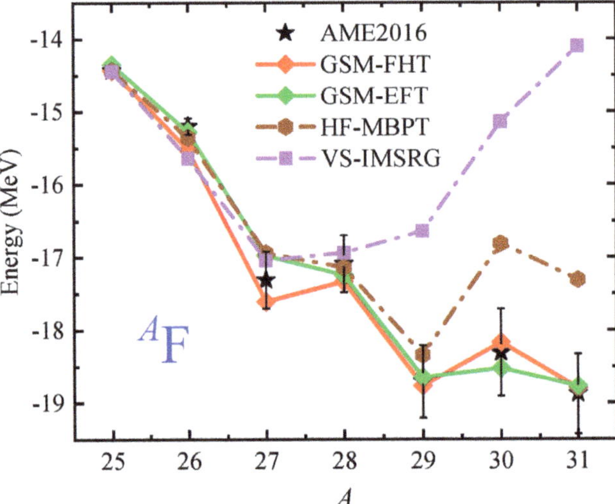


**Figure 6.** Energies of $^{25-31}$F with respect to the $^{24}$O core, calculated within different theoretical frameworks and compared to experimental data [78]. Besides the GSM calculations using FHT and EFT interactions, calculations in the Hartree-Fock many-body perturbation theory (HF-MBPT) [88] and VS-IMSRG [84] frameworks, utilizing the harmonic-oscillator (HO) basis, hence being without continuum coupling, are presented (with permissions from Ref. [18]).

Figure 6 shows the GSM calculations of the binding energies of fluorine isotopes using FHT and EFT interactions (see details in Ref. [18]), wich are compared with experimental data [78] and Hartree-Fock MBPT (HF-MBPT) [88] and VS-IMSRG [84] calculations, which are both performed in the HO basis. The energy of $^{25}$F has been fixed to its experimental datum in all used models in Figure 6. We can see that all calculations reproduce the

ground state energies of $^{25-28}$F isotopes well, situated in the well-bound region, whereas differences start after $^{29}$F, i.e., when one reaches the neutron drip line. Due to the absence of both multi-shell and continuum couplings, the VS-IMSRG calculations [84] provide visible differences, which are about 4- to 5-MeV in magnitude for $^{30,31}$F. When applying the HF-MBPT method [88], the cross-shell couplings generated by the *sd* and $pf$ shells are included, so that proper binding energies of up to $^{29}$F are predicted. However, due to the lack of continuum coupling, the binding energies of $^{30,31}$F are about 1 MeV away from experimental data. The GSM calculation performed with FHT and EFT interactions correctly provides binding energies of up to $^{31}$F. Moreover, the odd–even staggering encountered from the $^{28}$F isotope, typical of the presence of a strong proton–neutron interaction, is well reproduced, with $^{30}$F being unbound and $^{31}$F being loosely bound in our calculations.

Recent realistic shell model calculations [89] have pointed out that nuclear deformation plays an important role in the neutron drip line nuclei. Within GSM, deformation can be accounted for by configuration mixing using a cross-shell valence space [5]. Besides deformation, continuum coupling also gives important contributions in drip line nuclei. They are strongly coupled with continuum states near the particle-emission threshold, which provides additional binding energy [5]. This situation is unlike that occurring in well-bound systems, where one only has strong coupling with nearby deeply-bound single-particle states.

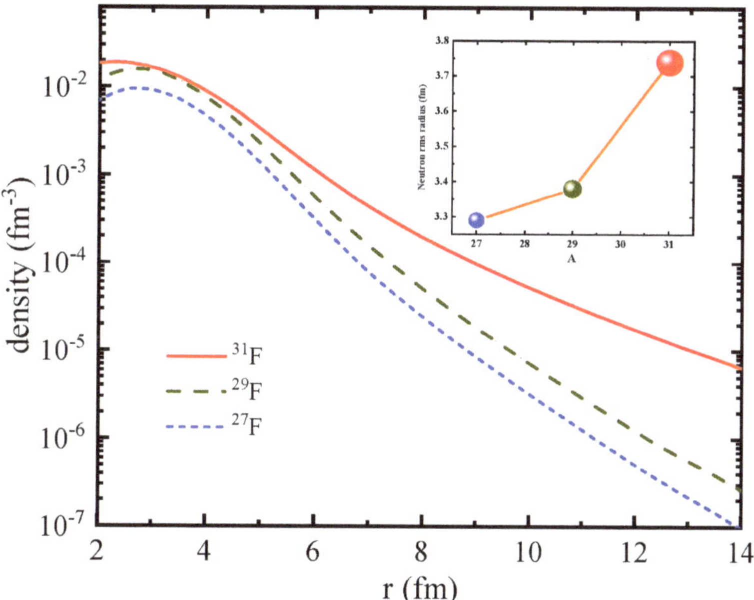


**Figure 7.** One-nucleon densities of the bound $^{27,29,31}$F isotopes calculated with the GSM using the EFT interaction in the valence space as a function of radii $r$, respectively, depicted by short-dashed, long-dashed, and solid lines. The rms radii of these isotopes are shown in the inset (with permissions from Ref. [18]).

The two-neutron separation energy $S_{2n}$ of $^{31}$F is about 170 keV [78], which is sufficiently small for sustaining a two-neutron halo. We calculated the one-nucleon densities and neutron rms radii of the neutron-bound $^{27,29,31}$F isotopes using GSM with the EFT interaction (see Figure 7). From our calculations, a halo clearly develops in the asymptotic region of $^{31}$F. Indeed, on the one hand, the one-nucleon density of $^{31}$F slowly decreases on the real axis and is about one to two orders of magnitude larger than those of $^{27,29}$F in

the asymptotic region. Added to that, on the other hand, the neutron rms radius of $^{31}$F does not follow the trend noticed in $^{27,29}$F, as the rms radius sharply increases compared to the associated values in $^{27}$F and $^{29}$F. Consequently, one can assume from these GSM calculations [18] that $^{31}$F is a two-neutron halo state.

### 3.2. Realistic Gamow Shell Model Calculations of Neutron-Rich Calcium Isotopes

The long chain of calcium isotopes provides an ideal laboratory for both theoretical and experimental investigations of unstable isotopes. With two typical doubly magic isotopes, $^{40}$Ca and $^{48}$Ca, and two new magic isotopes discovered in the neutron-rich region, $^{52}$Ca [90] and $^{54}$Ca [91], the calcium chain is speculated to end the $^{70}$Ca isotope. Its rich nuclear structure data [10] attract continued theoretical interest, especially using methods that include continuum coupling. The realistic GSM based on the realistic CD-Bonn [55] interaction has also been performed to investigate the properties of neutron-rich calcium isotopes up to the drip line.

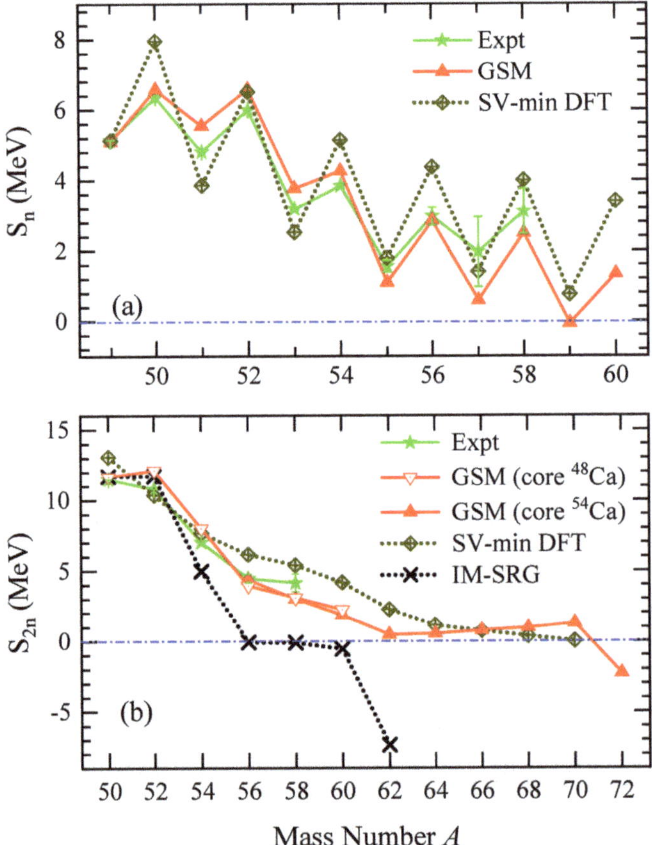


**Figure 8.** Calculated one-neutron separation energies $S_n$ (**a**) and two-neutron separation energies $S_{2n}$ (**b**), compared with data [78,92], and calculations obtained with SV-min density-functional theory (DFT) [93] and multireference IM-SRG ($S_{2n}$ only) [94]. The $S_n$ calculations end at $^{60}$Ca because odd isotopes heavier than $^{60}$Ca become unbound in our GSM calculations (with permissions from Ref. [47]).

The calculated one-neutron separation energies $S_n$ and two-neutron separation energies $S_{2n}$ are shown in Figure 8 and compared with experimental data [78,92], DFT [93], and IM-SRG [94] calculations. The calculated one-neutron separation energies $S_n$ show that $^{57}$Ca is the heaviest odd-mass bound calcium isotope, which is consistent with MBPT calculations [95]. $^{59}$Ca is weakly unbound with a small one-neutron separation energy $S_n = -326$ keV in our GSM calculations. For the two-neutron separation energy $S_{2n}$, the GSM calculations are performed with two different cores, $^{48}$Ca and $^{54}$Ca. For $^{56,58,60}$Ca, the two calculations give similar results. The calculated two-neutron separation energy $S_{2n}$ is in good agreement with experimental data [78,92] and other theoretical calculations, e.g., with DFT [93] and IM-SRG [94] calculations. The large decrease in $S_{2n}$ at neutron number $N = 32$ and 34 indicate that subshell closures occur therein, which has also been suggested from experiments [90,91] and theoretical calculations [94–96]. Moreover, the calculated two-neutron separation energy $S_{2n}$ using GSM predicts that the two-neutron drip line of the calcium isotopes should be located at $^{70}$Ca. This is consistent with the recent mean-field calculations of Ref. [97].

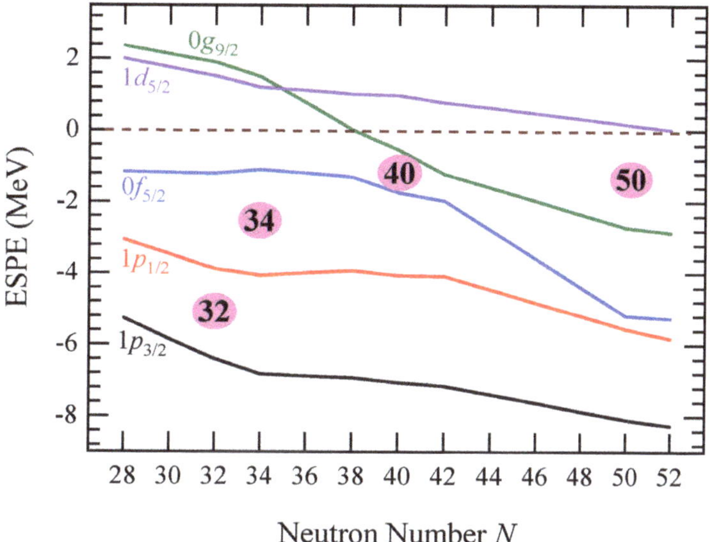


**Figure 9.** Neutron effective single-particle energies (ESPE) with respect to the $^{48}$Ca core, as a function of neutron number. The $V_{low-k}$ $\Lambda = 2.6$ fm$^{-1}$ CD-Bonn interaction is utilized (with permissions from Ref. [47]).

In order to see the shell evolution of the calcium isotopes around the neutron number $N = 32$, 34, 40, and 50, we calculated effective single-particle energies (ESPE) based on the GSM effective Hamiltonian. Figure 9 shows the evolution of the valence neutron ESPEs when increasing the neutron number. The calculations show that large shell gaps between $1p_{3/2}$ and $1p_{1/2}$ and between $1p_{1/2}$ and $0f_{5/2}$ exist, indicating that shell closures occur at $N = 32$ and 34, respectively. These results are consistent with experimental observations [90,91] and theoretical calculations [94–96,98]. The shell gap above the $0f_{5/2}$ orbit is reduced at around $N = 40$, implying a weakening of the $N = 40$ shell closure in the calcium chain. The $0g_{9/2}$ shell becomes bound at $N \geq 40$, which can enhance the stability of the heavy calcium isotopes. The observed $^{60}$Ca isotope in experiments [99] may be an indication of this enhanced stability. Moreover, the calculated ESPEs show a significant shell gap at $N = 50$, implying a shell closure at $^{70}$Ca.

### 3.3. One-Proton and Two-Proton Decays in $^{16}$Ne and $^{18}$Mg Unbound Nuclei

Two-proton decay is one of the most important drip-line phenomena. It occurs in proton drip line nuclei, such as $^{48}$Ni, $^{54}$F, $^{54}$Zn, $^{76}$K, $^{16}$Ne, and $^{19}$Mg (see a review of this topic in Ref. [4] ). While $^{18}$Mg has not been observed, it can decay in principle by proton and/or two-proton emissions. The GSM is then a suitable method to study these types of particle emissions. We carried out GSM calculations of the proton-rich carbon isotones of $^{14}$O, which are all resonance [10], using $^{14}$O as an inner core. The obtained energy spectra of carbon isotones are depicted in Figure 10 with respect to the ground state of $^{14}$O. One can see that both the energies and widths of experimentally known eigenstates are well reproduced for the low-lying states in $^{15}$F and $^{16}$Ne [10]. We also provide predictions for the $^{17}$Na and $^{18}$Mg nuclear spectra, of which, there are no experimental data. Our calculations show that the $^{16}$Ne and $^{18}$Mg isotopes are unbound nuclei, where both one-proton separation energies $S_p$ and two-proton separation energies $S_{2p}$ are negative, thereby indicating that two different particle-emission channels are open therein.

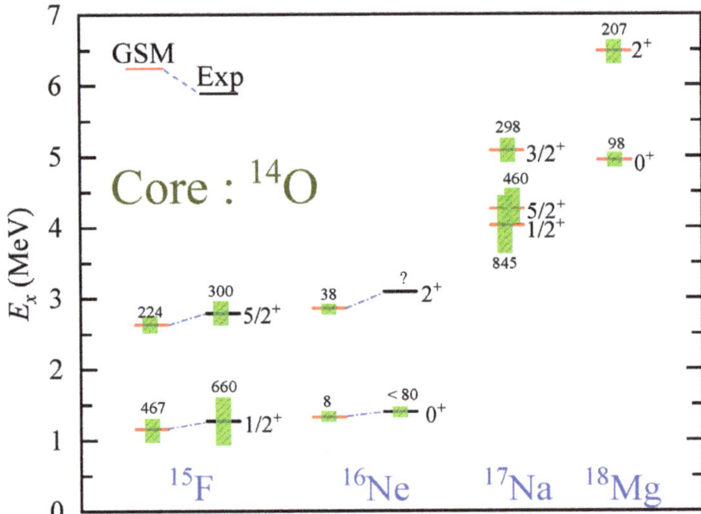


**Figure 10.** Excitation energies, $E_x$, and widths (in keV) of the ground and excited states of carbon isotones. The GSM calculations are compared to available experimental data [10,100–102]. Energies are given with respect to the $^{14}$O core. Widths are represented by green striped squares, and their explicit values are written above (with permissions from Ref. [48]).

To evaluate one-proton and two-proton decay widths, we changed the central potential depth $V_0$ of the WS core potential in order for the $S_p$ to become positive or very negative (see details in Ref. [48]). Consequently, it is possible to find a central potential depth for which only the two-proton decay channel is open, so that the obtained width is that of the two-proton emission. The obtained results are shown in Figure 11. As $^{15}$F and $^{17}$Ne are one-proton resonances, their width increases steadily with the Hamiltonian central potential depth. In contrast, one can see that the widths of $^{16}$Ne and $^{18}$Mg increase abruptly when the one-proton channel opens. The width of two-proton decay is almost constant with respect to the central potential depth below the one-proton emission threshold, and is also about 500 keV to 1 MeV above (see Figure 11). It is reasonable to assume that the two-proton decay width is almost independent of energy. Therefore, the GSM results shown in Figure 11, where only the two-proton channel is open, can be extrapolated to the physical case (indicated by an arrow in Figure 11). This two-proton decay width is about 10-15 keV for both $^{16}$Ne and $^{18}$Mg nuclei. The one-proton width can be assumed as

the difference between the total width and the two-proton emission width of 10–15 keV. Then, our calculations show that one-proton emission is negligible for $^{16}$Ne, whereas the one-proton decay width in $^{18}$Mg is estimated to be about 85-90 keV. The obtained data for $^{16}$Ne are also in agreement with experimental data [10,100–102].

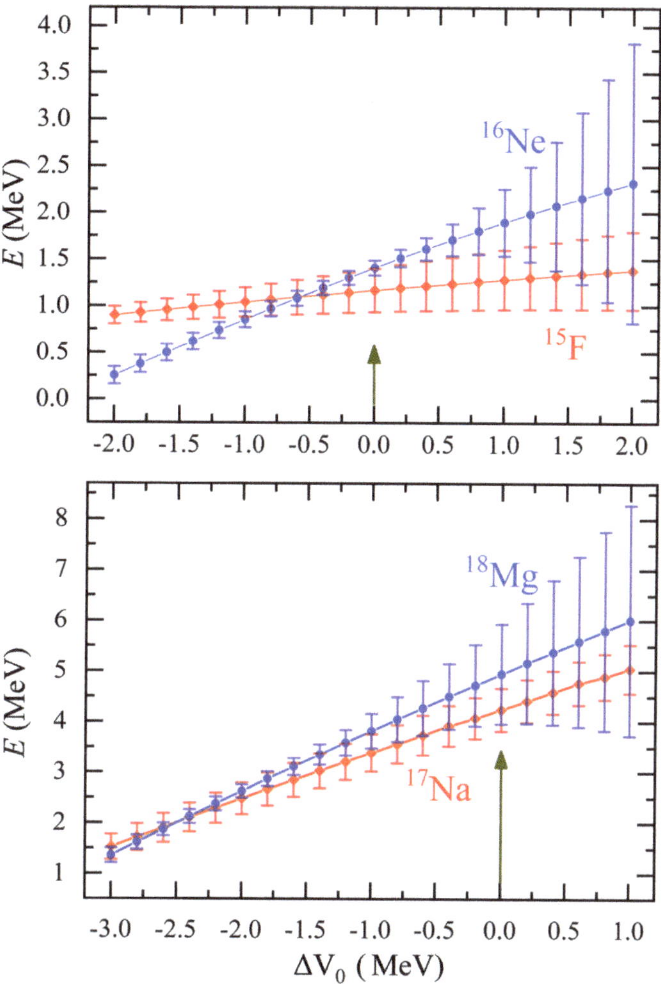


**Figure 11.** Calculated energies and widths (in MeV) of $^{15}$F, $^{16}$Ne (**upper panel**), $^{17}$Na, and $^{18}$Mg (**lower panel**) as a function of the difference $\Delta V_0 = V_0 - V_0^{(\text{fit})}$(fit) of the WS central potential depths (see details in Ref. [48]). Energies are depicted by blue disks and red lozenges for even and odd nuclei, respectively. Widths are represented by segments centered on disks and lozenges. The widths of $^{16}$Ne and $^{18}$Mg have been multiplied by 20 for readability. Energies are given with respect to the $^{14}$O core. The physical GSM calculation, for which $V_0 = V_0^{(\text{fit})}$, is indicated by an arrow (with permissions from Ref. [48]).

## 4. Summary

The Gamow shell model (GSM) is a powerful method for the description of the weakly bound and resonance properties of drip line nuclei. In the present review, we presented several recent applications of GSM dedicated to the study of drip line nuclei, including

GSM calculations of neutron-rich oxygen and fluorine drip line nuclei, of the long chain of neutron-rich calcium isotopes, and of the unbound proton-rich $^{16}$Ne and $^{18}$Mg isotopes. For the neutron-rich oxygen and fluorine drip line nuclei, both the realistic GSM and GSM with phenomenological forces have been utilized. Our calculations have described the weakly-bound and unbound properties of drip line nuclei well. Furthermore, the unbound properties of the $^{28}$O are obtained within the two both types of GSM calculations, and the two-neutron halo property of $^{31}$F has been predicted in GSM calculations as well. The realistic GSM calculations provide good agreements of the neutron-rich calcium isotopes with experimental data, as GSM calculations predict that the one- and two-neutron drip line nuclei of calcium isotopes are $^{57}$Ca and $^{70}$Ca, respectively. For the unbound proton-rich $^{16}$Ne and $^{18}$Mg nuclei, GSM calculations provide calculations and predictions for their low-lying spectra. Added to that, the one- and two-proton emission widths could be estimated for $^{16}$Ne and $^{18}$Mg isotopes. Our calculations have shown that $^{16}$Ne decays only by two-proton emission, whereas $^{18}$Mg can decay through both one- and two-proton emission channels, whose widths are estimated to be about 85–90 and 10–15 keV, respectively.

GSM has thus been shown to be the tool of choice for the study of drip line nuclei. Many challenges remain to be overcome for the future applications of GSM:

- Due to the large computational cost of the GSM calculations, the GSM has been applied for the neutron-rich nuclei with only one or two valence protons, and, for proton-rich nuclei, with only one or two valence neutrons in the non-resonant continuum. For example, the model space dimension of $^{31}$F is about $10^7$ with two valence particles in the continuum. It can easily reach $10^{10}$ without truncations, which is untractable numerically. In the nuclear chart, most of the drip line nuclei need to be described with many valence particles (protons and neutrons). One can think of the neutron-rich Ne and Mg drip line isotopes, where both continuum coupling and strong internucleon correlations must be treated properly. These isotopes will provide a challenge for future GSM calculations, due to the large dimensions;
- The diagonalization of the GSM Hamiltonian in order to obtain eigenstates of large resonance widths, such as the second $0_2^+$ state in $^8$He, is very difficult from a numerical point of view;
- The dimensions of the GSM Hamiltonian matrices increase extremely quickly when one adds valence particles, and thus the treatment of the many-body Hamiltonian is difficult when using the configuration interaction framework. Other kinds of many-body methods are urgently needed. The two-particle reduced density matrix method is one of the promising methods to solve the dimensionality problem of the GSM many-body Hamiltonian [103];
- The unbound single-particle states of *s* waves in neutron-rich nuclei are anti-bound states, which are difficult to include in many-body GSM calculations. The consideration of many-body anti-bound states in GSM (the ground state of $^{10}$Li is supposed to be anti-bound, for example [104]) is thus also a challenge for future applications of GSM.

**Author Contributions:** Writing—original draft preparation, J.L.; writing—review and editing, F.X. and N.M.; critically review, Y.M., B.H., Z.S. and W.Z. All authors have read and agreed to the published version of the manuscript.

**Funding:** This work has been funded by the National Key R&D Program of China under Grant No. 2018YFA0404401; the National Natural Science Foundation of China under Grants No. 11835001, NO. 11921006, NO. 12035001, and NO. 11975282; the State Key Laboratory of Nuclear Physics and Technology, Peking University under Grant NPT2020KFY13; the China Postdoctoral Science Foundation under Grant No. BX20200136 and 2020M682747; the Strategic Priority Research Program of Chinese Academy of Sciences under Grant No. XDB34000000; the Key Research Program of the Chinese Academy of Sciences under Grant No. XDPB15; and the CUSTIPEN (China-U.S. Theory Institute for Physics with Exotic Nuclei) funded by the U.S. Department of Energy, Office of Science

under Grant No. de-sc0009971. The High-Performance Computing Platform of Peking University is acknowledged.

**Data Availability Statement:** All of the relevant data are available from the corresponding author upon reasonable request.

**Conflicts of Interest:** The authors declare no conflict of interest.

## References


1. Tanihata, I.; Savajols, H.; Kanungo, R. Recent experimental progress in nuclear halo structure studies. *Prog. Part. Nucl. Phys.* **2013**, *68*, 215–313. [CrossRef]
2. Motobayashi, T. World new facilities for radioactive isotope beams. *EPJ Web Conf.* **2014**, *66*, 01013. [CrossRef]
3. Blank, B.; Borge, M. Nuclear structure at the proton drip line: Advances with nuclear decay studies. *Prog. Part. Nucl. Phys.* **2008**, *60*, 403–483. [CrossRef]
4. Pfützner, M.; Karny, M.; Grigorenko, L.V.; Riisager, K. Radioactive decays at limits of nuclear stability. *Rev. Mod. Phys.* **2012**, *84*, 567–619. [CrossRef]
5. Michel, N.; Nazarewicz, W.; Płoszajczak, M.; Vertse, T. Shell model in the complex energy plane. *J. Phys. G Nucl. Part. Phys.* **2009**, *36*, 013101. [CrossRef]
6. Riisager, K. Nuclear halo states. *Rev. Mod. Phys.* **1994**, *66*, 1105–1116. [CrossRef]
7. Al-Khalili, J. An introduction to halo nuclei. In *The Euroschool Lectures on Physics with Exotic Beams, Vol. I*; Al-Khalili, J., Roeckl, E., Eds.; Springer: Berlin/Heidelberg, Germany, 2004; pp. 77–112. [CrossRef]
8. Kondo, Y.; Nakamura, T.; Tanaka, R.; Minakata, R.; Ogoshi, S.; Orr, N.A.; Achouri, N.L.; Aumann, T.; Baba, H.; Delaunay, F.; et al. Nucleus $^{26}$O: A barely unbound system beyond the drip Line. *Phys. Rev. Lett.* **2016**, *116*, 102503. [CrossRef]
9. Michel, N.; Nazarewicz, W.; Płoszajczak, M.; Okołowicz, J. Gamow shell model description of weakly bound nuclei and unbound nuclear states. *Phys. Rev. C* **2003**, *67*, 054311. [CrossRef]
10. Available online: https://www.nndc.bnl.gov/ensdf/ (accessed on 10 August 2021).
11. Renzi, F.; Raabe, R.; Randisi, G.; Smirnov, D.; Angulo, C.; Cabrera, J.; Casarejos, E.; Keutgen, T.; Ninane, A.; Charvet, J.L.; et al. Spectroscopy of $^{7}$**He** using the $^{9}$**Be**($^{6}$**He**,$^{8}$ **Be**) transfer reaction. *Phys. Rev. C* **2016**, *94*, 024619. [CrossRef]
12. Votaw, D.; DeYoung, P.A.; Baumann, T.; Blake, A.; Boone, J.; Brown, J.; Chrisman, D.; Finck, J.E.; Frank, N.; Gombas, J.; et al. Low-lying level structure of the neutron-unbound $N = 7$ isotones. *Phys. Rev. C* **2020**, *102*, 014325. [CrossRef]
13. Alkhazov, G.D.; Andronenko, M.N.; Dobrovolsky, A.V.; Egelhof, P.; Gavrilov, G.E.; Geissel, H.; Irnich, H.; Khanzadeev, A.V.; Korolev, G.A.; Lobodenko, A.A.; et al. Nuclear matter distributions in $^{6}$He and $^{8}$He from small angle p-He scattering in inverse kinematics at intermediate energy. *Phys. Rev. Lett.* **1997**, *78*, 2313–2316. [CrossRef]
14. Wang, L.B.; Mueller, P.; Bailey, K.; Drake, G.W.F.; Greene, J.P.; Henderson, D.; Holt, R.J.; Janssens, R.V.F.; Jiang, C.L.; Lu, Z.T.; et al. Laser spectroscopic determination of the $^{6}$He nuclear charge radius. *Phys. Rev. Lett.* **2004**, *93*, 142501. [CrossRef]
15. Mueller, P.; Sulai, I.A.; Villari, A.C.C.; Alcántara-Núñez, J.A.; Alves-Condé, R.; Bailey, K.; Drake, G.W.F.; Dubois, M.; Eléon, C.; Gaubert, G.; et al. Nuclear charge radius of $^{8}$He. *Phys. Rev. Lett.* **2007**, *99*, 252501. [CrossRef]
16. Michel, N.; Nazarewicz, W.; Płoszajczak, M.; Rotureau, J. Antibound states and halo formation in the Gamow shell model. *Phys. Rev. C* **2006**, *74*, 054305. [CrossRef]
17. Papadimitriou, G.; Kruppa, A.T.; Michel, N.; Nazarewicz, W.; Płoszajczak, M.; Rotureau, J. Charge radii and neutron correlations in helium halo nuclei. *Phys. Rev. C* **2011**, *84*, 051304. [CrossRef]
18. Michel, N.; Li, J.G.; Xu, F.R.; Zuo, W. Two-neutron halo structure of $^{31}$F. *Phys. Rev. C* **2020**, *101*, 031301. [CrossRef]
19. Ma, Y.Z.; Xu, F.R.; Michel, N.; Zhang, S.; Li, J.G.; Hu, B.S.; Coraggio, L.; Itaco, N.; Gargano, A. Continuum and three-nucleon force in Borromean system: The $^{17}$Ne case. *Phys. Lett. B* **2020**, *808*, 135673. [CrossRef]
20. Barrett, B.R.; Navrátil, P.; Vary, J.P. Ab initio no core shell model. *Prog. Part. Nucl. Phys.* **2013**, *69*, 131–181. [CrossRef]
21. Somà, V.; Cipollone, A.; Barbieri, C.; Navrátil, P.; Duguet, T. Chiral two- and three-nucleon forces along medium-mass isotope chains. *Phys. Rev. C* **2014**, *89*, 061301. [CrossRef]
22. Jansen, G.R.; Engel, J.; Hagen, G.; Navratil, P.; Signoracci, A. Ab initio coupled-cluster effective interactions for the shell model: Application to neutron-rich oxygen and carbon isotopes. *Phys. Rev. Lett.* **2014**, *113*, 142502. [CrossRef]
23. Hagen, G.; Hjorth-Jensen, M.; Jansen, G.; Papenbrock, T. Emergent properties of nuclei from ab initio coupled-cluster calculations. *Phys.Scr.* **2016**, *91*, 063006. [CrossRef]
24. Stroberg, S.R.; Calci, A.; Hergert, H.; Holt, J.D.; Bogner, S.K.; Roth, R.; Schwenk, A. Nucleus-dependent valence-space approach to nuclear structure. *Phys. Rev. Lett.* **2017**, *118*, 032502. [CrossRef]
25. Caurier, E.; Martínez-Pinedo, G.; Nowacki, F.; Poves, A.; Zuker, A.P. The shell model as a unified view of nuclear structure. *Rev. Mod. Phys.* **2005**, *77*, 427–488. [CrossRef]
26. Otsuka, T.; Gade, A.; Sorlin, O.; Suzuki, T.; Utsuno, Y. Evolution of shell structure in exotic nuclei. *Rev. Mod. Phys.* **2020**, *92*, 015002. [CrossRef]
27. Papadimitriou, G.; Rotureau, J.; Michel, N.; Płoszajczak, M.; Barrett, B.R. Ab initio no-core Gamow shell model calculations with realistic interactions. *Phys. Rev. C* **2013**, *88*, 044318. [CrossRef]

28. Li, J.G.; Michel, N.; Hu, B.S.; Zuo, W.; Xu, F.R. Ab initio no-core Gamow shell-model calculations of multineutron systems. *Phys. Rev. C* **2019**, *100*, 054313. [CrossRef]
29. Li, J.G.; Michel, N.; Zuo, W.; Xu, F.R. Resonances of $A = 4$ $T = 1$ isospin triplet states within the ab initio no-core Gamow shell model. *Phys. Rev. C* **2021**, *104*, 024319. [CrossRef]
30. Navrátil, P.; Quaglioni, S.; Hupin, G.; Romero-Redondo, C.; Calci, A. Unified ab initio approaches to nuclear structure and reactions. *Phys. Scr.* **2016**, *91*, 053002. [CrossRef]
31. Hagen, G.; Hjorth-Jensen, M.; Jansen, G.R.; Machleidt, R.; Papenbrock, T. Continuum Effects and Three-Nucleon Forces in Neutron-Rich Oxygen Isotopes. *Phys. Rev. Lett.* **2012**, *108*, 242501. [CrossRef]
32. Hagen, G.; Hjorth-Jensen, M.; Jansen, G.R.; Machleidt, R.; Papenbrock, T. Evolution of Shell Structure in Neutron-Rich Calcium Isotopes. *Phys. Rev. Lett.* **2012**, *109*, 032502. [CrossRef]
33. Hu, B.S.; Wu, Q.; Sun, Z.H.; Xu, F.R. Ab initio Gamow in-medium similarity renormalization group with resonance and continuum. *Phys. Rev. C* **2019**, *99*, 061302. [CrossRef]
34. Volya, A.; Zelevinsky, V. Discrete and Continuum Spectra in the Unified Shell Model Approach. *Phys. Rev. Lett.* **2005**, *94*, 052501. [CrossRef]
35. Rotureau, J.; Okołowicz, J.; Płoszajczak, M. Microscopic theory of the two-proton radioactivity. *Phys. Rev. Lett.* **2005**, *95*, 042503. [CrossRef]
36. Id Betan, R.; Liotta, R.J.; Sandulescu, N.; Vertse, T. Two-particle resonant states in a many-body mean field. *Phys. Rev. Lett.* **2002**, *89*, 042501. [CrossRef]
37. Michel, N.; Nazarewicz, W.; Płoszajczak, M.; Bennaceur, K. Gamow shell model description of neutron-rich nuclei. *Phys. Rev. Lett.* **2002**, *89*, 042502. [CrossRef]
38. Sun, Z.H.; Wu, Q.; Zhao, Z.H.; Hu, B.S.; Dai, S.J.; Xu, F.R. Resonance and continuum Gamow shell model with realistic nuclear forces. *Phys. Lett. B* **2017**, *769*, 227–232. [CrossRef]
39. Ma, Y.Z.; Xu, F.R.; Coraggio, L.; Hu, B.S.; Li, J.G.; Fukui, T.; De Angelis, L.; Itaco, N.; Gargano, A. Chiral three-nucleon force and continuum for dripline nuclei and beyond. *Phys. Lett. B* **2020**, *802*, 135257. [CrossRef]
40. Berggren, T. On the use of resonant states in eigenfunction expansions of scattering and reaction amplitudes. *Nucl. Phys. A* **1968**, *109*, 265–287. [CrossRef]
41. Hu, B.S.; Wu, Q.; Li, J.G.; Ma, Y.Z.; Sun, Z.H.; Michel, N.; Xu, F.R. An ab initio Gamow shell model approach with a core. *Phys. Lett. B* **2020**, *802*, 135206. [CrossRef]
42. Li, J.G.; Michel, N.; Zuo, W.; Xu, F.R. Unbound spectra of neutron-rich oxygen isotopes predicted by the Gamow shell model. *Phys. Rev. C* **2021**, *103*, 034305. [CrossRef]
43. Jaganathen, Y.; Id Betan, R.M.; Michel, N.; Nazarewicz, W.; Płoszajczak, M. Quantified Gamow shell model interaction for *psd*-shell nuclei. *Phys. Rev. C* **2017**, *96*, 054316. [CrossRef]
44. Hagen, G.; Hjorth-Jensen, M.; Michel, N. Gamow shell model and realistic nucleon-nucleon interactions. *Phys. Rev. C* **2006**, *73*, 064307. [CrossRef]
45. Tsukiyama, K.; Hjorth-Jensen, M.; Hagen, G. Gamow shell-model calculations of drip-line oxygen isotopes. *Phys. Rev. C* **2009**, *80*, 051301. [CrossRef]
46. Furnstahl, R.J.; Hagen, G.; Papenbrock, T. Corrections to nuclear energies and radii in finite oscillator spaces. *Phys. Rev. C* **2012**, *86*, 031301. [CrossRef]
47. Li, J.G.; Hu, B.S.; Wu, Q.; Gao, Y.; Dai, S.J.; Xu, F.R. Neutron-rich calcium isotopes within realistic Gamow shell model calculations with continuum coupling. *Phys. Rev. C* **2020**, *102*, 034302. [CrossRef]
48. Michel, N.; Li, J.G.; Xu, F.R.; Zuo, W. Proton decays in $^{16}$Ne and $^{18}$Mg and isospin-symmetry breaking in carbon isotopes and isotones. *Phys. Rev. C* **2021**, *103*, 044319. [CrossRef]
49. Michel, N.; Płoszajczak, M. *The Gamow Shell Model: The Unified Theory of Nuclear Structure and Reactions*; Springer: Berlin/Heidelberg, Germany, 2021. [CrossRef]
50. Dudek, J.; Szymański, Z.; Werner, T. Woods-Saxon potential parameters optimized to the high spin spectra in the lead region. *Phys. Rev. C* **1981**, *23*, 920–925. [CrossRef]
51. Gyarmati, B.; Vertse, T. On the normalization of Gamow functions. *Nucl. Phys. A* **1971**, *160*, 523–528. [CrossRef]
52. Newton, R. *Scattering Theory of Waves and Particles*; Dover Publications; New York, NY, USA, 2013. [CrossRef]
53. Michel, N.; Aktulga, H.; Jaganathen, Y. Toward scalable many-body calculations for nuclear open quantum systems using the Gamow Shell Model. *Comp. Phys. Comm.* **2020**, *247*, 106978. [CrossRef]
54. Michel, N.; Li, J.G.; Xu, F.R.; Zuo, W. Description of proton-rich nuclei in the $A \approx 20$ region within the Gamow shell model. *Phys. Rev. C* **2019**, *100*, 064303. [CrossRef]
55. Machleidt, R. High-precision, charge-dependent Bonn nucleon-nucleon potential. *Phys. Rev. C* **2001**, *63*, 024001. [CrossRef]
56. Machleidt, R.; Entem, D. Chiral effective field theory and nuclear forces. *Phys. Rep.* **2011**, *503*, 1–75. [CrossRef]
57. Hjorth-Jensen, M.; Kuo, T.T.; Osnes, E. Realistic effective interactions for nuclear systems. *Phys. Rep.* **1995**, *261*, 125–270. [CrossRef]
58. Coraggio, L.; Covello, A.; Gargano, A.; Itaco, N.; Kuo, T. Effective shell-model hamiltonians from realistic nucleon–nucleon potentials within a perturbative approach. *Ann. Phys.* **2012**, *327*, 2125–2151. [CrossRef]

59. Bogner, S.; Furnstahl, R.; Schwenk, A. From low-momentum interactions to nuclear structure. *Prog. Part. Nucl. Phys.* **2010**, *65*, 94–147. [CrossRef]
60. Moshinsky, M. Transformation brackets for harmonic oscillator functions. *Nucl. Phys.* **1959**, *13*, 104–116. [CrossRef]
61. Kuo, T.; Lee, S.; Ratcliff, K. A folded-diagram expansion of the model-space effective hamiltonian. *Nucl. Phys. A* **1971**, *176*, 65–88. [CrossRef]
62. Suzuki, K.; Kumagai, H.; Okamoto, R.; Matsuzaki, M. Recursion method for deriving an energy-independent effective interaction. *Phys. Rev. C* **2014**, *89*, 044003. [CrossRef]
63. Takayanagi, K. Effective Hamiltonian in the extended Krenciglowa–Kuo method. *Nucl. Phys. A* **2011**, *864*, 91–112. [CrossRef]
64. Suzuki, Y.; Ikeda, K. Cluster-orbital shell model and its application to the He isotopes. *Phys. Rev. C* **1988**, *38*, 410–413. [CrossRef]
65. Thompson, D.; Lemere, M.; Tang, Y. Systematic investigation of scattering problems with the resonating-group method. *Nucl. Phys. A* **1977**, *286*, 53–66. [CrossRef]
66. Furutani, H.; Horiuchi, H.; Tamagaki, R. Cluster-model study of the T=1 states in A=4 system: 3He+p scattering. *Prog. Theor. Phys.* **1979**, *62*, 981–1001. [CrossRef]
67. Furutani, H.; Kanada, H.; Kaneko, T.; Nagata, S.; Nishioka, H.; Okabe, S.; Saito, S.; Sakuda, T.; Seya, M. Study of non-alpha-nuclei based on the viewpoint of cluster correlations. *Prog. Theor. Phys. Supp.* **1980**, *68*, 193–302. [CrossRef]
68. Brown, B.A.; Richter, W.A. New "USD" Hamiltonians for the *sd* shell. *Phys. Rev. C* **2006**, *74*, 034315. [CrossRef]
69. Bogner, S.K.; Hergert, H.; Holt, J.D.; Schwenk, A.; Binder, S.; Calci, A.; Langhammer, J.; Roth, R. Nonperturbative shell-model interactions from the in-medium similarity renormalization group. *Phys. Rev. Lett.* **2014**, *113*, 142501. [CrossRef]
70. Becheva, E.; Blumenfeld, Y.; Khan, E.; Beaumel, D.; Daugas, J.M.; Delaunay, F.; Demonchy, C.E.; Drouart, A.; Fallot, M.; Gillibert, A.; et al. $N = 14$ shell closure in $^{22}$O viewed through a neutron sensitive probe. *Phys. Rev. Lett.* **2006**, *96*, 012501. [CrossRef]
71. Stanoiu, M.; Azaiez, F.; Dombrádi, Z.; Sorlin, O.; Brown, B.A.; Belleguic, M.; Sohler, D.; Saint Laurent, M.G.; Lopez-Jimenez, M.J.; Penionzhkevich, Y.E.; et al $N = 14$ and 16 shell gaps in neutron-rich oxygen isotopes. *Phys. Rev. C* **2004**, *69*, 034312. [CrossRef]
72. Hoffman, C.R.; Baumann, T.; Bazin, D.; Brown, J.; Christian, G.; DeYoung, P.A.; Finck, J.E.; Frank, N.; Hinnefeld, J.; Howes, R.; et al Determination of the $N = 16$ shell closure at the oxygen drip line. *Phys. Rev. Lett.* **2008**, *100*, 152502. [CrossRef]
73. Tshoo, K.; Satou, Y.; Bhang, H.; Choi, S.; Nakamura, T.; Kondo, Y.; Deguchi, S.; Kawada, Y.; Kobayashi, N.; Nakayama, Y.; et al. $N = 16$ spherical shell closure in $^{24}$**O**. *Phys. Rev. Lett.* **2012**, *109*, 022501. [CrossRef]
74. Kohley, Z.; Baumann, T.; Bazin, D.; Christian, G.; DeYoung, P.A.; Finck, J.E.; Frank, N.; Jones, M.; Lunderberg, E.; Luther, B.; et al. Study of two-neutron radioactivity in the decay of $^{26}$**O**. *Phys. Rev. Lett.* **2013**, *110*, 152501. [CrossRef]
75. Ahn, D.S.; Fukuda, N.; Geissel, H.; Inabe, N.; Iwasa, N.; Kubo, T.; Kusaka, K.; Morrissey, D.J.; Murai, D.; Nakamura, T.; et al. Location of the neutron dripline at fluorine and neon. *Phys. Rev. Lett.* **2019**, *123*, 212501. [CrossRef]
76. Bagchi, S.; Kanungo, R.; Tanaka, Y.K.; Geissel, H.; Doornenbal, P.; Horiuchi, W.; Hagen, G.; Suzuki, T.; Tsunoda, N.; Ahn, D.S.; et al. Two-neutron halo is unveiled in $^{29}$F. *Phys. Rev. Lett.* **2020**, *124*, 222504. [CrossRef] [PubMed]
77. Revel, A.; Sorlin, O.; Marqués, F.M.; Kondo, Y.; Kahlbow, J.; Nakamura, T.; Orr, N.A.; Nowacki, F.; Tostevin, J.A.; Yuan, C.X.; et al. Extending the southern shore of the island of inversion to $^{28}$F. *Phys. Rev. Lett.* **2020**, *124*, 152502. [CrossRef] [PubMed]
78. Wang, M.; Audi, G.; Kondev, F.; Huang, W.; Naimi, S.; Xu, X. The AME2016 atomic mass evaluation (II). Tables, graphs and references. *Chin. Phys. C* **2017**, *41*, 030003. [CrossRef]
79. Ekström, A.; Baardsen, G.; Forssén, C.; Hagen, G.; Hjorth-Jensen, M.; Jansen, G.R.; Machleidt, R.; Nazarewicz, W.; Papenbrock, T.; Sarich, J.; et al. Optimized chiral nucleon-nucleon interaction at next-to-next-to-leading order. *Phys. Rev. Lett.* **2013**, *110*, 192502. [CrossRef] [PubMed]
80. Fossez, K.; Rotureau, J.; Michel, N.; Nazarewicz, W. Continuum effects in neutron-drip-line oxygen isotopes. *Phys. Rev. C* **2017**, *96*, 024308. [CrossRef]
81. Stroberg, S.R.; Hergert, H.; Holt, J.D.; Bogner, S.K.; Schwenk, A. Ground and excited states of doubly open-shell nuclei from ab initio valence-space Hamiltonians. *Phys. Rev. C* **2016**, *93*, 051301. [CrossRef]
82. Otsuka, T.; Suzuki, T.; Holt, J.D.; Schwenk, A.; Akaishi, Y. Three-body forces and the limit of oxygen isotopes. *Phys. Rev. Lett.* **2010**, *105*, 032501. [CrossRef]
83. Utsuno, Y.; Otsuka, T.; Mizusaki, T.; Honma, M. Varying shell gap and deformation in $N \sim 20$ unstable nuclei studied by the Monte Carlo shell model. *Phys. Rev. C* **1999**, *60*, 054315. [CrossRef]
84. Stroberg, S.R.; Holt, J.D.; Schwenk, A.; Simonis, J. Ab initio limits of atomic nuclei. *Phys. Rev. Lett.* **2021**, *126*, 022501. [CrossRef]
85. Hammer, H.W.; Nogga, A.; Schwenk, A. Colloquium: Three-body forces: From cold atoms to nuclei. *Rev. Mod. Phys.* **2013**, *85*, 197–217. [CrossRef]
86. Hammer, H.W.; König, S.; van Kolck, U. Nuclear effective field theory: Status and perspectives. *Rev. Mod. Phys.* **2020**, *92*, 025004. [CrossRef]
87. Grigorenko, L.V.; Mukha, I.G.; Scheidenberger, C.; Zhukov, M.V. Two-neutron radioactivity and four-nucleon emission from exotic nuclei. *Phys. Rev. C* **2011**, *84*, 021303. [CrossRef]
88. Wu, Q.; Xu, F.R.; Hu, B.S.; Li, J.G. Perturbation calculations of nucleon–nucleon effective interactions in the Hartree–Fock basis. *J. Phys. G Nucl. Part. Phys.* **2019**, *46*, 055104. [CrossRef]
89. Tsunoda, N.; Otsuka, T.; Takayanagi, K.; Shimizu, N.; Suzuki, T.; Utsuno, Y.; Yoshida, S.; Ueno, H. The impact of nuclear shape on the emergence of the neutron dripline. *Nature* **2020**, *587*, 66–71. [CrossRef] [PubMed]

90. Wienholtz, F.; Beck, D.; Blaum, K.; Borgmann, C.; Breitenfeldt, M.; Cakirli, R.B.; George, S.; Herfurth, F.; Holt, J.; Kowalska, M.; et al. Masses of exotic calcium isotopes pin down nuclear forces. *Nature* **2013**, *498*, 346. [CrossRef] [PubMed]
91. Steppenbeck, D.; Takeuchi, S.; Aoi, N.; Doornenbal, P.; Matsushita, M.; Wang, H.; Baba, H.; Fukuda, N.; Go, S.; Honma, M.; et al. Evidence for a new nuclear 'magic number'from the level structure of 54 Ca. *Nature* **2013**, *502*, 207. [CrossRef]
92. Michimasa, S.; Kobayashi, M.; Kiyokawa, Y.; Ota, S.; Ahn, D.S.; Baba, H.; Berg, G.P.A.; Dozono, M.; Fukuda, N.; Furuno, T.; et al. Magic nature of neutrons in $^{54}$Ca: First mass measurements of $^{55-57}$Ca. *Phys. Rev. Lett.* **2018**, *121*, 022506. [CrossRef]
93. Riley, L.A.; McPherson, D.M.; Agiorgousis, M.L.; Baugher, T.R.; Bazin, D.; Bowry, M.; Cottle, P.D.; DeVone, F.G.; Gade, A.; Glowacki, M.T.; et al. Octupole strength in the neutron-rich calcium isotopes. *Phys. Rev. C* **2016**, *93*, 044327. [CrossRef]
94. Hergert, H.; Bogner, S.K.; Morris, T.D.; Binder, S.; Calci, A.; Langhammer, J.; Roth, R. Ab initio multireference in-medium similarity renormalization group calculations of even calcium and nickel isotopes. *Phys. Rev. C* **2014**, *90*, 041302. [CrossRef]
95. Holt, J.D.; Menéndez, J.; Simonis, J.; Schwenk, A. Three-nucleon forces and spectroscopy of neutron-rich calcium isotopes. *Phys. Rev. C* **2014**, *90*, 024312. [CrossRef]
96. Coraggio, L.; Covello, A.; Gargano, A.; Itaco, N. Realistic shell-model calculations for isotopic chains "north-east" of $^{48}$Ca in the $(N,Z)$ plane. *Phys. Rev. C* **2014**, *89*, 024319. [CrossRef]
97. Neufcourt, L.; Cao, Y.; Nazarewicz, W.; Olsen, E.; Viens, F. Neutron drip line in the Ca region from Bayesian model averaging. *Phys. Rev. Lett.* **2019**, *122*, 062502. [CrossRef] [PubMed]
98. Coraggio, L.; De Gregorio, G.; Gargano, A.; Itaco, N.; Fukui, T.; Ma, Y.Z.; Xu, F.R. Shell-model study of calcium isotopes toward their drip line. *Phys. Rev. C* **2020**, *102*, 054326. [CrossRef]
99. Tarasov, O.B.; Ahn, D.S.; Bazin, D.; Fukuda, N.; Gade, A.; Hausmann, M.; Inabe, N.; Ishikawa, S.; Iwasa, N.; Kawata, K.; et al. Discovery of $^{60}$Ca and implications for the stability of $^{70}$Ca. *Phys. Rev. Lett.* **2018**, *121*, 022501. [CrossRef]
100. Woodward, C.J.; Tribble, R.E.; Tanner, D.M. Mass of $^{16}$Ne. *Phys. Rev. C* **1983**, *27*, 27–30. [CrossRef]
101. KeKelis, G.J.; Zisman, M.S.; Scott, D.K.; Jahn, R.; Vieira, D.J.; Cerny, J.; Ajzenberg-Selove, F. Masses of the unbound nuclei $^{16}$Ne, $^{15}$F, and $^{12}$O. *Phys. Rev. C* **1978**, *17*, 1929–1938. [CrossRef]
102. Brown, K.W.; Charity, R.J.; Sobotka, L.G.; Chajecki, Z.; Grigorenko, L.V.; Egorova, I.A.; Parfenova, Y.L.; Zhukov, M.V.; Bedoor, S.; Buhro, W.W.; et al. Observation of long-range three-body coulomb effects in the decay of $^{16}$Ne. *Phys. Rev. Lett.* **2014**, *113*, 232501. [CrossRef]
103. Li, J.G.; Michel, N.; Zuo, W.; Xu, F.R. Reexamining the variational two-particle reduced density matrix for nuclear systems. *Phys. Rev. C* **2021**, *103*, 064324. [CrossRef]
104. Aksyutina, Y.; Johansson, H.; Adrich, P.; Aksouh, F.; Aumann, T.; Boretzky, K.; Borge, M.; Chatillon, A.; Chulkov, L.; Cortina-Gil, D.; et al. Lithium isotopes beyond the drip line. *Phys. Lett. B* **2008**, *666*, 430–434. [CrossRef]

 *physics* 

*Review*

# Recent Progress of Shell-Model Calculations, Monte Carlo Shell Model, and Quasi-Particle Vacua Shell Model

**Noritaka Shimizu** [1,2]

[1] Center for Computational Sciences and Faculty of Pure and Applied Sciences, University of Tsukuba, 1-1-1 Tennodai, Tsukuba 305-8577, Japan; shimizu@nucl.ph.tsukuba.ac.jp
[2] Center for Nuclear Study, The University of Tokyo, 7-3-1, Hongo, Bunkyo-ku, Tokyo 113-0033, Japan

**Abstract:** Nuclear shell model is a powerful approach to investigate nuclear structure microscopically. However, the computational cost of shell-model calculations becomes huge in medium-heavy nuclei. I briefly review the theoretical framework and the code developments of the conventional Lanczos diagonalization method for shell-model calculations. In order to go beyond the conventional diagonalization method, the Monte Carlo shell model and the quasiparticle-vacua shell model were introduced. I present some benchmark examples of these models.

**Keywords:** nuclear shell model calculations; eigenvalue problem; Monte Carlo shell model; quasiparticle-vacua shell model

**Citation:** Shimizu, N. Recent Progress of Shell-Model Calculations, Monte Carlo Shell Model, and Quasi-Particle Vacua Shell Model. *Physics* **2022**, *4*, 1081–1093. https://doi.org/10.3390/physics4030071

Received: 31 July 2022
Accepted: 25 August 2022
Published: 9 September 2022

**Publisher's Note:** MDPI stays neutral with regard to jurisdictional claims in published maps and institutional affiliations.

**Copyright:** © 2022 by the author. Licensee MDPI, Basel, Switzerland. This article is an open access article distributed under the terms and conditions of the Creative Commons Attribution (CC BY) license (https://creativecommons.org/licenses/by/4.0/).

## 1. Introduction

Nuclear shell-model calculation is often called the "configuration interaction" (CI) method in analogy to the CI method in quantum chemistry. It provides us with a detailed description of the ground and low-lying excited states of medium-mass nuclei [1]. A shell-model study is an excellent theoretical tool to discuss exotic structures of nuclei, such as the shape coexistence, shape phase transition, emergence of new magic numbers in neutron-rich nuclei, and so on [1,2]. In addition, it enables us to predict nuclear data required for astrophysical applications [3,4] and for elementary particle physics [5–8]. In the shell-model framework, one separates a nuclear wave function into two parts: an inert core and active particles in the model space. Usually, an inert core is taken as a doubly magic nucleus closest to the Fermi level. For example, in the case of $^{48}$Ca, $^{40}$Ca ($N = Z = 20$, where $N$ denotes the number of neutrons and $Z$ the number of protons in a nucleus) is taken as a frozen inert core and eight valence neutrons are actively occupying the $pf$-shell orbits. Its wave function is written as a linear combination of the Slater determinants each of which represents how the active particles occupy single-particle orbits. This model has achieved successful description of $p$-shell [9], $sd$-shell [10], $pf$-shell, [1,11,12], and $f_5pg_9$-shell [13] nuclei with the conventional diagonalization method.

In most of the conventional shell-model studies, the shell-model Hamiltonian is constructed by the many-body perturbation theory [14] with minor phenomenological corrections to fit the experimental data. Its ab initio derivation has been recently developed by the valence-space in-medium similarity renormalization group (VS-IMSRG) method [15], the coupled-cluster method [16], the many-body perturbation theory [17], the extended Krenciglowa–Kuo method [18], and Okubo–Lee–Suzuki approach [19], while ab initio description of strongly quadrupole deformed states is still a challenge [15,20,21]. The shell-model calculation is now applied to ab initio theory and its importance increases now.

However, solving the eigenvalue problem of the shell-model Hamiltonian matrix often requires huge computational resources, which hampers us from performing the shell-model study in the whole nuclear region. Moreover, in the case of neutron-rich nuclei, since the proton and neutron Fermi levels locate in different shells, beyond one-major-shell model space is required for such model space and the exact diagonalization is often infeasible. To

circumvent this problem, Tokyo shell-model group introduced the Monte Carlo shell model (MCSM), which provides us with successful description of medium-mass nuclei [22,23]. For heavy nuclei in which the treatment of the pairing correlation is important, the MCSM is not sufficiently efficient and its extension, quasi-particle vacua shell model (QVSM), was introduced [24].

In this paper, I briefly describe the framework of the shell-model calculations and the developments of the shell-model codes in Section 2. In Section 3, various approximation methods going beyond the limitation of the conventional shell-model diagonalization are reviewed. Among them, the MCSM is discussed in Section 3.1. Section 3.2 is devoted to the description of the QVSM framework and its capability. This paper is summarized in Section 4.

## 2. Conventional Diagonalization Method for Shell-Model Calculations

In this Section, I briefly review computational aspects of conventional shell-model calculations with the Lanczos method and the developments of shell-model codes.

### 2.1. Shell-Model Hamiltonian Matrix and Its Dimension

A shell-model Hamiltonian consists of one-body and two-body interactions as

$$H = \sum_i \epsilon_i c_i^\dagger c_i + \sum_{ijkl} v_{ijkl} c_i^\dagger c_j^\dagger c_l c_k , \tag{1}$$

where $c_i$ ($c_i^\dagger$) denote the annihilation (creation) operator of the single-particle state $i$. $\epsilon_i$ and $v_{ijkl}$ are the single-particle energies (SPEs) and the two-body matrix elements (TBMEs), respectively, which are fixed so that rotational and parity symmetries are kept. In a conventional way, the SPEs and TBMEs are determined based on the many-body perturbation theory and are corrected phenomenologically so that the experimental binding energies and excitation energies are reproduced. In the so-called $M$-scheme framework, the shell-model wave function is written as a linear combination of the vast number of the $M$-scheme basis states as

$$|\Psi\rangle = \sum_{\alpha=1}^{D} v_\alpha |M_\alpha\rangle, \tag{2}$$

$$|M_\alpha\rangle = \prod_{i=1}^{A} c_{\alpha_i}^\dagger |-\rangle, \tag{3}$$

where $\{\alpha_i\}$ is a set of the single-particle orbits occupied by the active particles. The symbol $|-\rangle$ denotes the wave function of the inert core. $D$ is the number of the basis states allowed with a specified $z$-component of the angular momentum ($M$) and parity ($\pi$), and called the "$M$-scheme dimension". The Schrödinger equation is reduced to the eigenvalue problem:

$$\sum_{\beta=1}^{D} \langle M_\alpha|H|M_\beta\rangle v_\beta = E v_\alpha . \tag{4}$$

Here, $E$ denotes an eigenvalue of the Hamiltonian matrix, $\langle M_\alpha|H|M_\beta\rangle$.

What a shell-model code does is to solve the eigenvalue problem in Equation (4) and obtain a few numbers of the lowest eigenvalues and their eigenvectors with the Lanczos method. Since the $M$-scheme basis state $|M_\alpha\rangle$ is not an eigenstate of the total angular momentum, $J^2$, the wave function in Equation (2) is generally not an eigenstate of $J^2$. However, since the shell-model Hamiltonian is commutable with $J^2$, the solution obtained by the Lanczos method automatically becomes an eigenstate of $J^2$ in the case without any degeneracy. On the other hand, a $jj$-coupled basis state with a good $J^2$ eigenvalue can be constructed with additional angular-momentum algebra. Such basis state is called a

"*J*-scheme basis state" and the number of the allowed *J*-scheme basis states is called the "*J*-scheme dimension".

Table 1 shows the shell-model dimensions of several example nuclei with one-major-shell model space. Since the current limitation of the *M*-scheme dimension is around $10^{11} \sim 10^{12}$, nuclei with the $f_5pg_9$ model space and nuclei in the smaller mass region are able to be handled. Most open-shell heavy nuclei with the atomic mass number $A > 100$ are still far beyond the current capability. Since the numerical size of a vector with the *M*-scheme dimension $D$ is $8D$ Byte for the double precision, the capacity of the disk and memory restricts the capability. The dimensions of the *J*-scheme basis states are also shown in Table 1. It is two orders of magnitude smaller than the corresponding *M*-scheme dimensions. However, the *M*-scheme Hamiltonian matrix is simpler and more sparse than the *J*-scheme Hamiltonian matrix and then the *M*-scheme basis state is used for many numerical computations.

**Table 1.** *M*-scheme and *J*-scheme dimensions of several nuclei. The dimension of the $M^\pi = 0^+$ ($J^\pi = 0^+$) subspace is shown. See text for details.

| Nuclide | Model Space | *M*-Scheme Dim | *J*-Scheme Dim |
|---|---|---|---|
| $^{12}$C | $2 \leq N, Z \leq 8$ ($p$ shell) | 51 | 9 |
| $^{28}$Si | $8 \leq N, Z \leq 20$ ($sd$ shell) | 93,710 | 3372 |
| $^{56}$Ni | $20 \leq N, Z \leq 40$ ($pf$ shell) | $1.08 \times 10^9$ | $1.54 \times 10^7$ |
| $^{78}$Y | $28 \leq N, Z \leq 50$ ($f_5pg_9$, $jj44$) | $1.31 \times 10^{10}$ | $1.11 \times 10^8$ |
| $^{130}$Sm | $50 \leq Z, N \leq 82$ ($jj55$) | $2.06 \times 10^{15}$ | $9.58 \times 10^{12}$ |
| $^{172}$Dy | $50 \leq Z \leq 82, 82 \leq N \leq 126$ ($jj56$) | $1.71 \times 10^{19}$ | $5.54 \times 10^{16}$ |

In order to perform shell-model calculations in medium-heavy nuclei with huge dimensions, various shell-model codes were developed. One of the most famous, popular shell-model codes is the MSU (Michigan State University) version of the OXBASH code [25,26], which was published in the 1980s. Its algorithm is a hybrid method combining *M*-scheme and *J*-scheme basis states. It is equipped with a dialogue-type user-friendly interfaced and had been most widely used in the nuclear physics community. It was replaced in the 2010s by its successor, NuShellX@MSU [27]. On the other hand, the *M*-scheme ANTOINE code and its cousin, the *J*-scheme NATHAN code were developed by Strasbourg-Madrid shell-model group [28,29]. The ANTOINE code has also been used in various studies. It is equipped with an efficient algorithm based on on-the-fly generation of the coupling of the proton and neutron basis states [1] and heavily relies on the FORTRAN 77 language. The MSHELL and MSHELL64 codes were developed mainly by T. Mizusaki [30,31].

A recent major trend in the high-performance computing environment is massively parallel computation. In order to perform shell-model calculations on modern massively parallel computers efficiently, several shell-model codes have been developed recently, such as the MFDn [32], BIGSTICK [33], and KSHELL [34] codes. Among them, the KSHELL code was developing, which is equipped with MPI (Message Passing Interface)–OpenMP (Open Multi-Processing) hybrid parallel computation and can be used both in a personal computer and state-of-the-art supercomputer in the same manner. It shows good parallel scaling up to around $10^4$ threads [34]. The MFDn and BIGSTICK codes are mainly oriented to the no-core shell-model approach by including three-body forces explicitly.

### 2.2. Lanczos Method

The Lanczos method [35] is one of the simplest Krylov subspace methods, in which the low-lying exact eigenvalues are approximated by the Ritz values of the Krylov subspace. The Krylov subspace is spanned by an initial vector, $v_0$, and its vector multiplied by the shell-model Hamiltonian matrix as

$$\mathcal{K}_n(H, v_0) = \mathrm{span}\{v_0, Hv_0, H^2v_0, \cdots, H^{n-1}v_0\}. \quad (5)$$

The exact eigenvalues are approximated by the eigenvalues of the Hamiltonian matrix in the subspace $\mathcal{K}_n(H, v_0)$. In the Lanczos method one performs the orthonormalization of the vector at every $n$, which makes the method numerically stable. While in the limit of $n \to D$ the approximated eigenvalues agree with the exact ones, a few of the smallest eigenvalues converge with small $n$, typically $n \simeq 300$ to obtain the 10 lowest eigenvalues [36]. In practical codes, its extension, the thick-restart Lanczos method, is widely used [37].

### 2.3. Block Lanczos Method

In a large-scale problem, the number of the Hamiltonian matrix elements is too huge to be stored in the memory. In the KSHELL code, the product of the Hamiltonian matrix and a vector is the most time-consuming part of the Lanczos algorithm, since the Hamiltonian matrix elements are generated on-the-fly for every matrix-vector product. In order to reduce the time of the on-the-fly generation, the block algorithm to the thick-restart Lanczos method, named the thick-restart block Lanczos method [34].

In the block algorithm, the on-the-fly generation is performed on every product of the matrix and block vectors, not a vector. Since the number of the products in the block Lanczos method is usually much smaller than that in the Lanczos method, the number of the generations is reduced in comparison with the Lanczos method. The block Lanczos method is one of the block Krylov subspace methods. The block Krylov subspace is defined with a block of the initial vectors $V_0 = (v_{0,1}, v_{0,2}, \cdots v_{0,q})$ as

$$\mathcal{K}_n(H, V_0) \quad = \quad \mathrm{span}\{V_0,\ HV_0,\ H^2V_0,\ \cdots,\ H^{n-1}V_0\}, \tag{6}$$

where $q$ is the block size. Although this subspace is spanned by the $nq$ vectors the number of on-the-fly generation of the matrix elements is $n$. The eigenvalues of the Hamiltonian in the Krylov block subspace are a good approximation to the exact ones. In practical code, the thick-restart block Lanczos method was introduced in the KSHELL code [34].

As other eigensolvers for the shell-model calculations, the LOBPCG method [38] and the Sakurai–Sugiura method [39] have been tested. In our KSHELL code, the block Sakurai–Sugiura method and its variant, stochastic estimation of eigenvalue density, were also implemented [40,41]. To study resonance states such as the Gamow shell model, the energies of the target many-body resonant states correspond to interior eigenvalues, not the lowest one, and are complex numbers. Since the simple Lanczos method cannot be used in such a case, the Jacobi-Davidson method was adopted in the Gamow shell model [42] and the Sakurai–Sugiura method was tested in the complex scaling method [43].

## 3. Approximation Methods to Exact Diagonalization

While shell-model codes for large-scale calculations have been developed vigorously, various approximation schemes to the full model space were proposed. I briefly review these attempts hereafter.

The auxiliary-field Monte Carlo was applied to shell-model calculations and became one of the successful methods in describing $pf$-shell nuclei in the 1990s [44]. However, the notorious sign problem in the quantum Monte Carlo method hampers us from utilizing realistic effective interactions. As a possible solution to the sign problem, the extrapolation method [44], the complex Langevin method [45], and a constrained-path approximation [46] were suggested. Rather schematic interaction without the sign problem has been adopted for the study of nuclear level density [47].

The density matrix renormalization group (DMRG) method is known to be an efficient variational method to solve a one-dimensional quantum many-body problem. It was applied to shell-model calculations coupled to continuum states [48,49]. Its advanced one, the tensor network method, was applied later [50]. The projected shell model has been a useful model with rather schematic interaction to describe high-spin states [51]. The angular-momentum projected CI is another variational method whose basis states are the angular-momentum projected deformed Slater determinants with particle-hole excitation [52]. On top of that, a lot of effort has been paid to developing efficient truncation

schemes, such as the correlated basis method [53], the variational Monte Carlo method with random walkers in $M$-scheme basis states [54,55], nucleon-pair truncation [56,57], generator coordinate method (GCM) [58] and so on.

The variation after mean-field projection in realistic model space (VAMPIR) method is one of the successful methods to describe the $N \simeq Z$ medium-mass nuclei [59,60]. In the VAMPIR method, the nuclear wave function is expressed as a linear combination of the parity, angular-momentum, number projected quasiparticle vacua. Each basis state is determined one by one by minimizing the energy expectation value of this wave function. In that sense, this method is the variation after the projection and superposition. In the hybrid multi-determinant (HMD) method, the number-projected quasiparticle vacua are replaced by the deformed Slater determinants [61,62]. The HMD has been used mainly for the no-core shell-model approach [63].

These truncation methods provide us with the variational upper limit, which may be controlled by a parameter to define the truncation. The exact energy can be estimated by the extrapolation of the energy function of this parameter. The exponential convergence method was suggested, the energy eigenvalue of the truncated subspace is expected to converge as a function of its dimension [64]. As an alternative, the energy-variance expectation value of the variational wave function can be used as a parameter for extrapolation. The energy function of the energy variance can be fitted as a polynomial function, which enables us to estimate the exact energy precisely. It was first introduced into the shell-model calculation with particle-hole truncation in Ref. [65] and applied also to no-core shell model calculations [66], the MCSM [67], QVSM [24], HMD [68], and the importance-truncated shell model (ITSM) [69].

The ITSM was introduced firstly into the no-core shell model approach by R. Roth and his collaborators successfully [70], and later applied also to the conventional shell model approach [69]. The $M$-scheme basis states are truncated by the importance measure, which is estimated by the many-body perturbation theory. The energy is calculated as a function of the criteria of the importance measure, and it is extrapolated to the full space. One of its results is shown in Section 3.2.

Among these efforts, the MCSM and the QVSM have been introduced. These models are discussed in the following Subsections.

### *3.1. Monte Carlo Shell Model*

Among all the efforts to develop various approximation schemes, the MCSM was suggested in the 1990s [22,71,72] and revised in an advanced form [23,73] in the 2010s. It has been used for extensive studies of neutron-rich nuclei. In this Section, I briefly review the theoretical framework of the MCSM.

In the MCSM the nuclear wave function is written as a linear combination of the parity, angular-momentum projected deformed Slater determinants:

$$|\Psi^{(N_b)}\rangle = \sum_{n=1}^{N_b} \sum_{K=-J}^{J} f_{nK} P^{J\pi}_{MK} |\phi_n\rangle. \tag{7}$$

$|\phi_n\rangle$ is a deformed Slater determinant given as

$$|\phi_n\rangle = \prod_{\alpha} \left( \sum_k D^{(n)}_{k\alpha} c^\dagger_k \right) |-\rangle, \tag{8}$$

which is parametrized by the complex matrix $D^{(n)}_{k\alpha}$. The energy $E^{(N_b)}$ is provided by solving the following generalized eigenvalue problem of the Hamiltonian and norm matrices:

$$\begin{aligned} \sum_{nK} H_{mM,nK} f_{nK} &= E^{(N_b)} \sum_{nK} H_{mM,nK} f_{nK}, \\ H_{mM,nK} &= \langle \phi_m | H P^{J\pi}_{MK} | \phi_n \rangle, \\ N_{mM,nK} &= \langle \phi_m | P^{J\pi}_{MK} | \phi_n \rangle, \end{aligned} \tag{9}$$

where $P^{J\pi}_{MK}$ is the angular-momentum ($J$), parity ($\pi$) projector. $f_{iK}$ is the corresponding eigenvector. In the early stage of the history of the MCSM, many candidates of $D_{k\alpha}$ are generated by employing the auxiliary-field Monte Carlo technique and highly selected to lower the energy expectation value $E^{(N_b)}$ [22]. In the advanced version of the MCSM [23], one uses such selected basis states as initial states and optimize $D_{k\alpha}$ by minimizing the energy expectation value utilizing the conjugate gradient method.

The angular-momentum projection in Equation (9) is performed by a three-fold integral of the Euler angles and is time-consuming. The computation time of this part is almost proportional to the number of the discretized mesh points of this integral. Ref. [74] proposed that the Lebedev quadrature would reduce the number of the mesh points and thus the amount of the computation to 2/3. The practical code was tuned employing the technique in Ref. [75].

As a benchmark test, the ground-state energy of $^{56}$Ni with the FPD6 interaction [12] and the $pf$-shell model space is shown in Figure 1. It has been used as a target of the benchmark tests in various studies since the exact diagonalization is infeasible in the 1990s and its structure is interesting in terms of the soft closed core and shape coexistence. Its exact diagonalization was achieved in 2006 [76].

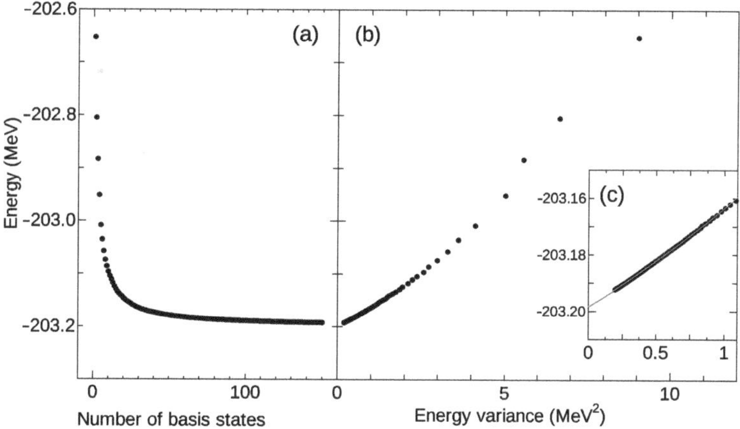


**Figure 1.** MCSM results of $^{56}$Ni with the FPD6 interaction. (**a**): MCSM energy against the number of the MCSM basis states. (**b**): extrapolation plot with the energy variance. (**c**): magnified view of (**b**). The red is the fitted curve with a second-order polynomial. See text for details.

Figure 1a shows the MCSM energy $E^{(N_b)}$ of the $^{56}$Ni as a function of $N_b$. The energy drops rapidly in the region $N_b < 20$ and it gradually converges giving the variational upper limit, $-203.192$ MeV ($N_b = 150$), which is only 6 keV higher than the exact one, $-203.198$ MeV. In our preceding works, the MCSM gave $-203.161$ MeV with $N_b = 150$ in 2010 [67], $-203.152$ MeV in 2001 [22], $-203.100$ MeV in 1998 [77]. Thus, the methodological development and progress in computational resources steadily improve the precision of the MCSM.

# The Nuclear Shell Model 70 Years after Its Advent: Achievements and Prospects

AF595452

www.ingramcontent.com/pod-product-compliance
Lightning Source LLC
LaVergne TN
LVHW071540170726
843515LV00008B/2183

MDPI
St. Alban-Anlage 66
4052 Basel
Switzerland
www.mdpi.com

*Physics* Editorial Office
E-mail: physics@mdpi.com
www.mdpi.com/journal/physics

Disclaimer/Publisher's Note: The statements, opinions and data contained in all publications are solely those of the individual author(s) and contributor(s) and not of MDPI and/or the editor(s). MDPI and/or the editor(s) disclaim responsibility for any injury to people or property resulting from any ideas, methods, instructions or products referred to in the content.

In this benchmark test, still the 6 keV difference with the exact value remains. In order to fill this small gap, the extrapolation method utilizing the energy variance was introduced. Figure 1b shows the energy against the corresponding energy variance, $\Delta E^{(N_b)} = \langle \Psi^{(N_b)} | H^2 | \Psi^{(N_b)} \rangle - \langle \Psi^{(N_b)} | H | \Psi^{(N_b)} \rangle^2$. The energy variance is a good measure of how well the approximation works since the energy variance of the exact eigenstate is zero. As $N_b$ increases, the energy and energy variance gradually decrease and the point smoothly approaches $y$ axis. Figure 1c shows the magnified view of Figure 1b. The red line in Figure 1c is a second order polynomial fitted for these points. The $y$-intercept of the fitted line becomes the extrapolated value, which agrees with the exact one within a keV.

Figure 2 shows the overlap probability between the MCSM wave function and the exact wave function. Even at $N_b = 1$ the probability is 0.95. As $N_b$ increases it approaches the unit smoothly and surpasses 0.99 at $N_b = 7$. The final value at $N_b = 150$ is 0.9998. Thus, the ground-state wave function of $^{56}$Ni is approximated by the MCSM in high precision, and, moreover, the estimation of its eigenenergy is improved by the energy-variance extrapolation. The MCSM method has been quite successful in medium-heavy nuclei mainly in $pf$-shell and neutron-rich nuclei in the medium-heavy mass region. However, the MCSM tends to underestimate the pairing correlation and $2^+$ excitation energies. In order to treat the pairing correlation properly, the QVSM was introduced and is discussed in the next Subsection.

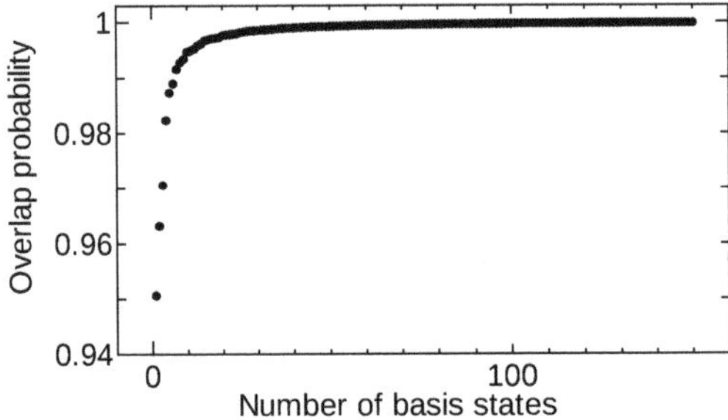


**Figure 2.** Overlap probability between the MCSM wave function and the exact wave function by the Lanczos method against the number of the MCSM basis states $N_b$. These wave functions were computed for the ground state of $^{56}$Ni with the FPD6 interaction [12].

### 3.2. *Quasiparticle Vacua Shell Model*

The MCSM wave function is a linear combination of the Slater determinants, which are not suitable for treating the pairing correlation in the heavy-mass region. In order to include the pairing correlation efficiently, one can replace the Slater-determinant basis by the number projected quasi-particle vacua. This framework is called the "QVSM" [24]. The QVSM wave function is defined as

$$|\Psi^{(N_b)}\rangle = \sum_{n=1}^{N_b} \sum_{K=-J}^{J} f_{iK} P^{J\pi}_{MK} P^Z P^N |\phi_n\rangle \tag{10}$$

where $P^Z$ and $P^N$ are the proton and neutron number projectors, respectively. $|\phi_n\rangle$ is a quasiparticle vaccum and is given as

$$\begin{aligned} \beta_k^{(n)} |\phi_n\rangle &= 0 \ \text{ for any } k, \\ \beta_k^{(n)} &= \sum_i (V_{ik}^{(n)*} c_i^\dagger + U_{ik}^{(n)*} c_i), \end{aligned} \tag{11}$$

where $\beta_k$ denotes a quasi-particle annihilation operator and $c_i^\dagger$ is the creation operator of the single-particle orbit $i$. Thus, the basis state is parametrized by the complex matrices $U$ and $V$ which keep the orthogonalization condition [78]. In the present work, this quasiparticle

does not mix the proton and neutron spaces. The energy is obtained in the same manner to the MCSM by solving Equation (9). $U^{(n)}$ and $V^{(n)}$ matrices are determined utilizing the conjugate gradient method to minimize $E^{(n)}$.

As a benchmark test, shell-model calculations of $^{132}$Ba were performed with the SN100PN interaction [79] and the *jj*55 model space. Figure 3 represents the results of various approximation frameworks. In Figure 3, (a)–(d) present the results of the generator coordinate methods (GCM) discussed in [58] in comparison with the MCSM result (e), the QVSM result (f), and the exact shell-model energy (g). Figure 3A shows that the two GCM results with the Slater determinants, (a) and (b), have 2 MeV or a larger deviation from the exact one. The GCM results with the quasiparticle vacua basis, the MCSM result, and the QVSM result show the deviation smaller than 1 MeV. The QVSM is the best result, the deviation of which is only 45 keV.

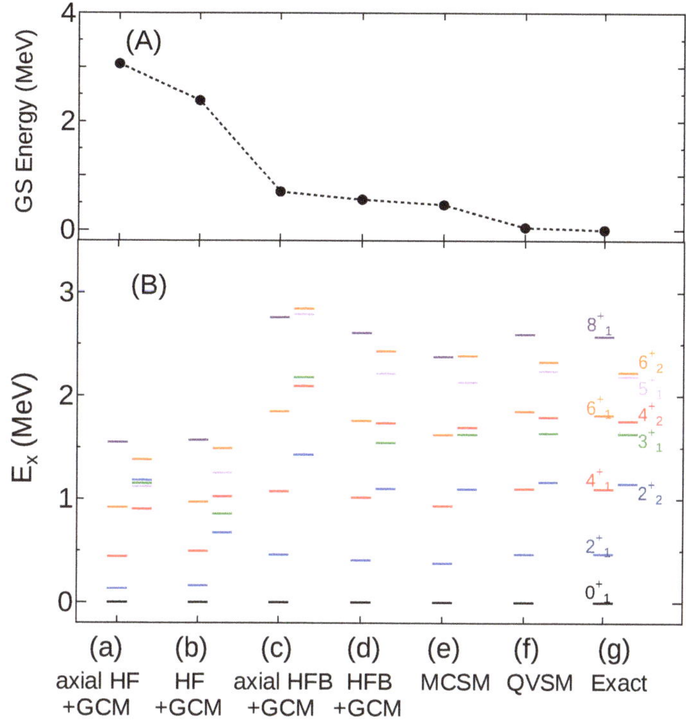


**Figure 3.** Shell-model results of $^{132}$Ba with the SN100PN interaction by various approximation methods. (**A**): the difference of shell-model energies from the exact one. (**B**): the excitation energies of the ground-state band ($2^+_1, 4^+_1, 6^+_1$ and $8^+_1$) and the quasi-$\gamma$ band ($2^+_2, 3^+_1, 4^+_2, 5^+_1$, and $6^+_2$). The results are shown for: (a) generator-coordinate method (GCM) with the Hartree-Fock (HF) calculations assuming axial symmetry, (b) GCM with the HF calculations without assuming axial symmetry, (c) GCM with the Hartree-Fock-Bogoliubov (HFB) calculations assuming axial symmetry, (d) GCM with the HFB calculations without assuming axial symmetry, (e) MCSM with 50 basis states without variance extrapolation, (f) QVSM with 30 basis states without variance extrapolation, and (g) exact shell-model result by the Lanczos diagonalization. Numerical data of (a–d) and (g) are taken from Ref. [58].

Figure 3B shows the excitation energies of the yrast band ($2^+_1$, $4^+_1$, $6^+_1$, and $8^+_1$) and the quasi-gamma band ($2^+_2$, $3^+_1$, $4^+_2$, $5^+_1$, and $6^+_2$) of $^{132}$Ba. The two GCM methods with the Slater determinant basis states, (a) and (b), present too low excitation energies of the yrast band because of the underestimation of the pairing correlation. The two GCM methods

with the quasiparticle vacua basis, (c) and (d) show the correct excitation energies of the yrast band. Two GCM methods assuming axial symmetry, (a) and (c), apparently failed to reproduce the quasi-gamma band. The GCM method with the quasiparticle vacua basis is referred to as the HFB+GCM ("HFB" stands for "Hartree-Fock-Bogoliubov") in Figure 3. The HFB+GCM result shows reasonable agreement with the exact one. The MCSM also shows the reasonable agreement with the exact one, the moment of inertia is slightly overestimated because of the underestimation of the pairing correlation. The QVSM result shows the almost perfect agreement with the exact one.

This benchmark test confirms that the QVSM outperforms the GCM and the MCSM in this mass region. Indeed, the QVSM wave function is superior to the MCSM wave function of the same $N_b$ by including the pairing correlation efficiently. However, the computation time of the QVSM is longer than the MCSM with the same $N_b$ mainly due to the number projection. If the difference between the MCSM and QVSM energies with the same $N_b$ is small, the MCSM can surpass the QVSM by increasing $N_b$ within the same computation time. In Ref. [24], we demonstrated that in the case of $^{68}$Ni in the $pfg_{9/2}d_{5/2}$ model space the MCSM and QVSM energies have small deviation with the same $N_b$, and thus the MCSM is efficient in terms of the computation time. In practice, one can try both the QVSM and the MCSM with a small $N_b$ and identify which one is efficient before performing heavy calculations. Empirically, it was found that the QVSM is more efficient in nuclei heavier than tin isotopes, in which the pairing correlation becomes important.

As another test of the QVSM, one can perform shell-model calculation of $^{101}$Sn with the $0g_{9/2}, 0g_{7/2}, 1d_{5/2}, 1d_{3/2}$, and $2s_{1/2}$ orbits as the model space. One adopts an effective interaction derived in an ab initio way, the VS-IMSRG method [15,80]. In the derivation, the chiral N$^3$LO 1.8/2.0(EM) (see details in Ref. [81]) was adopted for the two-body and three-body forces with a similarity-renormalization-group evolution. Figure 4 shows the results of the $7/2^+_1$ and $5/2^+_1$ energies of $^{101}$Sn provided by ITSM (Figure 4a), the Lanczos diagonalization with particle-hole truncation (Figure 4b), and the QVSM (Figure 4c).

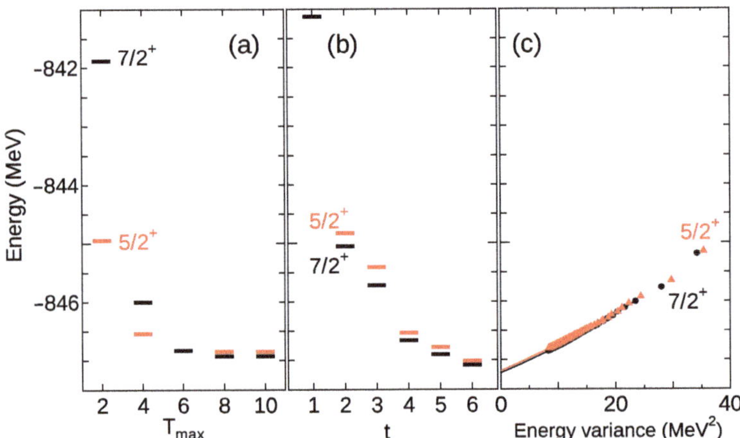


**Figure 4.** Energies of the $5/2^+$ (red) and $7/2^+$ (black) states of $^{101}$Sn. (**a**): ITSM result against $T_{\rm max}$ from Ref. [80]. (**b**): Result of the Lanczos diagonalization with $t$-particle $t$-hole truncation. (**c**): QVSM with 60 basis states. The solid lines in (**c**) are fitted for the extrapolation. See text for details.

Figure 4a shows the ITSM result as a function of the number of the allowed particle-hole excitation across the $N = Z = 50$ gap, $T_{\rm max}$ [80]. The result shows good convergence as a function of $T_{\rm max}$ and predicts the ground $7/2^+$ state and the small excitation energy of the $5/2^+$ state. Note that it does not mean a variational upper limit, since these results are extrapolated values as a function of the importance measure. Figure 4b shows the Lanczos diagonalization result with restricting $t$-particle $t$-hole excitation across the $N = Z = 50$ gap.

The $M$-scheme dimension of $t = 6$ is $9.6 \times 10^9$, which is quite large. The energies gradually lower as a function of $t$, but still it does not reach sufficient convergence. Figure 4c presents the QVSM results against the energy variance. as $N_b$ increases the energy and energy variance decreases smoothly and approaches the $y$-axis or zero energy variance. The $y$-intercepts of the fitted curve become the extrapolated values, which predict the $7/2^+$ ground state with small $5/2^+$ excitation energy.

Thus, these three methods predict the $7/2^+$ ground state and small excitation energy of the $5/2^+$ state consistently. The extrapolated value of the QVSM seems consistent with the behavior of and the diagonalization result with the $t$-particle $t$-hole truncation.

## 4. Summary

I reviewed the current status of the shell-model calculations, and our developments to overcome the limitation of the conventional Lanczos diagonalization method. One of the frontiers of shell-model study is to study neutron-rich nuclei towards the neutron drip line, in which a larger model space is required. To perform shell-model calculations with such a large model space, the MCSM was proposed and demonstrated in Section 3.1. The MCSM has been applied to various studies of exotic nuclei [23]. Another frontier is to go heavier open-shell nuclei, in which pairing correlation is essential to be treated efficiently. For such purpose, the QVSM has been developed and its feasibility was demonstrated in Section 3.2. Several benchmark tests to demonstrate the capabilities of the MCSM and QVSM methods are presented.

Other frontiers for the shell-model study are the microscopic description of giant resonance and statistical properties of the highly excited region. The Lanczos strength function method [1] is a solution to this problem, but it is still trapped by the rapid growth of the shell-model dimension. Although several attempts were performed to go beyond this limit, which shows promising results [82], further study is required.

**Funding:** This research was supported by "Program for Promoting Researches on the Supercomputer Fugaku" (JPMXP1020200105) and JICFuS, and KAKENHI grant (17K05433, 20K03981).

**Data Availability Statement:** The data used in this work are available from the author upon reasonable request.

**Acknowledgments:** The author acknowledges Takashi Abe, Michio Honma, Takaharu Otsuka, Takayuki Miyagi, Takahiro Mizusaki, Tomoaki Togashi, Yusuke Tsunoda, and Yutaka Utsuno for fruitful collaborations. I especially thank Takayuki Miyagi for providing us with the VS-IMSRG interaction for $^{101}$Sn. This research used computational resources of the supercomputer Fugaku (hp220174, hp210165, hp200130) at the RIKEN Center for Computational Science, Oakforest-PACS supercomputer, Wisteria supercomputer (CCS-Tsukuba MCRP xg18i035 and wo22i022) and Oakbridge-CX supercomputer.

**Conflicts of Interest:** The author declares no conflict of interest.

## References

1. Caurier, E.; Martinez-Pinedo, G.; Nowacki, F.; Poves, A.; Zuker, A.P. The shell model as unified view of nuclear structure. *Rev. Mod. Phys.* **2005**, *77*, 427–488. [CrossRef]
2. Otsuka, T.; Gade, A.; Sorlin, O.; Suzuki, T.; Utsuno, Y. Evolution of shell structure in exotic nuclei. *Rev. Mod. Phys.* **2020**, *92*, 015002. [CrossRef]
3. Martínez-Pinedo, G.; Langanke, K. Shell model applications in nuclear astrophysics. *Physics* **2022**, *4*, 677–689. [CrossRef]
4. Shimizu, N.; Togashi, T.; Utsuno, Y. Gamow-Teller transitions of neutron-rich $N = 82, 81$ nuclei by shell-model calculations. *Prog. Theor. Exp. Phys.* **2021**, *2021*, 033D01. [CrossRef]
5. Menéndez, J.; Poves, A.; Caurier, E.; Nowacki, F. Disassembling the nuclear matrix elements of the neutrinoless $\beta\beta$ decay. *Nucl. Phys. A* **2009**, *818*, 139–151. [CrossRef]
6. Coraggio, L.; Gargano, A.; Itaco, N.; Mancino, R.; Nowacki, F. Calculation of the neutrinoless double-$\beta$ decay matrix element within the realistic shell model. *Phys. Rev. C* **2020**, *101*, 044315. [CrossRef]
7. Shimizu, N.; Menéndez, J.; Yako, K. Double Gamow-Teller transitions and its relation to neutrinoless $\beta\beta$ decay. *Phys. Rev. Lett.* **2018**, *120*, 142502. [CrossRef]

8. Yanase, K.; Shimizu, N. Large-scale shell-model calculations of nuclear Schiff moments of $^{129}$Xe and $^{199}$Hg. *Phys. Rev. C* **2020**, *102*, 065502. [CrossRef]
9. Cohen, S.; Kurath, D. Effective interactions for the 1p shell. *Nucl. Phys.* **1965**, *73*, 1–24. [CrossRef]
10. Brown, B.A.; Richter, W. New "USD" Hamiltonians for the *sd* shell. *Phys. Rev. C* **2006**, *74*, 034315. [CrossRef]
11. Honma, M.; Otsuka, T.; Brown, B.A.; Mizusaki, T. New effective interaction for $pf$ shell nuclei and its implications for the stability of the $N = Z = 28$ closed core. *Phys. Rev. C* **2004**, *69*, 034335. [CrossRef]
12. Richter, W.A.; Van Der Merwe, M.G.; Julies, R.E.; Brown, B.A. New effective interactions for the 0f1p shell. *Nucl. Phys. A* **1991**, *523*, 325–353. [CrossRef]
13. Honma, M.; Otsuka, T.; Mizusaki, T.; Hjorth-Jensen, M. New effective interaction for $f_5pg_9$-shell nuclei. *Phys. Rev. C* **2009**, *80*, 064323. [CrossRef]
14. Hjorth-Jensen, M.; Kuo, T.T.S.; Osnes, E. Realistic effective interactions for nuclear systems. *Phys. Rep.* **1995**, *261*, 125–270. [CrossRef]
15. Stroberg, S.R.; Henderson, J.; Hackman, G.; Ruotsalainen, P.; Hagen, G.; Holt, J.D. Systematics of $E2$ strength in the *sd* shell with the valence-space in-medium similarity renormalization group. *Phys. Rev. C* **2022**, *105*, 034333. [CrossRef]
16. Jansen, G.R.; Engel, J.; Hagen, G.; Navratil, P.; Signoracci, A. Ab-initio coupled-cluster effective interactions for the shell model: Application to neutron-rich oxygen and carbon isotopes. *Phys. Rev. Lett.* **2014**, *113*, 142502. [CrossRef] [PubMed]
17. Coraggio, L.; De Gregorio, G.; Gargano, A.; Itaco, N.; Fukui, T.; Ma, Y.Z.; Xu, F.R. Shell-model study of calcium isotopes toward their drip line. *Phys. Rev. C* **2020**, *102*, 054326. [CrossRef]
18. Tsunoda, N.; Otsuka, T.; Shimizu, N.; Hjorth-Jensen, M.; Takayanagi, K.; Suzuki, T. Exotic neutron-rich medium-mass nuclei with realistic nuclear forces. *Phys. Rev. C* **2017**, *95*, 021304. [CrossRef]
19. Smirnova, N.A.; Barrett, B.R.; Kim, Y.; Shin, I.J.; Shirokov, A.M.; Dikmen, E.; Maris, P.; Vary, J.P. Effective interactions in the *sd* shell. *Phys. Rev. C* **2019**, *100*, 054329. [CrossRef]
20. Miyagi, T.; Stroberg, S.R.; Holt, J.D.; Shimizu, N. Ab initio multishell valence-space Hamiltonians and the island of inversion. *Phys. Rev. C* **2020**, *102*, 034320. [CrossRef]
21. Tsunoda, N.; Otsuka, T.; Takayanagi, K.; Shimizu, N.; Suzuki, T.; Utsuno, Y.; Yoshida, S.; Ueno, H. The impact of nuclear shape on the emergence of the neutron dripline. *Nature* **2020**, *587*, 66–71. [CrossRef] [PubMed]
22. Otsuka, T.; Honma, M.; Mizusaki, T.; Shimizu, N.; Utsuno, Y. Monte Carlo shell model for atomic nuclei. *Prog. Part. Nucl. Phys.* **2001**, *47*, 319–400. [CrossRef]
23. Shimizu, N.; Abe, T.; Honma, M.; Otsuka, T.; Togashi, T.; Tsunoda, Y.; Utsuno, Y.; Yoshida, T. Monte Carlo shell model studies with massively parallel supercomputers. *Phys. Scr.* **2017**, *92*, 063001. [CrossRef]
24. Shimizu, N.; Tsunoda, Y.; Utsuno, Y.; Otsuka, T. Variational approach with the superposition of the symmetry-restored quasiparticle vacua for nuclear shell-model calculations. *Phys. Rev. C* **2021**, *103*, 014312. [CrossRef]
25. Brown, B.A.; Etchegoyen, A.; Rae, W.D.M.; Godwin, N.S. *The Computer Code OXBASH*; MSU-NSCL Report 524; National Superconducting Cyclotron Laboratory: East Lansing, MI, USA, 1988.
26. Brown, B.A.; Etchegoyen, A.; Godwin, N.S.; Rae, W.D.M.; Richter, W.A.; Ormand, W.E.; Warburton, E.K.; Winfield, J.S.; Zhao, L.; Zimmerman, C.H. *Oxbash for Windows PC*; MSU-NSCL Report 1289; National Superconducting Cyclotron Laboratory: East Lansing, MI, USA, 2004.
27. Brown, B.A.; Rae, W.D.M. The shell-model code NuShellX@ MSU. *Nucl. Data Sheets* **2014**, *120*, 115–118. [CrossRef]
28. Caurier, E. *Computer Code ANTOINE*; Centre de Recherches Nucléaires: Strasbourg, France, 1989.
29. Caurier, E.; Nowacki, F. *Coupled Code NATHAN*; Centre de Recherches Nucléaires: Strasbourg, France, 1995.
30. Mizusaki, T. Development of a new code for large-scale shell-model calculations using a parallel computer. *RIKEN Accel. Prog. Rep.* **1999**, *33*, 14. Available online: https://www.nishina.riken.jp/researcher/APR/Document/ProgressReport_vol_33.pdf (accessed on 24 August 2020).
31. Mizusaki, T.; Shimizu, N.; Utsuno, Y.; Honma, M. The MSHELL64 code. 2013, unpublished.
32. Shao, M.; Aktulga, H.M.; Yang, C.; Ng, E.G.; Maris, P.; Vary, J.P. Accelerating nuclear configuration interaction calculations through a preconditioned block iterative eigensolver. *Comput. Phys. Commun.* **2018**, *222*, 1–13. [CrossRef]
33. Johnson, C.W.; Ormand, W.E.; Krastev, P.G. Factorization in large-scale many-body calculations. *Comput. Phys. Commun.* **2013**, *184*, 2761–2774. [CrossRef]
34. Shimizu, N.; Mizusaki, T.; Utsuno, Y.; Tsunoda, Y. Thick-restart block Lanczos method for large-scale shell-model calculations. *Comput. Phys. Commun.* **2019**, *244*, 372–384. [CrossRef]
35. Lanczos, C. An iteration method for the solution of the eigenvalue problem of linear differential and integral operators. *J. Res. Nat. Bur. Standards* **1950**, *45*, 255–282. [CrossRef]
36. Shimizu, N.; Utsuno, Y.; Futamura, Y.; Sakurai, T.; Mizusaki, T.; Otsuka, T. Stochastic estimation of level density in nuclear shell-model calculations. *EPJ Web Conf.* **2016**, *122*, 02003. [CrossRef]
37. Wu, K.; Simon, H. Thick-restart Lanczos method for large symmetric eigenvalue problems. *SIAM J. Matrix Anal. Appl.* **2000**, *22*, 602–616. [CrossRef]
38. Aktulga, H.M.; Afibuzzaman, M.; Williams, S.; Buluç, A.; Shao, M.; Yang, C.; Ng, E.G.; Maris, P.; Vary, J.P. A high performance block eigensolver for nuclear configuration interaction calculations. *IEEE Trans. Parallel Distrib. Syst.* **2017**, *28*, 1550–1563. [CrossRef]

39. Mizusaki, T.; Kaneko, K.; Honma, M.; Sakurai, T. Filter diagonalization of shell-model calculations. *Phys. Rev. C* **2010**, *82*, 024310. [CrossRef]
40. Ikegami, T.; Sakurai, T.; Nagashima, U. A filter diagonalization for generalized eigenvalue problems based on the Sakurai—Sugiura projection method. *J. Comput. Appl. Math* **2010**, *233*, 1927–1936. [CrossRef]
41. Shimizu, N.; Utsuno, Y.; Futamura, Y.; Sakurai, T.; Mizusaki, T.; Otsuka, T. Stochastic estimation of nuclear level density in the nuclear shell model: An application to parity-dependent level density in $^{58}$Ni. *Phys. Lett. B* **2016**, *753*, 13–17. [CrossRef]
42. Michel, N.; Aktulga, H.M.; Jaganathen, Y. Toward scalable many-body calculations for nuclear open quantum systems using the Gamow Shell Model. *Comput. Phys. Commun.* **2020**, *247*, 106978. [CrossRef]
43. Mizusaki, T.; Myo, T.; Katō, K. A new approach for many-body resonance spectroscopy with the complex scaling method. *Prog. Theor. Exp. Phys.* **2014**, *2014*, 091D01. [CrossRef]
44. Koonin, S.E.; Dean, D.J.; Langanke, K. Shell model Monte Carlo methods. *Phys. Rep.* **1997**, *278*, 1–77. [CrossRef]
45. Adami, C.; Koonin, S.E. Complex Langevin equation and the many fermion problem. *Phys. Rev. C* **2001**, *63*, 034319. [CrossRef]
46. Bonnard, J.; Juillet, O. Constrained-path quantum Monte Carlo approach for the nuclear shell model. *Phys. Rev. Lett.* **2013**, *111*, 012502. [CrossRef] [PubMed]
47. Alhassid, Y.; Bonett-Matiz, M.; Liu, S.; Nakada, H. Direct microscopic calculation of nuclear level densities in the shell model Monte Carlo approach. *Phys. Rev. C* **2015**, *92*, 024307. [CrossRef]
48. Dukelsky, J.; Pittel, S.; Dimitrova, S.; Stoitsov, M. Density matrix renormalization group method and large-scale nuclear shell-model calculations. *Phys. Rev. C* **2002**, *65*, 054319. [CrossRef]
49. Pittel, S.; Sandulescu, N. Density matrix renormalization group and the nuclear shell model. *Phys. Rev. C* **2006**, *73*, 014301. [CrossRef]
50. Legeza, Ö.; Veis, L.; Poves, A.; Dukelsky, J. Advanced density matrix renormalization group method for nuclear structure calculations. *Phys. Rev. C* **2015**, *92*, 051303. [CrossRef]
51. Hara, K.; Sun, Y. Projected shell model and high-spin spectroscopy. *Int. J. Mod. Phys. E* **1995**, *4*, 637–785. [CrossRef]
52. Gao, Z.-C.; Horoi, M. Angular momentum projected configuration interaction with realistic Hamiltonians. *Phys. Rev. C* **2009**, *79*, 014311. [CrossRef]
53. Jiao, L.F.; Sun, Z.H.; Xu, Z.X.; Xu, F.R.; Qi, C. Correlated-basis method for shell-model calculations. *Phys. Rev. C* **2014**, *90*, 024306. [CrossRef]
54. Mizusaki, T.; Shimizu, N. New variational Monte Carlo method with an energy variance extrapolation for large-scale shell-model calculations. *Phys. Rev. C* **2012**, *85*, 021301. [CrossRef]
55. Shimizu, N.; Mizusaki, T. Variational Monte Carlo method for shell-model calculations in odd-mass nuclei and restoration of symmetry. *Phys. Rev. C* **2018**, *98*, 054309. [CrossRef]
56. Yoshinaga, N.; Higashiyama, K. Systematic studies of nuclei around mass 130 in the pair-truncated shell model. *Phys. Rev. C* **2004**, *69*, 054309. [CrossRef]
57. Zhao, Y.M.; Arima, A. Nucleon-pair approximation to the nuclear shell model. *Phys. Rep.* **2014**, *545*, 1–45. [CrossRef]
58. Shimizu, N.; Mizusaki, T.; Kaneko, K.; Tsunoda, Y. Generator-coordinate methods with symmetry-restored Hartree-Fock-Bogoliubov wave functions for large-scale shell-model calculations. *Phys. Rev. C* **2021**, *103*, 064302. [CrossRef]
59. Schmid, K.W. On the use of general symmetry-projected Hartree–Fock–Bogoliubov configurations in variational approaches to the nuclear many-body problem. *Prog. Part. Nucl. Phys.* **2004**, *52*, 565–633. [CrossRef]
60. Schmid, K.W.; Grümmer, F.; Kyotoku, M.; Faessler, A. Selfconsistent description of non-yrast states in nuclei: The excited VAMPIR approach. *Nucl. Phys. A* **1986**, *452*, 493–512. [CrossRef]
61. Puddu, G. Hybrid schemes for the calculation of low-energy levels of shell model Hamiltonians. *J. Phys. G Nucl. Part. Phys.* **2006**, *32*, 321–331. [CrossRef]
62. Puddu, G. Calculation of energy levels with the hybrid multi-determinant method in the fp region. *Eur. Phys. J. A* **2007**, *34*, 413–415. [CrossRef]
63. Puddu, G. A study of open shell nuclei using chiral two-body interactions. *J. Phys. G Nucl. Part. Phys.* **2021**, *48*, 045105. [CrossRef]
64. Horoi, M.; Brown, B.A.; Zelevinsky, V. Exponential convergence method: Nonyrast states, occupation numbers, and a shell-model description of the superdeformed band in $^{56}$Ni. *Phys. Rev. C* **2003**, *67*, 034303. [CrossRef]
65. Mizusaki, T.; Imada, M. Extrapolation method for shell model calculations. *Phys. Rev. C* **2002**, *65*, 064319. [CrossRef]
66. Zhan, H.; Nogga, A.; Barrett, B.R.; Vary, J.P.; Navratil, P. Extrapolation method for the no-core shell model. *Phys. Rev. C* **2004**, *69*, 034302. [CrossRef]
67. Shimizu, N.; Utsuno, Y.; Mizusaki, T.; Otsuka, T.; Abe, T.; Honma, M. Novel extrapolation method in the Monte Carlo shell model. *Phys. Rev. C* **2010**, *82*, 061305. [CrossRef]
68. Puddu, G. An efficient method to evaluate energy variances for extrapolation methods. *J. Phys. G Nucl. Part. Phys.* **2012**, *39*, 085108. [CrossRef]
69. Stumpf, C.; Braun, J.; Roth, R. Importance-truncated large-scale shell model. *Phys. Rev. C* **2016**, *93*, 021301(R). [CrossRef]
70. Roth, R.; Navrátil, P. Ab initio study of $^{40}$Ca with an importance-truncated no-core shell model. *Phys. Rev. Lett.* **2007**, *99*, 092501. [CrossRef]

71. Honma, M.; Mizusaki, T.; Otsuka, T. Diagonalization of Hamiltonians for many-body systems by auxiliary field quantum Monte Carlo technique. *Phys Rev. Lett.* **1995**, *75*, 1284–1287. [CrossRef]
72. Honma, M.; Mizusaki, T.; Otsuka, T. Nuclear shell model by the quantum Monte Carlo diagonalization method. *Phys Rev. Lett.* **1996**, *77*, 3315–3318. [CrossRef]
73. Shimizu, N.; Abe, T.; Tsunoda, Y.; Utsuno, Y.; Yoshida, T.; Mizusaki, T.; Honma, M.; Otsuka, T. New-generation Monte Carlo shell model for the K computer era. *Prog. Theor. Exp. Phys.* **2012**, *2012*, 01A205. [CrossRef]
74. Shimizu, N.; Tsunoda, Y. $SO(3)$ quadratures in angular-momentum projection. *arXiv* **2022**, arXiv:2205.04119. [CrossRef]
75. Utsuno, Y.; Shimizu, N.; Otsuka, T.; Abe, T. Efficient computation of Hamiltonian matrix elements between non-orthogonal Slater determinants. *Comput. Phys. Commun.* **2013**, *184*, 102–108. [CrossRef]
76. Horoi, M.; Brown, B.A.; Otsuka, T.; Honma, M.; Mizusaki, T. Shell model analysis of the $^{56}$Ni spectrum in the full $pf$ model space. *Phys. Rev. C* **2006**, *73*, 061305(R). [CrossRef]
77. Otsuka, T.; Honma, M.; Mizusaki, T. Structure of the $N = Z = 28$ closed shell studied by Monte Carlo shell model calculation. *Phys. Rev. Lett.* **1998**, *81*, 1588–1591. [CrossRef]
78. Ring, P.; Schuck, P. *The Nuclear Many-Body Problem*; Springer: Berlin/Heidelberg, Germany, 2004.
79. Brown, B.A.; Stone, N.J.; Stone, J.R.; Towner, I.S.; Hjorth-Jensen, M. Magnetic moments of the $2_1^+$ states around $^{132}$Sn. *Phys. Rev. C* **2005**, *71*, 044317. [CrossRef]
80. Morris, T.D.; Simonis, J.; Stroberg, S.R.; Stumpf, C.; Hagen, G.; Holt, J.D.; Jansen, G.R.; Papenbrock, T.; Roth, R.; Schwenk, A. Structure of the lightest tin isotopes. *Phys. Rev. Lett.* **2018**, *120*, 152503. [CrossRef] [PubMed]
81. Hebeler, K.; Bogner, S.; Furnstahl, R.J.; Nogga, A.; Schwenk, A. Improved nuclear matter calculations from chiral low-momentum interactions. *Phys. Rev. C* **2011**, *83*, 031301(R). [CrossRef]
82. Togashi, T.; Otsuka, T.; Shimizu, N.; Utsuno, Y. $E1$ strength function in the Monte Carlo shell model. *JPS Conf. Proc.* **2018**, *23*, 012031. [CrossRef]

*Article*

# Low-Energy Coulomb Excitation for the Shell Model

**Marco Rocchini [1,*] and Magda Zielińska [2]**

[1] Department of Physics, University of Guelph, Guelph, ON N1G 2W1, Canada
[2] IRFU, CEA, Université Paris-Saclay, F-91191 Gif-sur-Yvette, France; magda.zielinska@cea.fr
* Correspondence: mrocchin@uoguelph.ca

**Abstract:** Low-energy Coulomb excitation is capable of providing unique information on static electromagnetic moments of short-lived excited nuclear states, including non-yrast states. The process selectively populates low-lying collective states and is, therefore, ideally suited to study phenomena such as shape coexistence and the development of exotic deformation (triaxial or octupole shapes). Historically, these experiments were restricted to stable isotopes. However, the advent of new facilities providing intense beams of short-lived radioactive species has opened the possibility to apply this powerful technique to a much wider range of nuclei. The paper discusses the observables that can be measured in a Coulomb-excitation experiment and their relation to the nuclear structure parameters with an emphasis on the nuclear shape. Recent examples of Coulomb-excitation studies that provided outcomes relevant for the Shell Model are also presented.

**Keywords:** nuclear structure; low-energy Coulomb excitation; Shell Model

**Citation:** Rocchini, M.; Zielińska, M. Low-Energy Coulomb Excitation for the Shell Model. *Physics* **2021**, *3*, 1237–1253. https://doi.org/10.3390/physics3040078

Received: 7 August 2021
Accepted: 28 September 2021
Published: 15 December 2021

**Publisher's Note:** MDPI stays neutral with regard to jurisdictional claims in published maps and institutional affiliations.

**Copyright:** © 2021 by the authors. Licensee MDPI, Basel, Switzerland. This article is an open access article distributed under the terms and conditions of the Creative Commons Attribution (CC BY) license (https://creativecommons.org/licenses/by/4.0/).

## 1. Introduction

Among the multitude of experimental techniques used in nuclear-structure physics, low-energy Coulomb excitation is one of the oldest and, still to this day, one of the most widely employed. The reason for its success is twofold. On the one hand, this technique requires ion beams with relatively low energy (a few MeV per nucleon) and the large cross sections of the Coulomb-excitation process can compensate for low beam intensity. For these reasons, it was widely used for experimental nuclear-structure studies in their early days and, at present, leads the way at new-generation radioactive ion beam (RIB) facilities. On the other hand, low-energy Coulomb excitation is particularly sensitive to nuclear collective properties, such as the shape. Specifically, this method can be used to determine reduced transition probabilities between low-lying states and their spectroscopic quadrupole moments. As it relies on the well-known electromagnetic interaction, all these observables can be extracted in a model-independent way. Furthermore, the unique and model-independent information on relative signs of $E2$ matrix elements, achievable solely with this technique, makes it possible to link transitional and diagonal $E2$ matrix elements to Hill–Wheeler parameters $(\beta_2, \gamma)$ describing a quadrupole shape, via non-energy weighted quadrupole sum rules [1]. Hence, low-energy Coulomb excitation constitutes a powerful tool to study phenomena such as shape coexistence, shape transitions, superdeformation, and octupole collectivity (see [2–4] for recent examples).

This paper aims to outline how the results of low-energy Coulomb-excitation measurements can be used to benchmark the Shell Model and inspire further theoretical developments. In the next Section the method is briefly introduced, and first-order and higher-order effects, giving rise to sensitivity to transitional and diagonal electromagnetic matrix elements, are discussed. The following Section presents examples of low-energy Coulomb-excitation experiments that provided outcomes particularly relevant for the Shell Model. The aim of this paper is not to provide a comprehensive review of low-energy Coulomb-excitation studies, as these can be found elsewhere (see, for instance, [5,6]).

## 2. Basics of Low-Energy Coulomb Excitation

Coulomb excitation is an inelastic scattering process, in which the two colliding nuclei are excited via a mutually-generated, time-dependent electromagnetic field. If the distance between the collision partners is sufficiently large, the short-range nuclear interaction has a negligible influence on the excitation process, which is governed solely by the well-known electromagnetic interaction. This condition can be quantified using the Cline's safe distance criterion [7], appropriate for heavy nuclei, which states that if the distance of the closest approach between the surfaces of the collision partners exceeds 5 fm, contributions from the nuclear interaction to the observed excitation cross sections are below 0.5%.

The excitation cross sections depend on electromagnetic matrix elements coupling the low-lying states in the nucleus of interest, including diagonal $E2$ matrix elements related to spectroscopic quadrupole moments. The decay of Coulomb-excited states is governed by the same set of electromagnetic matrix elements, although the influence of specific matrix elements on the excitation and decay processes may be very different as illustrated by Figure 1. Namely, low-energy Coulomb excitation favours the population of collective states through $E2$ and $E3$ transitions, while other multipolarities typically have a small impact on the measured cross sections (see [8] for further details). The $M1$ and $E1$ multipolarities, however, remain important in the de-excitation process. The quantities measured in low-energy Coulomb-excitation experiments are, most commonly, $\gamma$-ray yields in coincidence with at least one of the collision partners. It is, however, also possible to measure Coulomb-excitation cross sections by detecting only scattered particles or only $\gamma$ rays.

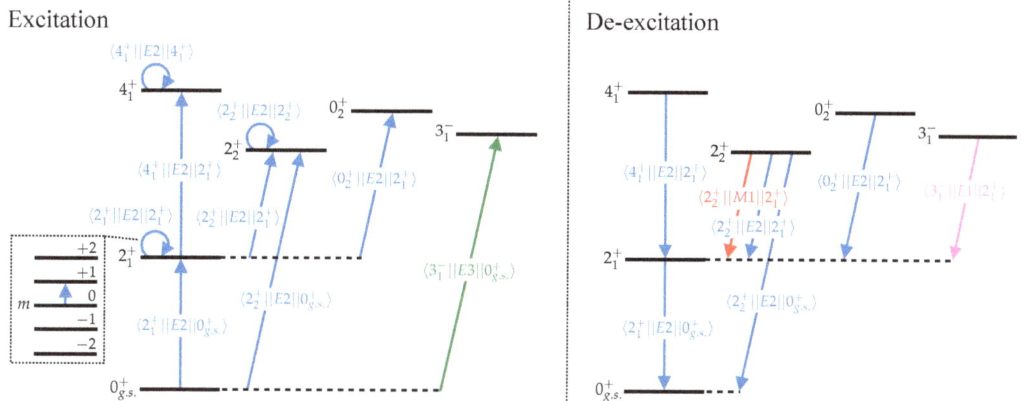


**Figure 1.** Low-lying level scheme of a fictitious even–even nucleus outlining dominant excitation (**left**) and de-excitation (**right**) patterns in low-energy Coulomb excitation. The transitions are labelled with the corresponding matrix elements. The inset on the left depicts the magnetic substates $m$ of the $2_1^+$ state and illustrates the reorientation effect. Some allowed transitions are neglected for simplicity.

While Coulomb-excitation cross sections can be calculated using a full quantum-mechanical treatment, a semi-classical approach is typically employed to overcome difficulties arising from the long-range of the Coulomb interaction and complex level schemes of the colliding nuclei. In this approach, introduced by K. Alder and A. Winther [9], the relative motion of collision partners is described using classical equations, and the quantal treatment is limited to the excitation process. The validity of this procedure, which provides a significant simplification of the calculations without a relevant loss of accuracy, stems from the fact that the interaction in the Coulomb-excitation process is dominated by the Rutherford term. For the semi-classical approximation to be valid, the de Broglie wavelength associated with the projectile must be small compared to the distance of closest approach, and the energy transferred in the excitation process must be small

compared with the total kinetic energy in the centre-of-mass reference system. These two conditions are well satisfied in low-energy Coulomb-excitation experiments involving heavy ions, but when light nuclei are involved (i.e., protons, deuterons, $\alpha$ particles), a full quantum-mechanical analysis is required.

### 2.1. First-Order Effects

If the interaction between the colliding nuclei is weak, i.e., the excitation probability is $\ll$1, Coulomb-excitation amplitudes can be calculated within the first-order perturbation theory. In the first order, the cross section for the excitation of a final state $I_f$ from the ground state $I_{\text{g.s.}}$ is proportional to the square of the transitional matrix element $\langle I_f||EL||I_{\text{g.s.}}\rangle$ , where $L = 2, 3$. Therefore, from the measured $I_{\text{g.s.}} \rightarrow I_f$ Coulomb-excitation cross section, it is possible to extract the reduced transition probability $B(EL; I_{\text{g.s.}} \rightarrow I_f)$.

The excitation process strongly depends on the kinematics and the mass es $A$ and atomic numbers $Z$ of the target and projectile nuclei. The first-order approximation is usually sufficiently accurate to describe the population of excited states from the ground state in experiments employing a light beam or a light target, or when small centre-of-mass scattering angles are used; examples of such recent studies are presented in Section 3.4. Larger kinetic energy, larger atomic numbers of the collision partners, and lower excitation energies enhance the excitation probability, leading to the appearance of higher-order effects in the excitation process.

### 2.2. Higher-Order Effects

If the electromagnetic field acting between the collision partners is strong enough and the collision process lasts a sufficiently long time, multi-step excitation becomes a possibility and higher-order contributions have to be taken into account. These contributions give rise to the experimental sensitivity to relative signs of transitional matrix elements and spectroscopic quadrupole moments of excited states, as described in the following.

#### 2.2.1. Multi-Step Excitation and Relative Signs

To understand the importance of multi-step excitation, it is useful to consider the population of two excited states , $I^\pi = 0_2^+, 4_1^+$, in an even–even nucleus (see Figure 1). As Coulomb excitation via an $E0$ transition is strictly forbidden, two-step excitation is the only way to populate the $0_2^+$ state. The $4_1^+$ state can be Coulomb-excited in two ways: directly from the ground state, via an $E4$ excitation, or with an $E2$ two-step excitation through the first excited state. Since the probability of Coulomb-exciting a given state through an $E4$ transition is much smaller than through the $E2$ excitation [8], the two-step excitation is typically dominant. Consequently, by measuring the intensities of the $4_1^+ \rightarrow 2_1^+, 0_2^+ \rightarrow 2_1^+$ $\gamma$-ray transitions with respect to the $2_1^+ \rightarrow 0_1^+$ decay, and relating them to excitation cross sections, it is possible to extract the $B(E2; 4_1^+ \rightarrow 2_1^+)$ and $B(E2; 0_2^+ \rightarrow 2_1^+)$ values.

In some cases, single-step and multi-step excitations are comparable in magnitude; an example is the $2_2^+$ state in an even–even nucleus (see Figure 1). This state can be populated by a direct $E2$ transition from the ground state and by a two-step excitation through the first excited state. The total excitation probability for the $2_2^+$ state can be written as:

$$P(0_{\text{g.s.}}^+ \rightarrow 2_2^+) = |a^{(1)}(0_{\text{g.s.}}^+ \rightarrow 2_2^+) + a^{(2)}(0_{\text{g.s.}}^+ \rightarrow 2_1^+ \rightarrow 2_2^+)|^2, \quad (1)$$

where $a^{(1)}$, $a^{(2)}$ are first-order and second-order excitation amplitudes. Consequently, $P(0_{\text{g.s.}}^+ \rightarrow 2_2^+)$ includes a term related to one-step excitation ($\langle 2_2^+||E2||0_{\text{g.s.}}^+\rangle^2$), one related to two-step excitation ($\langle 2_2^+||E2||2_1^+\rangle^2\langle 2_1^+||E2||0_{\text{g.s.}}^+\rangle^2$) and the interference term

$$\langle 2_2^+||E2||0_{\text{g.s.}}^+\rangle\langle 2_2^+||E2||2_1^+\rangle\langle 2_1^+||E2||0_{\text{g.s.}}^+\rangle. \quad (2)$$

In this last term, at variance with all the others, the matrix elements are not squared. As the total Coulomb-excitation cross section will be different for a negative (destructive) and a positive (constructive) interference term, its sign becomes an observable.

More complex interference terms can influence the Coulomb-excitation cross sections if states are populated through several excitation patterns involving multiple intermediate states. As such terms include non-squared matrix elements, their appearance leads to the experimental sensitivity to relative signs of transitional matrix elements. A sign convention should be adopted to ensure consistent analysis and facilitate a comparison with model predictions. Usually, signs of all in-band transitional $E2$ matrix elements are assumed to be positive, and, for each band head, a positive sign is imposed for one of the transitions linking it with a state in the ground-state band. The signs of all remaining matrix elements can be determined relative to those.

The probability of exciting a state via a process involving two or more steps can be comparable to that of one-step excitation, depending, for instance, on the magnitude of the involved matrix elements. Multi-step excitation is enhanced for larger scattering angles and masses of the collision partners. Experiments aiming at extracting reduced transition probabilities between the ground state and an excited state are typically performed in conditions reducing multi-step excitations, by limiting the scattering angle in the forward direction and selecting a light collision partner. In contrast, if the relative signs of transitional matrix elements and reduced transition probabilities between excited states are the objective of the experiment, the detection of scattered particles at backward angles and the use of a heavy collision partner is preferable.

#### 2.2.2. Reorientation Effect and Spectroscopic Quadrupole Moments

The reorientation effect [10] is another second-order effect in Coulomb excitation, which provides experimental sensitivity to spectroscopic quadrupole moments ($Q_s$) of excited nuclear states. This effect essentially consists in a double-step excitation, in which the intermediate state is identical to the final state, but the magnetic substate is different (see Figure 1). For a given state $I^\pi$, reorientation produces a second-order correction to its Coulomb-excitation cross section, which is proportional to the diagonal matrix element $\langle I^\pi||E2||I^\pi\rangle$, i.e., to $Q_s(I^\pi)$. Since this matrix element, and not its square, appears in the expression for cross section, its sign is also an observable. In favourable conditions, the reorientation effect may have a considerable influence on the measured $\gamma$-ray intensities. For example, in a recent study of $^{74}$Kr Coulomb-excited on $^{208}$Pb [11], a change of sign of the $Q_s(2_1^+)$ from negative to positive resulted in a 1.8-fold increase of the $4_1^+ \to 2_1^+/2_1^+ \to 0_1^+$ intensity ratio measured in coincidence with Kr nuclei scattered at 130° in the centre-of-mass frame.

The influence of the reorientation effect on Coulomb-excitation cross sections is often comparable to that of multi-step excitations. Consequently, the impact of the spectroscopic quadrupole moment can compete with, for instance, that of the sign of an interference term. This is why in early low-energy Coulomb-excitation measurements two values of the spectroscopic quadrupole moment were often reported: one corresponding to a positive sign of the $\langle 0_1^+||E2||2_1^+\rangle\langle 2_1^+||E2||2_2^+\rangle\langle 2_2^+||E2||0_1^+\rangle$ interference term, and the other one for a negative sign. This ambiguity can be solved by measuring $\gamma$-ray yields as a function of the scattering angle, thus exploiting the different angular dependence of the two effects [5,12]. This approach, typically referred to as a differential Coulomb-excitation measurement, is often employed in modern Coulomb-excitation studies. Alternatively, the use of different beam-target combinations in the same experiment can also help to disentangle competing contributions to the cross sections, and more constraints can be provided by including known spectroscopic data (lifetimes, branching ratios, and $E2/M1$ branching ratios) in the Coulomb-excitation data analysis.

### 2.3. *Quadrupole Sum Rules*

The nuclear shape can be inferred indirectly from transition probabilities or spectroscopic quadrupole moments, but this approach is not always unambiguous and generally depends on comparisons with models. An alternative model-independent approach, proposed by K. Kumar [1] and D. Cline [7], exploits the specific properties of the electro-

magnetic multipole operators. As these operators are spherical tensors, their zero-coupled products are rotationally invariant. The expectation values of these products are observables, and they are strictly related to the parameters describing the shape of the charge distribution.

The electric quadrupole operator in the principal axis system can be represented using the variables $Q$ and $\delta$, whose expectation values are equivalent to the Hill–Wheeler parameters $(\beta_2, \gamma)$ describing the quadrupole shape [1,7]. The simplest invariants read:

$$\{E2 \times E2\}^0 = \frac{1}{\sqrt{5}} Q^2, \tag{3}$$

$$\left\{[E2 \times E2]^2 \times E2\right\}^0 = -\sqrt{\frac{2}{35}} Q^3 \cos 3\delta. \tag{4}$$

The expectation values of these invariants for a state $I_n$ can be expressed through $E2$ matrix elements defined in the laboratory system. For instance:

$$\langle I_n | Q^2 | I_n \rangle = \frac{\sqrt{5}(-1)^{2I_n}}{\sqrt{2I_n + 1}} \sum_m M_{nm} M_{mn} \begin{Bmatrix} 2 & 2 & 0 \\ I_n & I_n & I_m \end{Bmatrix}, \tag{5}$$

$$\langle I_n | Q^3 \cos 3\delta | I_n \rangle = -\sqrt{\frac{35}{2}} \frac{(-1)^{2I_n}}{2I_n + 1} \sum_{ml} M_{nl} M_{lm} M_{mn} \begin{Bmatrix} 2 & 2 & 2 \\ I_n & I_m & I_l \end{Bmatrix}, \tag{6}$$

where $M_{ab} \equiv \langle I_a || E2 || I_b \rangle$ and the expression in curly brackets is a $6j$ coefficient. Higher-order invariants can be defined, such as $\langle Q^4 \rangle$, which can be linked to the dispersion in $\langle Q^2 \rangle$ via

$$\sigma(Q^2) = \sqrt{\langle Q^4 \rangle - (\langle Q^2 \rangle)^2}. \tag{7}$$

A similar definition applies to $\sigma(Q^3 \cos 3\delta)$. In principle, this approach can be extended to more complex, non-quadrupole shapes.

The invariants obtained from quadrupole sum rules provide a model-independent description of the nuclear shape in the intrinsic reference system. However, the experimental determination of such invariants requires numerous matrix elements to be known. While for the lowest-order shape invariant, $\langle Q^2 \rangle$, all matrix elements enter the sum in squares, this is not true for most higher-order invariants. In particular, the $\langle Q^3 \cos 3\delta \rangle$ invariant is constructed from triple products of $E2$ matrix elements, $\langle I_n || E2 || I_l \rangle \langle I_l || E2 || I_m \rangle \langle I_m || E2 || I_n \rangle$, where $|I_n\rangle$ is the state in question and $|I_l\rangle$ and $|I_m\rangle$ are the intermediate states. The diagonal matrix elements (i.e., $|I_l\rangle = |I_m\rangle$ ) and their signs are necessary to extract this invariant, as well as the relative signs of all relevant transitional matrix elements.

While the sums in Equations (5) and (6) formally run over all intermediate states that can be reached from the state in question via a single $E2$ transition, usually only a few key states contribute to the invariant. In particular, for the ground state of an even–even nucleus, the contributions to $\langle Q^2 \rangle$ are dominated by the coupling to the $2_1^+$ state, which typically amounts to well over 90% of the total. Similarly, the largest contributions to $\langle Q^3 \cos 3\delta \rangle$ for the ground state come from the $\langle 0_1^+ || E2 || 2_1^+ \rangle \langle 2_1^+ || E2 || 2_1^+ \rangle \langle 2_1^+ || E2 || 0_1^+ \rangle$ and $\langle 0_1^+ || E2 || 2_1^+ \rangle \langle 2_1^+ || E2 || 2_2^+ \rangle \langle 2_2^+ || E2 || 0_1^+ \rangle$ products. The situation becomes much more complicated for excited states, and the number of intermediate states that need to be included in the sum rules varies from one case to another. While theoretical approaches can, in principle, provide a complete set of electromagnetic matrix elements, this is not always true for experiments. Systematic studies employing the Shell Model addressed this convergence issue [13–15]. The contributions of individual products of matrix elements to the experimentally determined invariants have also been analysed in some cases [14,16–18].

## 3. Examples of Recent Low-Energy Coulomb-Excitation Studies Relevant for the Shell Model

The examples of experimental studies presented in this Section illustrate the variety of nuclear-structure questions, relevant for the Shell Model, that can be addressed using low-energy Coulomb excitation. The discussion is focused on the region of mid-mass nuclei between Ca and Sm and the reader is directed to [6] for other noteworthy examples, particularly concerning lighter nuclei important for astrophysical processes.

### 3.1. Superdeformation in $^{42}$Ca

The potential of Coulomb excitation as a tool to study superdeformation has been demonstrated in the very first experiment using the AGATA $\gamma$-ray tracking array [19]. The superdeformed (SD) structure in $^{42}$Ca was populated following Coulomb excitation of a $^{42}$Ca beam on $^{208}$Pb and $^{197}$Au targets [3,18]. From the measured $\gamma$-ray intensities, magnitudes and relative signs of numerous $E2$ matrix elements coupling the low-lying states in $^{42}$Ca were determined. In particular, two key pieces of information were obtained for the first time, which confirm that the band built on the $0_2^+$ state in $^{42}$Ca has a SD character at low spin: the spectroscopic quadrupole moment of the $2_2^+$ state, which corresponds to $\beta_2 = 0.48(16)$, as well as the enhanced $B(E2; 2_2^+ \to 0_2^+) = 15^{+6}_{-4}$ W.u. value. As discussed in [12], even though the $2_2^+ \to 0_2^+$ transition is too weak to be observed and, prior to the study of K. Hadyńska-Klęk et al. [3,18] only an upper limit for the branching ratio was known, the corresponding matrix element has a strong influence on excitation cross sections of the observed states, and hence it could be determined from the intensities of other transitions measured in the Coulomb-excitation experiment.

The obtained transitional and diagonal $E2$ matrix elements were further interpreted in terms of quadrupole invariants of the $0_{1,2}^+$ and $2_{1,2}^+$ states, leading to the conclusion that the spherical ground state of $^{42}$Ca exhibits large fluctuations into the $\beta_2$–$\gamma$ plane, while the excited structure has a large quadrupole deformation of $\beta_2 = 0.43(4)$ for the $0_2^+$ state, comparable to those deduced from lifetime measurements for other SD bands in this mass region. The important increase of the $\langle Q^2 \rangle$ quadrupole invariant for the $2_1^+$ state with respect to that for the ground state was attributed to the mixing of the $2^+$ states. Additionally, the triaxiality parameter $\langle \cos 3\delta \rangle$ obtained for the $0_2^+$ state, corresponding to $\gamma = (13^{+5}_{-6})^\circ$, provided the first experimental evidence for the non-axial character of SD structures around $A \approx 40$. The value obtained for the ground state, $\gamma = 28(3)^\circ$, was interpreted as resulting from its softness.

This experimental study triggered new Large-Scale Shell Model (LSSM) calculations for $^{42}$Ca [3,18]. They were performed using the SDPF.MIX interaction in the $sdpf$ model space for neutrons and protons, with a virtual $^{28}$Si core, as in the earlier study [20] that successfully described properties of the deformed $4p$–$4h$ and $8p$–$8h$ structures in $^{40}$Ca. Up to six particle –hole excitations from the $2s_{1/2}$ and $1d_{3/2}$ orbitals into the $pf$ orbitals were allowed, and the electric effective charges were $1.5e$ for protons and $0.5e$ for neutrons. The overall agreement of the calculations with the experimental level energies and decay patterns is remarkable, see Figure 2. The experimental values of the $\langle Q^2 \rangle$ and $\langle Q^3 \cos 3\delta \rangle$ invariants for the $0_{1,2}^+$ states were also well reproduced. The only notable systematic difference is the overestimation of $E2$ matrix elements in the SD band and underestimation of those in the yrast band as well as intra-band ones, which suggests that the mixing between the two bands is not fully reproduced by the calculations.

The LSSM results provide insight into the configurations of normal-deformed and SD states in $^{42}$Ca: the $0_2^+$ and $2_2^+$ states are predicted to be dominated by the $6p$–$4h$ excitation s, while the ground-state band has a predominantly two-particle configuration, with considerable $4p$–$2h$ and $6p$–$4h$ admixtures. Furthermore, they suggest that the experimentally known $2_3^+$ state is the band head of a $K = 2$ $\gamma$ band related to the SD structure, with the configuration dominated by almost equal contributions of $6p$–$4h$ and $8p$–$6h$ excitations ($\approx$40% each). This gives further support for the slightly triaxial shape of the SD band

in $^{42}$Ca, while the identification of higher-spin members of the predicted $K = 2$ $\gamma$ band represents a challenge for future experimental studies.

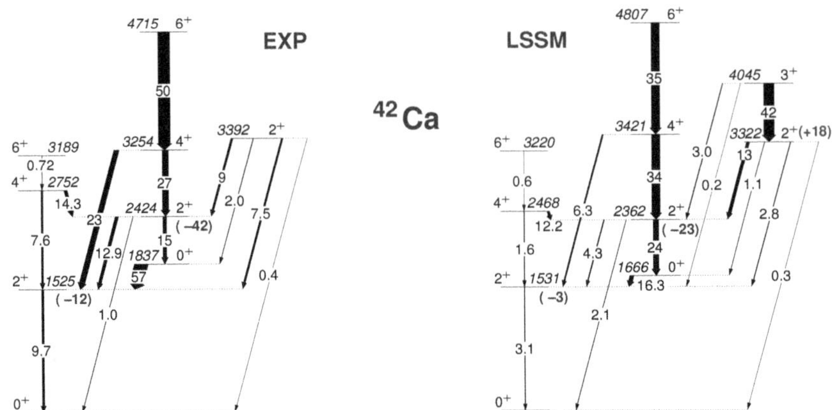


**Figure 2.** Comparison of the experimental low-energy part of the $^{42}$Ca level scheme with that calculated using Large-Scale Shell Model (LSSM) [3,18]. States are labelled with their energies in keV, transitions with $E2$ transition probabilities in Weisskopf units and spectroscopic quadrupole moments for the $2^+$ states, expressed in $e$fm$^2$, are reported in brackets.

*3.2. Shape Coexistence, Triaxiality, and the N = 50 Shell Closure in Germanium and Zinc Isotopes*

Detailed low-energy Coulomb-excitation studies were performed to investigate quadrupole properties of stable and exotic Ge and Zn isotopes, which are important in the context of the numerous Shell-Model calculations developed for this region. While extensive sets of electromagnetic matrix elements were extracted for the stable nuclei and interpreted within the quadrupole sum rules approach, in neutron-rich isotopes these measurements provided the first access to $B(E2)$ values and, in some cases, also excitation energies.

In the stable Ge isotopes, the $\langle Q^2 \rangle$ invariants extracted for the ground state and the $0^+_2$ state via low-energy Coulomb excitation represent one of the strongest signatures of shape coexistence [21,22]. As shown in Figure 3a, the ground-state $\langle Q^2 \rangle$ values in $^{70-76}$Ge are similar, 0.2–0.3 $e^2$b$^2$, while those of the $0^+_2$ states evolve as a function of the neutron number. The $0^+_2$ state in $^{70}$Ge is more deformed than the ground state [23], in $^{72}$Ge both states seem to have comparable overall deformations and considerable triaxiality [24], while those for the $0^+_2$ states in $^{74,76}$Ge point to nearly spherical shapes [25,26]. Based on the similarity of the $0^+_2$ energy systematics in Ge and Zn nuclei (see Figure 3b), one could speculate that shape coexistence is present also in the latter isotopic chain.

The first hints of the intruder character of the $0^+_2$ states in the Zn isotopes came from $E0$ measurements in the stable even–even $^{64-68}$Zn isotopes [27], a feature further supported by the results of multi-step Coulomb-excitation experiments on $^{66,68}$Zn [14,28]. However, only for $^{68}$Zn has the key $\langle 2^+_3 \| E2 \| 0^+_2 \rangle$ matrix element been determined, which, when combined with other matrix elements involving the $0^+_2$ state, leads to a $\langle Q^2 \rangle$ invariant significantly different from that of the ground state [28]. On the other hand, multiple low-energy Coulomb-excitation studies of stable Ge and Zn isotopes [14,25,28] demonstrated the importance of the triaxial degree of freedom in their structure, which was also evoked for the neighbouring $^{76,78}$Se nuclei [29,30]. Particularly relevant is the study of $^{76}$Ge [31], which yielded $(\beta_2, \gamma)$ parameters for the $0^+_1$, $2^+_1$ and $2^+_2$ states and their dispersions, which are consistent with rigid triaxial deformation. This is particularly important considering that $^{76}$Ge is a candidate for searches of neutrinoless double-$\beta$ decay, and the nuclear shape is predicted to play a significant role in this process [32,33].

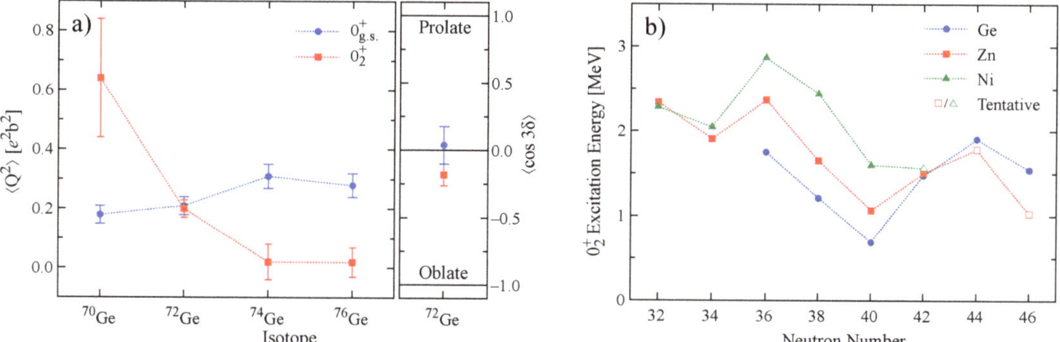


**Figure 3.** (**a**) $\langle Q^2 \rangle$ and $\langle \cos 3\delta \rangle$ quantities for the $0^+_{g.s.}$ and $0^+_2$ states in Ge isotopes. Data are taken from [21,23–26]. (**b**) Systematics of excitation energies for the $0^+_2$ states in Ge, Zn and Ni isotopes from neutron number $N = 32$ to $N = 46$. Tentative spin assignments are shown by open symbols. Data are taken from the ENSDF database [34–37]. See text for details.

Shell-model calculations focusing on the Ge, Zn, and Se isotopes well reproduced the features related to their triaxial shapes [14,29,38,39], even though the degree of $\gamma$ softness and the presence of static triaxial deformation are still debated [38]. The V-shaped pattern of the $0^+_2$ excitation energies in the Ge isotopes between the neutron numbers $N = 36$ and $N = 44$ (see Figure 3b) was related to shape coexistence by Shell-Model calculations [39] using the JUN45 effective interaction in a model space consisting of the $^{56}$Ni inert core and up to the $1g_{9/2}$ orbital for both neutrons and protons. The collectivity of the deformed ground states was linked to strong correlations (arising from pairing and the quadrupole –quadrupole force), which offset the $N = 40$ gap and lead to the enhanced occupation of the $1g_{9/2}$ neutron orbital that has a maximum predicted for $N = 40$. In contrast, the role of neutron excitations from the $pf$ shell into the $1g_{9/2}$ orbital is smaller for the $0^+_2$ states, with, on average, two additional neutrons promoted through the $N = 40$ gap with respect to the normal-order configuration. In particular, the wave function of the $0^+_2$ state in $^{72}$Ge is dominated by the normal-order configuration, i.e., neutrons completely filling the $pf$ shell, with a contribution of 37%, which suggests a nearly spherical shape.

As shown in Figure 3b, a decrease of the $0^+_2$ state energy between $N = 36$ and $N = 40$, similar to those observed in the Ge and Zn chains, is evident also in the Ni isotopes. According to Monte-Carlo Shell-Model (MCSM) calculations with the A3DA effective interaction in the $pfg_{9/2}d_{5/2}$ model space [40], the $0^+_2$ states in $^{64,66,68}$Ni are oblate deformed and result from neutron $2p$–$2h$ excitation across the $N = 40$ gap, similar to their counterparts in the Ge isotones. The V-shaped trend of the $0^+_2$ excitation energies with the vertex at $N = 40$ does not persist for $^{70}$Ni and beyond, as different configurations start to appear at low excitation energy. Specifically, proton $2p$–$2h$ excitations across the energy gap at $Z = 28$ are suggested [40,41] to dominate the structure of the $0^+_4$ state in $^{64,66}$Ni, the $0^+_3$ state in $^{68}$Ni and the $0^+_2$ state in $^{70}$Ni. MCSM calculations predict that these predominantly $\pi(2p$–$2h)$ states have well-deformed prolate shapes, resulting from an interplay of type-I and type-II shell evolution. The experimental verification of this multiple shape-coexistence scenario through the quadrupole sum rules approach represents a challenge for future low-energy Coulomb-excitation studies. Unfortunately, the population of excited $0^+$ states in both stable and radioactive Ni nuclei will be severely limited due to the high excitation energies involved, which is further complicated by the prohibitively low intensities of radioactive Ni beams that are currently available at energies suitable for low-energy Coulomb excitation.

On the neutron-rich side, low-energy Coulomb excitation has provided valuable structure information in the Ge and Zn isotopes. Experiments at ISOLDE identified the first excited $2^+_1$ state in $^{78,80}$Zn and yielded the $B(E2; 2^+_1 \to 0^+_{g.s})$ values in $^{74-80}$Zn and the $B(E2; 4^+_1 \to 2^+_1)$ values in $^{74,76}$Zn [42,43]. The obtained $B(E2)$ values hint at the importance

of triaxiality also in neutron-rich Zn isotopes, whose ground states were suggested to be rather diffuse in the $\gamma$ degree of freedom [42]. Furthermore, the energy of the first excited state in $^{80}$Zn confirms the persistence of the $N = 50$ shell closure two protons away from the doubly-magic $^{78}$Ni. The same conclusion was reached for the neutron-rich Ge isotopes from the $B(E2; 2_1^+ \to 0_{g.s}^+)$ values of the radioactive $^{78,80}$Ge measured using low-energy Coulomb excitation at ORNL [44].

The measured $B(E2; 2_1^+ \to 0_1^+)$ values in $^{74\text{–}80}$Zn were found in good agreement with those deduced from the experimental $2_1^+$ excitation energies via the Grodzins rule [45], provided that a renormalization factor (0.92) was applied to the calculated values [42]. The experimental results for $^{74\text{–}80}$Zn and $^{78,80}$Ge were compared with Shell-Model calculations comprising the $2p_{3/2}$, $1f_{5/2}$, $2p_{1/2}$, and $1g_{9/2}$ orbitals for both protons and neutrons outside of an inert $^{56}$Ni core. Effective charges significantly different from the standard $e_\nu = 0.5e$, $e_\pi = 1.5e$ values were adopted to compensate for the enhanced $^{56}$Ni core polarization reported in [46,47]. The persistence of the $N = 50$ shell closure in neutron-rich Zn and Ge isotopes, emerging from the experimental and calculated $B(E2)$ values and excitation energies, anticipated the more recent results for $^{78}$Ni , in which the first excited $2_1^+$ state was ultimately identified [48].

### 3.3. Shape Coexistence in Z ≈ 40 Nuclei

The sudden onset of deformation at $N = 60$ observed in the Zr and Sr isotopic chains has attracted a lot of attention, both from theoretical and experimental points of view. While the energies of the $2_1^+$ states in $^{90\text{–}100}$Zr were well reproduced by the LSSM calculations reported in [49], the required truncations of the model space made it impossible to account for the enhanced transition probability in $^{100}$Zr. Recently, the rapidity of the shape transition in the Zr isotopes has been reproduced, for the first time both in terms of level energies and transition probabilities, using the MCSM [50]. The calculations [50] also predict that $^{94,96,98,100}$Zr would present a multitude of low-lying states with various quadrupole shapes. A Coulomb-excitation study of $^{94}$Zr aiming to verify this scenario was performed at INFN-LNL [51], and its analysis is in progress. There exists, however, strong experimental evidence for the coexistence of deformed and spherical structures in $^{96,98}$Sr, recently reinforced by the results of Coulomb-excitation experiments performed at ISOLDE [2,52]. The rich set of transitional and diagonal $E2$ matrix elements determined in this study provides a consistent picture of a prolate-deformed ground-state band in $^{98}$Sr that coexists with an almost spherical structure built on the $0_2^+$ state. Similarity of the $B(E2; 2_2^+ \to 0_2^+) = 13(2)$ W.u. value in $^{98}$Sr with the $B(E2; 2_1^+ \to 0_1^+) = 17_{-3}^{+4}$ W.u. value in $^{96}$Sr, as well as of the quadrupole moments of the $2_2^+$ state in $^{98}$Sr and the $2_1^+$ state in $^{96}$Sr (both compatible with zero), suggest that the spherical and deformed structures interchange at $N = 60$. Contrary to what is observed in most known cases of shape coexistence, these two structures mix very weakly. This feature is in line with the type-II shell-evolution scenario proposed in [50] that links particular multiparticle–multihole excitations to significant reorganisations of the shell structure, which hinders configuration mixing.

A notable result of [2,52] is the observed reduction of the $Q_s(2_1^+)$ value in $^{98}$Sr with respect to the rotational estimate. This feature may indicate triaxiality of this state, which gives way to a more prolate deformation for higher-spin members of the ground-state band. Detailed Coulomb-excitation studies of $^{96,98,100}$Mo [17,53] yielded $\langle Q^2 \rangle$ and $\langle Q^3 \cos 3\delta \rangle$ invariants for the ground states and the low-lying $0_2^+$ states, demonstrating their different shapes and confirming that triaxiality is also a key feature of Mo nuclei with $A \approx 100$. The obtained invariant quantities indicate that, in $^{96}$Mo, an almost spherical $0_2^+$ state coexists with a triaxial ground state, while, in $^{98}$Mo, both the $0_1^+$ and $0_2^+$ states have approximately the same values of $\langle Q^2 \rangle$. However, the $\langle \cos 3\delta \rangle$ values suggest that the ground state in $^{98}$Mo is triaxial and the $0_2^+$ state has a prolate shape. The same pattern of a prolate $0_2^+$ state coexisting with a triaxial ground state appears in $^{100}$Mo, but the $\langle Q^2 \rangle$ invariants obtained for both the $0_1^+$ and $0_2^+$ states in this nucleus are significantly greater than those for $^{98}$Mo,

with that for the $0_2^+$ state being much larger. Given also that the proton vacancies and neutron occupancies for the ground states of $^{98,100}$Mo were recently extracted from an extensive series of single-proton and single-neutron transfer reactions [54], these nuclei would represent a stringent test for Shell-Model calculations. Such investigation would also be relevant in the context of neutrinoless double-$\beta$ decay studies, as $^{100}$Mo is one of candidate nuclei for this process.

### 3.4. Evolution of Collectivity in $Z \approx 50$ Nuclei

The tin nuclei, forming the longest chain of experimentally accessible isotopes between two doubly-magic nuclei, have traditionally been considered a prime example of the seniority scheme. While this description is supported by the almost constant energies of the $2_1^+$ states in the even–even Sn nuclei from $^{102}$Sn to $^{130}$Sn, the corresponding $B(E2; 2_1^+ \rightarrow 0_1^+)$ values seem to deviate from the expected parabolic behaviour (see Figure 4). Extensive Coulomb-excitation studies of stable [55–57] and exotic [58–62] Sn nuclei yielded $B(E2; 2_1^+ \rightarrow 0_1^+)$ values for $^{106-134}$Sn that were discussed in the context of Shell-Model calculations. In the Coulomb-excitation campaigns aiming at high-precision measurements of the $B(E2; 2_1^+ \rightarrow 0_1^+)$ values in stable Sn isotopes, the experimental conditions minimised the role of multi-step excitation and the reorientation effect. The experiments at ORNL [55] were performed in strongly inverse kinematics, with a $^{12}$C target bombarded by $^{112,114,116,118,120,122,124}$Sn beams; a $^{nat}$Ti target was also used for complementary $Q_s(2_1^+)$ measurements.

In the IUAC campaign [56,57], a reaction partner with a much higher $Z$ was used: a $^{58}$Ni beam impinged on $^{112,116,118,120,122,124}$Sn targets. However, due to the selection of events with the Ni beam particles scattered at forward angles, no excitation of higher-lying states was observed, although their possible weak influence on the $2_1^+$ excitation process was taken into account in the data analysis. The $B(E2; 2_1^+ \rightarrow 0_1^+)$ values were obtained with relative uncertainties of 5% or less in all cases, and the results of the two campaigns agreed within $3\sigma$ for $^{120,122,124}$Sn and within $1\sigma$ for the other isotopes, demonstrating the level of accuracy and precision that can be achieved (see Figure 4).

Low-energy Coulomb-excitation experiments on neutron-deficient Sn isotopes were performed at ISOLDE [58,59] with 2.8-MeV/$A$ $^{106,108,110}$Sn beams bombarding $^{58}$Ni targets. On the neutron-rich side, a campaign was performed at ORNL [60,61] to study $^{126,128,130,134}$Sn in very similar experimental conditions as those used for stable isotopes in [55]. In order to increase the excitation cross section for the $2_1^+$ state in $^{132}$Sn, located at 4.04-MeV excitation energy, targets of $^{48}$Ti and $^{206}$Pb were used in the ORNL [60] and HIE-ISOLDE [62] measurements, respectively.

While certain discrepancies with the values obtained using other methods exist (see e.g., [63] for a compilation of experimental data), the ensemble of experimental results points to an asymmetric shape of the $B(E2; 2_1^+ \rightarrow 0_1^+)$ distribution as a function of $N$, with a plateau extending towards lighter nuclei. The reproduction of this plateau represented a challenge for model calculations. Recently, its appearance has been discussed [64,65] in the context of pseudo-SU(3) symmetry acting in the space of $gds$ orbitals excluding $1g_{9/2}$. The calculations were performed using $V_{\mathrm{low}-k}$ variants of the realistic N3LO interaction, with the monopole part of the interaction replaced by a Hamiltonian provided by the GEMO code [66], adding the single-particle energies for $^{101}$Sn. They successfully reproduced the evolution of the $B(E2; 2_1^+ \rightarrow 0_1^+)$ values in $^{104-114}$Sn [64,65] (see Figure 4) and demonstrated that modifications of the pairing strength had a negligible effect on the calculated $B(E2; 2_1^+ \rightarrow 0_1^+)$ values, in contrast to what was observed for the $B(E2; 4_1^+ \rightarrow 2_1^+)$ strengths [65].

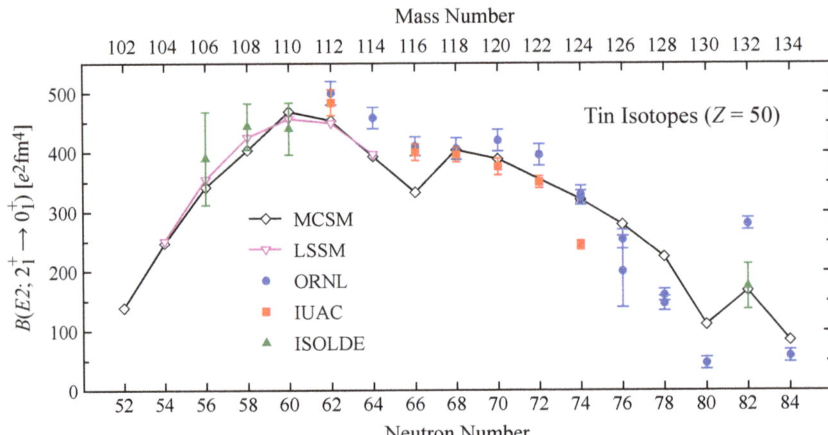


**Figure 4.** Reduced transition probabilities $B(E2; 2_1^+ \rightarrow 0_1^+)$ in the Sn isotopic chain determined from low-energy Coulomb-excitation measurements. The experimental results obtained at ORNL [55,60,61], IUAC [56,57], and ISOLDE [58,59,62] are compared with predictions from the Monte-Carlo Shell Model (MCSM) [63] and LSSM [64,65].

An alternative explanation was offered by the MCSM calculations [63] performed in the full *gds* model space complemented by the $1h_{11/2}$, $2f_{7/2}$, and $3p_{3/2}$ orbitals for protons and neutrons. These calculations provide good reproduction of all measured $B(E2; 2_1^+ \rightarrow 0_1^+)$ values in the Sn chain, including the local increase observed for $^{132}$Sn (see Figure 4), and link their enhancement for $^{108-114}$Sn to the development of quadrupole deformation driven by proton excitations from the $1g_{9/2}$ orbital. This scenario is consistent with the observed increase of the $Q_s(2_1^+)$ values at mid shell [55], which was suggested to be due to the mixing with a deformed configuration, resulting in the presence of proton $2p$–$2h$ and $4p$–$4h$ components in the $2_1^+$ wave function [55]. Low-lying states of predominantly proton $2p$–$2h$ character have been identified in $^{114,116,118}$Sn via two-proton transfer reactions [67], and later also in $^{110,112}$Sn and $^{120,122,124}$Sn, although at higher excitation energies. The MCSM calculations [63] predicted indeed that the ground states of Sn nuclei involve a significant promotion of protons across the $Z = 50$ gap, with the largest $2d_{5/2}$ occupation predicted at $N = 60$. The occupation of proton orbitals above the $Z = 50$ gap becomes even larger for the $2_1^+$ states, and the corresponding T-plots indicate deformed shapes [63], in line with the measured non-zero quadrupole moments. Multi-step Coulomb-excitation studies aiming at the determination of deformation parameters of the deformed structures built on the $0_2^+$ states, as well as their mixing with the ground-state configurations, would be of much interest. One should note here that the quadrupole invariants for the $0_{1,2}^+$ states in $^{110}$Cd were measured in a recent Coulomb-excitation experiment [68].

The $B(E2; 2_1^+ \rightarrow 0_1^+)$ and $B(E2; 4_1^+ \rightarrow 2_1^+)$ patterns in $^{100-110}$Cd nuclei closely resemble that of the $B(E2; 2_1^+ \rightarrow 0_1^+)$ values in the corresponding Sn isotones. They were well reproduced by the calculation of [65], and found almost independent of the assumed pairing strength. This was linked [65] to their static quadrupole deformation, consistent with non-zero quadrupole moments measured for the $^{102,104}$Cd isotopes in a Coulomb-excitation experiment at ISOLDE [69]. Interestingly, the obtained $Q_s(2_1^+)$ values are positive, in contrast to those measured for stable Cd nuclei. Unfortunately, they are subject to large uncertainties, and the $Q_s(2_1^+)$ value for $^{104}$Cd significantly changes if a previously measured lifetime of the $2_1^+$ state is used as an additional constraint in the Coulomb-excitation data analysis.

Quadrupole deformation of light Cd isotopes was explored in an LSSM study [15] using a modified *v3sb* effective interaction [70] in the $\pi(2p_{1/2}, 1g_{9/2})$, $\nu(2d_{5/2}, 3s_{1/2}, 2d_{3/2}, 1g_{7/2}, 1h_{11/2})$ model space. The calculated $E2$ matrix elements provide a good reproduction

of the experimental $B(E2;2_1^+ \to 0_1^+)$ and $B(E2;4_1^+ \to 2_1^+)$ values, and were analysed in terms of quadrupole invariants $\langle Q^2 \rangle$ and $\langle Q^3 \cos 3\delta \rangle$ pointing to a predominantly prolate character of $^{100-108}$Cd with both $\beta$ and $\gamma$ increasing with $N$. Very recently, Coulomb excitation of $^{106}$Cd was performed [71] at the NSCL ReA3 facility. Quadrupole moments of the $2_1^+, 4_1^+, 6_1^+$ and $2_2^+$ states were obtained, as well as the $\langle Q^2 \rangle$ and $\langle Q^3 \cos 3\delta \rangle$ invariants for the ground state, which suggest its considerable triaxiality. This feature does not emerge from the LSSM calculations reported in [71], which also used a G-matrix-renormalized CD-Bonn nucleon –nucleon potential and the same model space as those of [15], but allowed at most two neutrons in the $1h_{11/2}$ orbital. While they well reproduced the experimental $\langle Q^2 \rangle$ invariant for the ground state, the shapes that they predict for light Cd isotopes are decidedly prolate. The difference with respect to a more $\gamma$-soft behaviour suggested by [15] was attributed to the different $1h_{11/2}$ single-particle energies, as well as the adopted truncation. However, none of these calculations are able to explain the observed pattern of spectroscopic quadrupole moments in the light Cd nuclei, which will hopefully trigger future experimental and theoretical investigations aiming at understanding their quadrupole properties.

### 3.5. Heavier Collective Nuclei: Triaxiality in $^{130}$Xe and $^{140}$Sm

The $^{130}$Xe and $^{140}$Sm isotopes are examples of relatively heavy nuclei, probed with low-energy Coulomb excitation, for which extensive Shell-Model calculations have been performed [16,72]. Both isotopes were studied at ISOLDE, with the measurement for the stable $^{130}$Xe being a by-product of a radioactive beam experiment. Beam energies were 4.2 MeV/$A$ and 2.8 MeV/$A$, respectively, and states up to $I^\pi = 6_1^+$ were observed in $^{130}$Xe, while the $2_1^+, 4_1^+$ and $2_2^+$ states were populated in $^{140}$Sm. The results point to the importance of the triaxial degree of freedom in the structure of low-lying levels in both nuclei.

The extracted transitional and diagonal $E2$ matrix elements indicate that $^{130}$Xe and $^{140}$Sm are collective, and their ground states are characterized by $\beta_2 \approx 0.15$ and $\gamma \approx 30°$. For $^{130}$Xe, this conclusion was drawn on the basis of the determined quadrupole invariants, while, for $^{140}$Sm, it results from the measured $Q_s(2_1^+) = -0.06^{+0.41}_{-0.15}$ $e$b, compatible with zero, and the enhanced $B(E2;2_1^+ \to 0^+_{\text{g.s.}}) = 53(5)$ W.u. value. Shell-Model calculations for $^{130}$Xe and $^{140}$Sm were performed in a large model space consisting of the $^{100}$Sn inert core and all orbitals up to $N = Z = 82$. The GCN50:82 effective interaction [73] was employed for both cases, complemented by the SN100PN effective interaction [74] for $^{130}$Xe. The experimental and theoretical results showed good agreement (see Figure 5), which is remarkable considering the evident collective nature of the two nuclei and the relatively high number of allowed valence particles in the Shell-Model calculations. However, for both $^{130}$Xe and $^{140}$Sm, effective charges larger than the standard $e_\nu = 0.5e$, $e_\pi = 1.5e$ values were needed to reproduce the measured $B(E2)$ values. For $^{130}$Xe, $e_\nu = 0.945e$, $e_\pi = 1.53e$ and $e_\nu = 0.84e$, $e_\pi = 1.68e$ were adopted for the GCN50:82 and SN100PN interactions, respectively, while $e_\nu = 0.64e$, $e_\pi = 1.65e$ were used for the GCN50:82 interaction in the case of $^{140}$Sm. The need for increasing the effective charges in this mass region with respect to the standard values is known [75,76], and it suggests that a further expansion of the model space is necessary.

Despite the good reproduction of the experimental results by state-of-the-art Shell-Model calculations, further developments are needed to properly describe the structure of $A \approx 130$–140 nuclei within this theoretical approach. This is particularly relevant for $^{130}$Xe, which would be the daughter of the $^{130}$Te neutrinoless double-$\beta$ decay. If this process is observed at ongoing experiments, such as CUORE [77] and SNO+ [78], the relevant $\beta\beta$ nuclear matrix elements will need to be calculated in order to extract the Majorana mass. Such calculations are under way, also within the Shell Model [79], with important experimental constraints coming from recent measurements of valence proton and neutron occupations in $^{130}$Te and $^{130}$Xe [80,81]. Further low-energy Coulomb-excitation studies should help to elucidate the nuclear structure at $A \approx 130$–140. A $^{130}$Xe beam could be delivered by a stable ion beam facility with a much higher intensity than that available

in [16], and the use of a heavier target (e.g., $^{208}$Pb) would increase the excitation cross sections. For $^{140}$Sm, an experiment with a higher beam energy would be beneficial. Under favourable conditions, such experiments should be capable of extracting higher-order quadrupole invariants related to the dispersions in $\beta_2$ and $\gamma$ for the ground state.

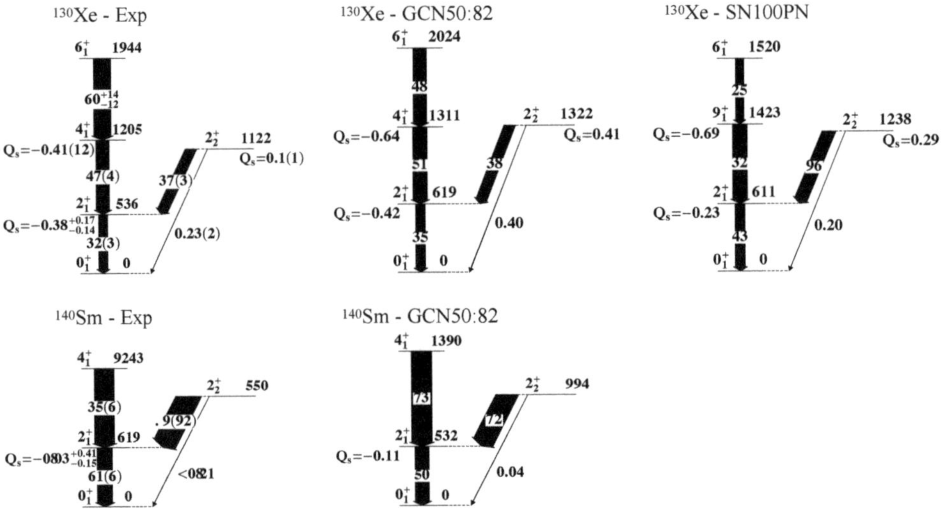


**Figure 5.** Comparison of low-energy parts of the experimental $^{130}$Xe and $^{140}$Sm level schemes with Shell-Model calculations using GCN50:82 and SN100PN interactions [16,72]. The states are labelled with their spin and parity $I^{\pi}$ and excitation energy in keV. Transitions are labelled with reduced transition probabilities expressed in Weisskopf units. Spectroscopic quadrupole moments are reported in $e$b. See text for further details about the calculations.

## 4. Summary and Outlook

In parallel to recent advances in accelerator and ion-source technologies, and the construction of new-generation high-resolution $\gamma$-ray tracking arrays as AGATA [19] and GRETINA [82], noteworthy developments have taken place in nuclear-structure theory. The state-of-the-art calculations, some of which were discussed in the preceding sections, are now able to predict the properties of nuclei with an unprecedented level of detail, particularly concerning the nuclear shape. Within the Shell Model, quadrupole shapes of ground and excited states can be inferred using T-plots [40] and the quadrupole sum rules approach [38]. Due to the large model spaces involved, Shell-Model studies of octupole collectivity are more rare, and one may hope that the availability of precise experimental data on $E3$ strengths will trigger further efforts in this direction.

The ongoing experimental and theoretical developments will bring forward our understanding of nuclear structure, while also being relevant for cross-disciplinary fields, such as astrophysics, neutrino physics, and physics of (and beyond) the Standard Model [4,33,83]. In this context, a precise understanding of the nuclear shape can bring us closer to answering long-standing questions in physics, such as how heavy elements originate in cataclysmic stellar events and the reason for the matter-antimatter asymmetry in the universe.

Thanks to the constant development of powerful computational resources, and refinements of Shell-Model codes and methods, this theoretical approach can now be extended to vast regions of the nuclear chart. It can be anticipated that this progress will be complemented and inspired by the availability of high-precision spectroscopic data and that low-energy Coulomb excitation will continue to play an important role in future studies throughout the nuclear chart. Let us emphasize, however, as in the cases of $^{98,100}$Mo and $^{130}$Xe, that the combination of data from a variety of techniques that probe both collective

and single-particle degrees of freedom will provide perhaps the most demanding tests of Shell-Model calculations, and studies in that direction should be pursued.

**Author Contributions:** The authors contributed equally to all aspects of this work. All authors have read and agreed to the published version of the manuscript.

**Funding:** This research was funded in part through the Natural Sciences and Engineering Research Council (NSERC) Canada.

**Data Availability Statement:** Data sharing not applicable.

**Acknowledgments:** We thankfully acknowledge P.E. Garrett for fruitful discussions and for the careful reading of the manuscript, and M. Siciliano for his contribution to Section 3.4 and Figure 4.

**Conflicts of Interest:** The authors declare no conflict of interest.

## Abbreviations

The following abbreviations are used in this manuscript:

| | |
|---|---|
| AGATA | Advanced GAmma Tracking Array |
| CUORE | Cryogenic Underground Laboratory for Rare Events |
| ENSDF | Evaluated Nuclear Structure Data File |
| GRETINA | Gamma-Ray Energy Tracking In-beam Nuclear Array |
| HIE-ISOLDE | High Intensity and Energy ISOLDE |
| INFN | Istituto Nazionale di Fisica Nucleare (National Institute for Nuclear Physics) |
| ISOLDE | Isotope Separator On-Line DEvice |
| IUAC | Inter-University Accelerator Centre |
| LNL | Legnaro National Laboratories |
| LSSM | Large-Scale Shell Model |
| MCSM | Monte-Carlo Shell Model |
| NSCL | National Superconducting Cyclotron Laboratory |
| ORNL | Oak Ridge National Laboratory |
| ReA3 | Re-accelerator facility |
| RIB | Radioactive Ion Beam |
| SD | Superdeformed |
| SNO+ | Sudbury Neutrino Observatory Plus |
| T-Plot | Tsunoda-Plot |

## References

1. Kumar, K. Intrinsic quadrupole moments and shapes of nuclear ground states and excited states. *Phys. Rev. Lett.* **1972**, *28*, 249. [CrossRef]
2. Clément, E.; Zielińska, M.; Görgen, A.; Korten, W.; Péru, S.; Libert, J.; Goutte, H.; Hilaire, S.; Bastin, B.; Bauer, C.; et al. Spectroscopic quadrupole moments in $^{96,98}$Sr: Evidence for shape coexistence in neutron-rich strontium isotopes at $N = 60$. *Phys. Rev. Lett.* **2016**, *116*, 022701. [CrossRef]
3. Hadyńska-Klęk, K.; Napiorkowski, P.J.; Zielińska, M.; Srebrny, J.; Maj, A.; Azaiez, F.; Valiente Dobón, J.J.; Kicińska-Habior, M.; Nowacki, F.; Naïdja, H.; et al. Superdeformed and triaxial states in $^{42}$Ca. *Phys. Rev. Lett.* **2016**, *117*, 062501. [CrossRef]
4. Gaffney, L.P.; Butler, P.A.; Scheck, M.; Hayes, A.B.; Wenander, F.; Albers, M.; Bastin, B.; Bauer, C.; Blazhev, A.; Bönig, S.; et al. Studies of pear shaped nuclei using accelerated radioactive beams. *Nature* **2013**, *497*, 199. [CrossRef]
5. Zielińska, M.; Gaffney, L.P.; Wrzosek-Lipska, K.; Clément, E.; Grahn, T.; Kesteloot, N.; Napiorkowski, P.J.; Pakarinen, J.; Van Duppen, P.; Warr, N. Analysis methods of safe Coulomb-excitation experiments with radioactive ion beams using the GOSIA code. *Eur. Phys. J. A* **2016**, *52*, 99. [CrossRef]
6. Görgen, A.; Korten, W. Coulomb excitation studies of shape coexistence in atomic nuclei. *J. Phys. G Nucl. Part. Phys.* **2010**, *43*, 024002. [CrossRef]
7. Cline, D. Nuclear shapes studied by Coulomb excitation. *Annu. Rev. Nucl. Part. Sci.* **1986**, *36*, 683. [CrossRef]
8. Srebrny, J.; Czosnyka, T.; Karczmarczyk, W.; Napiorkowski, P.; Droste, C.; Wollersheim, H.-J.; Emling, H.; Grein, H.; Kulessa, R.; Cline, D.; et al. $E1$, $E2$, $E3$ and $M1$ information from heavy ion Coulomb excitation. *Nucl. Phys. A* **1993**, *557*, 663c. [CrossRef]
9. Alder, K.; Winther, A. *Electromagnetic Excitation*; North-Holland: Amsterdam, The Netherlands, 1975.
10. De Boer, J.; Eichler, J. The Reorientation Effect. In *Advances in Nuclear Physics*; Plenum Press: New York, NY, USA, 1968; Volume 1, Chapter 1, pp. 1–65.

11. Clément, E.; Görgen, A.; Korten, W.; Bouchez, E.; Chatillon, A.; Delaroche, J.-P.; Girod, M.; Goutte, H.; Hürstel, A.; Le Coz, Y.; et al. Shape coexistence in neutron-deficient krypton isotopes. *Phys. Rev. C.* **2007**, *75*, 054313. [CrossRef]
12. Zielińska, M.; Hadyńska-Klęk, K. Nuclear shapes studied with low-energy Coulomb excitation. *EPJ Web Conf.* **2018**, *178*, 02014. [CrossRef]
13. Henderson, J. Convergence of electric quadrupole rotational invariants from the nuclear shell model. *Phys. Rev. C* **2020**, *102*, 054306. [CrossRef]
14. Rocchini, M.; Hadyńska-Klęk, K.; Nannini A.; Goasduff, A.; Zielińska, M.; Testov, D.; Rodriguez, T.R.; Gargano, A.; Nowacki, F.; De Gregorio, G.; et al. Onset of triaxial deformation in $^{66}$Zn and properties of its first excited $0^+$ state studied by means of Coulomb excitation. *Phys. Rev. C* **2021**, *103*, 014311. [CrossRef]
15. Schmidt, T.; Heyde, K.L.G.; Blazhev, A.; Jolie, J. Shell-model-based deformation analysis of light cadmium isotopes. *Phys. Rev. C* **2017**, *94*, 014302. [CrossRef]
16. Morrison, L.; Hadyńska-Klęk, K.; Podolyák, Z.; Doherty, D.T.; Gaffney, L.P.; Kaya, L.; Próchniak, L.; Samorajczyk-Pyśk, J.; Srebrny, J.; Berry, T.; et al. Quadrupole deformation of $^{130}$Xe measured in a Coulomb-excitation experiment. *Phys. Rev. C* **2020**, *102*, 054304. [CrossRef]
17. Wrzosek-Lipska, K.; Próchniak, L.; Zielińska, M.; Srebrny, J.; Hadyńska-Klęk, K.; Iwanicki, J.; Kisieliński, M.; Kowalczyk, M.; Napiorkowski, P.J.; Piętak, D.; et al. Electromagnetic properties of $^{100}$Mo: Experimental results and theoretical description of quadrupole degrees of freedom. *Phys. Rev. C* **2012**, *86*, 064305. [CrossRef]
18. Hadyńska-Klęk, K.; Napiorkowski, P.J.; Zielińska, M.; Srebrny, J.; Maj, A.; Azaiez, F.; Valiente Dobón, J.J.; Kicińska-Habior, M.; Nowacki, F.; Naïdja, H.; et al. Quadrupole collectivity in $^{42}$Ca from low-energy Coulomb excitation with AGATA. *Phys. Rev. C* **2018**, *97*, 024326. [CrossRef]
19. Akkoyun, S.; Algora, A.; Alikhani, B.; Ameil, F.; de Angelis, G.; Arnold, L.; Astier, A.; Ataç, A.; Aubert, Y.; Aufranc, C.; et al. AGATA—Advanced GAmma Tracking Array. *Nucl. Instrum. Methods Phys. Res. A* **2012**, *668*, 26. [CrossRef]
20. Caurier, E.; Menéndez, J.; Nowacki, F.; Poves, A. Coexistence of spherical states with deformed and superdeformed bands in doubly magic $^{40}$Ca: A shell-model challenge. *Phys. Rev. C* **2007**, *75*, 054317. [CrossRef]
21. Toh, Y.; Czosnyka, T.; Oshima, M.; Hayakawa, T.; Hatsukawa, Y. ; Matsuda, M.; Katakura, J.; Shinohara, N.; Sugawara, M.; Kusakari, H. Shape coexistence in even–even Ge isotopes—Complete spectroscopy with Coulomb excitation. *J. Nucl. Sci. Technol.* **2002**, *39*, 497. [CrossRef]
22. Heyde, K.; Wood, J.L. Shape coexistence in atomic nuclei. *Rev. Mod. Phys.* **2011**, *83*, 1467. [CrossRef]
23. Sugawara, M.; Toh, Y.; Czosnyka, T.; Oshima, M.; Hayakawa, T.; Kusakari, H.; Hatsukawa, Y.; Katakura, J.; Shinohara, N.; Matsuda, M.; et al. Multiple Coulomb excitation of a $^{70}$Ge beam and the interpretation of the $0_2^+$ state as a deformed intruder. *Eur. Phys. J. A* **2003**, *16*, 409. [CrossRef]
24. Ayangeakaa, A.D.; Janssens, R.V.F.; Wu, C.Y.; Allmond, J.M.; Wood, J.L.; Zhu, S.; Albers, M.; Almaraz-Calderon, S.; Bucher, B.; Carpenter, M.P.; et al. Shape coexistence and the role of axial asymmetry in $^{72}$Ge. *Phys. Lett. B* **2016**, *754*, 254. [CrossRef]
25. Toh, Y.; Czosnyka, T.; Oshima, M.; Hayakawa, T.; Kusakari, H.; Sugawara, M.; Hatsukawa, Y.; Katakura, J.; Shinohara, N.; Matsuda, M. Coulomb excitation of $^{74}$Ge beam. *Eur. Phys. J. A* **2000**, *9*, 353. [CrossRef]
26. Toh, Y.; Czosnyka, T.; Oshima, M.; Hayakawa, T.; Kusakari, H.; Sugawara, M.; Osa, A.; Koizumi, M.; Hatsukawa, Y.; Katakura, J.; et al. Multiple Coulomb excitation of a $^{76}$Ge beam. *J. Phys. G Nucl. Part. Phys.* **2001**, *27*, 1475. [CrossRef]
27. Passoja, A.; Julin, R.; Kantele, J.; Luontama, M.; Vergnes, M. $E0$ transitions in $^{70}$Ge and shape-coexistence interpretation of even-mass Ge isotopes. *Nucl. Phys. A* **1985**, *438*, 413. [CrossRef]
28. Koizumi, M.; Seki, A.; Toh, Y.; Osa, A.; Utsuno, Y.; Kimura, A.; Oshima, M.; Hayakawa, T.; Hatsukawa, Y.; Katakura, J.; et al. Multiple Coulomb excitation experiment of $^{68}$Zn. *Nucl. Phys. A* **2004**, *730*, 46. [CrossRef]
29. Henderson, J.; Wu, C.Y.; Ash, J.; Brown, B.A.; Bender, P.C.; Elder, R.; Elman, B.; Gade, A.; Grinder, M.; Iwasaki, H.; et al. Triaxiality in selenium-76. *Phys. Rev. C* **2019**, *99*, 054313. [CrossRef]
30. Hayakawa, T.; Toh, Y.; Oshima, M.; Osa, A.; Koizumi, M.; Hatsukawa, Y.; Utsuno, Y.; Katakura, J.; Matsuda, M.; Morikawa, T.; et al. Projectile Coulomb excitation of $^{78}$Se. *Phys. Rev. C* **2003**, *67*, 064310. [CrossRef]
31. Ayangeakaa, A.D.; Janssens, R.V.F.; Zhu, S.; Little, D.; Henderson, J.; Wu, C.Y.; Hartley, D.J.; Albers, M.; Auranen, K.; Bucher, B.; et al. Evidence for rigid triaxial deformation in $^{76}$Ge from a model-independent analysis. *Phys. Rev. Lett.* **2019**, *123*, 102501. [CrossRef] [PubMed]
32. Caurier, E.; Nowacki, F.; Poves, A. Nuclear-structure aspects of the neutrinoless $\beta\beta$-decays. *Eur. Phys. J. A* **2008**, *36*, 195. [CrossRef]
33. Rodríguez, T.R.; Martínez-Pinedo, G. Energy density functional study of nuclear matrix elements for neutrinoless $\beta\beta$ decay. *Phys. Rev. Lett.* **2010**, *105*, 252503. [CrossRef]
34. ENSDF, NNDC, Brookhaven National Laboratory. Available online: https://www.nndc.bnl.gov/ensdf/ (accessed on 1 June 2021).
35. Mărginean, N.; Little, D.; Tsunoda, Y.; Leoni, S.; Janssens, R.V.F.; Fornal, B.; Otsuka, T.; Michelagnoli, C.; Stan, L.; Crespi, F.C.L.; et al. Shape coexistence at zero spin in $^{64}$Ni driven by the monopole tensor interaction. *Phys. Rev. Lett.* **2020**, *125*, 102502. [CrossRef] [PubMed]
36. Tracy, J.L., Jr.; Winger, J.A.; Rasco, B.C.; Silwal, U.; Siwakoti, P.; Rykaczewski, K.P.; Grzywacz, R.; Batchelder, J.C.; Bingham, C.R.; Brewer, N.T.; et al. Updated $\beta$-decay measurement of neutron-rich $^{74}$Cu. *Phys. Rev. C* **2018**, *98*, 034309. [CrossRef]

37. Recchia, F.; Chiara, C.J.; Janssens, R.V.F.; Weisshaar, D.; Gade, A.; Walters, W.B.; Albers, M.; Alcorta, M.; Bader, V.M.; Baugher, T.; et al. Configuration mixing and relative transition rates between low-spin states in $^{68}$Ni. *Phys. Rev. C* **2013**, *88*, 041302(R). [CrossRef]
38. Poves, A.; Nowacki, F.; Alhassid, Y. Limits on assigning a shape to a nucleus. *Phys. Rev. C* **2020**, *101*, 054307. [CrossRef]
39. Honma, M.; Otsuka, T.; Mizusaki, T.; Hjorth-Jensen, M. New effective interaction for $f_5pg_9$-shell nuclei. *Phys. Rev. C* **2009**, *80*, 064323. [CrossRef]
40. Tsunoda, Y.; Otsuka, T.; Shimizu, N.; Honma, M.; Utsuno, Y. Novel shape evolution in exotic Ni isotopes and configuration-dependent shell structure. *Phys. Rev. C* **2014**, *89*, 031301(R). [CrossRef]
41. Leoni, S.; Fornal, B.; Marginean, N.; Sferrazza, M.; Tsunoda, Y.; Otsuka, T. Shape coexistence and shape isomerism in the Ni isotopic chain. *Acta Phys. Pol. B* **2019**, *50*, 605. [CrossRef]
42. Van de Walle, J.; Aksouh, F.; Behrens, T.; Bildstein, V.; Blazhev, A.; Cederkäll, J.; Clément, E.; Cocolios, T.E.; Davinson, T.; Delahaye, P.; et al. Low-energy Coulomb excitation of neutron-rich zinc isotopes. *Phys. Rev. C* **2009**, *79*, 014309. [CrossRef]
43. Van de Walle, J.; Aksouh, F.; Ames, F.; Behrens, T.; Bildstein, V.; Blazhev, A.; Cederkäll, J.; Clément, E.; Cocolios, T.E.; Davinson, T.; et al. Coulomb excitation of neutron-rich Zn isotopes: First observation of the $2_1^+$ state in $^{80}$Zn. *Phys. Rev. Lett.* **2007**, *99*, 142501. [CrossRef] [PubMed]
44. Padilla-Rodal, E.; Galindo-Uribarri, A.; Baktash, C.; Batchelder, J.C.; Beene, J.R.; Bijker, R.; Brown, B.A.; Castaños, R.; Fuentes, B.; Gomez del Campo, J.; et al. $B(E2)\uparrow$ measurements for radioactive neutron-rich Ge isotopes: Reaching the $N = 50$ closed shell. *Phys. Rev. Lett.* **2005**, *94*, 122501. [CrossRef]
45. Grodzins, L. The uniform behaviour of electric quadrupole transition probabilities from first $2^+$ states in even–even nuclei. *Phys. Lett.* **1962**, *2*, 88. [CrossRef]
46. Honma, M.; Otsuka, T.; Brown, B.A.; Mizusaki, T. New effective interaction for $pf$-shell nuclei and its implications for the stability of the $N = Z = 28$ closed core. *Phys. Rev. C* **2004**, *69*, 034335. [CrossRef]
47. Otsuka, T.; Honma, M.; Mizusaki, T. Structure of the $N = Z = 28$ closed shell studied by Monte Carlo Shell Model calculation. *Phys. Rev. Lett.* **1998**, *81*, 1588. [CrossRef]
48. Taniuchi, R.; Santamaria, C.; Doornenbal, P.; Obertelli, A.; Yoneda, K.; Authelet, G.; Baba, H.; Calvet, D.; Château3, F.; Corsi, A.; et al. $^{78}$Ni revealed as a doubly magic stronghold against nuclear deformation. *Nature* **2019**, *569*, 52. [CrossRef]
49. Sieja, K.; Nowacki, F.; Langanke, K.; Martínez-Pinedo, G. Shell model description of zirconium isotopes. *Phys. Rev. C* **2009**, *79*, 064310. [CrossRef]
50. Togashi, T.; Tsunoda, Y.; Otsuka, T.; Shimizu, N. Quantum Phase Transition in the shape of Zr isotopes. *Phys. Rev. Lett.* **2016**, *117*, 172502. [CrossRef]
51. Marchini, N.; Rocchini, M.; Nannini, A.; Doherty, D.T.; Zielińska, M.; Garrett, P.E.; Hadyńska-Klęk, K.; Testov, D.; Goasduff, A.; Benzoni, G.; et al. Shape coexistence in $^{94}$Zr studied via Coulomb excitation. *Eur. Phys. J. Web Conf.* **2019**, *223*, 01038. [CrossRef]
52. Clément, E.; Zielińska, M.; Péru, S.; Goutte, H.; Hilaire, S.; Görgen, A.; Korten, W.; Doherty, D.T.; Bastin, B.; Bauer, C.; et al. Low-energy Coulomb excitation of $^{96,98}$Sr beams. *Phys. Rev. C* **2016**, *94*, 054326. [CrossRef]
53. Zielińska, M.; Czosnyka, T.; Choiński, J.; Iwanicki, J.; Napiorkowski, P.; Srebrny, J.; Toh, Y.; Oshima, M.; Osa, A.; Utsuno, Y.; et al. Electromagnetic structure of $^{98}$Mo. *Nucl. Phys. A* **2002**, *712*, 3. [CrossRef]
54. Freeman, S.J.; Sharp, D.K.; McAllister, S.A.; Kay, B.P.; Deibel, C.M.; Faestermann, T.; Hertenberger, R.; Mitchell, A.J.; Schiffer, J.P.; Szwec, S.V.; et al. Experimental study of the rearrangements of valence protons and neutrons amongst single-particle orbits during double-$\beta$ decay in $^{100}$Mo. *Phys. Rev. C* **2017**, *96*, 054325. [CrossRef]
55. Allmond, J.M.; Stuchbery, A.E.; Galindo-Uribarri, A.; Padilla-Rodal, E.; Radford, D.C.; Batchelder, J.C.; Bingham, C.R.; Howard, M.E.; Liang, J.F.; Manning, B.; et al. Investigation into the semimagic nature of the tin isotopes through electromagnetic moments. *Phys. Rev. C* **2015**, *92*, 041303(R). [CrossRef]
56. Kumar, R.; Saxena, M.; Doornenbal, P.; Jhingan, A.; Banerjee, A.; Bhowmik, R.K.; Dutt, S.; Garg, R.; Joshi, C.; Mishra, V.; et al. No evidence of reduced collectivity in Coulomb-excited Sn isotopes. *Phys. Rev. C* **2017**, *96*, 054318. [CrossRef]
57. Kumar, R.; Doornenbal, P.; Jhingan, A.; Bhowmik, R.K.; Muralithar, S.; Appannababu, S.; Garg, R.; Gerl, J.; Górska, M.; Kaur, J.; et al. Enhanced $0^+_{\mathrm{g.s.}} \rightarrow 2^+_1$ $E2$ transition strength in $^{112}$Sn. *Phys. Rev. C* **2010**, *81*, 024306. [CrossRef]
58. Cederkäll, J.; Ekström, A.; Fahlander, C.; Hurst, A.M.; Hjorth-Jensen, M.; Ames, F.; Banu, A.; Butler, P.A.; Davinson, T.; Datta Pramanik, U.; et al. Sub-barrier Coulomb excitation of $^{110}$Sn and its implications for the $^{100}$Sn shell closure. *Phys. Rev. Lett.* **2007**, *98*, 172501. [CrossRef]
59. Ekström, A.; Cederkäll, J.; Fahlander, C.; Hjorth-Jensen, M.; Ames, F.; Butler, P.A.; Davinson, T.; Eberth, J.; Fincke, F.; Görgen, A.; et al. $0^+_{\mathrm{g.s.}} \rightarrow 2^+_1$ transition strengths in $^{106}$Sn and $^{108}$Sn. *Phys. Rev. Lett.* **2008**, *101*, 012502. [CrossRef]
60. Radford, D.C.; Baktash, C.; Beene, J.R.; Fuentes, B.; Galindo-Uribarri, A.; Gomez del Campo, J.; Gross, C.J.; Halbert, M.L.; Larochelle, Y.; Lewis, T.A.; et al. Nuclear structure studies with heavy neutron-rich RIBS at the HRIBF. *Nucl. Phys. A* **2004**, *746*, 83c. [CrossRef]
61. Allmond, J.M.; Radford, D.C.; Baktash, C.; Batchelder, J.C.; Galindo-Uribarri, A.; Gross, C.J.; Hausladen, P.A.; Lagergren, K.; Larochelle, Y.; Padilla-Rodal, E.; et al. Coulomb excitation of $^{124,126,128}$Sn. *Phys. Rev. C* **2012**, *84*, 061303(R). [CrossRef]
62. Rosiak, D.; Seidlitz, M.; Reiter, P.; Naïdja, H.; Tsunoda, Y.; Togashi, T.; Nowacki, F.; Otsuka, T.; Colò, G.; Arnswald, K.; et al. Enhanced quadrupole and octupole strength in doubly magic $^{132}$Sn. *Phys. Rev. Lett.* **2018**, *121*, 252501. [CrossRef]

63. Togashi, T.; Tsunoda, Y.; Otsuka, T.; Shimizu, N.; Honma, M. Novel shape evolution in Sn isotopes from magic numbers 50 to 82. *Phys. Rev. Lett.* **2018**, *121*, 062501. [CrossRef]
64. Siciliano, M.; Valiente-Dobón, J.J.; Goasduff, A.; Nowacki, F.; Zuker, A.P.; Bazzacco, D.; Lopez-Martens, A.; Clément, E.; Benzoni, G.; Braunroth, T.; et al. Pairing-quadrupole interplay in the neutron-deficient tin nuclei: First lifetime measurements of low-lying states in $^{106,108}$Sn. *Phys. Lett. B* **2020**, *806*, 135474. [CrossRef]
65. Zuker, A.P. Quadrupole dominance in the light Sn and in the Cd isotopes. *Phys. Rev. C* **2021**, *103*, 024322. [CrossRef]
66. Dufflo, J.; Zuker, A.P. The nuclear monopole Hamiltonian. *Phys. Rev. C* **1999**, *59*, R2347. [CrossRef]
67. Fielding, H.W.; Anderson, R.E.; Zafiratos, C.D.; Lind, D.A.; Cecil, F.E.; Wieman, H.H.; Alford, W.P. $0^+$ states observed in Cd and Sn nuclei with the ($^3$He, $n$) reaction. *Nucl. Phys. A* **1977**, *281*, 389. [CrossRef]
68. Wrzosek-Lipska, K.; Próchniak, L.; Garrett, P.E.; Yates, S.W.; Wood, J.L.; Napiorkowski, P.J.; Abraham, T.; Allmond, J.M.; Bello Garrote, F.L.; Bidaman, H.; et al. Quadrupole deformation of $^{110}$Cd studied with Coulomb excitation. *Acta Phys. Pol. B* **2020**, *51*, 789. [CrossRef]
69. Ekström, A.; Cederkäll, J.; DiJulio, D.D.; Fahlander, C.; Hjorth-Jensen, M.; Blazhev, A.; Bruyneel, B.; Butler, P.A.; Davinson, T.; Eberth, J.; et al. Electric quadrupole moments of the $2_1^+$ states in $^{100,102,104}$Cd. *Phys. Rev. C* **2009**, *80*, 054302. [CrossRef]
70. Boelaert, N.; Smirnova, N.; Heyde, K.; Jolie, J. Shell model description of the low-lying states of the neutron deficient Cd isotopes. *Phys. Rev. C* **2007**, *75*, 014316. [CrossRef]
71. Rhodes, D.; Brown, B.A.; Henderson, J.; Gade, A.; Ash, J.; Bender, P.C.; Elder, R.; Elman, B.; Grinder, M.; Hjorth-Jensen M.; et al. Exploring the role of high- $j$ configurations in collective observables through the Coulomb excitation of $^{106}$Cd. *Phys. Rev. C* **2021**, *103*, L051301. [CrossRef]
72. Klintefjord, M.; Hadyńska-Klęk, K.; Görgen, A.; Bauer, C.; Bello Garrote, F.L.; Bönig, S.; Bounthong, B.; Damyanova, A.; Delaroche, J.-P.; Fedosseev, V.; et al. Structure of low-lying states in $^{140}$Sm studied by Coulomb excitation. *Phys. Rev. C* **2016**, *93*, 054303. [CrossRef]
73. Caurier, E.; Nowacki, F.; Poves, A.; Sieja, K. Collectivity in the light xenon isotopes: A shell model study. *Phys. Rev. C* **2010**, *82*, 064304. [CrossRef]
74. Brown, B.A.; Stone, N.J.; Stone, J.R.; Towner, I.S.; Hjorth-Jensen, M. Magnetic moments of the $2_1^+$ states around $^{132}$Sn. *Phys. Rev. C* **2005**, *71*, 044317. [CrossRef]
75. Kaia, L.; Vogt, A.; Reiter, P.; Müller-Gatermann, C.; Siciliano, M.; Coraggio, L.; Itaco, N.; Gargano, A.; Arnswald, K.; Bazzacco, D.; et al. Millisecond $23/2^+$ isomers in the $N = 79$ isotones $^{133}$Xe and $^{135}$Ba. *Phys. Rev. C* **2018**, *98*, 054312. [CrossRef]
76. Vogt, A.; Birkenbach, B.; Reiter, P.; Blazhev, A.; Siciliano, M.; Hadyńska-Klęk, K.; Valiente-Dobón, J.J.; Wheldon, C.; Teruya, E.; Yoshinaga, N.; et al. Isomers and high-spin structures in the $N = 81$ isotones $^{135}$Xe and $^{137}$Ba. *Phys. Rev. C* **2017**, *95*, 024316. [CrossRef]
77. CUORE Collaboration. Arnaboldi, C.; Avignone, F.T., III; Beeman, J.; Barucci, M.; Balata, M.; Brofferio, C.; Bucci, C.; Cebrian, S.; Creswick, R.J.; et al. CUORE: A cryogenic underground observatory for rare events. *Nucl. Instrum. Methods Phys. Res. A* **2004**, *518*, 775.
78. Albanese, V.; Alves, R.; Anderson, M.R.; Andringa, S.; Anselmo, L.; Arushanova, E.; Asahi, S.; Askins, M.; Auty, D.J.; Back, A.R.; et al. The SNO+ experiment. *J. Instrum.* **2021**, *16*, P08059. [CrossRef]
79. Coraggio, L.; De Angelis, L.; Fukui, T.; Gargano, A.; Itaco, N. Calculation of Gamow-Teller and two-neutrino double-$\beta$ decay properties for $^{130}$Te and $^{136}$Xe with a realistic nucleon-nucleon potential. *Phys. Rev. C* **2017**, *95*, 064324. [CrossRef]
80. Entwisle, J.P.; Kay, B.P.; Tamii, A.; Adachi, S.; Aoi, N.; Clark, J.A.; Freeman, S.J.; Fujita, H.; Fujita, Y.; Furuno, T.; et al. Change of nuclear configurations in the neutrinoless double-$\beta$ decay of $^{130}\text{Te} \rightarrow {}^{130}\text{Xe}$ and $^{136}\text{Xe} \rightarrow {}^{136}\text{Ba}$. *Phys. Rev. C* **2016**, *93*, 064312. [CrossRef]
81. Kay, B.P.; Bloxham, T.; McAllister, S.A.; Clark, J.A.; Deibel, C.M.; Freedman, S.J.; Freeman, S.J.; Han, K.; Howard, A.M.; Mitchell, A.J.; et al. Valence neutron properties relevant to the neutrinoless double-$\beta$ decay of $^{130}$Te. *Phys. Rev. C* **2013**, *87*, 011302(R). [CrossRef]
82. Paschalis, S.; Lee, I.Y.; Macchiavelli, A.O.; Campbell, C.M.; Cromaz, M.; Gros, S.; Pavin, J.; Qian, J.; Clark, R.M.; Crawford, H.L.; et al. The performance of the Gamma-Ray Energy Tracking In-beam Nuclear Array GRETINA. *Nucl. Instrum. Methods Phys. Res. A* **2013**, *709*, 44. [CrossRef]
83. Surman, R.; Engel, J.; Bennett, J.R.; Meyer, B.S. Source of the rare-earth element peak in r-process nucleosynthesis. *Phys. Rev. Lett.* **1997**, *79*, 1809. [CrossRef]

*Article*

# Probing Different Characteristics of Shell Evolution Driven by Central, Spin-Orbit, and Tensor Forces

**Yutaka Utsuno** [1,2]

1 Advanced Science Research Center, Japan Atomic Energy Agency, Tokai, Ibaraki 319-1195, Japan; utsuno.yutaka@jaea.go.jp
2 Center for Nuclear Study, University of Tokyo, Hongo, Bunkyo-ku, Tokyo 113-0033, Japan

**Abstract:** In this paper, the validity of the shell-evolution picture is investigated on the basis of shell-model calculations for the atomic mass number $25 \lesssim A \lesssim 55$ neutron-rich nuclei. For this purpose, the so-called SDPF-MU interaction is used. Its central, two-body spin–orbit, and tensor forces are taken from a simple Gaussian force, the M3Y (Michigan 3-range Yukawa) interaction, and a $\pi + \rho$ meson exchange force, respectively. Carrying out almost a complete survey of the predicted effective single-particle energies, it is confirmed here that the present scheme is quite effective for describing shell evolution in exotic nuclei.

**Keywords:** shell evolution; exotic nuclei; shell model; effective interaction; tensor force; spectroscopic factor; effective single-particle energy

**Citation:** Utsuno, Y. Probing Different Characteristics of Shell Evolution Driven by Central, Spin-Orbit, and Tensor Forces. *Physics* **2022**, *4*, 185–201. https://doi.org/10.3390/physics4010014

Received: 29 November 2021
Accepted: 18 January 2022
Published: 9 February 2022

**Publisher's Note:** MDPI stays neutral with regard to jurisdictional claims in published maps and institutional affiliations.

**Copyright:** © 2022 by the author. Licensee MDPI, Basel, Switzerland. This article is an open access article distributed under the terms and conditions of the Creative Commons Attribution (CC BY) license (https://creativecommons.org/licenses/by/4.0/).

## 1. Introduction

One of the most important results obtained by investigating exotic nuclei (those from the $\beta$ stability) is the evolution of the shell structure, which is often called the shell evolution [1]. The evolution sometimes occurs in a more drastic way than as predicted by the standard Woods–Saxon potential model: some of the conventional neutron magic numbers, such as $N = 8$, 20, and 28, disappear, and new magic numbers, such as $N = 16$ and 34, appear.

These phenomena indicate a mechanism of shell evolution beyond the potential models, and the role of effective interactions has recently received much attention. Historically, this idea was developed in the context of the shell model, dating back to 1960 when Talmi and Unna accounted for the inversion of single-particle levels in the $p$-shell nuclei [2]. Later, a similar expression was derived in Ref. [3], in which the effect of two-body interactions was formulated with what is now called the monopole interaction [4].

The impact of the monopole interaction on nuclear structure has been investigated with the development of large-scale shell-model calculations [4–6], in which $pf$-shell nuclei are very successfully described by using Kuo–Brown interactions with a few monopole matrix elements appropriately modified. The single-particle energy that includes the effect of the monopole interaction is often referred to as the effective single-particle energy [7,8].

One of the remaining issues concerning shell evolution is the general properties of the monopole interaction and their origin. One of the earliest attempts in this direction was carried out by Federman and Pittel [9], who indicated that the central force causes a sharp drop of the neutron $1g_{7/2}$ orbital with the proton $1g_{9/2}$ orbital occupied. With more data on exotic nuclei accumulated in the 1990s, the spin-isospin dependence of the effective interaction was highlighted in Ref. [10]. This property well accounts for the monopole interaction that was phenomenologically introduced in Ref. [11] to describe the shifting magic number from $N = 16$ to 20. Finally, Otsuka et al. demonstrated [12] that the tensor force significantly increases or decreases spin–orbit splitting depending on the relative direction of the spin and orbital angular momenta that the last nucleons have.

For a unified description of the shell evolution, in [13], it was proposed that the central and tensor forces are the major sources of shell evolution, whereas the two-body spin–orbit

force plays a unique role in the monopole matrix elements between specific orbitals [14]. The same conclusion was drawn from the spin-tensor decomposition of an effective interaction fitted to the experimental data [15]. In Ref. [13], shell evolution is described by an interaction that consists of a simple Gaussian central force and a $\pi + \rho$ meson exchange tensor force, whose choice is supported by "renormalization persistency" [16]. This interaction, named the monopole-based universal interaction, $V_{\rm MU}$, and its variant were successfully applied to constructing effective interactions for shell-model calculations [17,18], whose focuses were placed on many-body properties, such as the onset of deformation due to the tensor force.

The aim of the present study is to quantitatively examine to what extent the shell evolution is described by such a simple scheme. To this end, the SDPF-MU interaction [18] is employed here whose cross-shell part is made of a variant of the $V_{\rm MU}$ interaction with the two-body spin–orbit force included, and the validity of its shell evolution is carefully examined by comparing with the relevant experimental data.

In this paper, neutron-rich nuclei with the atomic mass number $25 \lesssim A \lesssim 55$ are considered, where several doubly-closed-shell nuclei are known, including $^{24}$O, $^{34}$Si, $^{36}$S, $^{40}$Ca, $^{48}$Ca, $^{52}$Ca, and $^{54}$Ca. Hence, configuration mixing within the major shell is relatively suppressed along the atomic number, $Z$, and $N = 20$ chains, for instance, which makes easier to identify the monopole matrix element most relevant to the shell evolution under debate. Here, a rather complete survey that covers both the proton and neutron shell evolution is conducted, thus, enabling to separate the unique roles of the central, spin–orbit, and tensor forces.

This paper is organized as follows. In Section 2, the $V_{\rm MU}$ interaction is introduced as used in the SDPF-MU interaction, and the different characteristics of the central, spin–orbit, and tensor forces are quantitatively presented with regard to the monopole matrix element. Section 3 dicusses how the shell evolution, caused by this interaction, can be validated by experimental data. Sections 3.1 and 3.2 are devoted to proton shell evolution with varying neutron number and neutron shell evolution with varying proton numbers, respectively. Section 4 gives conclusions of the study.

## 2. Shell Evolution Caused by the SDPF-MU Interaction

### 2.1. Monopole Matrix Elements

The SDPF-MU interaction was constructed in Ref. [18] to describe the structure of neutron-rich nuclei around $N = 28$ whose Fermi surface is located in the *sd* shell for protons and the *pf* shell for neutrons. Hence, the proton–neutron cross-shell interaction, i.e., the part of the interaction that is relevant to both the sd shell and the pf shell is responsible for the shell evolution occurring in this region.

The cross-shell part of the SDPF-MU interaction is provided by a minor modification of the $V_{\rm MU}$ interaction [13]. The $V_{\rm MU}$ interaction was proposed to give a universal behavior of shell evolution over the nuclear chart, consisting of a Gaussian central force and a $\pi + \rho$ meson exchange tensor force. In the SDPF-MU interaction, the following refinements to the original $V_{\rm MU}$ interaction are introduced:

1. the central force includes density dependence;
2. the two-body spin–orbit force is included in addition.

The central force of the shell-model effective interaction is subject to complicated renormalization and many-body effects. The basic strategy of $V_{\rm MU}$ is to determine the central force so that its monopole matrix elements are close to those of a reliable effective interaction. Here, the monopole matrix element between the orbitals $j_1$ and $j_2$ is defined by

$$V_T^{\rm m}(j_1, j_2) = \frac{\sum_J (2J+1)\langle j_1 j_2; JT|V|j_1 j_2; JT\rangle}{\sum_J (2J+1)}, \tag{1}$$

where $J$ runs over all the possible angular-momentum coupling that the Pauli principle allows, and $T$ is the isospin coupling. In constructing the original $V_{\rm MU}$ interaction, the GXPF1A interaction [19] was used as a reference, and a reasonable but not perfect agreement

was achieved. Namely, while most of the monopole matrix elements agree within 0.2 MeV, a few matrix elements differ by 0.5 MeV or more; see Figure 1 of Ref. [13]. To obtain a better result, the central force of the SDPF-MU interaction has the form of

$$V_C(1,2) = D(R) \sum_{S,T} f_{S,T} P_{S,T} \exp\left(-(r/\mu)^2\right), \tag{2}$$

where $S$ and $T$ denote the spin and isospin coupling, respectively, and $P_{S,T}$ is the projection operator onto a given $(S, T)$. The $\vec{r}$ and $\vec{R}$ are the relative and center-of-mass coordinates, respectively: $\vec{r} = \vec{r_1} - \vec{r_2}$ and $\vec{R} = (\vec{r_1} + \vec{r_2})/2$. The $D(R)$ is the density dependent part that was newly introduced in the refined $V_{\mathrm{MU}}$, and its form was taken from the FPD6 interaction [20] as

$$D(R) = 1 + A_d\{1 + \exp((R - R_0)/a)\}^{-1} \tag{3}$$

with $R_0 = 1.2A^{-1/3}$ MeV and $a = 0.6$ fm. The interaction, thus, has six free parameters, $f_{S,T}$, $\mu$, and $A_d$. They were chosen to be $f_{0,0} = -140$ MeV, $f_{1,0} = 0$, $f_{0,1} = 0.6f_{0,0}$, $f_{1,1} = -0.6f_{0,0}$, $\mu = 1.2$ fm, and $A_d = -0.4$. The resulting agreement with the monopole matrix elements of the central force of GXPF1B is quite good, as illustrated in Figure 1 of Ref. [21].

The two-body spin–orbit force in the SDPF-MU interaction was taken from that of the M3Y (Michigan 3-range Yukawa) interaction [22]. The two-body spin-orbit force plays a minor role on shell evolution compared with the central and tensor forces, as far as a restricted region of the nuclear chart is considered: see Table 1 and discussion below. However, some specific evolutions of shell gaps are dominated by the two-body spin–orbit force, thus, included here for completeness.

The overall strength of the SDPF-MU interaction is scaled by a factor $A^{-0.3}$ in the same way as the USD (Universal *sd*) [23] and GXPF1 [24] interactions.

**Table 1.** Proton–neutron monopole matrix elements between the *sd* and *pf* orbitals obtained by the SDPF-MU interaction for the atomic mass number $A = 42$. The second to the fifth columns list the central (C), tensor (T), spin–orbit (LS), and the total values (in MeV), respectively. The sixth to ninth columns indicate the hierarchy of the C + T monopole matrix elements. The texts in red (blue) are to highlight the correspondence between the most attractive matrix elements of the central (tensor) force and $\Delta n = 0$ (spin direction). See text for details.

| | C | T | LS | Total | $\Delta n$ | Spin Direction | Label | C + T |
|---|---|---|---|---|---|---|---|---|
| $1d_{5/2}$-$1f_{5/2}$ | −1.10 | −0.19 | +0.05 | −1.24 | 0 | antiparallel | −− | −1.29 |
| $1d_{3/2}$-$1f_{7/2}$ | −1.10 | −0.21 | −0.04 | −1.34 | 0 | antiparallel | −− | −1.31 |
| $2s_{1/2}$-$2p_{3/2}$ | −1.15 | 0 | −0.09 | −1.24 | 0 | no direction | −0 | 1.15 |
| $2s_{1/2}$-$2p_{1/2}$ | −1.15 | 0 | +0.17 | −0.98 | 0 | no direction | −0 | −1.15 |
| $1d_{5/2}$-$1f_{7/2}$ | −1.16 | +0.14 | −0.03 | −1.05 | 0 | parallel | −+ | −1.02 |
| $1d_{3/2}$-$1f_{5/2}$ | −1.18 | +0.28 | +0.04 | −0.86 | 0 | parallel | −+ | −0.91 |
| $1d_{5/2}$-$2p_{1/2}$ | −0.68 | −0.06 | −0.05 | −0.78 | 1 | antiparallel | +(−) | −0.74 |
| $1d_{3/2}$-$2p_{3/2}$ | −0.68 | −0.05 | +0.06 | −0.66 | 1 | antiparallel | +(−) | −0.72 |
| $2s_{1/2}$-$1f_{7/2}$ | −0.88 | 0 | −0.02 | −0.90 | 1 | no direction | +0 | −0.88 |
| $2s_{1/2}$-$1f_{5/2}$ | −0.88 | 0 | +0.03 | −0.84 | 1 | no direction | +0 | −0.88 |
| $1d_{5/2}$-$2p_{3/2}$ | −0.69 | +0.03 | −0.03 | −0.70 | 1 | parallel | +(+) | −0.66 |
| $1d_{3/2}$-$2p_{1/2}$ | −0.71 | +0.09 | +0.05 | −0.57 | 1 | parallel | +(+) | −0.61 |

Table 1 presents the proton–neutron cross-shell monopole matrix elements, calculated with the SDPF-MU interaction, for central, tensor, and spin–orbit forces. The proton–neutron monopole matrix element for a pair with $(n_1, l_1) \neq (n_2, l_2)$ is given by

$$V^{\mathrm{m}}_{pn}(j_1, j_2) = \frac{1}{2}\{V^{\mathrm{m}}_{T=0}(j_1, j_2) + V^{\mathrm{m}}_{T=1}(j_1, j_2)\}. \tag{4}$$

The second column of Table 1 indicates that the strengths of the central matrix elements can be grouped into two categories: one has $\sim -1.1$ MeV, and the other has much weaker strengths. As explained [1,13], this difference occurs because two orbitals with the difference

of the number of nodes, $\Delta n = 0$, have a large spatial overlap, thus, gaining much attraction through short-range forces. Comparing the second and the sixth columns, one finds a good correspondence between $\Delta n$ and the strength of the central matrix elements.

The monopole matrix elements of the tensor force are characterized by the relative spin direction between the two orbitals considered, as pointed out in Ref. [12]. When the spins of two orbitals (with $l > 0$) are parallel, i.e., $j_>$ - $j'_>$ or $j_<$ - $j'_<$ ($j_>$ and $j_<$ stand for $j = l + 1/2$ and $j = l - 1/2$, respectively), the tensor monopole matrix element is positive and otherwise negative. The third and the seventh columns of Table 1 exactly point to this property. This fact is accepted now, [1,12], and quantitative aspects of the tensor monopole matrix elements are as follows.

1. Similar to the central force, the strengths for the $\Delta n \neq 0$ orbitals are weaker than those of $\Delta n = 0$.
2. Although the absolute values of the tensor matrix elements are much smaller than those of the central force, the difference between the largest matrix element and the smallest one reaches $\sim 0.5$ MeV, equivalent to that of the central force.

From the point 2, one concludes that the tensor force plays a role as important as the central force in shell evolution.

On the basis of the above arguments, let us label the orbital pairs to simply estimate the strengths of the monopole matrix elements due to the central and tensor forces without numerical calculations.

- The label consists of two characters: the first and the second ones are intended to grade the central and tensor monopole matrix elements, respectively. The net effect of these two characters stands for a rough estimate of the total monopole matrix element.
- Each part is evaluated on a scale of five levels defined by $-$, $(-)$, 0, $(+)$, and $+$, to indicate relative attraction within each type of force. The "$-$" character is given to the most attractive (i.e., largest negative) pairs, and the "$+$" character is given to the least attractive (or most repulsive) pairs among the whole monopole matrix elements of the central or tensor force.
- The first character gets "$-$" for $\Delta n = 0$, or "$+$" for $\Delta n \neq 0$.
- When the first character is "$-$", the second character gets either "$-$", 0 or "$+$" depending on the relative spin direction mentioned above. When the first character is "$+$", the second character is replaced by $(-)$, 0, or $(+)$.

These labels are listed in the eighth column of Table 1. The actual sum of the central and tensor monopole matrix elements shown in the ninth column of Table 1 rather well follows this ordering, except for a few cases with $\Delta n = 1$ in which the tensor force is less dominant.

Next, the two-body spin–orbit force is examined whose monopole matrix elements are presented in the fourth column of Table 1. The strengths of the elements are usually rather weak (see details in Supplemental Material in Ref. [1]), and the typical order of the monopole matrix elements is $\sim 20A^{-5/3}$ MeV $\approx 0.04$ MeV at $A = 42$. The signs of the elements are determined so that the inner nucleon (usually with lower orbital angular momentum, $l$) produces the normal spin–orbit splitting to the outer orbitals. Namely, when the inner and the outer orbitals are labeled $i$ and $j$, respectively, their monopole matrix elements satisfy $V^{\rm m}_{pn}(i,j) < 0$ for $j = l + 1/2$ and $V^{\rm m}_{pn}(i,j) > 0$ for $j = l - 1/2$.

More specifically, when monopole matrix elements between the *sd* and $1f$ orbitals are considered, the *sd* orbitals are located closer to the center and thus can be regarded as the inner orbitals. Hence, this rule causes negative and positive monopole matrix elements for the $1f_{7/2}$ and $1f_{5/2}$ orbitals, respectively. One can also find that the monopole matrix elements between the $2s$ and $2p$ orbitals are much larger than the others. This is because this pair, having a relative orbital angular momentum $L_{\rm rel} = 1$ alone, gains much energy due to the short-range nature of the two-body spin–orbit force.

### 2.2. Effective Single-Particle Energies

Once the cross-shell monopole matrix elements are determined the above-described way, one can obtain proton and neutron shell evolutions. The shell evolution is characterized by the effective-single-particle energy (ESPE), which includes the effects of valence nucleons on the single-particle energy. While the ESPEs can be defined for any wave function (see Ref. [1]), they are often estimated by filling configurations, so that one can directly connect monopole matrix elements to shell evolution. To simplify the discussion, a case of mass-independent two-body interactions is considered here. The ESPE of the neutron orbital, $j_n$, changes by filling protons in the orbital $j_p$ as

$$\varepsilon_{\nu j_n}(\pi j_p : \text{filled}) = \varepsilon_{\nu j_n}(\pi j_p : \text{empty}) + (2j_p + 1)V^{\text{m}}_{pn}(j_n, j_p). \quad (5)$$

When one defines the change of the ESPE of $\nu j_n$ with filling $\pi j_p$ as

$$\Delta_{\pi j_p}\varepsilon_{\nu j_n} \equiv \varepsilon_{\nu j_n}(\pi j_p : \text{filled}) - \varepsilon_{\nu j_n}(\pi j_p : \text{empty}), \quad (6)$$

the evolution of the *shell gap* between $\nu j_n$ and $\nu j'_n$ with filling $\pi j_p$ is expressed as

$$\Delta_{\pi j_p}(\varepsilon_{\nu j_n} - \varepsilon_{\nu j'_n}) = (2j_p + 1)\{V^{\text{m}}_{pn}(j_n, j_p) - V^{\text{m}}_{pn}(j'_n, j_p)\}. \quad (7)$$

Figure 1 provides a schematic illustration of what is represented in Equation (7). One of the most important properties of $\Delta_{\pi j_p}(\varepsilon_{\nu j_n} - \varepsilon_{\nu j'_n})$ is that this quantity does not depend on the choice of the core to define the ESPE. For example, the evolution of the $N = 34$ shell gap can be probed not only by the systems with the $N = 34$ core but also by those with the $N = 28$ core or the $N = 20$ core. This means that one can investigate a specific shell evolution for very neutron-rich isotones by using that of less neutron-rich ones, which will be utilized in some cases considered in Section 3.

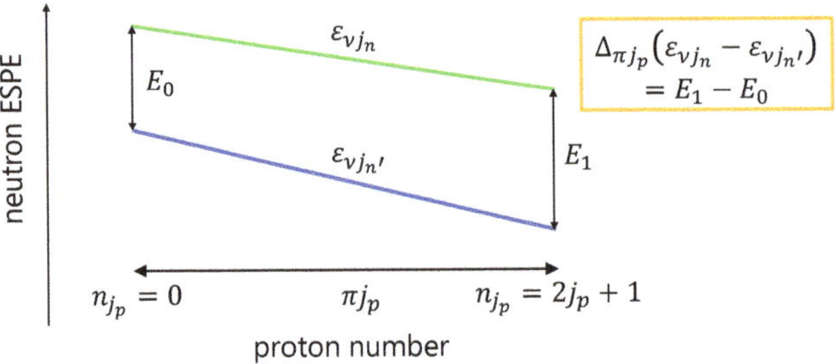


**Figure 1.** Schematic illustration of what is investigated in this paper. The blue and green lines are, respectively, the effective single-particle energies (ESPEs) of the neutron orbital, $j_n$ and $j'_n$, that change with the proton orbital, $j_p$, filled. The evolution of the shell gap, denoted as $\Delta_{\pi j_p}(\varepsilon_{\nu j_n} - \varepsilon_{\nu j'_n})$, is the main focus of this paper.

When one uses a mass-dependent interaction, Equation (7) is not exact but it is still useful for estimating shell evolution from monopole matrix elements.

## 3. Comparison to Experimental Data

The main objective of this paper is to examine how well the shell evolution described by Equation (7) is supported by experimental data. In Figure 2a,b, the proton shell evolution with neutrons occupying the $pf$ shell, and the neutron shell evolution with protons occupying the *sd* shell are plotted, respectively. The former and the latter are examined in Sections 3.1 and 3.2, respectively. In the following, for brevity, the quantum number $n$ are omitted and only the other quantum numbers like $d_{5/2}$ are given.

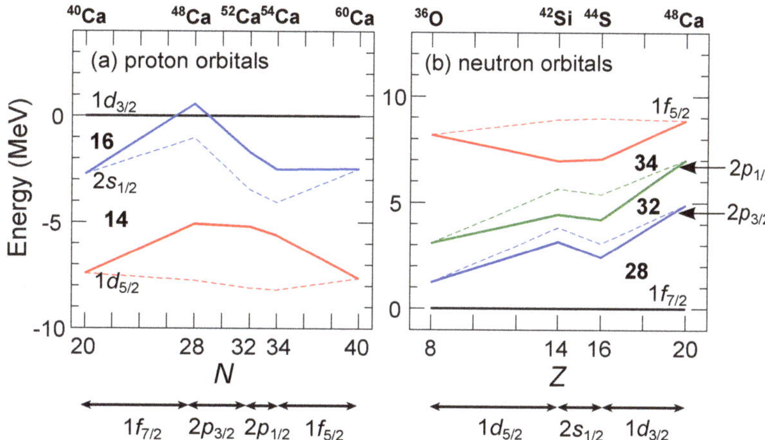


**Figure 2.** Evolution of the ESPEs calculated with the SDPF-MU interaction with the tensor force included (solid lines) and not included (dashed lines). (**a**) Proton orbitals measured from $1d_{3/2}$ for the atomic number $Z = 20$ isotopes and (**b**) neutron orbitals measured from $1f_{7/2}$ for the neutron number $N = 28$ isotones. The ESPEs are obtained by assuming filling configurations whose orders are indicated at the bottom of the figure.

In Figure 2, also the ESPE with the tensor force removed is plotted. One can immediately find that the proton $d_{5/2}$ orbital (Figure 2a) and the neutron $f_{5/2}$ orbital (Figure 2b) have the largest effect from the tensor force. Since the ESPEs shown are measured from the $d_{3/2}$ and $f_{7/2}$ orbitals, respectively, this result is a manifestation of a general property that the tensor force strongly affects the spin–orbit splitting (see Figure 1a of Ref. [12]).

To be more specific, when the proton orbital $j'$ is filled, the evolution of the neutron spin–orbit splitting between $j_<$ and $j_>$ is expressed, by using Equation (7), as $\Delta_{\pi j'}(\varepsilon_{\nu j_<} - \varepsilon_{\nu j_>}) = (2j'+1)\{V^{\mathrm{m}}_{pn}(j_<,j') - V^{\mathrm{m}}_{pn}(j_>,j')\}$. The $V^{\mathrm{m}}_{pn}(j_<,j')$ and $V^{\mathrm{m}}_{pn}(j_>,j')$ values for the tensor force are always of the opposite sign due to the identity $(2j_>+1)V^{\mathrm{m}}_{T}(j_>,j') + (2j_<+1)V^{\mathrm{m}}_{T}(j_<,j') = 0$ (valid for any isospin coupling $T$) [12], thus, magnifying the $\Delta_{\pi j'}(\varepsilon_{\nu j_<} - \varepsilon_{\nu j_>})$ value.

In addition to evaluating the ESPE, we conducted large-scale shell-model calculations to more directly compare to the data. The procedure of the calculation was the same as that employed earlier [18,25]. The valence shell consists of the full *sd* and *pf* shells. The basis states considered are truncated to allow only $0\hbar\omega$ (with $\hbar$ being the reduced Planck constant and $\omega$ the angular frequency) excitations for natural-parity states and to allow $1\hbar\omega$ excitations for unnatural-parity states. Note that, in the present case, $n\hbar\omega$ excitation is equivalent to $n$-particle-$n$-hole excitation across the $N = Z = 20$ shell gap.

Let us stress that this truncation scheme (restricted to the lowest $\hbar\omega$ space) is introduced not only to make numerical computation possible but also to be in accordance with the way how the SDPF-MU interaction is constructed (see also Section 2.2 of Ref. [21]): (i) the central force of the cross-shell interaction in SDPF-MU is fitted to the GXPF1B interaction and (ii) the intra-shell interactions employed in the SDPF-MU interaction are based on USD for the *sd* shell and GXPF1B for the *pf* shell. The USD and the GXPF1B interactions are intended for the use of the $0\hbar\omega$ model space. As shown in Ref. [18], the binding energies of neutron-rich nuclei in this region are well reproduced in this framework. The Hamiltonian matrices spanned by those basis states are numerically diagonalized by using the KSHELL code [26].

### 3.1. Proton Shell Evolution

#### 3.1.1. From $N = 20$ to $N = 28$

As shown in Figure 2a, the most distinct property by filling the $\nu f_{7/2}$ orbital is that the $Z = 16$ shell gap sharply diminishes and that the order of $\pi s_{1/2}$ and $\pi d_{3/2}$ is finally inverted at the $^{48}$Ca core. The change of this shell gap is expressed as $\Delta_{\nu f_{7/2}}(\varepsilon_{\pi d_{3/2}} - \varepsilon_{\pi s_{1/2}}) \approx 8\{V^{\rm m}_{pn}(d_{3/2}, f_{7/2}) - V^{\rm m}_{pn}(s_{1/2}, f_{7/2})\}$. Since the $d_{3/2}$-$f_{7/2}$ and $s_{1/2}$-$f_{7/2}$ pairs are labeled $\{--\}$ and $\{+0\}$, respectively, according to the rule, introduced in Section 2.1, this value is a large negative value. The actual number calculated with the SDPF-MU interaction is $-3.32$ MeV. If the tensor force is omitted from the interaction, this value decreases to $-1.71$ MeV, pointing to almost equal contributions of the central and tensor forces.

Experimentally, the evolution of the $Z = 16$ shell gap is well examined by the first excitation energies of $^{39}$K and $^{47}$K, which can be regarded as a proton hole in the $^{40}$Ca and $^{48}$Ca cores, respectively, from very large spectroscopic factors for the lowest two levels. The measured values of $E(1/2^+_1) - E(3/2^+_1)$ for $^{39}$K and $^{47}$K are 2.52 MeV and $-0.36$ MeV, respectively.

Hence, if one assumes the pure single-hole states for the $1/2^+_1$ and $3/2^+_1$ states in $^{39}$K and $^{47}$K, the $\Delta_{\nu f_{7/2}}(\varepsilon_{\pi d_{3/2}} - \varepsilon_{\pi s_{1/2}})$ value estimated from these experimental data is $-2.88$ MeV. The corresponding value obtained from large-scale shell-model calculations is $-3.33$ MeV, which is somewhat overestimated; however, the sharp decrease of $E(1/2^+_1) - E(3/2^+_1)$ in going from $^{39}$K to $^{47}$K is well explained. Note that this number is very close to that evaluated from the ESPE ($-3.32$ MeV; see the first paragraph of this Subsection) because the first two levels of $^{47}$K are very close to single-proton-hole states.

Another important property in filling the $\nu f_{7/2}$ orbital is that the proton spin–orbit splitting for the $d$ orbitals sharply decreases. This is caused almost solely by the tensor force (Figure 2a) because the central force gives similar monopole matrix elements between the $d_{3/2}$-$f_{7/2}$ and $d_{5/2}$-$f_{7/2}$ pairs: those are $\{--\}$ and $\{-+\}$ pairs, respectively. Hence, quantifying the spin–orbit splitting is the key to extracting the tensor-force driven shell evolution. By using the SDPF-MU interaction, the proton spin–orbit splittings for the $d$ orbital are obtained to be 7.42 and 5.05 MeV for the $^{40}$Ca and $^{48}$Ca cores, respectively, indicating a more than 2 MeV reduction.

Unlike the cases of $d_{3/2}$ and $s_{1/2}$, the $d_{5/2}$ proton hole does not appear as a nearly pure single-hole state because the excitation energy is much higher than other low-lying levels, making the hole state fragmented over many levels. For the present purpose, the distribution of spectroscopic factors provides crucial information. The one-proton removal spectroscopic factors from $^{40}$Ca and $^{48}$Ca were measured with reactions, such as $(d, {}^3\text{He})$ and $(e, e'p)$. Although the $(e, e'p)$ reaction gives more reliable spectroscopic factors, those measured for $^{40}$Ca are concerning only a few low-lying states. Thus, the $(d, {}^3\text{He})$ data were used to estimate the spin–orbit splitting for Ca isotopes from the centroid of the measured spectroscopic factors, as discussed in Refs. [27,28].

The centroid of the spectroscopic factors, actually, provides the exact single-particle energy. However, there are many energy levels that cannot be detected by the actual experiment because their spectroscopic factors are too small to be measured. Although each of these undetected levels has a tiny contribution to the centroid, the total effect is not negligible because the number of such levels is very large. In this sense, the centroid of the spectroscopic factors that is obtained from experiment cannot be free from uncertainty associated with the limited experimental sensitivity. Hence, in order to validate theoretical single-particle energies, it is rather helpful to compare between experiment and theory regarding how major peaks are distributed. The results are shown in Figure 3, in which the spectroscopic factor $C^2S(j)$ for the orbital $j$ is defined as

$$C^2S(j) = \frac{\left|\langle \Psi_B || a^\dagger_j || \Psi_A \rangle\right|^2}{2J_B + 1}, \tag{8}$$

where $\Psi_A$ and $\Psi_B$ are the wave functions of the nuclei $A$ and $B$, respectively (here, $A$ and $B$ correspond to Ca and K isotopes, respectively), and $J_B$ is the angular momentum of $B$.

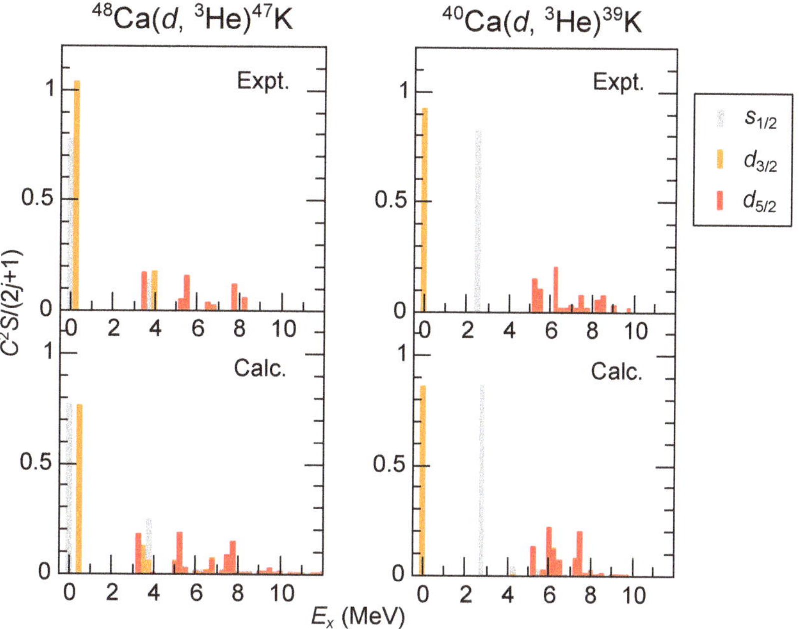


**Figure 3.** Distribution of the one-proton removal spectroscopic strengths (see Equation (8)) from $^{48}$Ca (**left**) and $^{40}$Ca (**right**) comparing experimentalal results ("Expt.") with shell-model calculations ("Calc."). The spectroscopic factors shown are divided by $2j+1$ to normalize to unity for fully occupied orbitals. The bin widths are 0.25 MeV. Data are from Refs. [29] ($^{48}$Ca) and [30] ($^{40}$Ca). See text for details.

For $^{48}$Ca, the calculations were carried out with the SDPF-MU interaction in the $0\hbar\omega$ model space [18]. The present calculation successfully captures the characteristics of the measured distribution. For $s_{1/2}$ and $d_{3/2}$, although the strengths are dominated by the lowest states, some strengths remain in the states slightly below 4 MeV due to the coupling to the $2_1^+$ state. Note that the sum of the experimental strengths for $d_{3/2}$ exceeds the sum-rule limit [29], indicating non-negligible uncertainties due to the reaction model employed. For $d_{5/2}$, the calculation well reproduces three major peaks located at 3–4, 5–6, and 7–8 MeV, although the calculated peaks are located a few hundred keV lower than those of the experiment. If the tensor force is omitted, the calculated weight of the $d_{5/2}$ strengths is shifted higher and fails to reproduce the data as presented in [18].

For $^{40}$Ca, as seen in Figure 3, the $d_{5/2}$ strengths are highly fragmented as in $^{48}$Ca. This property is impossible to reproduce with the same setup as $^{48}$Ca, since only one $5/2^+$ state appears in the $0\hbar\omega$ calculation. It is also found that the $2\hbar\omega$ calculation was not sufficient to obtain enough fragmentation because of much smaller level densities compared with the data. To resolve this problem, the large-scale shell-model calculations were done to allow many-particle many-hole excitations across the $N = Z = 20$ core. Since it is still difficult to perform such calculations in the full $sd$-$pf$ valence shell, the $p_{1/2}$ and $f_{5/2}$ orbitals are omitted from the valence shell, thus enabling $6\hbar\omega$ calculations with the KSHELL code [26].

The effective interaction is taken from Ref. [31], a modified SDPF-M interaction whose single-particle energies are fine-tuned to reproduce the correct one-neutron separation energies of $^{40,41}$Ca. Note that the original SDPF-M interaction [11] was designed for the full $sd + f_{7/2} + p_{3/2}$ model space. One expects that the $6\hbar\omega$ truncation is sufficient to achieve

convergent results. The resulting spin–orbit splitting of the *d* orbitals for the $^{40}$Ca core is close to that of SDPF-MU, 7.49 MeV, estimated from the ESPE.

Figure 3 presents the results of the calculations. Similar to $^{48}$Ca, the agreement with experiment is quite satisfactory. For $d_{3/2}$ and $s_{1/2}$, the strengths near the $2^+_1$ level of $^{40}$Ca ($\sim$4 MeV) are much smaller than those for $^{48}$Ca, in good accordance with the measured distribution [30]. For $d_{5/2}$, the calculated three major peaks at 5–6, $\sim$6, and 7–8 MeV well correspond to the measured peaks, although the highest peak is more fragmented in the experiment.

The above detailed comparisons of spectroscopic distributions confirm that a large reduction of the spin–orbit splitting, which amounts to $\sim$2 MeV, occurs in reality as a $\pi + \rho$ meson exchange tensor force produces.

3.1.2. From $N = 28$ to $N = 32$ and Beyond

As the neutron number increases from $N = 28$, the Fermi surface moves to $p_{3/2}$, which causes a different proton shell evolution from that for $20 \leq N \leq 28$. Figure 2a indicates that the most prominent is that $s_{1/2}$ goes down relative to $d_{3/2}$. This is caused by a positive $\Delta_{\pi p_{3/2}}(\varepsilon_{\nu d_{3/2}} - \varepsilon_{\nu s_{1/2}}) = 4\{V^{\rm m}_{pn}(d_{3/2}, p_{3/2}) - V^{\rm m}_{pn}(s_{1/2}, p_{3/2})\}$ because the $d_{3/2}$-$p_{3/2}$ and $s_{1/2}$-$p_{3/2}$ pairs are labeled "$\{+(-)\}$" and "$\{-0\}$". Although the tensor force causes attraction for the former pair, the central force that favors the latter surpasses this effect due to a larger spatial overlap. It is thus predicted that the $d_{3/2}$ orbital becomes the highest in the *sd* shell again at $N = 32$, leading to the reinversion of the $s_{1/2}$-$d_{3/2}$ level ordering.

Similar to that of Section 3.1.1, K isotopes play a key role in probing this level ordering from experiment. The observed hyperfine structure ruled out a $1/2^+$ ground state for the $N = 32$ isotope $^{51}$K [32], and its measured *g*-factor of $+0.3420(15)$ [32] is very close to that of the single-proton hole in $d_{3/2}$. From these data, it is concluded in Ref. [32] that the ground state of $^{51}$K must be a $3/2^+$ state that is dominated by the $\pi(d_{3/2})^{-1}$ configuration. The predicted reinversion has thus been confirmed by experiment.

A deeper understanding of shell evolution can be obtained from excitation energies. In Figure 4, the energies of the $1/2^+_1$ levels, measured from the $3/2^+_1$ levels in neutron-rich K isotopes, are compared to theory. Very recently, the first excited levels in $^{51,53}$K ($N = 32, 34$) were measured to be 0.74 and 0.84 MeV, respectively [33]. These states are assigned to be $1/2^+$ from the observed parallel momentum distributions of the $^{51,53}$K residues after $(p, 2p)$ reactions.

As shown in Figure 4, the measured values are lower than the shell-model results with the SDPF-MU interaction, 1.40 and 1.74 MeV, respectively. Although the calculated levels are located lower than those estimated from the ESPE, 1.71 and 2.50 MeV, respectively; the deviation from the experimental data may indicate the need of refining the monopole matrix elements, related to the shell evolution under discussion.

In Ref. [33], a modified SDPF-MU interaction was introduced (named SDPF-MUs) in which $V^{\rm m}_{T=0}(s, p)$ is shifted by $+0.4$ MeV, equivalent to a $+0.2$ MeV shift for the proton–neutron channel. The resulting $1/2^+_1$ levels in $^{51,53}$K are improved to be 0.85 and 0.79 MeV, respectively. These SDPF-MUs levels are also somewhat lower than those estimated from its ESPE, 0.95 and 1.38 MeV, respectively. This difference is caused by a many-body correlation, which makes single-hole strengths fragmented. Experimentally, three more levels are observed from the $^{52}$Ca$(p, 2p)^{51}$K reaction [33], which may indicate some deviation from the single-hole nature for $1/2^+_1$ or proton $d_{5/2}$ hole states fragmented.

As shown in Figure 4, the $E(1/2^+_1) - E(3/2^+_1)$ value evolves in a non-monotonic way; that is, it decreases until $N = 28$ and then turns to increase. This evolution, following the ESPE, is caused by that of the ESPE of $\pi d_{3/2}$ measured from $\pi s_{1/2}$. The reinversion of the $1/2^+_1$-$3/2^+_1$ level ordering is a consequence of the non-monotonic evolution of single-particle level spacings.

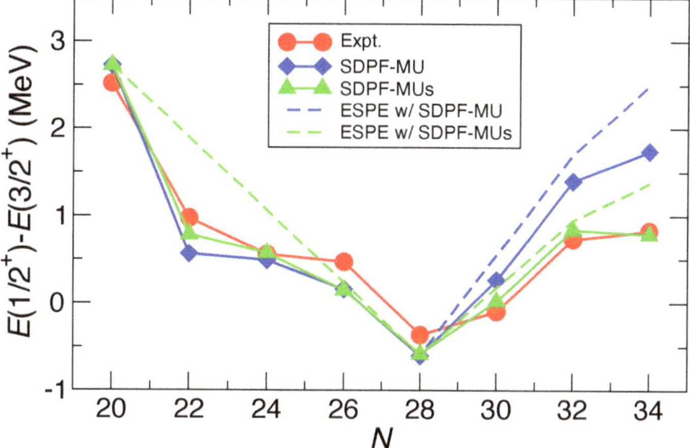


**Figure 4.** Comparison of the evolution of the energy difference $E(1/2_1^+) - E(3/2_1^+)$ in neutron-rich K isotopes between experiment and theory. The red circles represent experimental data, and the blue diamonds and the green triangles stand for the results of large-scale shell-model calculations with the SDPF-MU and the SDPF-MUs interactions, respectively. The dashed lines in blue and green are the corresponding values evaluated from the ESPE (i.e., $\varepsilon_{\pi(s_{1/2})^{-1}} - \varepsilon_{\pi(d_{3/2})^{-1}} = \varepsilon_{\pi d_{3/2}} - \varepsilon_{\pi s_{1/2}}$) for the SDPF-MU and the SDPF-MUs interaction, respectively.

Let us point out that such a non-monotonic evolution constitutes a strong evidence for the dominance of the effective interaction in shell evolution because simple one-body potential models like the Woods–Saxon ones always produce monotonic evolution of level spacings with changing mass number. Furthermore, in this particular case, the non-monotonic evolution is caused by the central force. To account for this, let us first remind one that the changes of $\varepsilon_{\pi d_{3/2}} - \varepsilon_{\pi s_{1/2}}$ for $N = 20$–28 and for $N = 28$–32 amounts, respectively, to $\Delta E_1 = 8\{V_{pn}^{\rm m}(d_{3/2}, f_{7/2}) - V_{pn}^{\rm m}(s_{1/2}, f_{7/2})\}$ and $\Delta E_2 = 4\{V_{pn}^{\rm m}(d_{3/2}, p_{3/2}) - V_{pn}^{\rm m}(s_{1/2}, p_{3/2})\}$.

For the tensor force, $V_{pn}^{\rm m}(s_{1/2}, f_{7/2}) = V_{pn}^{\rm m}(s_{1/2}, f_{7/2}) = 0$ holds, and only the first terms contribute to $\Delta E_1$ and $\Delta E_2$. As shown in Table 1, both of them are negative, and the $\varepsilon_{\pi d_{3/2}} - \varepsilon_{\pi s_{1/2}}$ value keeps decreasing. On the other hand, the central-force contributions to $\Delta E_1$ and $\Delta E_2$ are negative and positive, respectively, thus producing a kink in $E(1/2_1^+) - E(3/2_1^+)$ and $\varepsilon_{\pi d_{3/2}} - \varepsilon_{\pi s_{1/2}}$. Since this non-monotonic evolution is dominated by the central force, any microscopic model, with a reasonable two-body force, is able to describe that. In fact, both nonrelativistic and relativistic mean-field models produce similar effects [34,35].

Here, let us comment on the idea behind the empirical shift of monopole matrix elements employed in the SDPF-MUs interaction. As presented in Section 2.1, the cross-shell part of the SDPF-MU interaction consists of the central, two-body spin–orbit, and tensor terms. Among them, the tensor term is the most strongly supported by microscopic theories in terms of the "renormalization persistency", named in Ref. [16]. On the other hand, the central term is constructed in a fully phenomenological way. The two-body spin–orbit term is too small to tune.

On the basis of this general consideration, it seems that the most reasonable method of monopole tuning is for the central term alone, with the other terms untouched. The SDPF-MUs interaction is made to follow this policy. With respect to the cross-shell interaction, the difference between SDPF-MU and SDPF-MUs is the shift of $V_{T=0}^{\rm m}(s, p)$. The shift, $\Delta V_{T=0}^{\rm m}(s, p) = +0.4$ MeV, is applied not only to the $p_{3/2}$ orbital but also to the $p_{1/2}$ orbital. The latter change is needed to keep the tensor term unchanged after carrying out the spin-tensor decomposition [36].

Finally, let us mention that the $V^{\rm m}_{pn}(s,p)$ monopole matrix elements contain non-negligible contributions from the two-body spin–orbit force. This feature is discussed in Section 3.2.2.

### 3.2. Neutron Shell Evolution

In this Subsection, the neutron shells that change with the proton number are considered, as illustrated in Figure 2b. Let us start with the $^{48}$Ca core, since its neutron $pf$-shell energies are well established from the data. Next, the protons are removed from the $d_{3/2}$, $s_{1/2}$, and $d_{5/2}$ orbitals, and the relevant issues are discussed in Sections 3.2.1 to 3.2.3, respectively.

#### 3.2.1. From $Z = 20$ to $Z = 16$

As shown in Figure 2b, with protons removed from $d_{3/2}$, the $N = 28$ shell gap changes by $-\Delta_{\pi d_{3/2}}(\varepsilon_{\nu p_{3/2}} - \varepsilon_{\nu f_{7/2}}) \approx 4\{V^{\rm m}_{pn}(f_{7/2}, d_{3/2}) - V^{\rm m}_{pn}(p_{3/2}, d_{3/2})\}$. Note that the negative sign in $-\Delta_{\pi d_{3/2}}$ is needed because the shell evolution is considered with decreasing $Z$. Since $f_{7/2}$-$d_{3/2}$ and $p_{3/2}$-$d_{3/2}$ are $\{--\}$ and $\{+(-)\}$ pairs, respectively, this quantity should be negative. As discussed in Section 3.1.1, the strongly attractive monopole matrix element of $V^{\rm m}_{pn}(f_{7/2}, d_{3/2})$ causes the rapid decrease of the $Z = 16$ shell gap in going from $N = 20$ to $N = 28$.

The decrease of the $N = 28$ shell gap is difficult to evaluate from experimental data in the vicinity of $N = 28$ isotones because the corresponding sulfur isotopes are deformed. As emphasized in Section 2.2, however, this decrease can be probed from another isotone chain. In this case, $N = 20$ isotones provide useful information since both $^{36}$S and $^{40}$Ca are regarded as doubly-closed-shell nuclei with rather large first excitation energies (>3 MeV).

From the $^{36}$S$(d,p)^{37}$S reaction data, the $\nu f_{7/2}$ and $\nu p_{3/2}$ strengths are concentrated in the ground state and the 0.646 MeV state, respectively, while some strengths remain in the $3/2^-$ state at 3.263 MeV with $C^2S \approx 0.14$ and in the $7/2^-$ state at 3.443 MeV with $C^2S \approx 0.06$ [37]. The measured spectroscopic strengths, thus, indicate a small $N = 28$ shell gap that is less than 1 MeV on top of the $^{36}$S core. The shell-model calculation with the SDPF-MU interaction rather well reproduces this feature with $C^2S(f_{7/2}) = 0.86$ at $E_x = 0$ MeV, $C^2S(p_{3/2}) = 0.77$ at $E_x = 0.56$ MeV, $C^2S(p_{3/2}) = 0.19$ at $E_x = 2.97$ MeV, and $C^2S(f_{7/2}) = 0.07$ at $E_x = 3.21$ MeV. The calculated $N = 28$ shell gap for the $^{36}$S core is 0.32 MeV.

Similar data exist for the $^{40}$Ca core. The $p_{3/2}$ strengths are fragmented into the states at 1.94, 2.46, and 4.60 MeV, which is impossible to reproduce with the $0\hbar\omega$ calculations. The centroids of the spectroscopic factors measured with the $^{40}$Ca$(\vec{d},p)^{41}$Ca reaction [38] suggest that the $N = 28$ shell gap for $^{40}$Ca is 2.5 MeV. The SDPF-MU interaction produces the $N = 28$ shell gap of 2.94 MeV, which is slightly larger than this value. Hence, a large decrease of the $N = 28$ shell gap is confirmed, although the SDPF-MU interaction may overestimate this decrease by a few hundred keV.

The reduction of the $N = 28$ shell gap should have a significant impact on the $N = 28$ closed-shell structure. The breaking of the $N = 28$ closure can be probed with one-neutron removal spectroscopic strengths from $p_{3/2}$: if no $\nu p_{3/2}$ strengths are observed, then no neutrons occupy the $p_{3/2}$ orbital, implying a complete closure. Although summing up all the $p_{3/2}$ strengths are desirable for a quantitative evaluation, excited states available in neutron-rich nuclei are limited. For this purpose, the strengths of the first $3/2^-$ levels between experiment and theory are compared and the results are shown in Figure 5.

It is natural that the strength for $^{48}$Ca is very small. As the proton number is away from $Z = 20$, the strengths are naively expected to increase due to deformation caused by valence proton particles or holes. If deformation is controlled by the number of valence protons alone, those spectroscopic factors should be symmetric with respect to $Z = 20$. However, the observed spectroscopic factors are rather large for the $Z < 20$ isotones, whereas they remain small for the $Z > 20$ isotones.

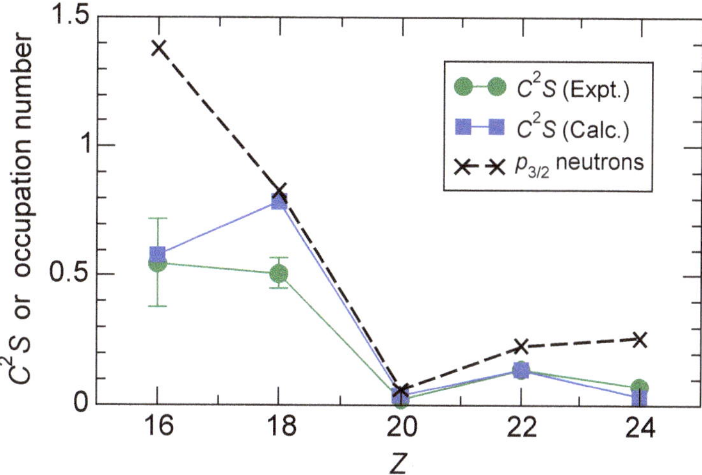


**Figure 5.** One-neutron removal spectroscopic factors of the $3/2_1^-$ states from the ground states of $N = 28$ isotones. The crosses denote the calculated neutron occupation numbers in the ground states of the $N = 28$ isotones. Data are from Refs. [39] ($Z = 20$, 22, and 24), [40] ($Z = 18$), and [41] ($Z = 16$).

This behavior is well reproduced by the shell-model calculations with the SDPF-MU interaction. The same trend is seen in the $\nu p_{3/2}$ occupation numbers, which are the upper limit of these spectroscopic factors. Hence, one concludes that the breaking of the $N = 28$ closure is much greater for $Z < 20$ than for $Z > 20$ and that the reduction of the $N = 28$ shell gap for lower $Z$ works to enhance this property.

From Figure 5, it may look unexpected that the $C^2S$ value for $Z = 16$ is only half the neutron $p_{3/2}$ occupation number of $^{44}$S unlike that for $^{46}$Ar. This is caused by a unique nuclear structure of $^{44}$S. As pointed out in Ref. [25], sulfur isotopes around $^{44}$S have two nearly degenerate deformed neutron orbitals on the Fermi surface with $\Omega^\pi = 7/2^-$ and $3/2^-$, which make the $K^\pi = 7/2^-$ and $3/2^-$ bands in $^{43}$S, respectively, by one-neutron occupation. Here, $\Omega$ and $K$ are, respectively, single-particle and total angular-momentum projection onto the symmetry axis, and $\pi$ is parity. The $3/2_1^-$ state in $^{43}$S is a $K^\pi = 3/2^-$ member. Due to the near degeneracy of the $\Omega^\pi = 7/2^-$ and $3/2^-$ orbitals, the ground state of $^{44}$S has a strongly mixed configuration with two neutrons in $\Omega^\pi = 7/2^-$ and those in $\Omega^\pi = 3/2^-$. As a result, about half the ground-state wave function of $^{44}$S, i.e., the part with two neutrons occupying the $\Omega^\pi = 3/2^-$, is able to contribute to populating the $K^\pi = 3/2^-$ band in $^{43}$S. The remaining fractions of $C^2S$ should be distributed to the excited $3/2^-$ states, which was indeed observed [41].

Let us comment on other shell gaps. The discussions of the $N = 32$ shell gap is given in Section 3.2.2, and here, just a brief remark to be made about the $N = 34$ shell gap. A recent $^{54}$Ca$(p, pn)^{53}$Ca measurement clarified that the $N = 34$ shell closure is rather good [42], while the $N = 34$ shell gap for the $^{54}$Ca core was estimated to be $\sim$ 2.5 MeV from the GXPF1Br interaction [43]. It was predicted that this shell gap enlarges with decreasing $Z$ and that the fingerprint of the enlargement can be seen in the $2_1^+$ energies of the $N = 34$ isotones with $Z < 20$ [44,45].

This prediction was confirmed later by measuring the $2_1^+$ level in $^{52}$Ar that is located at 1.656(18) MeV [46]. Interestingly, this level is even higher than that of the $N = 28$ isotope, $^{46}$Ar. The change of the $N = 34$ shell gap from $Z = 20$ to 16 is expressed as $-\Delta_{\pi d_{3/2}}(\varepsilon_{\nu f_{5/2}} - \varepsilon_{\nu p_{1/2}}) \approx 4\{V_{pn}^{\mathrm{m}}(d_{3/2}, p_{1/2}) - V_{pn}^{\mathrm{m}}(d_{3/2}, f_{5/2})\}$. The $d_{3/2}$-$p_{1/2}$ and $d_{3/2}$-$f_{5/2}$ are $\{+(+)\}$ and $\{-+\}$ pairs, respectively. Since the former pair is the most unfavored combination in energy in terms of both the central and tensor forces, this value is positive leading to the enlargement of the $N = 34$ shell gap. Experimental evaluation of this enhancement is difficult for $N = 34$ cores, but it is, however, possible for $N = 20$ cores

through spectroscopic strengths. Although the measured $f_{5/2}$ strengths are not complete for the $^{36}$S core, such an enlargement possibly occurs from the existing data (Figure 3 of [14]).

3.2.2. From $Z = 16$ to $Z = 14$

Two protons are removed from the $s_{1/2}$ orbital as moving from $Z = 16$ to $Z = 14$. Although the $N = 28$ Si isotope, $^{42}$Si, is strongly deformed, the $N = 20$ isotope, $^{34}$Si, can be regarded as a doubly closed-shell nucleus: its first excited state is $0^+$ (not $2^+$) and is located as high as 2.719(3) MeV [47]. In addition, a proton knockout experiment from $^{34}$Si [48] indicated small spectroscopic strengths of $s_{1/2}$ below $E_x \approx 4$ MeV, thus, suggesting a good $\pi(d_{5/2})^6$ closure in $^{34}$Si. For this reason, it is a good approximation to substitute the yrast levels in $^{35}$Si for the neutron effective single-particle energies on top of the $^{34}$Si core.

As shown in Figure 2b, the $N = 28$ shell gap changes by $-\Delta_{\pi s_{1/2}}(\varepsilon_{\nu p_{3/2}} - \varepsilon_{\nu f_{7/2}}) \approx 2\{V^{\rm m}_{pn}(s_{1/2}, f_{7/2}) - V^{\rm m}_{pn}(s_{1/2}, p_{3/2})\}$ from $Z = 16$ to $Z = 14$. The $s_{1/2}$-$f_{7/2}$ and $s_{1/2}$-$p_{3/2}$ pairs are $\{+0\}$ and $\{-0\}$, respectively, since the tensor force does not contribute to the monopole matrix elements for $s_{1/2}$ . As discussed next, the spin–orbit force also adds a negative value for the $s_{1/2}$-$p_{3/2}$, and therefore the $N = 28$ shell gap should enlarge. This enlargement is estimated from the yrast levels in $^{35}$Si and $^{37}$S to be $+0.264$ MeV. The shell-model calculations with the SDPF-MU interaction lead to $+0.667$ MeV, which is somewhat too large.

On the other hand, when one uses the SDPF-MUs interaction [33]—the one introduced in Section 3.1.2—to reproduce the $1/2^+$ levels in $^{51,53}$K, this value is modified to be $+0.317$ MeV. Note that the $-\Delta_{\pi s_{1/2}}(\varepsilon_{\nu p_{3/2}} - \varepsilon_{\nu f_{7/2}})$ values estimated from the ESPEs of SDPF-MU and SDPF-MUs are $+0.78$ and $+0.37$ MeV, respectively. These two independent experimental data—K isotopes and $N = 21$ isotones—consistently require about a $+0.2$ MeV modification of $V^{\rm m}_{pn}(s, p)$ matrix elements for the SDPF-MU interaction. This looks like due to the uncertainty of the central force that is determined empirically with a simple potential.

Next, the evolution of the $N = 32$ shell gap is discussed. The $^{34}$Si$(d, p)$ reaction experiment in inverse kinematics found two prominent $l = 1$ peaks at 0.910 and 2.044 MeV, the former and the latter of which should be the $3/2^-$ and $1/2^-$ levels, respectively [14]. The interval of these two levels, 1.134 MeV, is much smaller than the corresponding value of $^{37}$S, 1.911 MeV. If these values are identical with the spin–orbit splitting between the $p$ orbitals, the data point to a sharp reduction of 0.857 MeV. Since the matrix elements for the $s_{1/2}$-$p_{3/2}$ and $s_{1/2}$-$p_{1/2}$ pairs have no tensor contributions and the same central strengths (see Table 1), only the spin–orbit force can change this shell gap in terms of the shell model.

As pointed out in Section 2.1, the two-body spin orbit force produces particularly large monopole matrix elements between the $s$ and $p$ orbitals. The reduction of the $p$ orbital splitting is evaluated from the ESPEs of the SDPF-MU interaction to be 0.54 MeV, while the actual shell-model calculation produces a 0.758 MeV reduction of the $3/2^-_1$-$1/2^-_1$ level splitting in going from $^{37}$S to $^{35}$Si. Hence, although the two-body spin–orbit force is the dominant source of the observed reduction, correlation energy may account for the energy of a hundred keV order.

The origin of the observed reduction is still controversial. It is claimed [49] that Woods–Saxon potentials well account for the observed reduction of the spin–orbit splitting in going from the $^{40}$Ca to $^{34}$Si and that this occurs due to weak binding for lower $Z$ isotopes. This effect causes a gradual reduction with decreasing $Z$, whereas the two-body spin–orbit force affects the $p$ orbital splitting primarily with $s_{1/2}$ filled. Hence, one of the key issues to discriminate these effects is to establish how sharp this reduction occurs from the $^{36}$S to $^{34}$Si cores compared to that occurring from the $^{40}$Ca to $^{36}$S cores. Although one-neutron adding spectroscopic factors are measured for the $^{36}$S and $^{40}$Ca cores, the experimental uncertainty does not converge within the required accuracy (see the Supplemental Material of Ref. [1]).

3.2.3. From $Z = 14$ to $Z = 8$

Finally, protons in $d_{5/2}$ are removed from $Z = 14$ to $Z = 8$. As shown in Figure 2b, the $N = 28$ shell gap sharply decreases again. Note that Figure 2b presents neutron ESPEs for the $N = 28$ cores. When a similar figure is drawn for the $N = 20$ cores, the ESPE of $p_{3/2}$ shifts downward by $\sim$2 MeV , and the neutron $f_{7/2}$ and $p_{3/2}$ orbitals cross at around $Z = 11$. In this Section, the evolution of the $N = 28$ shell gap is examined; other gaps are difficult to access with the current experimental capability.

The change of the $N = 28$ shell gap with $Z$ decreasing from 14 to 8 is estimated to be $-\Delta_{\pi d_{5/2}}(\varepsilon_{\nu p_{3/2}} - \varepsilon_{\nu f_{7/2}}) \approx 6\{V^{\rm m}_{pn}(d_{5/2}, f_{7/2}) - V^{\rm m}_{pn}(d_{5/2}, p_{3/2})\}$. The $d_{5/2}$-$f_{7/2}$ and $d_{5/2}$-$p_{3/2}$ pairs are $\{-+\}$ and $\{+(+)\}$, respectively. Although the tensor force produces a slightly larger positive value for the former pair, the central attraction overrides this effect, thus, causing a negative value in total.

For such proton deficient isotopes, one cannot obtain sufficient experimental information from the nuclei around $N = 28$. Moreover, $N = 20$ isotones do not provide direct data for the present purpose because some isotopes in the "island of inversion" are strongly deformed. Hence, one relies on single-particle levels on top of the $N = 16$ cores, although $N = 16$ does not form a good closed shell except for with oxygen.

In Figure 6, the $3/2^-_1$ energy levels relative to $7/2^-_1$ are compared for experiment vs. theory. The data for $^{27}$Ne, $^{29}$Mg, and $^{31}$Si indicate a nearly linear change of these energies. Since the relevant one-neutron adding spectroscopic factors are not large, i.e., typically $\sim$0.5, as measured [50–52], these energy differences cannot be identified with the $N = 28$ shell gap. However, the linear evolution reminds one of the famous "Talmi plot" [2], which successfully predicted the $1/2^+_1$ level in $^{11}$Be from the linearity. Thus, this behavior is worthy of particular attention.

One can see from Figure 6 that the measurements are in a good agreement with the calculations based on the SDPF-MU and SDPF-MUs interactions. The SDPF-MUs interaction achieves better agreement because its $N = 28$ shell gap on for the $^{34}$Si core is improved (see Section 3.2.2). These two interactions are quite successful in reproducing the slope of $E_x(3/2^-_1) - E_x(7/2^-_1)$.

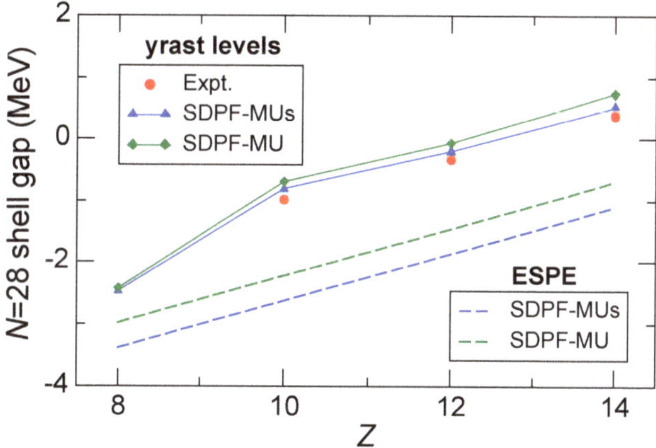


**Figure 6.** Evolution of the $N = 28$ shell gap in going from $Z = 8$ to 14 estimated from the $E_x(3/2^-_1) - E_x(7/2^-_1)$ values in the $N = 17$ isotones (solid lines) and from the ESPE calculations (dashed lines). Data are from Refs. [50–52].

As one can also see from Figure 6, the slope is quite similar to what the ESPE predicts. Since the $V^{\rm m}_{pn}(d_{5/2}, f_{7/2})$ and $V^{\rm m}_{pn}(d_{5/2}, p_{3/2})$ values are kept unchanged in making the SDPF-MUs interaction based on SDPF-MU, the $\nu p_{3/2}$ ESPEs are parallel. On the other hand, these ESPEs are shifted downward in parallel from $E_x(3/2^-_1) - E_x(7/2^-_1)$ by $\sim$1.5 MeV.

This difference arises from the assumption of the $\nu(d_{5/2})^6(s_{1/2})^2$ closure taken here to evaluate the ESPE.

However, in reality, a significant number of neutron excitations to $d_{3/2}$ occur in the $^{27}$Ne, $^{29}$Mg, and $^{31}$Si eigenstates. These neutron excitations attract a neutron in the $f_{7/2}$ orbital more than a one in $p_{3/2}$ because the $T = 1$ monopole matrix element of $d_{3/2}$-$f_{7/2}$ is more attractive than that of $d_{3/2}$-$p_{3/2}$, thus, shifting $E_x(3/2^-_1) - E_x(7/2^-_1)$ upward. Such significant neutron excitation to $d_{3/2}$ occurs similarly in the Ne, Mg, and Si isotopes. Hence, the evolution of $E_x(3/2^-_1) - E_x(7/2^-_1)$ is predominantly changed by the ESPE, providing evidence for the narrowing $N = 28$ shell gap caused by the monopole matrix element $V^{\rm m}_{pn}(d_{5/2}, f_{7/2}) - V^{\rm m}_{pn}(d_{5/2}, p_{3/2})$.

It should be noted that the predicted $E_x(3/2^-_1) - E_x(7/2^-_1)$ at $Z = 8$ is closer to the ESPE estimate than those of other isotopes. This is due to the fact that the assumed $N = 16$ closure works better at $Z = 8$ due to the occurrence of the $N = 16$ magic number.

## 4. Conclusions

In this paper, an almost complete survey of proton and neutron shell evolution for atomic mass number $25 \lesssim A \lesssim 55$ neutron-rich nuclei is performed on the basis of shell-model calculations, in order to understand how well the observed evolution is explained with a simple monopole-based universal interaction, $V_{\rm MU}$.

On the proton side, the observed one-proton removal spectroscopic distributions in $^{40,48}$Ca were very well reproduced with shell-model calculations, pointing to a ~2 MeV change of $\pi d_{5/2}$-$\pi d_{3/2}$ spin–orbit splitting. Since this change is caused almost solely by the tensor force, this agreement quantitatively confirms the validity of a $\pi + \rho$ meson exchange tensor force in the $V_{\rm MU}$ interaction. The $1/2^+_1$-$3/2^+_1$ level difference in K isotopes changes the sign twice, with $\nu f_{7/2}$ filled and with $\nu p_{3/2}$ filled.

As discussed, this change is caused by the "reinversion" of single-particle level ordering between $\pi d_{3/2}$ and $\pi s_{1/2}$ as a result of the non-monotonic evolution of these level spacings. Such a manner of evolution cannot be produced by one-body potential models, and therefore it is strong evidence for the dominant role of two-body forces in shell evolution. In this particular case, the non-monotonic evolution observed in K isotopes is driven by the central force.

On the neutron side, the neutron-number $N = 28$ shell gap is reduced with protons removed from the $d_{3/2}$ and $d_{5/2}$ orbitals, dominated by the central force. The relevant single-particle-like levels are well reproduced by the shell-model calculation. In addition, the central force causes the enhancement of the $N = 34$ shell gap for the atomic-number $Z < 20$ isotopes. This effect well accounts for the recently observed $2^+_1$ level in $^{52}$Ar.

In this way, the present scheme, based on $V_{\rm MU}$, provides a successful description of the shell evolution. Neutron shell evolution in exotic nuclei is often argued in the context of weak binding. In the present study, we were successful in obtaining not only neutron shells but also proton shells that were free from weak binding. Thus, such a unified description strongly indicates the dominance of the effective interaction in shell evolution, as far as the region of the present study is concerned, including the narrowing $N = 28$ shell gap toward a neutron-rich nuclei.

**Funding:** This work was supported in part by JSPS KAKENHI Grant Numbers 20K03981 and 15K05094.

**Data Availability Statement:** Not applicable.

**Acknowledgments:** The author thanks Takaharu Otsuka for valuable discussions on shell evolution and thank Noritaka Shimizu for his help with large-scale shell-model calculations.

**Conflicts of Interest:** The author declares no conflict of interest.

## References


1. Otsuka, T.; Gade, A.; Sorlin, O.; Suzuki, T.; Utsuno, Y. Evolution of shell structure in exotic nuclei. *Rev. Mod. Phys.* **2020**, *92*, 015002. [CrossRef]
2. Talmi, I.; Unna, I. Order of Levels in the Shell Model and Spin of $Be^{11}$. *Phys. Rev. Lett.* **1960**, *4*, 469–470. [CrossRef]
3. Bansal, R.; French, J. Even-parity-hole states in $f_{7/2}$-shell nuclei. *Phys. Lett.* **1964**, *11*, 145–148. [CrossRef]
4. Poves, A.; Zuker, A. Theoretical spectroscopy and the fp shell. *Phys. Rep.* **1981**, *70*, 235–314. [CrossRef]
5. McGrory, J.B.; Wildenthal, B.H.; Halbert, E.C. Shell-Model Structure of $^{42-50}$Ca. *Phys. Rev. C* **1970**, *2*, 186–212. [CrossRef]
6. Caurier, E.; Martínez-Pinedo, G.; Nowacki, F.; Poves, A.; Zuker, A.P. The shell model as a unified view of nuclear structure. *Rev. Mod. Phys.* **2005**, *77*, 427–488. [CrossRef]
7. Storm, M.H.; Watt, A.; Whitehead, R.R. Crossing of single-particle energy levels resulting from neutron excess in the sd shell. *J. Phys. Nucl. Phys.* **1983**, *9*, L165–L168. [CrossRef]
8. Warburton, E.K.; Becker, J.A.; Brown, B.A. Mass systematics for $A$ = 29–44 nuclei: The deformed $A$ ~32 region. *Phys. Rev. C* **1990**, *41*, 1147–1166. [CrossRef]
9. Federman, P.; Pittel, S. Towards a unified microscopic description of nuclear deformation. *Phys. Lett. B* **1977**, *69*, 385–388. [CrossRef]
10. Otsuka, T.; Fujimoto, R.; Utsuno, Y.; Brown, B.A.; Honma, M.; Mizusaki, T. Magic numbers in exotic nuclei and spin-isospin properties of the *NN* interaction. *Phys. Rev. Lett.* **2001**, *87*, 082502. [CrossRef]
11. Utsuno, Y.; Otsuka, T.; Mizusaki, T.; Honma, M. Varying shell gap and deformation in $N \sim 20$ unstable nuclei studied by the Monte Carlo shell model. *Phys. Rev. C* **1999**, *60*, 054315. [CrossRef]
12. Otsuka, T.; Suzuki, T.; Fujimoto, R.; Grawe, H.; Akaishi, Y. Evolution of nuclear shells due to the tensor force. *Phys. Rev. Lett.* **2005**, *95*, 232502. [CrossRef] [PubMed]
13. Otsuka, T.; Suzuki, T.; Honma, M.; Utsuno, Y.; Tsunoda, N.; Tsukiyama, K.; Hjorth-Jensen, M. Novel features of nuclear forces and shell evolution in exotic nuclei. *Phys. Rev. Lett.* **2010**, *104*, 012501. [CrossRef] [PubMed]
14. Burgunder, G.; Sorlin, O.; Nowacki, F.; Giron, S.; Hammache, F.; Moukaddam, M.; de Séréville, N.; Beaumel, D.; Càceres, L.; Clément, E.; et al. Experimental Study of the two-body spin-orbit force in nuclei. *Phys. Rev. Lett.* **2014**, *112*, 042502. [CrossRef] [PubMed]
15. Smirnova, N.; Bally, B.; Heyde, K.; Nowacki, F.; Sieja, K. Shell evolution and nuclear forces. *Phys. Lett. B* **2010**, *686*, 109–113. [CrossRef]
16. Tsunoda, N.; Otsuka, T.; Tsukiyama, K.; Hjorth-Jensen, M. Renormalization persistency of the tensor force in nuclei. *Phys. Rev. C* **2011**, *84*, 044322. [CrossRef]
17. Yuan, C.; Suzuki, T.; Otsuka, T.; Xu, F.; Tsunoda, N. Shell-model study of boron, carbon, nitrogen, and oxygen isotopes with a monopole-based universal interaction. *Phys. Rev. C* **2012**, *85*, 064324. [CrossRef]
18. Utsuno, Y.; Otsuka, T.; Brown, B.A.; Honma, M.; Mizusaki, T.; Shimizu, N. Shape transitions in exotic Si and S isotopes and tensor-force-driven Jahn-Teller effect. *Phys. Rev. C* **2012**, *86*, 051301. [CrossRef]
19. Honma, M.; Otsuka, T.; Brown, B.A.; Mizusaki, T. Shell-model description of neutron-rich pf-shell nuclei with a new effective interaction GXPF1. *Eur. Phys. J. Hadron. Nucl.* **2005**, *25*, 499–502. [CrossRef]
20. Richter, W.; Van Der Merwe, M.; Julies, R.; Brown, B. New effective interactions for the 0f1p shell. *Nucl. Phys. A* **1991**, *523*, 325–353. [CrossRef]
21. Utsuno, Y.; Otsuka, T.; Shimizu, N.; Honma, M.; Mizusaki, T.; Tsunoda, Y.; Abe, T. Recent shell-model results for exotic nuclei. *Epj Web Conf.* **2014**, *66*, 02106. [CrossRef]
22. Bertsch, G.; Borysowicz, J.; McManus, H.; Love, W. Interactions for inelastic scattering derived from realistic potentials. *Nucl. Phys. A* **1977**, *284*, 399–419. [CrossRef]
23. Brown, B.A.; Wildenthal, B.H. Status of the nuclear shell model. *Annu. Rev. Nucl. Part. Sci.* **1988**, *38*, 29–66. [CrossRef]
24. Honma, M.; Otsuka, T.; Brown, B.A.; Mizusaki, T. Effective interaction for pf-shell nuclei. *Phys. Rev. C* **2002**, *65*, 061301. [CrossRef]
25. Utsuno, Y.; Shimizu, N.; Otsuka, T.; Yoshida, T.; Tsunoda, Y. Nature of Isomerism in Exotic Sulfur Isotopes. *Phys. Rev. Lett.* **2015**, *114*, 032501. [CrossRef]
26. Shimizu, N.; Mizusaki, T.; Utsuno, Y.; Tsunoda, Y. Thick-restart block Lanczos method for large-scale shell-model calculations. *Comput. Phys. Commun.* **2019**, *244*, 372–384. [CrossRef]
27. Cottle, P.D.; Kemper, K.W. Persistence of the $N = 28$ shell closure in neutron-rich nuclei. *Phys. Rev. C* **1998**, *58*, 3761–3762. [CrossRef]
28. Sorlin, O.; Porquet, M.G. Nuclear magic numbers: New features far from stability. *Prog. Part. Nucl. Phys.* **2008**, *61*, 602–673. [CrossRef]
29. Banks, S.; Spicer, B.; Shute, G.; Officer, V.; Wagner, G.; Dollhopf, W.; Qingli, L.; Glover, C.; Devins, D.; Friesel, D. The $^{48}$Ca($\vec{d}$, $^{3}$He)$^{47}$K reaction at 80 MeV. *Nucl. Phys. A* **1985**, *437*, 381–396. [CrossRef]
30. Doll, P.; Wagner, G.; Knöpfle, K.; Mairle, G. The quasihole aspect of hole strength distributions in odd potassium and calcium isotopes. *Nucl. Phys. A* **1976**, *263*, 210–236. [CrossRef]
31. Go, S.; Ideguchi, E.; Yokoyama, R.; Aoi, N.; Azaiez, F.; Furutaka, K.; Hatsukawa, Y.; Kimura, A.; Kisamori, K.; Kobayashi, M.; et al. High-spin states in $^{35}$S. *Phys. Rev. C* **2021**, *103*, 034327. [CrossRef]

32. Papuga, J.; Bissell, M.L.; Kreim, K.; Blaum, K.; Brown, B.A.; De Rydt, M.; Garcia Ruiz, R.F.; Heylen, H.; Kowalska, M.; Neugart, R.; et al. Spins and Magnetic Moments of $^{49}$K and $^{51}$K: Establishing the $1/2^+$ and $3/2^+$ Level Ordering Beyond $N = 28$. *Phys. Rev. Lett.* **2013**, *110*, 172503. [CrossRef] [PubMed]
33. Sun, Y.; Obertelli, A.; Doornenbal, P.; Barbieri, C.; Chazono, Y.; Duguet, T.; Liu, H.; Navrátil, P.; Nowacki, F.; Ogata, K.; et al. Restoration of the natural $E(1/2_1^+) - E(3/2_1^+)$ energy splitting in odd-K isotopes towards $N = 40$. *Phys. Lett. B* **2020**, *802*, 135215. [CrossRef]
34. Grasso, M.; Ma, Z.Y.; Khan, E.; Margueron, J.; Giai, N.V. Evolution of the proton *sd* states in neutron-rich Ca isotopes. *Phys. Rev. C* **2007**, *76*, 044319. [CrossRef]
35. Nakada, H.; Sugiura, K.; Margueron, J. Tensor-force effects on single-particle levels and proton bubble structure around the $Z$ or $N = 20$ magic number. *Phys. Rev. C* **2013**, *87*, 067305. [CrossRef]
36. Kirson, M. Spin-tensor decomposition of nuclear effective interactions. *Phys. Lett. B* **1973**, *47*, 110–114. [CrossRef]
37. Piskoř, Š.; Franc, P.; Křemének, J.; Schäferlingová, W. Spectroscopic information on $^{35}$S and $^{37}$S from the $(d, p)$ reaction. *Nucl. Phys. A* **1984**, *414*, 219–239. [CrossRef]
38. Uozumi, Y.; Kikuzawa, N.; Sakae, T.; Matoba, M.; Kinoshita, K.; Sajima, S.; Ijiri, H.; Koori, N.; Nakano, M.; Maki, T. Shell-model study of $^{40}$Ca with the 56-MeV $(\vec{d}, p)$ reaction. *Phys. Rev. C* **1994**, *50*, 263–274. [CrossRef]
39. Evaluated Nuclear Structure Data File (ENSDF). Available online: http://www.nndc.bnl.gov/ensdf/ (accessed on 1 January 2022 ).
40. Lu, F.; Lee, J.; Tsang, M.B.; Bazin, D.; Coupland, D.; Henzl, V.; Henzlova, D.; Kilburn, M.; Lynch, W.G.; Rogers, A.M.; et al. Neutron-hole states in $^{45}$Ar from $^{1}$H($^{46}$Ar, $d$) $^{45}$Ar reactions. *Phys. Rev. C* **2013**, *88*, 017604. [CrossRef]
41. Momiyama, S.; Wimmer, K.; Bazin, D.; Belarge, J.; Bender, P.; Elman, B.; Gade, A.; Kemper, K.W.; Kitamura, N.; Longfellow, B.; et al. Shell structure of $^{43}$S and collapse of the $N = 28$ shell closure. *Phys. Rev. C* **2020**, *102*, 034325. [CrossRef]
42. Chen, S.; Lee, J.; Doornenbal, P.; Obertelli, A.; Barbieri, C.; Chazono, Y.; Navrátil, P.; Ogata, K.; Otsuka, T.; Raimondi, F.; et al. Quasifree neutron knockout from $^{54}$Ca corroborates arising $N = 34$ neutron magic number. *Phys. Rev. Lett.* **2019**, *123*, 142501. [CrossRef] [PubMed]
43. Steppenbeck, D.; Takeuchi, S.; Aoi, N.; Doornenbal, P.; Matsushita, M.; Wang, H.; Baba, H.; Fukuda, N.; Go, S.; Honma, M.; et al. Evidence for a new nuclear 'magic number' from the level structure of $^{54}$Ca. *Nature* **2013**, *502*, 207–210. [CrossRef] [PubMed]
44. Utsuno, Y.; Otsuka, T.; Tsunoda, Y.; Shimizu, N.; Honma, M.; Togashi, T.; Mizusaki, T. Recent advances in shell evolution with shell-model calculations. *JPS Conf. Proc.* **2015**, *6*, 010007. [CrossRef]
45. Steppenbeck, D.; Takeuchi, S.; Aoi, N.; Doornenbal, P.; Matsushita, M.; Wang, H.; Utsuno, Y.; Baba, H.; Go, S.; Lee, J.; et al. Low-Lying Structure of $^{50}$Ar and the $N = 32$ Subshell Closure. *Phys. Rev. Lett.* **2015**, *114*, 252501. [CrossRef] [PubMed]
46. Liu, H.N.; Obertelli, A.; Doornenbal, P.; Bertulani, C.A.; Hagen, G.; Holt, J.D.; Jansen, G.R.; Morris, T.D.; Schwenk, A.; Stroberg, R.; et al. How robust is the $N = 34$ subshell closure? First spectroscopy of $^{52}$Ar. *Phys. Rev. Lett.* **2019**, *122*, 072502. [CrossRef]
47. Rotaru, F.; Negoita, F.; Grévy, S.; Mrazek, J.; Lukyanov, S.; Nowacki, F.; Poves, A.; Sorlin, O.; Borcea, C.; Borcea, R.; et al. Unveiling the Intruder deformed $0_2^+$ state in $^{34}$Si. *Phys. Rev. Lett.* **2012**, *109*, 092503. [CrossRef]
48. Mutschler, A.; Lemasson, A.; Sorlin, O.; Bazin, D.; Borcea, C.; Borcea, R.; Dombrádi, Z.; Ebran, J.P.; Gade, A.; Iwasaki, H.; et al. A proton density bubble in the doubly magic $^{34}$Si nucleus. *Nat. Phys.* **2017**, *13*, 152–156. [CrossRef]
49. Kay, B.P.; Hoffman, C.R.; Macchiavelli, A.O. Effect of weak binding on the apparent spin-orbit splitting in nuclei. *Phys. Rev. Lett.* **2017**, *119*, 182502. [CrossRef]
50. Brown, S.M.; Catford, W.N.; Thomas, J.S.; Fernández-Domínguez, B.; Orr, N.A.; Labiche, M.; Rejmund, M.; Achouri, N.L.; Al Falou, H.; Ashwood, N.I.; et al. Low-lying neutron $fp$-shell intruder states in $^{27}$Ne. *Phys. Rev. C* **2012**, *85*, 011302. [CrossRef]
51. Matta, A.; Catford, W.N.; Orr, N.A.; Henderson, J.; Ruotsalainen, P.; Hackman, G.; Garnsworthy, A.B.; Delaunay, F.; Wilkinson, R.; Lotay, G.; et al. Shell evolution approaching the $N = 20$ island of inversion: Structure of $^{29}$Mg. *Phys. Rev. C* **2019**, *99*, 044320. [CrossRef]
52. Piskoř, Š.; Novák, J.; Šimečková, E.; Cejpek, J.; Kroha, V.; Dobeš, J.; Navrátil, P. A study of the $^{30}$Si$(d, p)^{31}$Si reaction. *Nucl. Phys. A* **2000**, *662*, 112–124. [CrossRef]

MDPI

Review

# Reaching into the $N = 40$ Island of Inversion with Nucleon Removal Reactions

**Alexandra Gade** [1,2]

1 FRIB Laboratory, Michigan State University, East Lansing, MI 48824, USA; gade@nscl.msu.edu
2 Department of Physics and Astronomy, Michigan State University, East Lansing, MI 48824, USA

**Abstract:** One ambitious goal of nuclear physics is a predictive model of all nuclei, including the ones at the fringes of the nuclear chart which may remain out of experimental reach. Certain regions of the chart are providing formidable testing grounds for nuclear models in this quest as they display rapid structural evolution from one nucleus to another or phenomena such as shape coexistence. Observables measured for such nuclei can confirm or refute our understanding of the driving forces of the evolution of nuclear structure away from stability where textbook nuclear physics has been proven to not apply anymore. This paper briefly reviews the emerging picture for the very neutron-rich Fe, Cr, and Ti isotopes within the so-called $N = 40$ island of inversion as obtained with nucleon knockout reactions. These have provided some of the most detailed nuclear spectroscopy in very neutron-rich nuclei produced at rare-isotope facilities. The results indicate that our current understanding, as encoded in large-scale shell-model calculations, appears correct with exciting predictions for the $N = 40$ island of inversion left to be proven in the experiment. A bright future emerges with predictions of continued shell evolution and shape coexistence out to neutron number $N = 50$, below $^{78}$Ni on the chart of nuclei.

**Keywords:** rare isotopes; shell evolution; $N = 40$ island of inversion; knockout reactions

**Citation:** Gade, A Reaching into the $N = 40$ Island of Inversion with Nucleon Removal Reactions. *Physics* **2021**, 3, 1226–1236. https://doi.org/10.3390/physics3040077

Received: 29 October 2021
Accepted: 2 December 2021
Published: 8 December 2021

**Publisher's Note:** MDPI stays neutral with regard to jurisdictional claims in published maps and institutional affiliations.

**Copyright:** © 2021 by the author. Licensee MDPI, Basel, Switzerland. This article is an open access article distributed under the terms and conditions of the Creative Commons Attribution (CC BY) license (https://creativecommons.org/licenses/by/4.0/).

## 1. Introduction

One of the challenging goals of the field of nuclear structure physics is to model atomic nuclei, including their properties and their reactions—rooted in the fundamental forces at play between protons and neutrons—with predictive power also for the shortest-lived nuclear species located near the driplines of the chart. With this ultimate vision to extrapolate towards the most neutron-rich nuclei that may elude experimental study in the near future, much can be gleaned from nuclei in regions that display the effects of structural evolution away from the valley of stability and so offer a window into the driving forces of structural change and our understanding of it [1–3]. Specifically, the complex interplay between single-particle and collective degrees of freedom can provide exciting experimental challenges and demanding theoretical benchmarks.

The region of rapid structural change of interest in this review is the so-called "$N = 40$ island of inversion" [4,5], where the neutron-rich Fe and Cr nuclei around neutron number 40 become the most deformed in the region. In nuclear models, this is theorized to be caused by the strong quadrupole-quadrupole interaction producing a nuclear shape transition in which highly-correlated many-particle–many-hole configurations become energetically more favored than the normal-order (spherical) ones [4]. Such islands of inversion are characterized by rapid structural changes and shape coexistence [5,6], providing insight into nuclear structure physics far from stability [7]. Large-scale shell-model calculations with the LNPS (Lenzi-Nowacki-Poves-Sieja) effective interaction [4] in the full $fp$ shell for protons and the $f_{5/2}$, $p_{3/2}$, $p_{1/2}$, $g_{9/2}$, and $d_{5/2}$ orbitals for neutrons have confirmed the picture described above, with many successful predictions that preceded experimental results [3].

A recent prediction extends this island of inversion to $N = 50$ [5] and includes nuclei that will only be reached at next-generation rare-isotope beam facilities. This exciting prospect of extending the island towards the magic neutron number $N = 50$ is based on extrapolations of calculations using the LNPS shell-model effective interaction and its monopole drifts [4,5]. These prediction together with advances in experimentation continue to push the field forward on the journey to the $N = 50$ island of inversion.

The furthest experimental reach into the Fe, Cr, and Ti isotopes has been afforded by inverse-kinematics nucleon removal studies induced by fast rare-isotope projectile beams [8,9] to probe the nuclei of interest via in-beam $\gamma$-ray spectroscopy [10]. Often, such reactions provide the first glimpse of the excitation level scheme [11] and, in some cases, the direct character of such reactions is used to conclude on wave function overlaps within the shell-model framework [8,9]. This paper reviews the recent results for the very neutron-rich nuclei $^{66,68,70,72}$Fe, $^{64,66}$Cr, and $^{60,62}$Ti, all located near the center of the $N = 40$ island of inversion or already on the path to $N = 50$, obtained with such experimental approaches which have provided pioneering information the furthest away from the line of stability; see Figure 1.

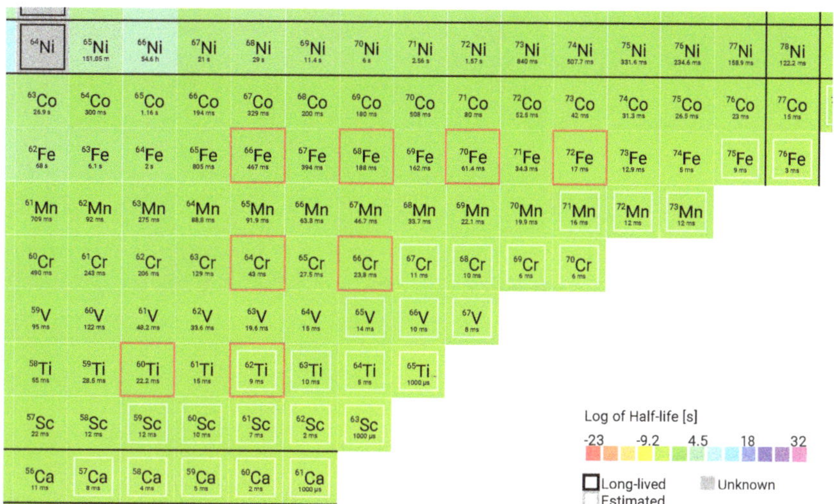


**Figure 1.** Portion of the nuclear chart that shows the $N = 40$ island of inversion. Nuclei discussed in this review are highlighted with a red outline. This chart was generated with [12] (half-life color coding based on NuBase2020 and corresponding extrapolations).

## 2. Experimental Approaches

Experimental techniques aimed at tracking the changes in the structure of nuclei are multi-pronged. They include measurements of ground-state properties such as masses, radii, $\beta$-decay properties, and electromagnetic moments as well as the study of properties of bound and unbound excited states. One way of probing specific nuclear structure aspects in quantitative ways is the use of nuclear reactions that selectively probe a specific degree of freedom. Inelastic scattering of nuclei, including Coulomb excitation, has long been used to probe nuclear collectivity, characterized by the coherent motion of several protons and neutrons. The single-particle degree of freedom, on the other hand, is commonly associated with the single-particle composition of the many-body wave function in a shell-model picture. Such single-particle properties can be studied rather selectively by using *direct* reactions that add or remove one or a few nucleons from the nucleus of interest. Intriguing possibilities now arise in the above mentioned islands of inversion, where the telltale onset of collectivity and the underlying migration of single-particle levels can be tracked to provide a consistent picture.

At in-flight rare-isotope facilities, short-lived nuclei away from stability can be efficiently produced by fragmentation (or fission) of stable, primary beams impinging upon stable targets at high beam energy. The resulting secondary beams of rare isotopes are then available for experiments at velocities typically exceeding a $v/c$ of 30%, where $c$ is the speed of light. Well-established experimental techniques used for decades to study stable nuclei are not readily applicable in inverse kinematics and at the low beam rates encountered for the shortest-lived nuclear species. Instead, powerful new experimental approaches have been developed to enable in-beam nuclear spectroscopy studies of fast rare-isotope beams with intensities that are several orders of magnitude less than needed for typical low-energy techniques.

The intensities of rare-isotope beams are lower than stable-beam rates by several orders of magnitude. However, the experimental approach of in-beam $\gamma$-ray spectroscopy compensates for the reduced intensities by enabling thick reaction targets, due to the high beam velocity, and realizing measurements with luminosities comparable to stable-beam experiments but at beam rates of up to a factor of $10^4$ less. Reactions such as nucleon removal are induced in thick reaction targets (several hundred mg/cm$^2$ to g/cm$^2$) and with the detection of $\gamma$ rays for the identification of the reaction residue's final state [10]. Since the residue's $\gamma$-ray emission occurs *in flight*, the $\gamma$-ray detection systems have to be granular or position-sensitive to allow for an angle-dependent event-by-event reconstruction of the Doppler-shifted $\gamma$-ray energies into the rest frame of the emitter. The choice of the target material depends on the desired reaction; one- and two-nucleon knockout reactions [8] are often induced by light targets, for example $^9$Be or $^{12}$C, while quasi-free scattering of the $(p, 2p)$ or $(p, pn)$ type are nowadays performed with MINOS, an extended liquid hydrogen target that allows reaction vertex-reconstruction and tracking following the concept of a time projection chamber [13]. The projectile-like reaction residues exiting the target has to be identified with magnetic spectrographs or advanced detector systems to cleanly select the reaction channel of interest. In-beam $\gamma$-ray spectroscopy programs with fast beams are pursued at a number of fragmentation facilities around the world, while the work using nucleon removal reactions in the $N = 40$ region has been performed largely at NSCL [14] in the US and RIBF/RIKEN [15] in Japan with the GRETINA [16,17] and DALI2 [18] arrays for $\gamma$-ray spectroscopy, respectively. A sketch of the experimental scheme is shown in Figure 2.

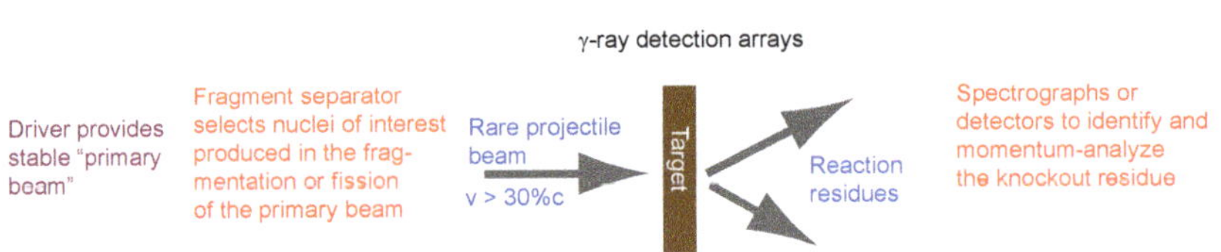


**Figure 2.** Experimental scheme for inverse-kinematics nucleon knockout reactions at rare-isotope beam facilities that provide fast beams of rare isotopes via projectile fragmentation or fission with velocities, $v$, exceeding 30% of the speed of light, $c$.

For $^9$Be- or $^{12}$C-induced one-nucleon knockout reactions, the exit channel of interest is one where—in a single step—one proton or neutron is removed from the fast rare-isotope beam and the projectile-like residue with one less nucleon survives in a bound final state. This channel is characterized by swift, surface-grazing collisions of the projectile and the target nuclei. From a large body of experiments performed at energies from 50 MeV/nucleon to more than 1 GeV/nucleon at (rare-isotope) facilities around the world, it has been established that, with large cross sections, the dominant single-hole states relative to the projectile ground state are populated in the projectile-like reaction residue, unambiguously demonstrating the unmatched sensitivity to the single-particle degree of freedom. The residue parallel momentum distributions encode in their shape and width the information of the orbital angular momentum $\ell$ and separation energy of the removed

nucleon [8]. The cross section of the selectively populated single-hole configurations scale with the respective spectroscopic factor or wave-function overlap in a shell-model picture. Statistical descriptions of these reactions will not capture these features. Comparisons of such one-nucleon removal data with nuclear structure calculations have been enabled by a direct reaction model [8,19,20] that uses the sudden (short interaction time) and the spectator-core approximation to many-body eikonal (forward-scattering) theory [19] with a detailed prescription provided in [21]. The single-particle nuclear structure information then enters the calculations through spectroscopic factors, or wave-function overlaps, that scale the calculated cross sections for the removal of one nucleon from the corresponding orbital. With that, the measured knockout cross sections can serve as formidable probes of shell-model interactions on the quest to identify the single-particle makeup of the projectile ground state and the residue final states [8,22].

It has been shown also that two-proton and two-neutron removal from neutron-rich and neutron-deficient projectiles, respectively, also proceed as direct reactions [23–25]. By combining eikonal reaction dynamics, that assumes a sudden single-step removal of two nucleons and shell-model calculations of the two-nucleon amplitudes (TNAs), the cross sections for two-nucleon knockout from the parent-nucleus ground state to each of the final states in the daughter nucleus can be calculated [26]. Also, it was shown that the shape of the parallel momentum distribution of the two-nucleon knockout residues depends strongly on the total angular momentum of the two removed nucleons, allowing spin values to be assigned to populated final states [27–29]. One step further, it was proposed and confirmed that since the two-nucleon overlaps contain components with different values of the total orbital angular momentum, information beyond the total angular momentum can be probed. This opens up the possibility to uniquely explore this composition and couplings within the wave functions of rare isotopes [30,31].

More recently, quasi-free $(p,2p)$ and $(p,pn)$ reactions, extensively used in normal kinematics with stable targets, have been successfully adapted for inverse-kinematics studies of rare-isotope beams on proton targets [9]. Just as the heavy-ion-induced knockout reactions sketched above, the proton-induced knockout reaction selectively probes the single-particle structure of the nucleus of interest. Also, the shape of the momentum distribution of the knockout residue is connected to the momentum distribution of the knocked-out nucleon. Protons are a penetrating probe that interrogate the nuclear interior, and their rescattering inside the nucleus has to be understood and modeled [9]. In heavy-ion induced knockout, the orbital radii need to be modeled precisely due to their surface localization [21]. This experimental approach has been used recently at RIBF/RIKEN for measurements reviewed here. Various reaction models have been developed and their consistency remains a challenge for the future [32]. The different nucleon removal reactions were described and confronted with each other recently and extensive details on sensitivities and model dependencies can be found in reference [32].

## 3. The Fe Isotopic Chain

The first in-beam nuclear spectroscopy of $^{66}$Fe and $^{68}$Fe was published in 2008 from a measurement performed at NSCL where these two Fe isotopes with 40 and 42 neutrons were populated each in $^{9}$Be-induced one- and two-proton knockout reactions, using the laboratory's S800 spectrograph for particle identification and SeGA for in-beam $\gamma$-ray spectroscopy [33]. For $^{66}$Fe, in addition to the tentative $2^+_1 \rightarrow 0^+_1$ and $4^+_1 \rightarrow 2^+_1$ transitions reported earlier from $\beta$ decay [34], a 957(10) keV $\gamma$-ray was observed in both reactions, while a 1310(15) keV transition was only seen in the one-proton knockout. Within a simple two-proton knockout picture, for two protons removed from the $f_{7/2}$ orbital, one would expect to populate states in $^{66}$Fe with spin-parity $6^+$, $4^+$, and $2^+$, suggesting that the 957-keV line depopulates the $6^+$ state of $^{66}$Fe. Subsequent $\beta$-decay work limited to lower-spin states seems to suggest that the 1310-keV transition could originate from the second $2^+$ level and feeds the first $2^+$ state [35]. This first in-beam work predates the publication of the LNPS effective interaction [4] and it is interesting to explore the suggestion that

the $6_1^+$ state was observed. Kotila and Lenzi [36] discuss collective phenomena in Fe and Cr and show that the $6_1^+$ level predicted by the LNPS calculations agrees with the tentatively assigned $(6^+)$ state proposed by Adrich et al. in $^{66}$Fe [36]. Subsequent in-beam spectroscopy work performed at NSCL explored the $2_1^+$ and $4_1^+$ states of $^{66}$Fe in $^9$Be-induced inelastic scattering [37] and the quadrupole collectivity in intermediate-energy Coulomb excitation [38] and excited-state lifetime measurements [39].

Beyond $N = 40$, in $^{68}$Fe, the first observation of $\gamma$-ray transitions was reported in [33] from $^9$Be-induced one- and two-proton removal reactions with proposed $2_1^+ \rightarrow 0_1^+$ and $4_1^+ \rightarrow 2_1^+$ decays, later supported by intermediate-energy Coulomb excitation measurements [38] as well as $\beta$ decay [40]. It turned out that the energy of the first $2^+$ state in $^{68}$Fe is lower than in $^{66}$Fe, indicating that the maximum collectivity is assumed beyond $N = 40$. Taking the $4_1^+$ assignment at face value, the $R_{4/2}$ energy ratio increases as well. The shell model calculations with the LNPS effective interaction are in good agreement with the energies and transition strengths, lending even more confidence that the shell evolution past $N = 40$ is captured by the incorporated driving forces [4]. Excited states beyond the tentatively assigned yrast $2^+$ and $4^+$ remained elusive until a $\beta$-decay study [41], where a number of low-spin states were proposed. Two candidates for the $6_1^+$ state just emerged recently from a $^9$Be-induced charge-exchange reaction on $^{68}$Co projectiles in the $(7^-)$ ground state and a low-spin isomer [42]. Governed by the charge-exchange selection rules, access to never-before observed states was provided, predominantly higher-spin states [42]. The calculations with the LNPS effective interaction show good agreement with the energies of the candidate yrast states up to the suggested candidate $6^+$ levels [42]. This reaction mechanism holds great promise to reach beyond the selectivity of knockout reactions and $\beta$ decay, depending on the spin and parity of the projectile initial state.

At $N = 44$, spectroscopy of $^{70}$Fe became first possible in 2015 at the RIBF facility at RIKEN using a $(p, 2p)$ reaction with the MINOS hydrogen target [13] and the DALI2 scintillator array [43] and in the same year with $\beta$ decay at RIKEN [41]. Two transitions were consistently identified in both measurements and proposed to correspond to the $2_1^+ \rightarrow 0_1^+$ and $4_1^+ \rightarrow 2_1^+$ decays, establishing the corresponding states. It took until 2019 to get beyond the yrast $4^+$ state and identify a transition on top of the $4^+$ level in a $^9$Be-induced one-proton knockout measurement performed at NSCL with GRETINA and the S800 spectrograph [44]. The measured and calculated partial one-proton removal cross sections were confronted and showed, at first glance, a striking disagreement with high-lying states populated more strongly than the yrast states observed in the measurement. The emerging picture is one that is not unlike the Pandemonium in $\beta$ decay [45], where indiscernible feeding from a multitude of higher-lying states funnels intensity into low-lying states which then appear prominent albeit carrying little direct feeding. This demonstrates that, while one-proton removal is a powerful experimental probe to reach nuclei more neutron-rich than the projectile, the collectivity prevalent in this region of the nuclear chart can lead to fragmentation of the single-particle strength which may then be thinly spread over many states in the reaction residue, leading to a Pandemonium-like feeding scheme when $\gamma$-ray spectroscopy is used [44].

The most neutron-rich Fe isotope with spectroscopic information is $^{72}$Fe, studied at RIBF/RIKEN in the same experiment and with the same approach as $^{70}$Fe [43] and two $\gamma$ rays were observed and proposed to correspond to the $2_1^+ \rightarrow 0_1^+$ and $4_1^+ \rightarrow 2_1^+$ transitions, establishing the corresponding states. It will likely take a next-generation rare-isotope facility to move beyond $^{72}$Fe with nuclear spectroscopy. A peculiar picture emerges where starting at $N = 40$, the evolution of the $2_1^+$ and $4_1^+$ excitation energies largely stays flat, as shown in Figure 3. Across the Fe isotopic chain, the LNPS shell-model calculations, using the slightly modified LNPS-m effective interaction, reproduce the measured excitation energies.

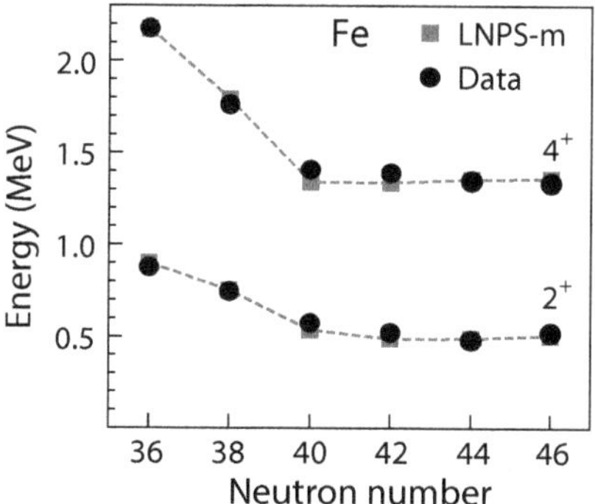


**Figure 3.** Evolution of the yrast $2^+$ and $4^+$ states in the Fe isotopic chain from $N = 36$ to 46, the most neutron-rich Fe isotope with spectroscopic information. The data is confronted with the results of LNPS-m (modified Lenzi-Nowacki-Poves-Sieja) shell-model calculations from reference [43]. LNPS-m is a slightly modified version of the original LNPS interaction as detailed in [43]. The calculations reproduce the signature drop in excitation energy at $N = 40$, corresponding to an onset of collectivity, and the subsequent flat evolution.

## 4. The Cr Isotopes

In the heart of the $N = 40$ island of inversion, the nucleus $^{64}$Cr eluded spectroscopy until 2010. In fact, Adrich et al. attempted to populate $^{64}$Cr in the two-proton removal from $^{66}$Fe. This measurement failed as the cross section turned out to be only 0.13(5) mb, an order of magnitude smaller than the cross section leading from $^{68}$Ni to $^{66}$Fe along $N = 40$ [33]. The conclusion at the time was that the cross section is small due to a structural mismatch between the $^{66}$Fe ground state and the bound states of $^{64}$Cr. This early idea was partially supported in 2010 through the LNPS effective interaction which predicts significant differences in the neutron 2p2h and 6p6h content in the ground states of $^{66}$Fe and $^{64}$Cr, hinting indeed at a potentially reduced overlap of the neutron wave functions [4]. The first spectroscopy of $^{64}$Cr was then accomplished via $^{9}$Be-induced inelastic scattering at NSCL where candidates for the $2^+_1 \rightarrow 0^+_1$ and $4^+_1 \rightarrow 2^+_1$ transitions were identified. While published in the same year, this data did not enter the development of the LNPS effective interaction and so the close match between experiment and calculation can be viewed as a stunningly successful prediction [4]. The $2^+_1$ state and the energy of the candidate $4^+_1$ level were since confirmed in intermediate-energy Coulomb excitation [38] and $\beta$ decay [46], respectively. The first one- and two-proton knockout study into $^{64}$Cr, using GRETINA and the S800, revealed a high $\gamma$-ray transition density, indicative of a rather complex and dense level scheme [47]. A quantitative knockout study was not possible as the knockout reaction channels may have contained small contaminations from $^{64}$Cr populated in fragmentation of other projectiles in the cocktail beam [47]. A study of the $B(E2)$ transition strength predicted by the LNPS shell-model calculations revealed several very interesting collective band structures, resembling a gamma and beta band but with deviations from the textbook expectations for such structures [47]. These proposed bands are barely linked via $E2$ transitions. Identifying these predicted collective structures in measurements has to remain a challenge for future studies and next-generation rare-isotope facilities where these states can be accessed with reactions at low beam energies such as deep-inelastic scattering [47].

The most neutron-rich Cr isotope with spectroscopic information is $^{66}$Cr studied at RIBF/RIKEN with $\gamma$-ray spectroscopy following a $(p, 2p)$ reaction [43]. Candidate transitions for the $2^+_1 \rightarrow 0^+_1$ and $4^+_1 \rightarrow 2^+_1$ decays were proposed, in agreement with slightly modified LNPS shell-model calculations, termed LNPS-m in [43]. The Cr isotopes mirror the observations for the Fe isotopic chain, with a rather flat evolution of the $2^+_1$ and $4^+_1$ energies, but starting at $N = 38$ already instead of at $N = 40$ as for the Fe chain; see Figure 4.

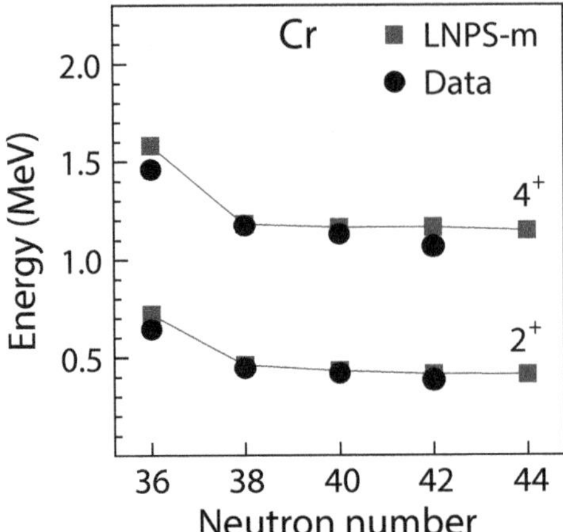


**Figure 4.** Evolution of the yrast $2^+$ and $4^+$ states in the Cr isotopic chain from $N = 36$ to 44, the most neutron-rich Cr isotope with spectroscopic information. The data is confronted with the results of LNPS-m shell-model calculations from reference [43]. LNPS-m is a slightly modified version of the original LNPS interaction as discussed in [43]. The calculations reproduce the signature drop in excitation energy at $N = 38$, corresponding to an onset of collectivity, and the subsequent flat evolution. Note that the onset of collectivity in Cr sets in already at $N = 38$, unlike for the Fe isotopes.

## 5. The Ti Isotopic Chain

The $N = 38$ Ti isotope $^{60}$Ti was studied in the $^{9}$Be-induced one-proton knockout at NSCL using GRETINA at the S800 spectrograph, providing the first spectroscopy of this nucleus [48]. One $\gamma$-ray peak was observed which was argued to be a doublet of two transitions corresponding to the $2_1^+ \rightarrow 0_1^+$ transition and perhaps the decay of the $4_1^+$ level. This measurement exploited the knockout reaction mechanism and compared calculated and measured partial cross sections [48]. The comparison supported the suggestion of a doublet as well as the spin assignments for the candidate states and the expectation for the inclusive cross section. This analysis provided a unique benchmark for the LNPS effective interaction that goes beyond excitation energies and includes wave-function overlaps, at the time the closest to $^{60}$Ca as possible.

At $N = 40$, $^{62}$Ti was accessed with $\gamma$-ray spectroscopy only recently, using a $(p, 2p)$ reaction with the MINOS target and the DALI2 scintillator array at SAMURAI [49]. Candidate $\gamma$-ray decays attributed to the $2_1^+ \rightarrow 0_1^+$ and $4_1^+ \rightarrow 2_1^+$ transitions were proposed. As in the work on $^{60}$Ti [48], also for $^{62}$Ti the direct nature of the reaction was exploited, comparing measured and calculated partial cross sections that probe the wave-function overlaps between projectile ground state and knockout residue final states. A $^{63}$V ground-state spin assignment $J = 3/2^-$ was found the most likely given the calculated cross section distributions for the other alternatives [49]. Along the $N = 40$ isotone line, $^{62}$Ti is the last extrapolation point towards the elusive $^{60}$Ca (see Figure 5), which was proven to exist only recently with implications for the dripline in the Ca isotopic chain [50].

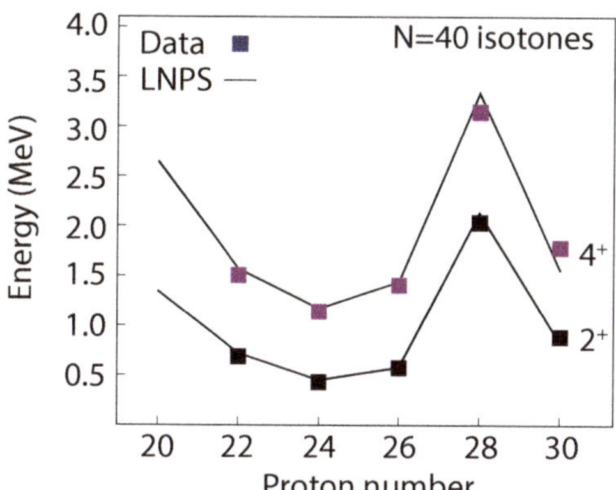


**Figure 5.** Evolution of the $2^+$ and $4^+$ energies in the $N = 40$ isotones from Ca to Zn as predicted by LNPS shell-model calculations presented in reference [49] in comparison to data, where $^{62}$Ti is the most neutron-rich in the chain. The excellent agreement lends confidence in the prediction for the elusive nucleus $^{60}$Ca which was only recently identified [50].

## 6. Complementary Descriptions of the Region

While this review focuses on the shell-model description of the nuclei above, the reader is referred to an interesting discussion of the Cr and Fe isotopic chains in the framework of the proton neutron interacting boson model (IBM-2) by Kotila and Lenzi [36]. Among the discussed collective observables, for example, the measured as well calculated energy ratios $R_{4/2}$ and $R_{6/4}$ within the shell-model and the IBM-2 are examined from $N = 30 - 40$ [36]. Complementary to the effective-interaction shell model and the IBM, Coraggio et al. [51] performed pioneering realistic shell-model calculations starting from a low-momentum potential derived from the high-precision CD-Bonn free nucleon-nucleon interaction. The energies of the first $2^+$ states and $B(E2)$ strengths are calculated inside the $N = 40$ island of inversion and the best agreement is reached with the largest possible model space [51]. These calculations were extended for $^{68,70}$Fe and confronted with experiment in [41]. The level structures of odd-Z $^{63,65,67}$Mn isotopes located on the nuclear chart just between the collective Cr and Fe isotopic chains were shown to be consistent with strongly coupled rotational bands built on a state with $K$ quantum number $K = 5/2$ [52], providing yet another means to characterize the collectivity that has become a hallmark of the region.

## 7. Summary and Conclusions

The $N = 40$ island of inversion, centered on $^{64}$Cr, has enjoyed intense attention from experimentalists and theorists alike. Experimental efforts at NSCL/MSU and RIBF/RIKEN have pushed the frontiers of spectroscopy by utilizing proton removal reactions which always lead to reaction residues more neutron-rich than the projectile. Technological advances, such as GRETINA at NSCL and MINOS at RIBF, allowed $\gamma$-ray spectroscopy, for the first time, at $N = 46$, $N = 42$ and $N = 40$ in the Fe, Cr, and Ti isotopic chains, respectively. On the theory side, the LNPS effective shell-model interaction continues to demonstrate not just the capability to reproduce the data but also to predict some of it. This close collaboration between experiment and theory has led to continuous refinements of the LNPS effective interaction and, in turn, motivated cutting-edge measurements at rare-isotope facilities around the world. Particularly exciting are predictions of an $N = 50$ island of inversion which rely on the monopole drifts from the LNPS interaction. This unmatched success of this effective interaction at $N = 40$ and beyond now lends confidence to these extrapolation and promises an exciting future for experiments at next generation rare-isotope facilities needed to reach these outskirts of the nuclear chart. In the $N = 40$ island of inversion itself, open challenges for experiment remain with respect to shape and configuration coexistence as predicted to be manifested in quadrupole-collective band

structures in $^{62,64}$Cr, for example. Also the study of such higher-lying off-yrast states has to remain a challenge for upcoming rare-isotope facilities. The future is exciting with experimental and theoretical advances in lockstep pushing the field forward to conquer the $N = 50$ island of inversion and fully characterize the $N = 40$ one. For this, more exotic fast-beam reactions such as the HI-induced charge exchange on high-spin projectiles [42] or multi-nucleon pickup reaction [53] may turn out to be promising tools in the arsenal of direct reactions.

**Funding:** Support from the U.S. Department of Energy, Office of Science, Office of Nuclear Physics, under Grant No. DE-SC0020451 is acknowledged.

**Conflicts of Interest:** The author declares no conflict of interest.

## References

1. Brown, B. The nuclear shell model towards the drip lines. *Prog. Part. Nucl. Phys.* **2001**, *47*, 517–599. [CrossRef]
2. Otsuka, T.; Gade, A.; Sorlin, O.; Suzuki, T.; Utsuno, Y. Evolution of shell structure in exotic nuclei. *Rev. Mod. Phys.* **2020**, *92*, 015002. [CrossRef]
3. Nowacki, F.; Obertelli, A.; Poves, A. The neutron-rich edge of the nuclear landscape: Experiment and theory. *Prog. Part. Nucl. Phys.* **2021**, *120*, 103866. [CrossRef]
4. Lenzi, S.M.; Nowacki, F.; Poves, A.; Sieja, K. Island of inversion around $^{64}$Cr. *Phys. Rev. C* **2010**, *82*, 054301. [CrossRef]
5. Nowacki, F.; Poves, A.; Caurier, E.; Bounthong, B. Shape coexistence in $^{78}$Ni as the portal to the fifth island of inversion. *Phys. Rev. Lett.* **2016**, *117*, 272501. [CrossRef] [PubMed]
6. Gade, A.; Liddick, S.N. Shape coexistence in neutron-rich nuclei. *J. Phys. G Nucl. Part. Phys.* **2016**, *43*, 024001. [CrossRef]
7. Brown, B.A. Islands of insight in the nuclear chart. *APS Physics* **2010**, *3*, 104. [CrossRef]
8. Hansen, P.; Tostevin, J. Direct reactions with exotic nuclei. *Annu. Rev. Nucl. Part. Sci.* **2003**, *53*, 219–261. [CrossRef]
9. Aumann, T.; Bertulani, C.A.; Ryckebusch, J. Quasifree ($p$,2$p$) and ($p$,$pn$) reactions with unstable nuclei. *Phys. Rev. C* **2013**, *88*, 064610. [CrossRef]
10. Gade, A.; Glasmacher, T. In-beam nuclear spectroscopy of bound states with fast exotic ion beams. *Prog. Part. Nucl. Phys.* **2008**, *60*, 161–224. [CrossRef]
11. Gade, A. Excitation energies in neutron-rich rare isotopes as indicators of changing shell structure. *Eur. Phys. J. A* **2015**, *51*, 118. [CrossRef]
12. Simpson, E. The Colourful Nuclide Chart. Available online: https://people.physics.anu.edu.au/~ecs103/chart/ (accessed on 1 November 2021).
13. Obertelli, A.; Delbart, A.; Anvar, S.; Audirac, L.; Authelet, G.; Baba, H.; Bruyneel, B.; Calvet, D.; Château, F.; Corsi, A.; et al. MINOS: A vertex tracker coupled to a thick liquid-hydrogen target for in-beam spectroscopy of exotic nuclei. *Eur. Phys. J. A* **2014**, *50*, 8. [CrossRef]
14. Gade, A.; Sherrill, B.M. NSCL and FRIB at Michigan State University: Nuclear science at the limits of stability. *Phys. Scr.* **2016**, *91*, 053003. [CrossRef]
15. Sakurai, H. Nuclear physics with RI Beam Factory. *Front. Phys.* **2018**, *13*, 132111. [CrossRef]
16. Paschalis, S.; Lee, I.; Macchiavelli, A.; Campbell, C.; Cromaz, M.; Gros, S.; Pavan, J.; Qian, J.; Clark, R.; Crawford, H.; et al. The performance of the Gamma-Ray Energy Tracking In-beam Nuclear Array GRETINA. *Nucl. Instrum. Meth. Phys. Res. Sect. Accel. Spectrometers Detect. Assoc. Equip.* **2013**, *709*, 44–55. [CrossRef]
17. Weisshaar, D.; Bazin, D.; Bender, P.; Campbell, C.; Recchia, F.; Bader, V.; Baugher, T.; Belarge, J.; Carpenter, M.; Crawford, H.; et al. The performance of the $\gamma$-ray tracking array GRETINA for $\gamma$-ray spectroscopy with fast beams of rare isotopes. *Nucl. Instrum. Meth. Phys. Res. Sect. A Accel. Spectrometers Detect. Assoc. Equip.* **2017**, *847*, 187–198. [CrossRef]
18. Takeuchi, S.; Motobayashi, T.; Togano, Y.; Matsushita, M.; Aoi, N.; Demichi, K.; Hasegawa, H.; Murakami, H. DALI2: A NaI(Tl) detector array for measurements of $\gamma$ rays from fast nuclei. *Nucl. Instrum. Meth. Phys. Res. Sect. A Accel. Spectrometers Detect. Assoc. Equip.* **2014**, *763*, 596–603. [CrossRef]
19. Tostevin, J. Single-nucleon knockout reactions at fragmentation beam energies. *Nucl. Phys. A* **2001**, *682*, 320–331. [CrossRef]
20. Tostevin, J. Core excitation in halo nucleus break-up. *J. Phys. G Nucl. Part. Phys.* **1999**, *25*, 735–739. [CrossRef]
21. Gade, A.; Adrich, P.; Bazin, D.; Bowen, M.D.; Brown, B.A.; Campbell, C.M.; Cook, J.M.; Glasmacher, T.; Hansen, P.G.; Hosier, K.; et al. Reduction of spectroscopic strength: Weakly-bound and strongly-bound single-particle states studied using one-nucleon knockout reactions. *Phys. Rev. C* **2008**, *77*, 044306. [CrossRef]
22. Tostevin, J.A.; Gade, A. Updated systematics of intermediate-energy single-nucleon removal cross sections. *Phys. Rev. C* **2021**, *103*, 054610. [CrossRef]
23. Bazin, D.; Brown, B.A.; Campbell, C.M.; Church, J.A.; Dinca, D.C.; Enders, J.; Gade, A.; Glasmacher, T.; Hansen, P.G.; Mueller, W.F.; et al. New direct reaction: Two-proton knockout from neutron-rich nuclei. *Phys. Rev. Lett.* **2003**, *91*, 012501. [CrossRef] [PubMed]

24. Yoneda, K.; Obertelli, A.; Gade, A.; Bazin, D.; Brown, B.A.; Campbell, C.M.; Cook, J.M.; Cottle, P.D.; Davies, A.D.; Dinca, D.C.; et al. Two-neutron knockout from neutron-deficient $^{34}$Ar, $^{30}$S, and $^{26}$Si. *Phys. Rev. C* **2006**, *74*, 021303. [CrossRef]
25. Tostevin, J.A.; Podolyák, G.; Brown, B.A.; Hansen, P.G. Correlated two-nucleon stripping reactions. *Phys. Rev. C* **2004**, *70*, 064602. [CrossRef]
26. Tostevin, J.A.; Brown, B.A. Diffraction dissociation contributions to two-nucleon knockout reactions and the suppression of shell-model strength. *Phys. Rev. C* **2006**, *74*, 064604. [CrossRef]
27. Simpson, E.C.; Tostevin, J.A.; Bazin, D.; Gade, A. Longitudinal momentum distributions of the reaction residues following fast two-nucleon knockout reactions. *Phys. Rev. C* **2009**, *79*, 064621. [CrossRef]
28. Simpson, E.C.; Tostevin, J.A.; Bazin, D.; Brown, B.A.; Gade, A. Two-nucleon knockout spectroscopy at the limits of nuclear stability. *Phys. Rev. Lett.* **2009**, *102*, 132502. [CrossRef] [PubMed]
29. Santiago-Gonzalez, D.; Wiedenhöver, I.; Abramkina, V.; Avila, M.L.; Baugher, T.; Bazin, D.; Brown, B.A.; Cottle, P.D.; Gade, A.; Glasmacher, T.; et al. Triple configuration coexistence in $^{44}$S. *Phys. Rev. C* **2011**, *83*, 061305. [CrossRef]
30. Simpson, E.C.; Tostevin, J.A. Correlations probed in direct two-nucleon removal reactions. *Phys. Rev. C* **2010**, *82*, 044616. [CrossRef]
31. Longfellow, B.; Gade, A.; Tostevin, J.A.; Simpson, E.C.; Brown, B.A.; Magilligan, A.; Bazin, D.; Bender, P.C.; Bowry, M.; Elman, B.; et al. Two-neutron knockout as a probe of the composition of states in $^{22}$Mg,$^{23}$ Al, and $^{24}$Si. *Phys. Rev. C* **2020**, *101*, 031303. [CrossRef]
32. Aumann, T.; Barbieri, C.; Bazin, D.; Bertulani, C.; Bonaccorso, A.; Dickhoff, W.; Gade, A.; Gomez-Ramos, M.; Kay, B.; Moro, A.; et al. Quenching of single-particle strength from direct reactions with stable and rare-isotope beams. *Prog. Part. Nucl. Phys.* **2021**, *118*, 103847. [CrossRef]
33. Adrich, P.; Amthor, A.M.; Bazin, D.; Bowen, M.D.; Brown, B.A.; Campbell, C.M.; Cook, J.M.; Gade, A.; Galaviz, D.; Glasmacher, T.; et al. In-beam $\gamma$-ray spectroscopy and inclusive two-proton knockout cross section measurements at $N \approx 40$. *Phys. Rev. C* **2008**, *77*, 054306. [CrossRef]
34. Hannawald, M.; Kautzsch, T.; Wöhr, A.; Walters, W.B.; Kratz, K.L.; Fedoseyev, V.N.; Mishin, V.I.; Böhmer, W.; Pfeiffer, B.; Sebastian, V.; et al. Decay of neutron-rich Mn nuclides and deformation of heavy Fe isotopes. *Phys. Rev. Lett.* **1999**, *82*, 1391–1394. [CrossRef]
35. Liddick, S.N.; Abromeit, B.; Ayres, A.; Bey, A.; Bingham, C.R.; Brown, B.A.; Cartegni, L.; Crawford, H.L.; Darby, I.G.; Grzywacz, R.; et al. Low-energy level schemes of $^{66,68}$Fe and inferred proton and neutron excitations across $Z = 28$ and $N = 40$. *Phys. Rev. C* **2013**, *87*, 014325. [CrossRef]
36. Kotila, J.; Lenzi, S.M. Collective features of Cr and Fe isotopes. *Phys. Rev. C* **2014**, *89*, 064304. [CrossRef]
37. Gade, A.; Janssens, R.V.F.; Baugher, T.; Bazin, D.; Brown, B.A.; Carpenter, M.P.; Chiara, C.J.; Deacon, A.N.; Freeman, S.J.; Grinyer, G.F.; et al. Collectivity at $N = 40$ in neutron-rich $^{64}$Cr. *Phys. Rev. C* **2010**, *81*, 051304. [CrossRef]
38. Crawford, H.L.; Clark, R.M.; Fallon, P.; Macchiavelli, A.O.; Baugher, T.; Bazin, D.; Beausang, C.W.; Berryman, J.S.; Bleuel, D.L.; Campbell, C.M.; et al. Quadrupole collectivity in neutron-rich Fe and Cr isotopes. *Phys. Rev. Lett.* **2013**, *110*, 242701. [CrossRef]
39. Rother, W.; Dewald, A.; Iwasaki, H.; Lenzi, S.M.; Starosta, K.; Bazin, D.; Baugher, T.; Brown, B.A.; Crawford, H.L.; Fransen, C.; et al. Enhanced quadrupole collectivity at $N = 40$: The case of neutron-rich Fe isotopes. *Phys. Rev. Lett.* **2011**, *106*, 022502. [CrossRef]
40. Daugas, J.M.; Matea, I.; Delaroche, J.P.; Pfützner, M.; Sawicka, M.; Becker, F.; Bélier, G.; Bingham, C.R.; Borcea, R.; Bouchez, E.; et al. $\beta$-decay measurements for $N > 40$ Mn nuclei and inference of collectivity for neutron-rich Fe isotopes. *Phys. Rev. C* **2011**, *83*, 054312. [CrossRef]
41. Benzoni, G.; Morales, A.; Watanabe, H.; Nishimura, S.; Coraggio, L.; Itaco, N.; Gargano, A.; Browne, F.; Daido, R.; Doornenbal, P.; et al. Decay properties of 68,69,70 Mn: Probing collectivity up to $N = 44$ in Fe isotopic chain. *Phys. Lett. B* **2015**, *751*, 107–112. [CrossRef]
42. Gade, A.; Janssens, R.V.F.; Brown, B.A.; Zegers, R.G.T.; Bazin, D.; Farris, P.; Hill, A.M.; Li, J.; Little, D.; Longfellow, B.; et al. In-beam $\gamma$-ray spectroscopy of $^{68}$Fe from charge exchange on $^{68}$Co projectiles. *Phys. Rev. C* **2021**, *104*, 024313. [CrossRef]
43. Santamaria, C.; Louchart, C.; Obertelli, A.; Werner, V.; Doornenbal, P.; Nowacki, F.; Authelet, G.; Baba, H.; Calvet, D.; Château, F.; et al. Extension of the $N = 40$ Island of inversion towards $N = 50$: Spectroscopy of $^{66}$Cr, $^{70,72}$Fe. *Phys. Rev. Lett.* **2015**, *115*, 192501. [CrossRef]
44. Gade, A.; Janssens, R.V.F.; Tostevin, J.A.; Bazin, D.; Belarge, J.; Bender, P.C.; Bottoni, S.; Carpenter, M.P.; Elman, B.; Freeman, S.J.; et al. Structure of $^{70}$Fe: Single-particle and collective degrees of freedom. *Phys. Rev. C* **2019**, *99*, 011301. [CrossRef]
45. Hardy, J.; Carraz, L.; Jonson, B.; Hansen, P. The essential decay of pandemonium: A demonstration of errors in complex beta-decay schemes. *Phys. Lett. B* **1977**, *71*, 307–310. [CrossRef]
46. Suchyta, S.; Liddick, S.N.; Chiara, C.J.; Walters, W.B.; Carpenter, M.P.; Crawford, H.L.; Grinyer, G.F.; Gürdal, G.; Klose, A.; McCutchan, E.A.; et al. $\beta$ and isomeric decay of $^{64}$V. *Phys. Rev. C* **2014**, *89*, 067303. [CrossRef]
47. Gade, A.; Janssens, R.V.F.; Bazin, D.; Farris, P.; Hill, A.M.; Lenzi, S.M.; Li, J.; Little, D.; Longfellow, B.; Nowacki, F.; et al. In-beam $\gamma$-ray spectroscopy of $^{62,64}$Cr. *Phys. Rev. C* **2021**, *103*, 014314. [CrossRef]
48. Gade, A.; Janssens, R.V.F.; Weisshaar, D.; Brown, B.A.; Lunderberg, E.; Albers, M.; Bader, V.M.; Baugher, T.; Bazin, D.; Berryman, J.S.; et al. Nuclear structure towards $N = 40$ $^{60}$Ca: In-beam $\gamma$-ray spectroscopy of $^{58,60}$Ti. *Phys. Rev. Lett.* **2014**, *112*, 112503. [CrossRef]

49. Cortes, M.; Rodriguez, W.; Doornenbal, P.; Obertelli, A.; Holt, J.; Lenzi, S.; Menendez, J.; Nowacki, F.; Ogata, K.; Poves, A.; et al. Shell evolution of $N = 40$ isotones towards $^{60}$Ca: First spectroscopy of $^{62}$Ti. *Phys. Lett. B* **2020**, *800*, 135071. [CrossRef]
50. Tarasov, O.B.; Ahn, D.S.; Bazin, D.; Fukuda, N.; Gade, A.; Hausmann, M.; Inabe, N.; Ishikawa, S.; Iwasa, N.; Kawata, K.; et al. Discovery of $^{60}$Ca and Implications For the Stability of $^{70}$Ca. *Phys. Rev. Lett.* **2018**, *121*, 022501. [CrossRef]
51. Coraggio, L.; Covello, A.; Gargano, A.; Itaco, N. Realistic shell-model calculations for isotopic chains "north-east" of $^{48}$Ca in the $(N,Z)$ plane. *Phys. Rev. C* **2014**, *89*, 024319. [CrossRef]
52. Liu, X.; Liu, Z.; Ding, B.; Doornenbal, P.; Obertelli, A.; Lenzi, S.; Walker, P.; Chung, L.; Linh, B.; Authelet, G.; et al. Spectroscopy of 2565,67Mn: Strong coupling in the $N = 40$ island of inversion. *Phys. Lett. B* **2018**, *784*, 392–396. [CrossRef]
53. Gade, A.; Weisshaar, D.; Brown, B.; Tostevin, J.; Bazin, D.; Brown, K.; Charity, R.; Farris, P.; Hill, A.; Li, J.; et al. In-beam $\gamma$-ray spectroscopy at the proton dripline: $^{40}$Sc. *Phys. Lett. B* **2020**, *808*, 135637. [CrossRef]

MDPI

*Review*

# Trends in the Structure of Nuclei near $^{100}$Sn

**Magdalena Górska**

GSI Helmholtzzentrum für Schwerionenforschung GmbH, D-64291 Darmstadt, Germany; m.gorska@gsi.de

**Abstract:** Inevitable progress has been achieved in recent years regarding the available data on the structure of $^{100}$Sn and neighboring nuclei. Updated nuclear structure data in the region is presented using selected examples. State-of-the-art experimental techniques involving stable and radioactive beam facilities have enabled access to those exotic nuclei. The analysis of experimental data has established the shell structure and its evolution towards $N = Z = 50$ of the number of neutrons, $N$, and the atomic number, $Z$, seniority conservation and proton–neutron interaction in the $g_{9/2}$ orbit, the super-allowed Gamow–Teller decay of $^{100}$Sn, masses and half-lives along the rapid neutron-capture process (r-process) path and super-allowed $\alpha$ decay beyond $^{100}$Sn. The status of theoretical approaches in shell model and mean-field investigations are discussed and their predictive power assessed. The calculated systematics of high-spin states for $N = 50$ isotopes including the $5^-$ state and $N = Z$ nuclei in the $g_{9/2}$ orbit is presented for the first time.

**Keywords:** nuclear structure; shell model; magic nuclei; gamma-ray spectroscopy

**Citation:** Górska, M. Trends in the Structure of Nuclei near $^{100}$Sn. *Physics* **2022**, *4*, 364–382. https://doi.org/10.3390/physics4010024

Received: 26 November 2021
Accepted: 2 March 2022
Published: 21 March 2022

**Publisher's Note:** MDPI stays neutral with regard to jurisdictional claims in published maps and institutional affiliations.

**Copyright:** © 2022 by the author. Licensee MDPI, Basel, Switzerland. This article is an open access article distributed under the terms and conditions of the Creative Commons Attribution (CC BY) license (https://creativecommons.org/licenses/by/4.0/).

## 1. Introduction and Ground-State Properties

The $N = Z = 50$ nucleus $^{100}$Sn, with $N$ being the number of neutrons and $Z$ being the atomic number, is the heaviest self-conjugate and doubly-magic nucleus that remains stable with respect to heavy-particle emission and thus provides an excellent opportunity for shell-model studies. In particular, its unique placement in the chart of nuclei makes it and its neighbors the most suitable to investigate neutron–proton correlations based on the coupling of single particle states with respect to a doubly-magic-core. However, in view of these advantages, the progress on the relevant experimental information in this region is moderate in spite of enormous efforts of physicists around the globe.

It directly relates to the accessibility of these nuclei in any known production reaction and therefore also to technical accelerator developments. Recent related technical developments have a large impact on this field. The history of the approaches to investigate $^{100}$Sn was addressed in the latest review [1] where the experimental and theoretical status of the region was summarized until 2013. The purpose of this work is to report on recent developments in this region relevant for the understanding of the nuclear force.

The $^{100}$Sn ground state is expected to be bound by about 3 MeV with respect to proton emission [2], which makes its yrast states accessible to $\gamma$-ray spectroscopy. The proton dripline was recently predicted [3] to be at $N = 47$ for the element of tin. The first ab initio prediction for the charge radius and density distribution of $^{100}$Sn was attempted in Ref. [4]. The latest example of experimental developments to study nuclear size in this region is given in Ref. [5]. In-gas-cell laser ionization spectroscopy and extraction of magnetic moments and mean-square charge radii of light Ag isotopes was presented in Ref. [6].

The strength of the Super Gamow-Teller transition [7], $B(GT)$ from the ground state of $^{100}$Sn, which could yield the largest value observed within the electron capture (EC) decay energy, $Q_{EC}$, window in the whole chart of nuclei, was originally predicted in Ref. [8]. It was measured for the first time by Hinke et al. [9]. Significant progress was obtained since then as the experimental value was revisited recently [10] and discussed theoretically in [11–14]. The two experimental $B(GT)$ values, originating from beta decay process, differ significantly.

This is mainly caused by the greatest deficiency of the beta-decay spectroscopy measurements that is the determination of $Q_{EC}$, i.e., $Q_{\beta,}$ which is needed for extraction of $B(GT)$ value. The reason for that is a strong dependence of phase-space factor $f$ on the decay energy. This calls for a high-precision mass measurement of $^{100}$Sn. Indeed, also here the progress is significant. Mass measurements in the region were recently extended [15,16], suggesting that the $Q_{EC}$ of Ref. [9] to be more consistent with those new results. The mass of $^{100}$Sn itself is likely within reach very soon.

The first possible excited yrast states in $^{100}$Sn are particle-hole excitations of the closed core. No excited states of $^{100}$Sn were reported thus far, as shown in Figure 1. To address directly the structure of $^{100}$Sn itself, an extended work has been invested into calculations of both $\alpha$-cluster formation and decay probabilities in ideal heavy $\alpha$ emitters $^{104}$Te and the $^{212}$Po for a direct comparison [16]. In this microscopic calculation of $\alpha$-cluster formation with an improved treatment of shell structure for the core nucleus, it was found out that the effective potential is sensitive to the contributing single-particle wave functions.

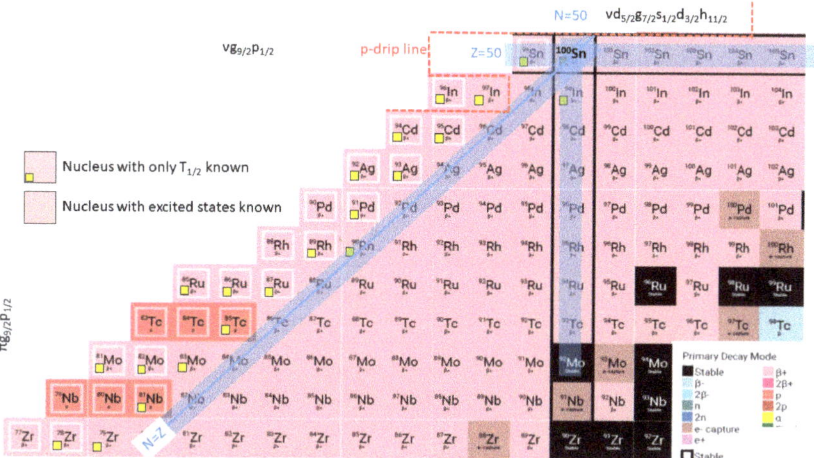


**Figure 1.** Present status of experimental information on the ground-state lifetime and excited states in the $^{100}$Sn region. Nuclei with only the ground-state lifetime known and no excited states reported are indicated with a small yellow square. The nuclei to the left of them were only produced in an experiment. The blue-shaded chains of nuclei highlight the main focus of this review.

Striking shell effects on the $\alpha$-cluster formation probabilities are shown for magic numbers 50, 82 and 126 by using the same nucleon-nucleon interaction. An enhanced $\alpha$-cluster formation probability was shown for both $^{104}$Te and $^{212}$Po as compared with their neighbors. In Ref. [17], a particular enhancement in the $^{100}$Sn region was suggested with respect to that in $^{208}$Pb. The analysis of the statistical significance of the neutron skin thickness to the symmetry energy in $^{132}$Sn and comparison to proton skin in $^{100}$Sn was performed by Muir et al. [18]. As in the first case, a clear correlation was observed for neutron skin, in the later no correlation could be deduced for $^{100}$Sn.

The nuclear structure of hole states in the region "southwest" of the shell closure at $^{100}$Sn, close to the $N = Z$ line is dominated by the $g_{9/2}$ intruder orbital from the $N = 4$ harmonic oscillator (HO) shell. This is well separated from the $N = 3$, *pf* orbitals, both energetically and by parity, allowing only $2p$-$2h$ excitations into the intruder orbital space. Dominated by the strong proton-neutron interaction, the $0g_{9/2}$ orbit gives rise to unique structural features [1] such as spin-gaps, seniority [19–21] and parity-changing isomerism [22] in addition to proton-neutron pairing correlations [23] and seniority-induced symmetries [24]. Moreover, when moving below $Z = 45$, deformation and shape coexistence of spherical and

deformed shapes start to appear. Therefore, the region "southwest" of $^{100}$Sn has become and remains, the subject of ever-increasing efforts both in experiments and theory.

The $N = Z$ nuclei provide the best quantum laboratory to investigate the characteristics of the neutron–proton ($np$) interaction, isospin symmetry and mixing, in addition to evolution of nuclear shapes. The $N \approx Z$ nuclei up to the $A = 60$ mass region have been intensively investigated during the past twenty years in various laboratories around the world. Here, the nuclei have been experimentally accessible as they are located only few neutrons away from their stable isotopes. From the nuclear theory perspective, especially regarding the nuclear shell model, the $N \approx Z$ nuclei between the $A = 40$–60 mass region have been an ideal subject to study since the valence nucleons occupy primarily the $f_{7/2}$ orbital making the calculations feasible due to the small valence space.

Currently, experimental ground-state decay and nuclear structure data, such as the level schemes and lifetimes of excited states for the $N \approx Z$ nuclei around the $A = 60$–90 mass region, are relatively scarce. This is due to the fact that these nuclei are located further away from the line of stability, near the proton-drip line. The production cross sections of these systems in nuclear reactions are very low (tens of nb to few μb). The missing experimental data from this region is naturally required in order to scrutinize and develop theoretical models operating in larger model spaces. However, radioactive ion beams (RIB) in this region are becoming gradually available for experiments, which can be utilized in various ways to search for new physics around the $N = Z$ line.

The evolution of nuclear shell structure in the vicinity of doubly-magic nuclei is of major importance in nuclear physics. The Sn isotopes provide a unique testing ground in this respect. The Sn isotopes represent the longest chain of semi-magic nuclei in nature, which makes them attractive for systematic investigations. How the shell structure evolves as a function of the number of protons and neutrons can be related to collective as well to single-particle effects. Unique correlation effects may be manifested at a self-conjugate shell-closure as the same spin-orbit partners for neutrons and protons reside just above and below the shell gap.

A sensitive probe for correlations of this kind is to measure transition probabilities for certain selected states. With this approach the results of large-scale shell-model (LSSM) calculations based on microscopically-derived interactions can be tested through direct comparison with experiment. The study of simple nuclear systems, with only a few nucleons outside a closed core, can thus provide insight into the underlying nucleon-nucleon interaction as applied to finite nuclei.

This paper summarizes briefly the new results on the structure of excited states of nuclei in $^{100}$Sn region and is organized as follows. After the general introduction including ground-state properties, the most successful experimental methods to obtain knowledge on excited states in nuclei in the region are elaborated in Section 2. Theoretical approaches are summarized briefly in Section 3.

The focus on new results of the shell-model calculations and their comparison to the recent experimental data is put in Section 4 for three sub regions describing certain symmetries shaded in blue in Figure 1. The $N = 50$ isotones are addressed in Section 4.1.1. The progress on $N = Z$ nuclear chain just below $^{100}$Sn is described in Section 4.1.2. The recent results on light Sn isotopes and selected nuclei with $N > 50$ below Sn are presented in Section 4.2. The choice of presented data from recent experimental and theoretical results is based on the author's subjective taste.

## 2. Experimental Methods

Although a comprehensive summary of the experimental status in the $^{100}$Sn region was reported in 2013 [1], the struggle to discover new aspects of structure of those nuclei continued. To assure progress in this very difficult to reach nuclear region, an enormous effort is devoted to the experimental techniques. Several aspects are crucial among the developments. The prerequisite is the availability of accelerators with beam parameters, i.e., energy and intensity etc., optimal for a given experimental apparatus.

Two types of experiments are distinguished with the highest impact on the progress obtained: low-energy facilities providing higher beam intensities for fusion and multi-nucleon transfer reactions and high energy facilities for fragmentation and spallation reactions producing radioactive beams, so called in-flight and isotope separation on-line (ISOL) facilities. While in earlier times the majority of experimental information was delivered from fusion-evaporation reactions at the low-energy facilities, recent years have shown the significance of the later ones.

### 2.1. *Low Energy Facilities*

Highly-intense beams of stable isotopes are available at several facilities such as GANIL (Large Heavy Ion National Accelerator, Caen, France) [25], Jyväskylä University [26], Argonne National Laboratory (ANL) [27] and INFN (National Institute for Nuclear Physics, Italy) Legnaro [28], to name only a few. These low-energy accelerators serving fusion-evaporation or multi-nucleon-transfer reactions newly-applied also in the $^{100}$Sn region [29], play a very important role when combined with highly-efficient detectors.

Such a combination was recently available at GANIL [25] with AGATA [30,31] and ancillary detectors as e.g., DIAMANT [32,33] and NEDA [34,35], which make the exit channel identification possible. Alternatively, a recoil spectrometer such as VAMOS [36–38], MARA [39], or the FMA [40] may serve for residue identification. A clear advantage of this method is prompt spectroscopy at the reaction target tagged with identified recoils or a decay particle (e.g. $\gamma$, $\beta$, $\alpha$).

Often, population of high-spin states, in particular in fusion-evaporation reactions, is considered advantageous. However, in the $^{100}$Sn region the most exotic nuclei that can be investigated are produced in 2-neutron ($2n$) or more and $2n\alpha$ exit channels [41]. Those residues are produced with the highest cross section at relatively low energies above the Coulomb barrier keeping the total reaction cross section low in order to avoid misidentification due to contaminants. At those lower energies, the reached spin values and residue excitation energy are reduced.

Impressive experiments of this type were performed in the last years leading to important discoveries, e.g., the delayed rotational alignment in a deformed $N = Z$ $^{88}$Ru [42]. Another way for the production of very neutron deficient Sn isotopes is alpha decay tagging measurements of the Te isotopes. The two leading groups at ANL with FMA [43,44] and at Oak Ridge National Laboratory (ORNL) [45,46] have been hunting for superallowed $\alpha$-decay signatures for the last 15 years. The burning question—whether it is energetically possible to produce excited states in $^{100}$Sn in this way—remains open [47].

### 2.2. *High-Energy Fragmentation and ISOL Facilities*

When the beam energy reaches the range suitable for fragmentation reactions, in-flight separation and identification of reaction products can be achieved. The lower beam intensities at high energies and lowered production cross sections are then compensated by the possibility of using thicker targets and higher efficiency of tracking and identification detectors. Unprecedented primary-beam intensities at relativistic energies have become available at the Radioactive Isotope Beam Factory (RIBF) at the RIKEN Nishina Center (Japan).

The reaction products are identified in the BigRIPS fragment separator [48] accompanied by an efficient $\gamma$-ray array, e.g., EURICA [49] (see Figure 2) consisting of EUROBALL cluster detectors for high-energy resolution or DALI2+ spectrometer [50] for low-resolution spectroscopy. Experiments of this type marked a new chapter in the available data in the $^{100}$Sn region including the discovery of new isotopes and proton emitters [51]. An important add-on was the new measurement of the *B(GT)* value for $^{100}$Sn $\beta$-decay measurement made with sufficient accuracy so that the interpretation allowed for distinguishing between different models used [10].

The existing data can be used for *B(GT)* re-determination once the mass measurement of $^{100}$Sn is available. Extensive data on lifetimes of $\beta$ decay and $\beta$-delayed proton

emission, as well as beta-delayed spectroscopy was published in Refs. [52,53]. Large progress on excited states in the region was obtained including gamma-gamma coincidence data for the $^{100}$In nucleus [10], identification of excited states in $^{96}$Cd [22,54] and others [55,56] based on isomer spectroscopy are mentioned in Section 4.1. The disadvantage of this method is that it does not allow for prompt gamma-ray spectroscopy at the (primary) target.

**Figure 2.** EURICA array consisting of 12 cluster detectors of EUROBALL at RIKEN RIBF in 2012 [49]. See text for details.

Two-step fragmentation is used to study prompt radiation from Coulomb excitation or knockout reactions with the DALI2+ spectrometer. From several experimental campaigns, extensive and spectacular data was collected and, to large extent, published as discussed in Section 4.2. Very recently, in 2020–2021 the HICARI (High-resolution in-beam gamma-ray spectroscopy at RIBF) project [57] used a Ge-detector array to perform several experiments addressing this region.

Efforts continue at NSCL (National Superconducting Cyclotron Laboratory, Michigan, USA) to contribute to the region [58], e.g., with spectroscopy using knock-out reactions [59], or the recent mass measurement of $^{80}$Zr [60]. At GSI, revisited isomeric decay in $^{102}$Sn and resulting effective neutron and proton charges based on state-of-the-art shell-model calculations were published [61], 13 years after the RISING [62] experiment.

In 2020, the GSI (Society for Heavy Ion Research, Darmstadt, Germany) facility came back into operation again after 6 years with the Fragement Separator (FRS) [63] and DEcay SPECtroscopy (DESPEC) [64] setup including the FATIMA [65] gamma-ray array and a Ge-detector array to address this region of nuclei again with lifetime measurement of intermediate states below isomers or states populated in beta decay.

To determine the excitation energy of long-lived isomeric states, a complementary technique employing the Multi-Reflection Time-of-Flight Mass Spectrometer (MR-TOF-MS) at the FRS Ion Catcher was recently used in this region of nuclei [15].

Alternatively, high-energy and high-intensity protons are used in spallation reaction and the radioactive beam is stored and separated in an ion source. Laser-ionized secondary beams are accelerated to fusion energies to impinge on a secondary target surrounded by a $\gamma$-ray array. The enormous success of this method was demonstrated at CERN (the European Organization for Nuclear Research) REX-ISOLDE [66] and continued with the HIE-ISOLDE project [67] where secondary beams were used for transfer and Coulomb-excitation measurements using the MINIBALL $\gamma$-ray array [68,69]. Several experiments were devoted to the study of neutron-deficient tin via transfer and Coulomb excitation measurements [70–74].

A similar principle was recently also applied at NSCL in [75] using the JANUS setup [76].

## 3. Theoretical Approaches

Similarly, theory activities in the $^{100}$Sn region did not lose momentum in the years since the last review [1]. Indeed, several approaches ([61,77–84]) could enlarge the treated configuration space (truncation level) and/or go to new regions of the nuclidic chart for The LSSM calculations, which helped to explain certain phenomena not clarified before and suggested experiments for future studies. These calculations use mostly realistic interactions derived from a nucleon-nucleon potential with various treatments to obtain two-body matrix elements and single-particle energies as described for example in [1,83] and references therein.

On the other hand, the comparison of those advanced calculation results to the ones with a smaller model space and empirical interactions, which are doable in the scope of this review work, can shed light on certain basic principles, which were not treated yet with large codes and computer power. For this purpose, the empirical GF [85] and SLGT [86] interactions in the $p_{1/2}g_{9/2}$ model space is used here, with single-particle level energies adjusted as given in [1], to guide the basic understanding of the underlying structure.

In the scope of this work, JUN45 interaction [78] results in the $\pi\nu(f_{5/2},p_{3/2},p_{1/2},g_{9/2})$ (or r3$g$) model space for high spin states are also presented in Section 4 for several $N$ = 50 isotones and $N$ = Z nuclei of the $g_{9/2}$ shell. The shell-model code NuShellX [87] was used for these computations. The MHJM interaction in the $\pi p_{1/2}g_{9/2}$ $\nu d_{5/2}g_{7/2}d_{3/2}s_{1/2}h_{11/2}$ model space originating from [88] was successfully used in the literature to describe the structure of nuclei with $Z \leq 50$ and $N \geq 50$ [89], see e.g., [1,53,90,91]. The LSSM calculations in the e.g., [61,92,93] with the SDG interaction [22] for the $\pi\upsilon(g_{9/2},d_{5/2},g_{7/2},s_{1/2},d_{3/2})$ model space (further referred as *gds*) are described in more detaill below.

Recently, new approaches were proposed in this region of nuclei promising further success. The role of 3-body residual interactions in nuclear chains in a single j-shell is discussed within the shell model [94]. Beyond the standard shell-model approach, the 3-body interaction was considered in calculations of the energies of excited states in $N$ = 50 isotones [95], which caused a significant improvement in reproducing experimental values. Furthermore, nuclear flied theory group investigated Sn isotopic chain using particle-vibration coupling [96]. The quadrupole-vibrational excitations in even–even Cd isotopes was revisited by the mean-field interacting boson model [97].

As already indicated in the introduction the ab initio methods are recently possible for this region of nuclei. The prediction of energies of excited states of nuclei in vicinity of $^{100}$Sn were very recently presented based on the particle-hole effective interaction derived from shell model couple cluster method [98].

## 4. Results

The results presented in this paper are narrowed to the three aspects of the structure of $^{100}$Sn with an idea of demonstrating the role of like-nucleon vs. proton-neutron coupling in the $g_{9/2}$ shell (see Figure 1). First, the evolution of excited states was examined along the $N$ = 50 line describing seniority in a $g_{9/2}{}^{n}$ and simple level coupling $g_{9/2}p_{1/2}$. Additionally, a JUN45 calculation and, even larger model spaces when available, are presented. Second, the $N$ = $Z$ line is analyzed below $^{100}$Sn where the shape change is expected along the $g_{9/2}{}^{n}$ orbital configuration. The recent developments in nuclear structure above $N$ = 50 is summarized in the last subsection.

### *4.1. Nuclei below $^{100}$Sn, $A \leq 100$*

#### 4.1.1. Even-Even $N$ = 50 Isotones of the $g_{9/2}$ Shell

The seniority quantum number $\upsilon$, which counts the number of unpaired nucleons for protons and neutrons occupying the same shell-model orbital, is very useful when

discussing the structure based on high-spin orbitals. Just below $^{100}$Sn, isotopes with even $Z$ form a long chain of a seniority isomers, which exhibit observable decays as the $g_{9/2}$ is well isolated from other high spin orbitals. A direct consequence of the short-range nature of the nucleon-nucleon interaction is the conservation of the seniority $\upsilon$ in any n-particle configuration $j^n$ of like particles [99].

As the mixing of states with different seniority is expected to be small, several symmetries are imposed [24,100] (and references therein), of which the constant excitation energies within the shell and symmetry against the mid shell of $B(E2)$ values for transitions with non-changing seniority, are addressed in this Section. In Figure 3 the experimental excitation energies for the $2^+$–$8^+$ levels, as well as $5^-$ are shown with the differently-colored symbols for each spin value. The lowering of the excitation energy seen for the states with even spin values is understood mostly by the increased binding of the $0^+$ ground-state when removing protons from $^{100}$Sn. This effect is caused by the contribution of lower shell orbitals such as $p_{3/2}$ and $f_{5/2}$, becoming closer to the Fermi level.

Most of the shell-model calculations, presented for $N = 50$ isotones, can reproduce level energies relatively well, with less accuracy for the $6^+$ and $8^+$ states. For the GF shell-model calculation (shown with continuous lines in Figure 3), the agreement is very good (note the expanded energy scale) in the lower shell e.g., for $^{92}$Mo, which is trivial as it was used to fit the two-body matrix elements (TBME) and therefore effectively includes the contribution of lower shells. It is also understandable that there is a trend of increased level deviation of the $g_{9/2}{}^n$ coupling towards $^{100}$Sn.

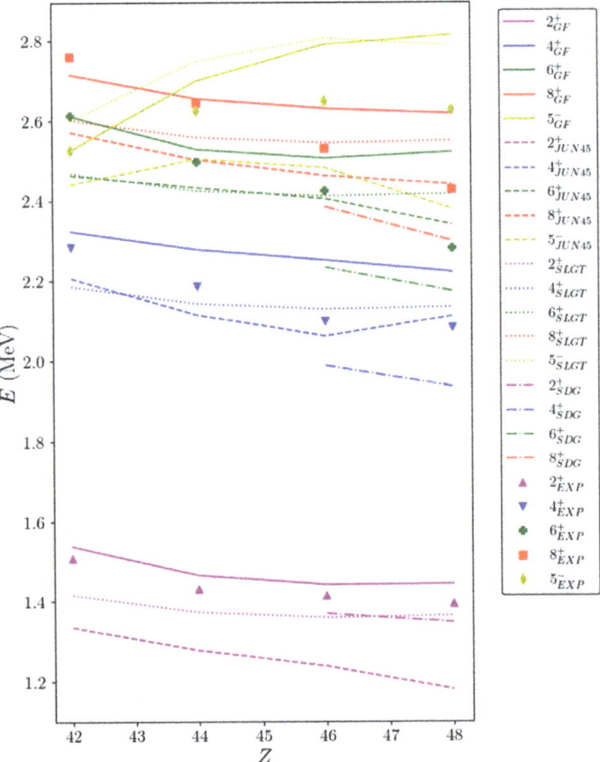


**Figure 3.** Systematics of the level energies in the $N$ (number of neutrons) = 50 isotonic chain, for even-$Z$ (atomic number) nuclei. Experimental data shown with colored symbols are taken from [101].

The energy of the $5^-$ state in $^{98}$Cd is taken from [102]. The lines shown in corresponding colors represent theoretical level energies. Calculations were done with the NuShellX code [87] using empirical interaction GF [85], SLGT [86] in the $\pi(g_{9/2}p_{1/2})$ and realistic effective interaction JUN45 [78] in the $\pi(f_{5/2},p_{3/2},p_{1/2},g_{9/2})$ and SDG in $\pi\upsilon(g_{9/2},d_{5/2},g_{7/2},s_{1/2},d_{3/2})$ model spaces [92]. The values of single-particle level energies were adopted from [1]. A pure $\pi g_{9/2}{}^{n}$ configuration explains the levels $2^+$–$8^+$. The $5^-$ states are obtained from the $\pi p_{1/2}{}^{-1}g_{9/2}{}^{n+1}$ coupling. The "EXP" denotes experimental values.

To note is that the newly-observed $5^-$ state in $^{98}$Cd [102] completes its systematics in those nuclei build of coupling $\pi p_{1/2}{}^{-1}g_{9/2}{}^{n+1}$ and resembles the trend of increasing deviation towards the full shell. The discrepancy of the calculated $5^-$ and $8^+$ states from the experimental values are almost identical, while their crossing is reproduced almost perfectly at $^{94}$Ru. For the calculations using SLGT interaction (dotted lines in Figure 3), instead, the absolute values of energies are improving with increasing $Z$ towards $^{96}$Pd and worsen again for $^{98}$Cd (except the $2^+$ state).

The energy trend of $5^-$ states along the shell is not reproduced, similarly as for GF interaction. While for the low-spin states the JUN45 [78] spectrum is compressed, the $6^+$ and $8^+$ are rather well reproduced, improving further towards the end of the shell (dashed line). The $5^-$ state is missed by about 200 keV, which indicates the $p_{1/2}$ single particle evolution correction needed for those isotopes.

The SDG spectrum is generally contracted, caused by the inclusion of core excitations, but it reproduces consistently the slope of the line between $^{96}$Pd and $^{98}$Cd for which the calculation is available [92]. Moreover, the second $4^+$ state (not shown in the figure for clarity) is predicted only 250 keV higher than the first one, which is approximately at the same energy as the first $6^+$ state. In contrast, the second $6^+$ state is predicted about 400 keV above the first one, which is considerably higher than the $8^+$ state [92].

More relevant for the wave functions, however, is the comparison of experimental reduced transition probabilities, *B(E2)* values in $N = 50$ isotones with the calculated values. In that case, all calculations including the GF calculations, presented here, can reproduce the data rather well for the $6^+$ and $8^+$ states as their wave functions are mostly not affected by other configurations; see [1,100] and references therein. The *B*(*E*2) for transition from the $6^+$ state in $^{98}$Cd became recently available and the accuracy of that from $8^+$ state was improved [56]. The lower lying yrast states, however, caused an extended discussion in recent years on seniority conservation and seniority mixing in the $g_{9/2}$ orbital; see [1,100,103] and references therein.

Particularly for the nonaligned $4^+$ systematics in the mid shell, of which two different seniority states are predicted in close vicinity, evidence is discussed [100] for seniority breakdown due to neutron excitations across the $N = 50$ shell gap. There, the clear advantage of LSSM calculation with core excitations included is evident. An extension of the experimental data to lower $Z$ for $N = 50$ isotones will be soon available [104,105] and therefore the discussion on this topic is not extended in this work.

#### 4.1.2. $N = Z$ Chain from $^{100}$Sn to $^{80}$Zr

The region of the heaviest and bound $N = Z$ nuclei is below $^{100}$Sn. This fact makes it particularly attractive for studies of proton-neutron correlations and their dependance on the spin of the nucleus. Already decades ago theoretical calculations predicted the existence of $I^\pi = 16^+$ and $I^\pi = 25/2^+$ high-spin isomers in $^{96,97}$Cd, respectively. Only much later, technical advances allowed the confirmation of these isomers experimentally [22,106].

Early shell-model studies employing empirical interactions in the $\pi\nu(p_{1/2}, g_{9/2})$ model space were reviewed in Ref. [1]. Realistic interactions with LSSM calculations were presented for the full $\pi\nu(f_{5/2},p_{3/2},p_{1/2},g_{9/2})$ model space [78,79,107–109] as well as for the upper $\pi\nu(gds)$ shell using the SDG interaction [110]. The strength of the $\pi\nu$ interaction in the $\pi\nu g_{9/2}$ orbits manifests itself best in the strongly-binding $T = 0$, $I^\pi = 9^+$ TBME, which is comparable in strength with the "normal", $T = 1$ pairing mode [85,86]. With the identification of excited states in $^{92}$Pd [23], the role of the $\pi\nu g_{9/2}$ pairs with maximum spin of $I^\pi = 9^+$

in the $N = Z$ nuclei $^{96}$Cd, $^{94}$Ag and $^{92}$Pd has been investigated in a series of multi-step shell model and IBM studies with respect to the "fully aligned" $9^+$ -TBME [24,111–114].

The experimental yrast states for the even–even $N = Z$ nuclei in the $g_{9/2}$ shell are shown with colored symbols in Figure 4. For the lower mass $N = Z$ even–even nuclei the level structure implies an onset of deformation. The spectrum of $^{88}$Ru [42] exhibits a rotational band similar to the known states of $^{84}$Mo [115]. This could intuitively understood, similarly to the $N = 50$ chain, as even stronger influence by the lower shell *fp*-shell therefore driving deformation. However, recent developments in the calculation power could disprove this hypothesis, as shown below.

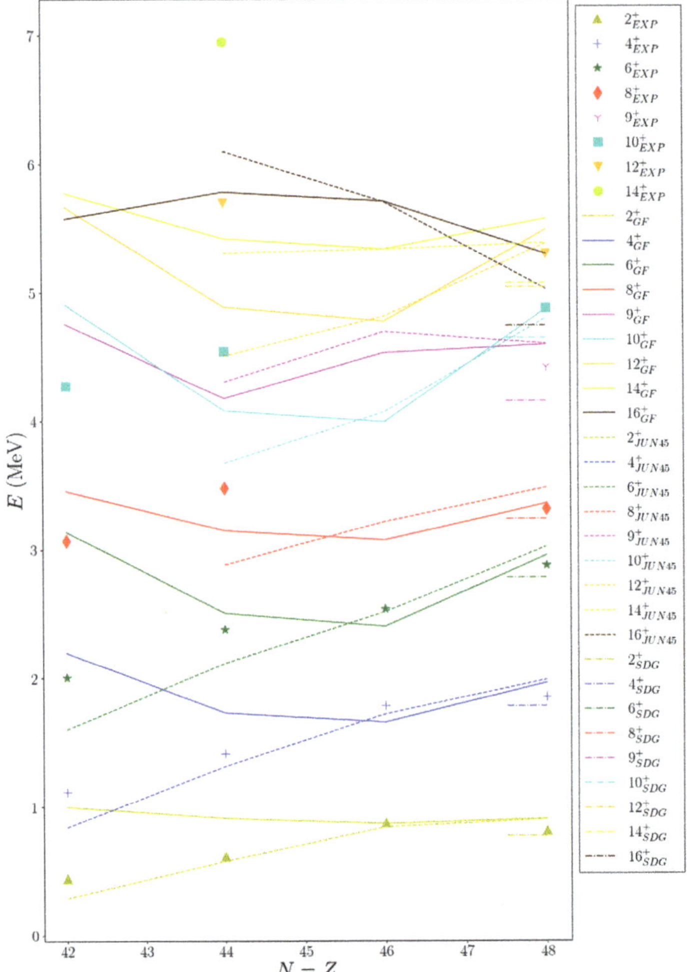


**Figure 4.** Excitation energies of the even–even $N = Z$ nuclei in the $g_{9/2}$ shell with $A$ (atomic mass number) = 84–96 for spins $I^\pi = 2^+$–$16^+$. The spins of excited states in all isotopes are assigned tentatively [101]. Experimental energies are given with colored symbols distinguished by their spin values. Shell-model calculations are shown with solid lines with different colors for each spin for the GF [85] calculation, dashed line for the JUN45 calculation [78,107] and present work for the higher-spin states and colored and short dashed-dotted colored lines for SDG [110] calculation for $^{96}$Cd.

Various shell-model approaches in different model spaces have been reported for this region. The calculation presented with solid colored lines is GF [85], which includes all yrast spins states up to the $16^+$ for all the $N = Z$ nuclei in $g_{9/2}$ shell nuclei and in particular an isomeric trap, known to exist experimentally in $^{96}$Cd [110]. Calculation using JUN45 interactions [78] are shown with dashed lines whenever possible for the computer power available within this work or in the literature [78,107].

The LSSM SDG [22] approach with up to $5p5h$ excitations (t = 5) across the Z, $N = 50$ closed shell are available only for $^{96}$Cd up to high spins and is shown with short, colored dashed-dotted lines. A consistent description for all the calculations is obtained in the upper $g_{9/2}$ shell, as shown for $^{96}$Cd. However, towards the mid shell, the $N = Z$ results are hampered by severe truncation. The advantage of including a larger model space, particularly the lower-lying orbitals, is evident. Nevertheless, with increasing spin the lower-shell nuclei follow very different trend than the predicted one.

According to the GF calculation presented in Figure 4, the spin trap at $I^\pi = 16^+$ in $^{84}$Mo will stay yrast even if the spherical $0^+$ state in the small model space of GF calculation is shifted by ~2 MeV up in energy with respect to the known deformed ground state and the rotational band member energies extrapolated with a constant moment of inertia to higher spins [115].

However, a common conclusion of various shell model approaches with $Z \geq 40$, $N = Z$ nuclei is the quest to include excitations across the Z, $N = 50$ shell closure [79,116–119] in order to attempt the description of the deformation. $^{84}$Mo marks the transitional region where the shape-driving role of the $r3g$ space is replaced/enhanced by the $gds$ space [79]. The spectra of $^{96}$Cd is a benchmark for both model spaces and yield similar results. In addition, the $^{84}$Mo spectrum was calculated with the help of the nucleon-pair approximation method [120], which also could be expanded to higher spins and other $N = Z$ nuclei in the $g_{9/2}$ shell in the future.

Recently, a new approach was proposed in the mean field [95], where within a simple SO(8) pairing model, it was shown that the symmetry-projected condensates of mixed isovector and isoscalar pairs very accurately describe properties of the exact solutions, including the coexistence of the isovector and isoscalar pairing. Lack of symmetry restoration thus explains the limited success in describing such a coexistence in the standard mean-field approaches to date. It was concluded that the future work investigating properties of the proton-neutron nuclear pairing can be carried out within the variation-after-projection approach to mean-field pairing methods.

### 4.2. Nuclei with $N > 50$

The Sn isotopes represent the longest chain of semi-magic nuclei, which makes them attractive for studies of shell-structure evolution as a function of the number of neutrons and how it can be related to collective as well to single-particle effects. The known, almost constant excitation energy of the first $2^+$ state in the Sn isotopic chain has been a textbook example of the seniority scheme for a long time.

A sensitive probe for correlations of this kind is to measure transition probabilities for first excited selected states, which will manifest the configuration content of those states. With this approach the results of LSSM calculations based on microscopically-derived interactions can be tested through direct comparison with experiment. This approached was recognized with the availability of radioactive beams, and new measurements or new theory values have been seen frequently in recent years. A recent update on $6^+$ states in Sn isotopes is reported in [61] for the $6^+$ state lifetime and effective charge analysis. The energy of the second $2^+$ state in $^{102}$Sn was recently claimed in Ref. [121].

However, the main focus in the studies of Sn isotopes is on the first $2^+$ states since the first measurement of radioactive isotopes 16 years ago [122], and the last ones being the theoretical study of Togashi et al. [77] and experimental study of Siciliano et al. [29]. The study of Togashi et al., represents the first approach in which it is possible to reproduce remarkably well the whole Sn chain (shown in Figure 2 of [77]) in same calculation.

The method used for this unified description of the detailed nuclear structure is Monte Carlo Shell Model including isospin conserving interaction calculation of the *gds* HO shell as well as the lower part of the neighboring HO shells [77] (and reference therein). An alternative calculation with a small modification is shown by the authors as to give an Ansatz to the experimentalists for a more precise experimental answer to the values in the mid-shell.

Within the generalized seniority scheme this mid-shell valley would be interpreted as changing single-particle orbitals filled along the Sn chain (see also [123]). In Figure 3 of [77] the authors describe the complex wave functions including core excitation (and deformation) for the $2^+$ as well as the $4^+$ states in Sn isotopes towards the mid shell modified by the quadrupole component of the proton-neutron interaction, which was first postulated in [124] by the LSSM analysis for the $2^+$ states. The quadrupole collectivity was also predicted in Ref. [12].

The results, presented in Figure 3 by Siciliano et al. [29], show an overview of experimental knowledge on $B(E2{:}2^+\rightarrow0^+)$ for the full Sn isotopic chain based on intense efforts of many laboratories and experimental groups. References [122,124–144] represent a complete up-to-date list. The usage of general Doppler methods, such as the Doppler shift attenuation (DSAM) and the recoil distance Doppler shift methods (RDDS) to measure the lifetimes and to extract the $B(E2)$ values, were hampered until recently by the existence of higher-lying isomeric states.

Indeed, the authors of Ref. [29] managed to overcome the problem by using multi-nucleon transfer reactions and adjusting the excitation energy of the final product such that the $6^+$ isomer feeding was minimized allowing for RDDS measurements. Moreover, the experimentalists harvested the first information on the lifetimes of $4^+$ states, which opened up the systematics of $B(E2{:}4^+\rightarrow2^+)$ states below $N = 60$. This pioneering experimental work was accompanied by LSSM calculations, which could well reproduce experimental data using the new realistic effective interaction in the *gds* model space with a proper monopole treatment.

Another calculation indicating for the first time the double-hump shape associated to the quadrupole dominance, as shown in Ref. [77], refer to the importance of further investigations of the $4^+$ states, where pairing effects related to single particle energies dominate instead. The new findings, together with the recent theory calculation [145] request for further experimental and theoretical effort in this direction.

The systematics of the reduced transition probabilities, the $B(E2{:}2^+\rightarrow0^+)$ values, is expected to be completed soon with the inclusion of the $^{102}$Sn value [146]. The review dedicated to nuclear collectivity is in preparation [147], where updated figure will be presented.

Alternatively, the collective properties of $^{100}$Sn can be approached again by studying nuclei with slightly lower $Z$. Several dedicated attempts for such experiments were undertaken at ISOLDE and NSCL. In particular, light Cd isotopes were addressed already in earlier days and Coulomb excitation transition probabilities and quadrupole moments were extracted up to $^{102}$Cd [70]. The latest update on this can be found in [148], where $B(E2{:}4^+\rightarrow2^+)$ values were also measured, and the accompanying theory work, which attempted to explain particular conditions for collectivity in light Cd isotopes [70].

The beyond-mean-field calculations, presented in Ref. [148], reproduce the cadmium systematics, but also predict rotational structures for all of the $Z = 48$ isotopes, breaking the common view of the textbook example of vibrational nuclei. In addition, there (and in Ref. [85] therein), Z = 48 isotopes are predicted to be semi-magic deformed nuclei. The lightest Cd isotope for which $B(E2{:}2^+ \rightarrow 0^+)$ value measurement was attempted thus far is $^{100}$Cd [70].

## 5. Summary and Outlook

Recent years have shown a great deal of interest from experimental and theoretical groups from all over the globe dedicated to investigations of the $^{100}$Sn region. This materialized in many publications (referred to here and with more to come) in the last decade, as well as active and waiting proposals and, presently, a large amount of as-yet unevaluated data. The primary reason for the particular excitement is the relevance of this region for understanding the nuclear force in general and various specific aspects that can be uniquely studied in this region of the heaviest doubly-magic $N = Z$ nucleus.

To further encourage a steady level of development and the need for new data of key nuclei and particular states, two examples are mentioned here. The first one is the search for excited states in $^{100}$Sn, which could be determined from the decay of the predicted isomeric $6^+$ state [93] (and references therein). The second is excited states in $^{98}$Sn (predictions presented in Figure 5), a mirror nucleus of $^{98}$Cd [149]. Those two nuclei likely constitute the heaviest possibly bound mirror pairs of all nuclei.

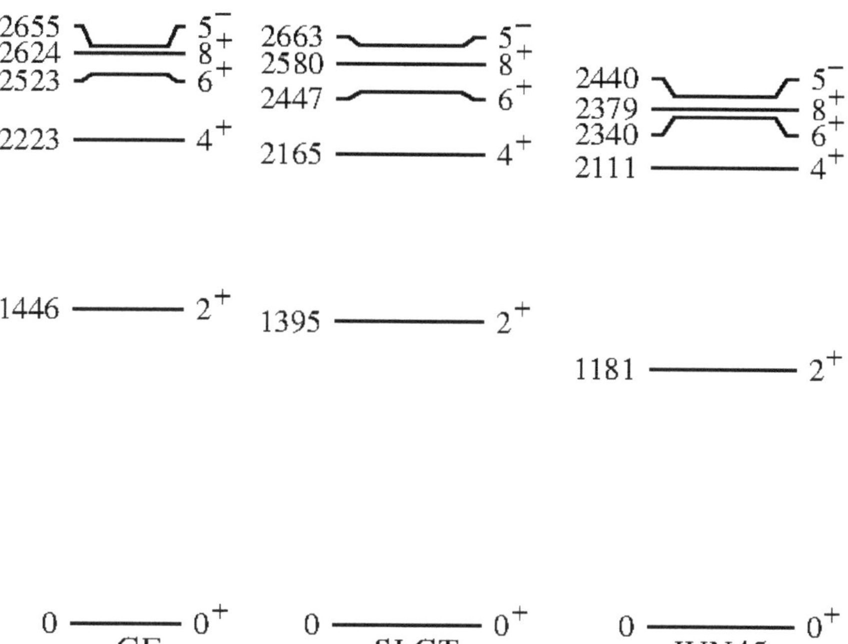


**Figure 5.** Predictions of the $^{98}$Sn excited states according to the available interactions. All of them suggest an $8^+$ isomeric state as the one known in the mirror nucleus $^{98}$Cd [149]. The JUN45 spectrum is identical to that one of $^{98}$Cd because of the isospin symmetry of this interaction.

**Funding:** This research received no external funding.

**Data Availability Statement:** No new data were created or analyzed in this study. Data sharing is not applicable to this article.

**Acknowledgments:** Hubert Grawe is gratefully acknowledged post mortem for his mentoring until his very last days. Without him, this paper would not be possible. Andrey Blazhev, Piet Van Isacker, Frédéric Nowacki and Taka Otsuka are acknowledged for useful discussions and support. Kathrin Wimmer is acknowledged for introducing the author to the basics of Python libraries as well as for running calculations with JUN45 interaction for high spin states in $N = Z$ nuclei on a larger computer than available for the author. Zsolt Podolyák and Helena May Albers are acknowledged for valuable comments and proofreading of this article.

**Conflicts of Interest:** The author declares no conflict of interest.

## References

1. Faestermann, T.; Górska, M.; Grawe, H. The structure of $^{100}$Sn and neighbouring nuclei. *Prog. Part. Nucl. Phys.* **2013**, *69*, 85–130. [CrossRef]
2. Dobaczewski, J.; Nazarewicz, W. Limits of proton stability near $^{100}$Sn. *Phys. Rev. C* **1995**, *51*, R1070–R1073. [CrossRef] [PubMed]
3. Neufcourt, L.; Cao, Y.; Giuliani, S.; Nazarewicz, W.; Olsen, E.; Tarasov, O.B. Beyond the proton drip line: Bayesian analysis of proton-emitting nuclei. *Phys. Rev. C* **2020**, *101*, 014319. [CrossRef]
4. Arthuis, P.; Barbieri, C.; Vorabbi, M.; Finelli, P. Ab Initio Computation of Charge Densities for Sn and Xe Isotopes. *Phys. Rev. Lett.* **2020**, *125*, 182501. [CrossRef] [PubMed]
5. Reponen, M.; de Groote, R.P.; Al Ayoubi, L.; Beliuskina, O.; Bissell, M.L.; Campbell, P.; Cañete, L.; Cheal, B.; Chrysalidis, K.; Delafosse, C.; et al. Evidence of a sudden increase in the nuclear size of proton-rich silver-96. *Nat. Commun.* **2021**, *12*, 4596. [CrossRef] [PubMed]
6. Ferrer, R.; Bree, N.; Cocolios, T.E.; Darby, I.; De Witte, H.; Dexters, W.; Diriken, J.; Elseviers, J.; Franchoo, S.; Huyse, M.; et al. In-gas-cell laser ionization spectroscopy in the vicinity of $^{100}$Sn: Magnetic moments and mean-square charge radii of N = 50–54 Ag. *Phys. Lett. B* **2014**, *728*, 191–197. [CrossRef]
7. Brown, B. The nuclear shell model towards the drip lines. *Prog. Part. Nucl. Phys.* **2001**, *47*, 517–599. [CrossRef]
8. Brown, B.A.; Rykaczewski, K. Gamow-Teller strength in the region of $^{100}$Sn. *Phys. Rev. C* **1994**, *50*, R2270–R2273. [CrossRef]
9. Hinke, C.B.; Böhmer, M.; Boutachkov, P.; Faestermann, T.; Geissel, H.; Gerl, J.; Gernhäuser, R.; Górska, M.; Gottardo, A.; Grawe, H.; et al. Superallowed Gamow–Teller decay of the doubly magic nucleus $^{100}$Sn. *Nature* **2012**, *486*, 341–345. [CrossRef]
10. Lubos, D.; Park, J.; Faestermann, T.; Gernhäuser, R.; Krücken, R.; Lewitowicz, M.; Nishimura, S.; Sakurai, H.; Ahn, D.S.; Baba, H.; et al. Improved Value for the Gamow-Teller strength of the $^{100}$Sn beta decay. *Phys. Rev. Lett.* **2019**, *122*, 222502. [CrossRef]
11. Konieczka, M.; Satuła, W.; Kortelainen, M. Gamow-Teller response in the configuration space of a density-functional-theory–rooted no-core configuration-interaction model. *Phys. Rev. C* **2018**, *97*, 034310. [CrossRef]
12. Morris, T.D.; Simonis, J.; Stroberg, S.R.; Stumpf, C.; Hagen, G.; Holt, J.D.; Jansen, G.R.; Papenbrock, T.; Roth, R.; Schwenk, A. Structure of the Lightest Tin Isotopes. *Phys. Rev. Lett.* **2018**, *120*, 152503. [CrossRef]
13. Gysbers, P.; Hagen, G.; Holt, J.D.; Jansen, G.R.; Morris, T.; Navrátil, P.; Papenbrock, T.; Quaglioni, S.; Schwenk, A.; Stroberg, S.R.; et al. Discrepancy between experimental and theoretical β-decay rates resolved from first principles. *Nat. Phys.* **2019**, *15*, 428–431. [CrossRef]
14. Mougeot, M.; Atanasov, D.; Karthein, J.; Wolf, R.N.; Ascher, P.; Blaum, K.; Chrysalidis, K.; Hagen, G.; Holt, J.D.; Huang, W.J.; et al. Mass measurements of 99–101In challenge ab initio nuclear theory of the nuclide $^{100}$Sn. *Nat. Phys.* **2021**, *17*, 1099–1103. [CrossRef]
15. Hornung, C.; Amanbayev, D.; Dedes, I.; Kripko-Koncz, G.; Miskun, I.; Shimizu, N.; Andrés, S.A.S.; Bergmann, J.; Dickel, T.; Dudek, J.; et al. Isomer studies in the vicinity of the doubly-magic nucleus $^{100}$Sn: Observation of a new low-lying isomeric state in $^{97}$Ag. *Phys. Lett. B* **2020**, *802*, 135200. [CrossRef]
16. Yang, S.; Xu, C.; Röpke, G.; Schuck, P.; Ren, Z.; Funaki, Y.; Horiuchi, H.; Tohsaki, A.; Yamada, T.; Zhou, B. α decay to a doubly magic core in the quartetting wave function approach. *Phys. Rev. C* **2020**, *101*, 024316. [CrossRef]
17. Clark, R.M.; Macchiavelli, A.O.; Crawford, H.L.; Fallon, P.; Rudolph, D.; Såmark-Roth, A.; Campbell, C.M.; Cromaz, M.; Morse, C.; Santamaria, C. Enhancement of α-particle formation near $^{100}$Sn. *Phys. Rev. C* **2020**, *101*, 034313. [CrossRef]
18. Muir, D.; Pastore, A.; Dobaczewski, J.; Barton, C. Bootstrap technique to study correlation between neutron skin thickness and the slope of symmetry energy in atomic nuclei. *Acta Phys. Pol. B* **2018**, *49*, 359. [CrossRef]
19. Blazhev, A.; Górska, M.; Grawe, H.; Nyberg, J.; Palacz, M.; Caurier, E.; Dorvaux, O.; Gadea, A.; Nowacki, F.; Andreoiu, C.; et al. Observation of a core-excited *E4* isomer in $^{98}$Cd. *Phys. Rev. C* **2004**, *69*, 064304. [CrossRef]

20. Blazhev, A.; Braun, N.; Grawe, H.; Boutachkov, P.; Singh, B.S.N.; Brock, T.; Liu, Z.; Wadsworth, R.; Gorska, M.; Jolie, J.; et al. High-energy excited states in98Cd. *J. Phys. Conf. Ser.* **2010**, *205*, 012035. [CrossRef]
21. Boutachkov, P.; Górska, M.; Grawe, H.; Blazhev, A.; Braun, N.; Brock, T.S.; Liu, Z.; Singh, B.S.N.; Wadsworth, R.; Pietri, S.; et al. High-spin isomers in $^{96}$Ag: Excitations across the Z = 38 and Z = 50, N = 50 closed shells. *Phys. Rev. C* **2011**, *84*, 044311. [CrossRef]
22. Davies, P.J.; Park, J.; Grawe, H.; Wadsworth, R.; Gernhäuser, R.; Krücken, R.; Nowacki, F.; Ahn, D.S.; Ameil, F.; Baba, H.; et al. Toward the limit of nuclear binding on the $N = Z$ line: Spectroscopy of $^{96}$Cd. *Phys. Rev. C* **2019**, *99*, 021302. [CrossRef]
23. Cederwall, B.; Moradi, F.G.; Bäck, T.; Johnson, A.; Blomqvist, J.; Clement, E.; de France, G.; Wadsworth, R.; Andgren, K.; Lagergren, K.; et al. Evidence for a spin-aligned neutron–proton paired phase from the level structure of $^{92}$Pd. *Nature* **2010**, *469*, 68–71. [CrossRef] [PubMed]
24. Zamick, L.; Escuderos, A. Single $j$-shell studies of cross-conjugate nuclei and isomerism: $(2j - 1)$ rule. *Nucl. Phys. A* **2012**, *889*, 8–17. [CrossRef]
25. Goutte, H.; Navin, A. Microscopes for the physics at the femtoscale: GANIL-SPIRAL2. *Nucl. Phys. News* **2021**, *31*, 5–12. [CrossRef]
26. Nuclear Physics Facilities—Department of Physics. Available online: https://www.jyu.fi/en/frontpage (accessed on 1 August 2021).
27. Argonne Tandem Linac Accelerator System—Argonne National Laboratory. Available online: https://www.anl.gov/ (accessed on 1 August 2021).
28. Home INFN Legnaro—National Institute for Nuclear Physics. Available online: https://www.lnl.infn.it/en/welcome-on-the-site-of-the-national-laboratories-of-legnaro/ (accessed on 1 August 2021).
29. Siciliano, M.; Valiente-Dobón, J.; Goasduff, A.; Nowacki, F.; Zuker, A.; Bazzacco, D.; Lopez-Martens, A.; Clément, E.; Benzoni, G.; Braunroth, T.; et al. Pairing-quadrupole interplay in the neutron-deficient tin nuclei: First lifetime measurements of low-lying states in $^{106,108}$Sn. *Phys. Lett. B* **2020**, *806*, 135474. [CrossRef]
30. Akkoyun, S.; Algora, A.; Alikhani, B.; Ameil, F.; de Angelis, G.; Arnold, L.; Astier, A.; Ataç, A.; Aubert, Y.; Aufranc, C.; et al. AGATA—Advanced GAmma Tracking Array. *Nucl. Instr. Meth. A* **2012**, *668*, 26. [CrossRef]
31. Clément, E.; Michelagnoli, C.; de France, G.; Li, H.J.; Lemasson, A.; Barthe Dejean, C.; Beuzard, M.; Bougault, P.; Cacitti, J.; Foucher, J.-L.; et al. Conceptual design of the AGATA $\pi$1 array at GANIL. *Nucl. Instr. Meth. A* **2017**, *855*, 1. [CrossRef]
32. Scheurer, J.; Aiche, M.; Aleonard, M.M.; Barreau, G.; Bourgine, F.; Boivin, D.; Cabaussel, D.; Chemin, J.F.; Doan, T.P.; Goudour, J.P.; et al. Improvements in the in-beam y-ray spectroscopy provided by an ancillary detector coupled to a Ge y-spectrometer: The Diamant-Eurogam II example. *Nucl. Instr. Meth. A* **1997**, *385*, 501. [CrossRef]
33. Gál, J.; Hegyesi, G.; Molnár, J.; Nyakó, B.; Kalinka, G.; Scheurer, J.; Aléonard, M.; Chemin, J.; Pedroza, J.; Juhász, K.; et al. The VXI electronics of the DIAMANT particle detector array. *Nucl. Instrum. Methods Phys. Res. Sect. A* **2004**, *516*, 502–510. [CrossRef]
34. Hüyük, T.; Di Nitto, A.; Jaworski, G.; Gadea, A.; Valiente-Dobón, J.J.; Nyberg, J.; Palacz, M.; Söderström, P.-A.; Aliaga-Varea, R.J.; De Angelis, G.; et al. Conceptual design of the early implementation of the NEutron Detector Array (NEDA) with AGATA. *Eur. Phys. J. A* **2016**, *52*, 55. [CrossRef]
35. Valiente-Dobón, J.J.; Jaworski, G.; Goasduff, A.; Egea, F.J.; Modamio, V.; Hüyük, T.; Triossi, A.; Jastrząb, M.; Söderström, P.A.; Di Nitto, A.; et al. NEDA—NEutron Detector Array. *Nucl. Instr. Meth. A* **2019**, *927*, 81. [CrossRef]
36. Pullanhiotan, S.; Chatterjee, A.; Jacquot, B.; Navin, A.; Rejmund, M. Improvement in the reconstruction method for VAMOS spectrometer. *Nucl. Instrum. Methods Phys. Res. Sect. B* **2008**, *266*, 4148–4152. [CrossRef]
37. Rejmund, M.; Lecornu, B.; Navin, A.; Schmitt, C.; Damoy, S.; Delaune, O.; Enguerrand, J.; Fremont, G.; Gangnant, P.; Gaudefroy, L.; et al. Performance of the improved larger acceptance spectrometer: VAMOS++. *Nucl. Instrum. Methods Phys. Res. Sect. A* **2011**, *646*, 184–191. [CrossRef]
38. Vandebrouck, M.; Lemasson, A.; Rejmund, M.; Fremont, G.; Pancin, J.; Navin, A.; Michelagnoli, C.; Goupil, J.; Spitaels, C.; Jacquot, B. Dual Position Sensitive MWPC for tracking reaction products at VAMOS++. *Nucl. Instrum. Methods Phys. Res. Sect. A* **2016**, *812*, 112–117. [CrossRef]
39. Mass Analysing Recoil Apparatus (MARA)—Department of Physics. Available online: https://www.jyu.fi/science/en/physics/research/infrastructures/accelerator-laboratory/nuclear-physics-facilities/recoil-separators/mara-mass-analysing-recoil-apparatus (accessed on 1 November 2021).
40. Davids, C.N.; Larson, J.D. The argonne fragment mass analyzer. *Nucl. Instr. Meth. B* **1989**, *40–41*, 1224. [CrossRef]
41. Kalandarov, S.A.; Adamian, G.G.; Antonenko, N.V.; Wieleczko, J.P. Production of the doubly magic nucleus $^{100}$Sn in fusion and quasifission reactions via light particle and cluster emission channels. *Phys. Rev. C* **2014**, *90*, 024609. [CrossRef]
42. Cederwall, B.; Liu, X.; Aktas, Ö.; Ertoprak, A.; Zhang, W.; Qi, C.; Clément, E.; de France, G.; Ralet, D.; Gadea, A.; et al. Isospin properties of nuclear pair correlations from the level structure of the self-conjugate nucleus $^{88}$Ru. *Phys. Rev. Lett.* **2020**, *124*, 062501. [CrossRef]
43. Auranen, K. Superallowed $\alpha$ decay to doubly magic $^{100}$Sn. *Phys. Rev. Lett.* **2018**, *121*, 182501. [CrossRef]
44. Seweryniak, D.; Carpenter, M.P.; Gros, S.; Hecht, A.A.; Hoteling, N.; Janssens, R.V.F.; Khoo, T.L.; Lauritsen, T.; Lister, C.J.; Lotay, G.; et al. Single-neutron states in $^{101}$Sn. *Phys. Rev. Lett.* **2007**, *99*, 022504. [CrossRef]

45. Xiao, Y.; Go, S.; Grzywacz, R.; Orlandi, R.; Andreyev, A.N.; Asai, M.; Bentley, M.A.; de Angelis, G.; Gross, C.J.; Hausladen, P.; et al. Search for $\alpha$ decay of $^{104}$Te with a novel recoil-decay scintillation detector. *Phys. Rev. C* **2019**, *100*, 034315. [CrossRef]
46. Darby, I.G.; Grzywacz, R.K.; Batchelder, J.C.; Bingham, C.R.; Cartegni, L.; Gross, C.J.; Hjorth-Jensen, M.; Joss, D.T.; Liddick, S.N.; Nazarewicz, W.; et al. Orbital dependent nucleonic pairing in the lightest known Isotopes of Tin. *Phys. Rev. Lett.* **2010**, *105*, 162502. [CrossRef] [PubMed]
47. Erler, J.; Birge, N.; Kortelainen, M.; Nazarewicz, W.; Olsen, E.; Perhac, A.M.; Stoitsov, M. The limits of the nuclear landscape. *Nature* **2012**, *486*, 509–512. [CrossRef] [PubMed]
48. Kubo, T.; Kameda, D.; Suzuki, H.; Fukuda, N.; Takeda, H.; Yanagisawa, Y.; Ohtake, M.; Kusaka, K.; Yoshida, K.; Inabe, N.; et al. BigRIPS separator and ZeroDegree spectrometer at RIKEN RI Beam Factory. *Prog. Theor. Exp. Phys.* **2012**, *2012*, 03C003. [CrossRef]
49. RIKEN. Euroball-RIKEN Cluster Array (EURICA) Project Unveiled. Available online: https://www.riken.jp/en/news_pubs/news/2012/20120326_4/index.html (accessed on 1 August 2021).
50. Takeuchi, S.; Motobayashi, T.; Togano, Y.; Matsushita, M.; Aoi, N.; Demichi, K.; Hasegawa, H.; Murakami, H. DALI2: A NaI(Tl) detector array for measurements of $\gamma$-rays from fast nuclei. *Nucl. Instr. Meth. A* **2014**, *763*, 596. [CrossRef]
51. Čeliković, I.; Lewitowicz, M.; Gernhäuser, R.; Krücken, R.; Nishimura, S.; Sakurai, H.; Ahn, D.; Baba, H.; Blank, B.; Blazhev, A.; et al. New isotopes and proton emitters–Crossing the drip line in the vicinity of $^{100}$Sn. *Phys. Rev. Lett.* **2016**, *116*, 162501. [CrossRef]
52. Park, J.; Krücken, R.; Lubos, D.; Gernhäuser, R.; Lewitowicz, M.; Nishimura, S.; Ahn, D.S.; Baba, H.; Blank, B.; Blazhev, A.; et al. New and comprehensive $\beta$- and $\beta$p-decay spectroscopy results in the vicinity of $^{100}$Sn. *Phys. Rev. C* **2019**, *99*, 034313. [CrossRef]
53. Park, J.; Krücken, R.; Blazhev, A.; Lubos, D.; Gernhäuser, R.; Lewitowicz, M.; Nishimura, S.; Ahn, D.; Baba, H.; Blank, B.; et al. Spectroscopy of $^{99}$Cd and $^{101}$In from $\beta$ decays of $^{99}$In and $^{101}$Sn. *Phys. Rev. C* **2020**, *102*, 014304. [CrossRef]
54. Davies, P.; Grawe, H.; Moschner, K.; Blazhev, A.; Wadsworth, R.; Boutachkov, P.; Ameil, F.; Yagi, A.; Baba, H.; Bäck, T.; et al. The role of core excitations in the structure and decay of the 16+ spin-gap isomer in $^{96}$Cd. *Phys. Lett. B* **2017**, *767*, 474–479. [CrossRef]
55. Häfner, G.; Moschner, K.; Blazhev, A.; Boutachkov, P.; Davies, P.J.; Wadsworth, R.; Ameil, F.; Baba, H.; Bäck, T.; Dewald, M.; et al. Properties of $\gamma$-decaying isomers in the $^{100}$Sn region populated in fragmentation of a $^{124}$Xe beam. *Phys. Rev. C* **2019**, *100*, 024302. [CrossRef]
56. Park, J.; Krücken, R.; Lubos, D.; Gernhäuser, R.; Lewitowicz, M.; Nishimura, S.; Ahn, D.S.; Baba, H.; Blank, B.; Blazhev, A.; et al. Properties of $\gamma$ -decaying isomers and isomeric ratios in the $^{100}$Sn region. *Phys. Rev. C* **2017**, *96*, 044311. [CrossRef]
57. Doornenbal, P. In-beam gamma-ray spectrometer at the RIBF. *Prog. Theor. Exp. Phys.* **2012**, *2012*, 03C004. [CrossRef]
58. Loruso, G.; Becerril, A.D.; Amthor, A.M.; Baumann, T. $\beta$-delayed proton emission in the $^{100}$Sn region. *Phys. Rev. C* **2012**, *86*, 014313. [CrossRef]
59. Cerizza, G.; Ayres, A.; Jones, K.L.; Grzywacz, R.; Bey, A.; Bingham, C.; Cartegni, L.; Miller, D.; Padgett, S.; Baugher, T.; et al. Structure of $^{107}$Sn studied through single-neutron knockout reactions. *Phys. Rev. C* **2016**, *93*, 021601. [CrossRef]
60. Hamaker, A.; Leistenschneider, E.; Jain, R.; Bollen, G.; Giuliani, S.A.; Lund, K.; Nazarewicz, W.; Neufcourt, L.; Nicoloff, C.R.; Puentes, D.; et al. Precision mass measurement of lightweight self-conjugate nucleus 80Zr. *Nat. Phys.* **2021**, *17 1408*, 1–5. [CrossRef]
61. Grawe, H.; Straub, K.; Faestermann, T.; Górska, M.; Hinke, C.; Krücken, R.; Nowacki, F.; Böhmer, M.; Boutachkov, P.; Geissel, H.; et al. The (6+) isomer in $^{102}$Sn revisited: Neutron and proton effective charges close to the double shell closure. *Phys. Lett. B* **2021**, *820*, 136591. [CrossRef]
62. Pietri, S.; Regan, P.H.; Podolyak, Z.; Rudolph, D.; Steer, S.; Garnsworthy, A.B.; Werner-Malento, E.; Hoischen, R.; Górska, M.; Gerl, J.; et al. Recent results in fragmentation isomer spectroscopy with rising. *Nucl. Instr. Met. B* **2007**, *261*, 1079. [CrossRef]
63. Geissel, G.; Armbruster, P.; Behr, K.H.; Brünle, A.; Burkard, K.; Chen, M.; Folger, H.; Franczak, B.; Keller, H.; Klepperet, O.; et al. The GSI projectile fragment separator (FRS): A versatile magnetic system for relativistic heavy ions. *Nucl. Instr. Meth. B* **1992**, *70*, 286. [CrossRef]
64. Mistry, A.K.; Albers, H.M.; Arıcı, T.; Banerjee, A.; Benzoni, G.; Cederwall, B.; Gerl, J.; Górska, M.; Hall, O.; Hubbard, N.; et al. The DESPEC setup for GSI and FAIR. *Nucl. Inst. Meth. B, in print.*
65. Technical Report for the Design, Construction and Commissioning of FATIMA, the FAst TIMing Array (2015). Available online: https://edms.cern.ch/document/1865981/1 (accessed on 1 August 2021).
66. Kester, O.; Sieber, T.; Emhofer, S.; Ames, F.; Reisinger, K.; Reiter, P.; Thirolf, P.; Lutter, R.; Habs, D.; Wolf, B.; et al. Accelerated radioactive beams from REX-ISOLDE. *Nucl. Instrum. Methods Phys. Res. Sect. B* **2002**, *204*, 20–30. [CrossRef]
67. Borge, M. Highlights of the ISOLDE facility and the HIE-ISOLDE project. *Nucl. Instrum. Methods Phys. Res. Sect. B* **2016**, *376*, 408–412. [CrossRef]

68. Eberth, J.; Pascovici, G.; Thomas, H.; Warr, N.; Weisshaar, D.; Habs, D.; Reiter, P.; Thirolf, P.; Schwalm, D.; Gund, C.; et al. MINIBALL A Ge detector array for radioactive ion beam facilities. *Prog. Part. Nucl. Phys.* **2001**, *46*, 389–398. [CrossRef]
69. Warr, N.; Van De Walle, J.; Albers, M.; Ames, F.; Bastin, B.; Bauer, C.; Bildstein, V.; Blazhev, A.; Bönig, S.; Bree, N.; et al. The Miniball spectrometer. *Eur. Phys. J. A* **2013**, *49*, 40. [CrossRef]
70. Ekström, A.; Cederkäll, J.; DiJulio, D.D.; Fahlander, C.; Hjorth-Jensen, M.; Blazhev, A.; Bruyneel, B.; Butler, P.A.; Davinson, T.; Eberth, J.; et al. Electric quadrupole moments of the 21+states in Cd100,102,104. *Phys. Rev. C* **2009**, *80*, 054302. [CrossRef]
71. Ekstrom, A.; Cederkall, J.; Fahlander, C.; Hjorth-Jensen, M.; Engeland, T.; Blazhev, A.; Butler, P.A.; Davinson, T.; Eberth, J.; Finke, F.; et al. Coulomb excitation of the odd-odd isotopes $^{106,108}$In. *Eur. Phys. J. A* **2010**, *44*, 355–361. [CrossRef]
72. DiJulio, D.D.; Cederkall, J.; Fahlander, C.; Ekstrom, A.; Hjorth-Jensen, M.; Albers, M.; Bildstein, V.; Blazhev, A.; Darby, I.; Davinson, T.; et al. Excitation strengths in $^{109}$Sn: Single-neutron and collective excitations near $^{100}$Sn. *Phys. Rev. C* **2012**, *86*, 031302. [CrossRef]
73. DiJulio, D.D.; Cederkall, J.; Fahlander, C.; Ekstrom, A.; Hjorth-Jensen, M.; Albers, M.; Bildstein, V.; Blazhev, A.; Darby, I.; Davinson, T.; et al. Coulomb excitation of $^{107}$Sn. *Eur. Phys. J. A* **2012**, *48*, 105. [CrossRef]
74. DiJulio, D.; Cederkall, J.; Fahlander, C.; Ekström, A. Coulomb excitation of $^{107}$In. *Phys. Rev. C* **2013**, *87*, 017301. [CrossRef]
75. Lunderberg, E.; Belarge, J.; Bender, P.; Bucher, B.; Cline, D.; Elman, B.; Gade, A.; Liddick, S.; Longfellow, B.; Prokop, C.; et al. JANUS—A setup for low-energy Coulomb excitation at ReA3. *Nucl. Instrum. Methods Phys. Res. Sect. A* **2018**, *885*, 30–37. [CrossRef]
76. Rhodes, D.; Brown, B.A.; Henderson, J.; Gade, A.; Ash, J.; Bender, P.C.; Elder, R.; Elman, B.; Grinder, M.; Hjorth-Jensen, M.; et al. Exploring the role of high-$j$ configurations in collective observables through the Coulomb excitation of $^{106}$Cd. *Phys. Rev. C* **2021**, *103*, L051301. [CrossRef]
77. Togashi, T.; Tsunoda, Y.; Otsuka, T.; Shimizu, N.; Honma, M. Novel shape evolution in Sn isotopes from magic numbers 50 to 82. *Phys. Rev. Lett.* **2018**, *121*, 062501. [CrossRef] [PubMed]
78. Honma, M.; Otsuka, T.; Mizusaki, T.; Hjorth-Jensen, M. New effective interaction for $f_5pg_9$-shell nuclei. *Phys. Rev. C* **2009**, *80*, 064323. [CrossRef]
79. Zuker, A.P.; Poves, A.; Nowacki, F.; Lenzi, S.M. Nilsson-SU3 self-consistency in heavy N = Z nuclei. *Phys. Rev. C* **2015**, *92*, 024320. [CrossRef]
80. Gargano, A.; Coraggio, L.; Covello, A.; Itaco, N. Investigating neutron-deficient tin isotopes via realistic shell-model calculations. *AIP Conf. Proc.* **2015**, *1681*, 020007.
81. Coraggio, L.; Gargano, A.; Itaco, N. Double-step truncation procedure for large-scale shell-model calculations. *Phys. Rev. C* **2016**, *93*, 064328. [CrossRef]
82. Zucker, A. Quadrupole dominance in the light Sn and in the Cd isotope. *Phys. Rev. C* **2021**, *103*, 024322. [CrossRef]
83. Coraggio, L.; Itaco, N. Perturbative approach to effective shell-model hamiltonians and operators. *Front. Phys.* **2020**, *8*, 345. [CrossRef]
84. Gargano, A.; Coraggio, L.; Itaco, N. Effectively-truncated large-scale shell-model calculations and nuclei around $^{100}$Sn. *Phys. Scr.* **2017**, *92*, 094003. [CrossRef]
85. Gross, R.; Frenkel, A. Effective interaction of protons and neutrons in the $2p_{12}$-$1g_{92}$ subshells. *Nucl. Phys. A* **1976**, *267*, 85. [CrossRef]
86. Serduke, F.J.D.; Lawson, R.D.; Gloeckner, D.H. Shell-modelstudy of the N = 49 isotones. *Nucl. Phys. A* **1976**, *256*, 45. [CrossRef]
87. Brown, B.A.; Rae, W.D.M. The Shell-Model Code NuShellX@MSU. *Nucl. Data Sheets* **2014**, *120*, 115. [CrossRef]
88. Hjorth-Jensen, M.; Kuo, T.T.; Osnes, E. Realistic effective interactions for nuclear systems. *Phys. Rep.* **1995**, *261*, 125–270. [CrossRef]
89. Grawe, H.; Blazhev, A.; Górska, M. Correspondence Affiliation: GSI Darmstadt, Darmstadt, Germany. Empirically-modified realistic intereraction for the $^{100}$Sn region. **2021**. to be published.
90. Yordanov, D.T.; Balabanski, D.L.; Bissell, M.L.; Blaum, K.; Blazhev, A.; Budinčević, I.; Frömmgen, N.; Geppert, C.; Grawe, H.; Hammen, M.; et al. Spins and electromagnetic moments of $^{101-109}$Cd. *Phys. Rev. C* **2018**, *98*, 011303. [CrossRef]
91. Coombes, B.J.; Stuchbery, A.E.; Blazhev, A.; Grawe, H.; Reed, M.W.; Akber, A.; Dowie, J.T.H.; Gerathy, M.S.M.; Gray, T.J.; Kibedi, T.; et al. Spectroscopy and excited-state g factors in weakly collective $^{111}$Cd: Confronting collective and micro-scopic models. *Phys. Rev. C* **2019**, *100*, 024322. [CrossRef]
92. Nowacki, F. (Institut Pluridisciplinaire Hubert Curien, Strasbourg, France). Updated calculations with t=5. Private communication, 2021.
93. Arnswald, K.; Blazhev, A.; Nowacki, F.; Petkov, P.; Reiter, P.; Braunroth, T.; Dewald, A.; Droste, M.; Fransen, C.; Hirsch, R.; et al. Enhanced quadrupole collectivity in doubly-magic $^{56}$Ni: Lifetime measurements of the $4_1{}^+$ and $6_1{}^+$ states. *Phys. Lett. B* **2021**, *820*, 136592. [CrossRef]

94. Derbali, E.; van Isacker, P.; Tellili, B.; Souga, C. Effective operators in a single-*j* orbital. *J. Phys. G Nucl. Part. Phys.* **2018**, *45*, 035102. [CrossRef]
95. Romero, A.; Dobaczewski, J.; Pastore, A. Symmetry restoration in the mean-field description of proton-neutron pairing. *Phys. Lett. B* **2019**, *795*, 177–182. [CrossRef]
96. Idini, A.; Potel, G.; Barranco, F.; Vigezzi, E.; Broglia, R.A. Interweaving of elementary modes of excitation in superfluid nuclei through particle-vibration coupling: Quantitative account of the variety of nuclear structure observables. *Phys. Rev. C* **2015**, *92*, 031304. [CrossRef]
97. Nomura, K.; Jolie, J. Structure of even-even cadmium isotopes from the beyond-mean-field interacting boson model. *Phys. Rev. C* **2018**, *98*, 024303. [CrossRef]
98. Sun, Z.H.; Hagen, G.; Jansen, G.R.; Papenbrock, T. Effective shell-model interaction for nuclei "southeast" of $^{100}$Sn. *Phys. Rev. C* **2021**, *104*, 064310. [CrossRef]
99. De-Shalit, A.; Talmi, I.; Wigner, E.P. Review of "Nuclear Shell Theory". In *Historical and Biographycal Reflections and Syntheses*; Mehra, J., Ed.; Springer: Berlin/Heidelberg, Germany, 2001; pp. 494–495. [CrossRef]
100. Mach, H.; Korgul, A.; Górska, M.; Grawe, H. Ultrafast-timing lifetime measurements in $^{94}$Ru and $^{96}$Pd: Breakdown of the seniority scheme in N = 50 iso-tones. *Phys. Rev. C* **2017**, *95*, 014313. [CrossRef]
101. National Nuclear Data Center, NuDat 3. Available online: https://www.bnl.gov/world/ (accessed on 1 August 2021).
102. Jin, S.Y.; Wang, S.T.; Lee, J.; Corsi, A.; Wimmer, K.; Browne, F.; Chen, S.; Cortés, M.L.; Doornenbal, P.; Koiwai, T.; et al. Spectroscopy of $^{98}$Cd by two-nucleon removal from $^{100}$In. *Phys. Rev. C* **2021**, *104*, 024302. [CrossRef]
103. Qi, C. Partial conservation of seniority and its unexpected influence on E2 transitions in $g_{9/2}$ nuclei. *Phys. Lett. B* **2017**, *773*, 616. [CrossRef]
104. Pérez-Vidal, R.M. Collectivity along *N*=50: Nuclear Structure studies on the neutron-magic nuclei $^{92}$Mo and $^{94}$Ru with AGATA and VAMOS++. Ph.D. Thesis, Universidad de Valencia, València, Spain, 2019. Available online: https://roderic.uv.es/handle/10550/72450 (accessed on 1 November 2021).
105. Das, B.; Cederwall, B.; Qi, C.; Górska, M.; Regan, P.H.; Aktas, Ö.; Albers, H.M.; Banerjee, A.; Chishti, M.M.R.; Gerl, J.; et al. Nature of seniority symmetry breaking in the semimagic nucleus $^{94}$Ru. *Phys. Rev. C, in print.*
106. Lorusso, G.; Becerril, A.; Amthor, A.; Baumann, T.; Bazin, D.; Berryman, J.; Brown, B.; Cyburt, R.; Crawford, H.; Estrade, A.; et al. Half-lives of ground and isomeric states in $^{97}$Cd and the astrophysical origin of $^{96}$Ru. *Phys. Lett. B* **2011**, *699*, 141–144. [CrossRef]
107. Robinson, S.J.Q.; Hoang, T.; Zamick, L.; Escuderos, A.; Sharon, Y.Y. Shell model calculations of *B(E2)* values, static quadrupole moments, and *g* factors for a number of *N* = *Z* nuclei. *Phys. Rev. C* **2014**, *89*, 014316. [CrossRef]
108. Kingan, A.; Zamick, L. Odd-J states of isospin zero and one for 4-nucleon systems: Near-degeneracies. *Phys. Rev. C* **2018**, *98*, 014301. [CrossRef]
109. Fu, G.J.; Cheng, Y.Y.; Zhao, Y.M.; Arima, A. Shell model study of $T = 0$ states for $^{96}$Cd by the nucleon-pair approximation. *Phys. Rev. C* **2016**, *94*, 024336. [CrossRef]
110. Nara Singh, B.S.; Liu, Z.; Wadsworth, R.; Grawe, H.; Brock, T.S.; Boutachkov, P.; Braun, N.; Blazhev, A.; Górska, M.; Pietri, S.; et al. 16+ spin-gap isomer in $^{96}$Cd. *Phys. Rev. Lett.* **2011**, *107*, 172502. [CrossRef]
111. Zerguine, S.; Van Isacker, P. Spin-aligned neutron-proton pairs in *N* = *Z* nuclei. *Phys. Rev. C* **2011**, *83*, 064314. [CrossRef]
112. Xu, Z.; Qi, C.; Blomqvist, J.; Liotta, R.; Wyss, R. Multistep shell model description of spin-aligned neutron–proton pair coupling. *Nucl. Phys. A* **2012**, *877*, 51–58. [CrossRef]
113. Qi, C.; Blomqvist, J.; Bäck, T.; Cederwall, B.; Johnson, A.; Liotta, R.J.; Wyss, R. Spin-aligned neutron-proton pair mode in atomic nuclei. *Phys. Rev. C* **2011**, *84*, 021301. [CrossRef]
114. Van Isacker, P. Neutron–proton pairs in nuclei. *Int. J. Mod. Phys.* **2013**, *E22*, 1330028. [CrossRef]
115. Marginean, N.; Bucurescu, D.; Alvarez, C.R.; Ur, C.A. Delayed alignments in the *N* = *Z* nuclei $^{84}$Mo and $^{88}$Ru. *Phys. Rev. C* **2002**, *65*, 051303. [CrossRef]
116. Sun, Y. Projected shell model study on nuclei near the *N* = *Z* line. *Eur. Phys. J. A* **2003**, *20*, 133–138. [CrossRef]
117. Sun, Y.; Sheikh, J. Anomalous rotational alignment in *N* = *Z* nuclei and residual neutron-proton interaction. *Phys. Rev. C* **2001**, *64*, 162031302. [CrossRef]
118. Kaneko, Y.S.K.; de Angelis, G. Enhancement of high-spin collectivity in *N* = *Z* nuclei by the isoscalar neutron–proton pairing. *Nucl. Phys. A* **2017**, *957*, 144. [CrossRef]
119. Kaneko, Y.S.K.; Mizusaki, T.; Sun, Y.; Tazaki, S. Toward a unified realistic shell-model Hamiltonian with the mono-pole-based universal force. *Phys. Rev. C* **2014**, *89*, 011302. [CrossRef]
120. Fu, G.J.; Johnson, C.W. Nucleon-pair approximation for nuclei from spherical to deformed regions. *Phys. Rev. C* **2021**, *104*, 024312. [CrossRef]
121. Corsi, A.; Obertelli, A.; Doornenbal, P.; Nowacki, F.; Sagawa, H.; Tanimura, Y.; Aoi, N.; Baba, H.; Bednarczyk, P.; Boissinot, S.; et al. Spectroscopy of nuclei around $^{100}$Sn populated via two-neutron knockout reactions. *Phys. Rev. C* **2018**, *97*, 044321. [CrossRef]

122. Banu, A.; Gerl, J.; Fahlander, C.; Górska, M.; Grawe, H.; Saito, T.R.; Wollersheim, H.-J.; Caurier, E.; Engeland, T.; Gniady, A.; et al. $^{108}$Sn studied with intermediate-energy Coulomb excitation. *Phys. Rev. C* **2005**, *72*, 061305. [CrossRef]
123. Maheshwari, B. A unified view of the first-excited $2^+$ and $3^-$ states of Cd, Sn and Te isotopes. *Eur. Phys. J. Spéc. Top.* **2020**, *229*, 2485–2495. [CrossRef]
124. Guastalla, G.; DiJulio, D.D.; Górska, M.; Cederkall, J.; Boutachkov, P.; Golubev, P.; Pietri, S.; Grawe, H.; Nowacki, F.; Sieja, K.; et al. Coulomb Excitation of $^{104}$Sn and the Strength of the $^{100}$Sn Shell Closure. *Phys. Rev. Lett.* **2013**, *110*, 172501. [CrossRef]
125. Jonsson, N.-G.; Bäcklin, A.; Kantele, J.; Julin, R.; Luontama, M.; Passoja, A. Collective states in even Sn nuclei. *Nucl. Phys. A* **1981**, *371*, 333–348. [CrossRef]
126. Vaman, C.; Andreoiu, C.; Bazin, D.; Becerril, A.; Brown, B.A.; Campbell, C.M.; Chester, A.; Cook, J.M.; Dinca, D.C.; Gade, A.; et al. Z = 50 shell gap near $^{100}$Sn from intermediate-energy Coulomb excitations in even-mass $^{106–112}$Sn isotopes. *Phys. Rev. Lett.* **2007**, *99*, 162501. [CrossRef]
127. Cederkall, J.; Ekstrom, A.; Fahlander, C.; Hurst, A.M.; Hjorth-Jensen, M.; Ames, F.; Banu, A.; Butler, P.A.; Davinson, T.; Pramanik, U.D.; et al. Sub-barrier Coulomb excitation of $^{110}$Sn and its implications for the $^{100}$Sn shell closure. *Phys. Rev. Lett.* **2007**, *98*, 172501. [CrossRef]
128. Ekström, A.; Cederkäll, J.; Fahlander, C.; Hjorth-Jensen, M.; Ames, F.; Butler, P.A.; Davinson, T.; Eberth, J.; Fincke, F.; Görgen, A.; et al. $0^+_{gs} \rightarrow 2^+_1$ transition strengths in $^{106}$Sn and $^{108}$Sn. *Phys. Rev. Lett.* **2008**, *101*, 012502. [CrossRef]
129. Orce, J.N.; Choudry, S.N.; Crider, B.; Elhami, E.; Mukhopadhyay, S.; Scheck, M.; McEllistrem, M.T.; Yates, S.W. $2^+_1 \rightarrow 0^+_1$ transition strengths in Sn nuclei. *Phys. Rev. C* **2007**, *76*, 021302. [CrossRef]
130. Bader, V.M.; Gade, A.; Weisshaar, D.; Brown, B.A.; Baugher, T.; Bazin, D.; Berryman, J.S.; Ekstrom, A.; Hjorth-Jensen, M.; Stroberg, R.; et al. Quadrupole collectivity in neutron-deficient Sn nuclei: $^{104}$Sn and the role of proton excitations. *Phys. Rev. C* **2013**, *88*, 051301. [CrossRef]
131. Doornenbal, P.; Reiter, P.; Grawe, H.; Wollersheim, H.J.; Bednarczyk, P.; Caceres, L.; Cederkäll, J.; Ekström, A.; Gerl, J.; Górska, M.; et al. Enhanced strength of the $2^+_1 \rightarrow 0^+_{gs}$ transition in $^{114}$Sn studied via Coulomb excitation in inverse kinematics. *Phys. Rev. C* **2008**, *78*, 031303. [CrossRef]
132. Kumar, R.; Doornenbal, P.; Jhingan, A.; Bhowmik, R.K.; Muralithar, S.; Appannababu, S.; Garg, R.; Gerl, J.; Górska, M.; Kaur, J.; et al. Enhanced $0^+_{gs} \rightarrow 2^+_1$ $E2$ transition strength in $^{112}$Sn. *Phys. Rev. C* **2010**, *81*, 024306. [CrossRef]
133. Kumar, R.; Saxena, M.; Doornenbal, P.; Jhingan, A.; Banerjee, A.; Bhowmik, R.K.; Dutt, S.; Garg, R.; Joshi, C.; Mishra, V.; et al. No evidence of reduced collectivity in Coulomb-excited Sn isotopes. *Phys. Rev. C* **2017**, *96*, 054318. [CrossRef]
134. Allmond, J.M.; Radford, D.C.; Baktash, C.; Batchelder, J.C.; Galindo-Uribarri, A.; Gross, C.J.; Hausladen, P.; Lagergren, K.; LaRochelle, Y.; Padilla-Rodal, E.; et al. Coulomb excitation of $^{124,126,128}$Sn. *Phys. Rev. C* **2011**, *84*, 061303. [CrossRef]
135. Allmond, J.M.; Stuchbery, A.E.; Galindo-Uribarri, A.; Padilla-Rodal, E.; Radford, D.C.; Batchelder, J.C.; Bingham, C.R.; Howard, M.E.; Liang, J.F.; Manning, B.; et al. Investigation into the semimagic nature of the tin isotopes through electromagnetic moments. *Phys. Rev. C* **2015**, *92*, 041303. [CrossRef]
136. Jungclaus, A.; Walker, J.; Leske, J.; Speidel, K.-H.; Stuchbery, A.; East, M.; Boutachkov, P.; Cederkall, J.; Doornenbal, P.; Egido, J.L.; et al. Evidence for reduced collectivity around the neutron mid-shell in the stable even-mass Sn isotopes from new lifetime measurements. *Phys. Lett. B* **2011**, *695*, 110–114. [CrossRef]
137. Kumbartzki, G.J. Transient field g factor and mean-life measurements with a rare isotope beam of $^{126}$Sn. *Phys. Rev. C* **2012**, *86*, 034319. [CrossRef]
138. Kumbartzki, G.J.; Benczer-Koller, N.; Speidel, K.-H.; Torres, D.A.; Allmond, J.M.; Fallon, P.; Abramovic, I.; Bernstein, L.A.; Bevins, J.E.; Crawford, H.; et al. Z = 50 core stability in $^{110}$Sn from magnetic-moment and lifetime measurements. *Phys. Rev. C* **2016**, *93*, 044316. [CrossRef]
139. Doornenbal, P.; Takeuchi, S.; Aoi, N.; Matsushita, M.; Obertelli, A.; Steppenbeck, D.; Wang, H.; Audirac, L.; Baba, H.; Bednarczyk, P.; et al. Intermediate-energy Coulomb excitation of $^{104}$Sn: Moderate $E2$ strength decrease approaching $^{100}$Sn. *Phys. Rev. C* **2014**, *90*, 061302. [CrossRef]
140. Radford, D.C.; Baktash, C.; Barton, C.J.; Batchelder, J.; Beene, J.R.; Bingham, C.R.; Caprio, M.A.; Danchev, M.; Fuentes, B.; Galindo-Uribarri, A.; et al. Coulomb excitation and transfer reactions with neutron-rich radioactive beams. *Int. Conf. Exot. Nucl. At. Masses* **2005**, *746*, 383–387. [CrossRef]
141. Radford, D.; Baktash, C.; Barton, C.; Batchelder, J.; Beene, J.; Bingham, C.; Caprio, M.; Danchev, M.; Fuentes, B.; Galindo-Uribarri, A.; et al. Coulomb excitation and transfer reactions with rare neutron-rich isotopes. *Nucl. Phys. A* **2005**, *752*, 264–272. [CrossRef]
142. Spieker, M.; Petkov, P.; Litvinova, E.; Müller-Gatermann, C. Shape coexistence and collective low-spin states in $^{112,114}$Sn studied with the (p, p′γ ) Doppler-shift at-tenuation coincidence technique. *Phys. Rev. C* **2018**, *97*, 054319. [CrossRef]
143. Berger, M. Kernstrukturuntersuchungen bis zur Teilchenseparationsschwelle mit der Methode der Kernresonanzfluoreszen. Ph.D. Thesis, Technical University of Darmstadt, Darmstadt, Germany, 2020. [CrossRef]
144. Rosiak, D.; Seidlitz, M.; Reiter, P.; Naïdja, H.; Tsunoda, Y.; Togashi, T.; Nowacki, F.; Otsuka, T.; Colò, G.; Arnswald, K.; et al. Enhanced quadrupole and octupole strength in doubly magic $^{132}$Sn. *Phys. Rev. Lett.* **2018**, *121*, 252501. [CrossRef]
145. Coraggio, L.; Covello, A.; Gargano, A.; Itaco, N.; Kuo, T.T.S. Shell-model study of quadrupole collectivity in light tin isotopes. *Phys. Rev. C* **2015**, *91*, 041301. [CrossRef]
146. Cortes, L.M. (INFN Legnro, Italy). Data analysis of the Coulomb excitations of $^{102}$Sn experiment. Private communication, 2021.

147. Wimmer, K. (GSI Darmstadt, Germany). Review submitted to Prog. Part. Nucl. Phys. Private communication, 2021.
148. Siciliano, M.; Valiente-Dobón, J.J.; Goasduff, A.; Rodríguez, T.R.; Bazzacco, D.; Benzoni, G.; Braunroth, T.; Cieplicka-Oryńczak, N.; Clément, E.; Crespi, F.C.L.; et al. Lifetime measurements in the even-even $^{102-108}$Cd isotopes. *Phys. Rev. C* **2021**, *104*, 034320. [CrossRef]
149. Górska, M.; Lipoglavsek, M.; Grawe, H.; Nyberg, J.; Atac, A.; Axelsson, A.; Bark, R.; Blomqvist, J.; Cederkäll, J.; Cederwall, B.; et al. $^{98}{}_{48}Cd_{50}$: The two-proton-hole spectrum in $^{100}{}_{50}Sn_{50}$. *Phys. Rev. Lett.* **1997**, *79*, 2415. [CrossRef]

MDPI

Review

# Excited States in Isobaric Multiplets—Experimental Advances and the Shell-Model Approach

**Michael A Bentley**

School of Physics, Engineering and Technology, University of York, York YO10 5DD, UK; michael.bentley@york.ac.uk

**Abstract:** A review of recent advances in the study of the energy splitting between excited isobaric analogue states is presented. Some of the experimental developments, and new approaches, associated with spectroscopy of the most proton-rich members of isobaric multiplets, are discussed. The review focuses on the immense impact of the shell-model in the analysis of energy differences and their interpretation in terms of nuclear structure phenomena.

**Keywords:** isospin symmetry; nuclear shell model; charge symmetry; charge independence; $\gamma$-ray spectroscopy; knockout reactions

**Citation:** Bentley, M.A. Excited States in Isobaric Multiplets—Experimental Advances and the Shell-Model Approach. *Physics* **2022**, *4*, 995–1011. https://doi.org/10.3390/physics4030066

Received: 12 July 2022
Accepted: 4 August 2022
Published: 5 September 2022

**Publisher's Note:** MDPI stays neutral with regard to jurisdictional claims in published maps and institutional affiliations.

**Copyright:** © 2022 by the author. Licensee MDPI, Basel, Switzerland. This article is an open access article distributed under the terms and conditions of the Creative Commons Attribution (CC BY) license (https://creativecommons.org/licenses/by/4.0/).

## 1. Introduction

The approximate charge symmetry and charge independence of the nucleon-nucleon (NN) interaction [1] results in elegant symmetries in the behaviour of the otherwise exceptionally complex nuclear system. Examining and exploiting these isospin-related symmetries, and determining the extent to which they are broken, has become a rich field of nuclear structure physics over the last 30 years. When the symmetries are slightly broken, this provides an opportunity to observe nuclear behaviour through the lens of the well-understood electromagnetic interaction, providing a probe of nuclear structure phenomena such as pairing, particle alignments, shape changes and radii. It may even be possible to learn about the charge-dependent components of the nuclear interaction itself. Moreover, it is possible to exploit the often near-perfect isospin symmetry between pairs of analogue states to extract information other phenomena; in this review such an example is provided in the study of neutron skins.

Wigner's isospin concept [2] provided the conceptual and mathematical foundation for describing these symmetries. All states are assigned an isospin,**T**, quantum number, $T$, with a projection defined by $T_z = \sum_i t_z(i) = (N - Z)/2$, where $N$ denotes the number of neutrons and $Z$ the number of protons in a nucleus. In this formalism, the nucleon is treated as two states of the same particle with quantum number $t$ and projection $t_z = \mp\frac{1}{2}$ for the proton/neutron respectively. With the concept of isospin established, we now have a powerful isospin classification scheme, which enables us to map out, in isospin space, the resulting symmetries—visualised in Figure 1. Crucially, the mathematical formalism of isospin enables the treatment of the two types of fermion in the same system, allowing predictions based on the assumption of pure isospin symmetry, and the tools to model the observed deviations from that symmetry.

This short review focuses on the energy differences between excited isobaric analogue states—i.e., analogue states of the same isospin $T$ in different members of an isobaric multiplet (different $T_z$). With perfect isospin symmetry, and in the absence of electromagnetic effects, the excitation energies would be identical. In reality the electromagnetic effects, and any other isospin-non conserving interactions, such as charge-dependent nuclear forces, lift the degeneracy. The study, and modelling, of these differences is discussed here. Two types of energy difference are usually measured: mirror energy differences (MED)

for mirror nuclei ($T_z = \pm T$) or triplet energy differences (TED) for isobaric triplets ($T = 1$, $T_z = 0, \pm 1$). MED and TED, the differences in excitation energy, $E^*$, are defined by:

$$\text{MED}_{J,T} = E^*_{J,T,T_z=-T} - E^*_{J,T,T_z=T}, \text{ and} \tag{1}$$

$$\text{TED}_{J,T=1} = E^*_{J,T_z=-1} + E^*_{J,T_z=1} - 2E^*_{J,T_z=0}, \tag{2}$$

respectively, with $J$ the total angular momentum quantum number.

Developments of experimental technique, especially in the $\gamma$-ray spectroscopy of excited states in proton-rich nuclei, have led to a wealth of new data in recent years, allowing for experimental measurements of MED, e.g., [3–11], and TED, e.g., [12–15]. It is, however, the interpretation of these observations through shell-model analysis that has energised this field of study (e.g., [16–21]). This has allowed detailed nuclear structure phenomena, and especially their evolution with angular momentum and excitation along the yrast line, to be investigated in detail. This review outlines some experimental advances in Section 2 including specific case studies. The shell-model approach is outlined in Section 3 and some recent advances, made through shell-model interpretation, are discussed in Section 4.

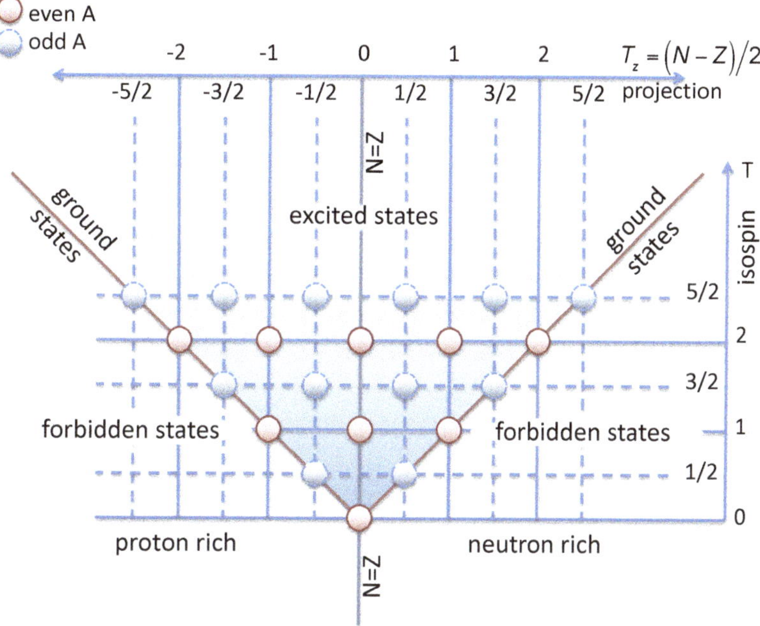


**Figure 1.** A schematic visualisation of the classification of nuclear states according to the total isospin quantum numbers $T, T_Z$. Each circle represents a set of states, of given isospin, which are allowed by the Pauli principle. Note that the diagram assumes that the lowest-energy set of states in any nucleus have the lowest allowed value of isospin. This is usually, but not always, true, e.g., odd-odd $N = Z$ nuclei (equal and odd numbers of neutrons, $N$, and protons, $Z$).

## 2. Advances in Experimental Techniques and Selected Case Studies

The key challenge, in experimental measurements of energy differences between excited states of isobaric multiplets, is the typically low cross sections for, or low production rates of, the required proton-rich (i.e., $Z \geq N$, $T_z \leq 0$) nuclei. Two reaction mechanisms are generally employed: fusion-evaporation reactions with stable beams at near Coulomb-barrier energies and knockout reactions from relativistic radioactive beams. For fusion-evaporation reactions, the major difficultly is the low production cross section

of the neutron-evaporation channels that lead to the required proton-rich systems, leading to cross sections often less than 1 μb—i.e., representing a fraction of $<1 \times 10^{-6}$ of the total reaction cross section. The experimental challenge is therefore the clean selection of the reaction channel to remove the huge background from proton-emission channels. In the second method, knockout from fast radioactive beams, the knockout cross sections are reasonable (∼few mb) and the identification of the desired proton-rich fragment is straightforwardly achieved with post-target magnetic spectrometers. However, here the experimental challenge comes from the potentially low secondary beam rates and from performing high-resolution $\gamma$-ray spectroscopy at high beam velocities, $v$ ($v/c \sim$0.35–0.55, where $c$ is the speed of light), with the associated Doppler-broadening issues. Since the last reviews of this topic, e.g., [3,4], progress has been made in addressing these two sets of challenges, which have in turn led to advances in our understanding of MED and TED.

In the following Sections 2.1–2.3, three example cases studies are presented which highlight the recent experimental advances. The impact of these case studies on our shell-model based interpretation of isospin-symmetry breaking, in mirror nuclei and $T = 1$ isobaric triplets, is discussed in Section 4.

### 2.1. Prompt Tagging of Fusion-Evaporation Channels and a Case Study: The A = 23, $T_z = \pm\frac{1}{2}$ Mirror Nuclei

In fusion-evaporation reactions, the required proton-rich nuclei are populated with low cross sections and, following the evaporation of at least one prompt neutron, often at the same time as evaporated charged particles. One method of selection of the desired reaction channel is to surround the target with high-efficiency neutron- and charged-particle detectors, in addition to the high-resolution $\gamma$-ray array. As a case study, we use the example of the mass number $A = 23$, $T_z = \pm\frac{1}{2}$ mirror nuclei $^{23}$Mg/$^{23}$Na [11]. This mirror pair was studied at GANIL (Grand Accélérateur National d'Ions Lourds), Caen, France, using an $^{16}$O beam on $^{12}$C target with the nuclei of interest populated through the $\alpha, n$ and $\alpha, p$ reaction channels, respectively. The prompt $\gamma$ rays were detected with the EXOGAM array [22]. The prompt evaporated neutrons were detected with the Neutron Wall [23], an array of 50 liquid scintillator detectors. The proton and alpha particles were detected with DIAMANT [24], an array of 80 CsI scintillators. These highly-efficient detectors enabled a clean channel selection through the full identification of all emitted particles, allowing for the event-by-event tagging of the $\gamma$ rays from the nuclei of interest. In this case study, the cleanliness of the channel selection allowed for the confident assignment of states in proton-rich $^{23}$Mg up to angular momentum/parity of $J^\pi = \frac{15}{2}^+$, through a $\gamma$–$\gamma$ coincidence analysis and using comparisons with the mirror nucleus, on which an identical analysis was performed. The identification of these states enabled MED to be determined up to high spin, and this proved crucial in the subsequent shell-model analysis. The impact of this measurement, and of the resulting shell-model analysis, connected to radii and neutron skins, is discussed in Section 4.2.

For the study of heavier proton-rich or $N = Z$ nuclei, and especially where $N = Z$ beam/target combinations are not possible, prompt particle tagging of the nuclei of interest becomes more challenging due to very low production cross sections and the need to identify more than one evaporated neutron. The development of more efficient, highly modular, neutron detector arrays such as NEDA [25], coupled to the improvements in high-resolution and high-efficiency $\gamma$-ray measurement afforded by the AGATA $\gamma$-ray array [26], provide exciting possibilities (e.g., [27]). The recent in-beam spectroscopy of $^{88}$Ru [28] through a $2n$ evaporation channel, using the AGATA, DIAMANT, Neutron Wall and NEDA arrays, provides a characteristic example.

### 2.2. Decay Tagging of Fusion-Evaporation Channels and a Case Study: The $T_z = -1$ Nucleus $^{66}$Se

Instead of tagging the prompt emitted $\gamma$ rays by the prompt evaporated particles, an alternative approach to select the low cross-section neutron-evaporation channel of interest is to tag the $\gamma$ rays by the ground-state decay emissions characteristic of the nucleus

of interest. For spectroscopy of $Z \geq N$ nuclei, a highly effective technique is recoil-beta tagging (RBT) [29,30], which takes advantage of cases where the ground state of the nucleus of interest $(N, Z)$ $\beta$-decays to its isobaric analogue state in the $N+1$, $Z-1$ neighbour. Such decays are characterised by fast, superallowed, $\beta$-decays, with high $\beta$ end-point energy.

In the RBT approach, outlined in Figure 2, a triggerless data acquisition system is used to enable temporal correlation between prompt $\gamma$-ray emission at the target and the subsequent decays of the residual nuclear ground state. The recoiling nuclei are separated using a magnetic spectrometer and implanted in a highly pixellated double-sided silicon strip detector (DSSSD). The subsequent decay of the ground state of the implanted nucleus is detected in the same position as the implantation within the DSSSD and a second detector (a planar Ge detector or plastic scintillator) is used to measure the remaining energy of the $\beta$-decay A correlation in time of the three events (prompt emission, implantation and $\beta$-decay) and in position using the pixellated DSSSD, allows the selection of the proton-rich nucleus when a short correlation time (few 10 s of ms, typically) is required as well as a high-energy $\beta$-decay

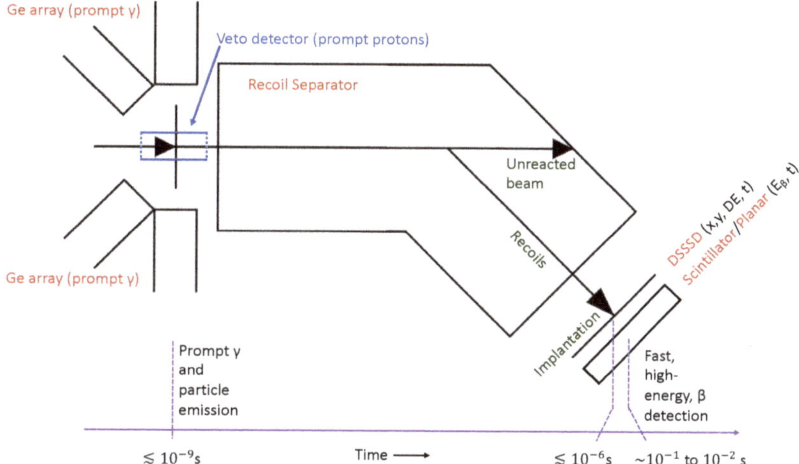


**Figure 2.** A schematic diagram summarising the recoil-beta-tagging technique, [29] used for identifying prompt $\gamma$ decays, emitted from proton-rich nuclei through tagging with the characteristic superallowed $\beta$-decay of the residue ground state. See text for details.

The example of spectroscopy of $T_z = -1$ $^{66}$Se [12] is chosen as the case study for this technique. The experiment was performed at the at the University of Jyväskylä (JYFL) using the JUROGAMII $\gamma$-ray array and the RITU gas filled separator [31,32], in which $^{66}$Se was populated through a $2n$ evaporation channel. The fusion products were implanted in the DSSSD, which was followed by a planar Ge detector for detection of the high-energy positrons from the fast superallowed $\beta$-decay. A key component of this experiment was the inclusion of a high-efficiency veto detector to measure prompt charged particles—the UoYTube [33] detector. This is essential to help identify, and remove, contamination in the final spectrum coming from reaction channels with evaporation of one or more charged particles. The resulting clean spectrum identified decays from states with $J^{\pi} = 2^+, 4^+$ and $6^+$, which in turn enabled the completion of the full set of $T = 1$ isobaric analogue states up to $6^+$ in the $A = 66$ $T = 1$ triplet, allowing the TED to be extracted. The impact of this result on the understanding of isotensor isospin non-conserving interactions, within the shell model description of TED, is discussed in Section 4.1.

Since the work on $^{66}$Se, the same RBT approach, including charged-particle vetoing, has been applied successfully at JYFL to identify the excited states in the $T_z = -1$ nuclei $^{70}$Kr [13] and $^{74}$Sr [15], and a programme using the same methodology is underway

using the new MARA spectrometer [34], which additionally allows for mass selection and identification.

### 2.3. *Knockout Reactions at Intermediate Energies and a Case Study: The $T_z = -2$ Nuclei $^{48}Fe$ and $^{56}Zn$*

Spectroscopy of the most proton-rich systems (i.e., $T_z \leq -\frac{3}{2}$) presents significant challenges for fusion reactions, since evaporation of at least three neutrons will be required to access the nuclei of interest. Indeed the majority of the recent in-beam $\gamma$-ray spectroscopic studies of $T_z \leq -\frac{3}{2}$ nuclei have been performed with one- (or two-) neutron knockout reactions from relativistic fragmentation beams. The knockout reaction, being a direct process, will populate specific, usually low-lying, states, those bound states for which there is a large spectroscopic overlap between the ground-state configuration of the beam and the final state of the residue, with respect to neutron removal from a specific orbital. Whilst the range of final states can be rather limited, compared with fusion reactions, the reactions (and final spectra) can be easier to interpret, especially when combined with cross-section calculations based on a reaction model using shell-model spectroscopic factors. This, in turn, helps give confidence to the $J^\pi$ assignment of the observed states, when comparing with the analogue states in the mirror nucleus. Moreover, population of high-$J$ states in proton-rich systems is possible in specific conditions, e.g., through knockout from isomeric states (e.g., [7]) or through two-neutron removal from a beam species with a $J \neq 0$ ground state (e.g., [5]).

The case studies discussed here are the very recent works related to the observation of excited states in $T_z = -2$ nuclei $^{56}$Zn [9] and $^{48}$Fe [8]. These studies have enabled the examination of $T = 2$, $T_z = \pm 2$, mirror pairs, providing stringent tests of the shell-model prescription for "distant" mirror pairs (large difference in $T_z$). In both of these examples, one-neutron knockout reactions were performed on odd-$A$ relativistic fragmentation beams. For $^{56}$Zn [9], the experiment was performed at the RIBF facility (Radioactive Isotope Beam Factory), at the RIKEN Nishina Center, Japan. Fragmentation of a beam of $^{78}$Kr at 345MeV/u produced a secondary beam of $^{57}$Zn fragments, separated and identified using the BigRIPS spectrometer [35]. The Be reaction target was surrounded by the DALI2+ NaI $\gamma$-ray array [36] and the final knockout residues identified by the Zero Degree Spectrometer [35]. For $^{48}$Fe [8], the experiment was performed at NCSL (National Superconducting Cyclotron Laboratory, East Lansing, MI, USA). A primary beam of $^{58}$Ni at 160 MeV/u was used to create a $^{49}$Fe fragment beam, separated using the A1900 spectrometer [37]. The reaction target was surrounded by the GRETINA Ge $\gamma$-ray tracking array [38] and the final knockout residues identified by the S800 Spectrograph [39].

In both the above reactions, the ground-state of the beam species was $J^\pi = \frac{7}{2}^-$, where the Fermi-level for the odd, unpaired, neutron was in the $f_{\frac{7}{2}}$ shell. In both cases, excited states of $J^\pi = 2^+, 4^+$, and $6^+$ were observed ($6^+$ is the highest-$J$ state that can be populated directly). The predicted spectroscopic factors for both reactions suggest that the yrast and yrare states of $J^\pi = 2^+$, $4^+$, and $6^+$ are expected to be directly populated, with strong populations of $6^+$ states, which matched the experimental observations [8,9]. It was not possible to identify decays from the yrare states in $^{56}$Zn, but the higher resolution of the $\gamma$-ray array in the $^{48}$Fe study enabled the yrare state decays to be tentatively identified.

In the $^{48}$Fe case, the experiment also used the "mirrored knockout" technique, which has proven to be especially powerful for the observation and assignment of analogue states in mirror pairs. In this approach, as well as using the $^{49}$Fe$-1n$ reaction, the mirror partner to $^{48}$Fe, $^{48}$Ti, was studied through a $^{49}$V$-1p$ reaction (this required a separate setting of the A1900 spectrometer). Since the two beam species, $^{49}$Fe and $^{49}$V, are also mirror nuclei, these reactions comprise a complete pair of "analogue" knockout reactions—i.e., reflected around the $N = Z$ line. Isospin symmetry also implies that the spectroscopic factor for each specific knockout path (removal from a specific orbital to a specific final state) should be essentially identical in both mirror nuclei, and this should, in turn, lead to very similar distributions of knockout strength when the mirror nuclei are studied in the same experimental conditions.

Since the scheme of $^{48}$Ti is known, this helps considerably in the assignment of their analogue states in $^{48}$Fe. The mirrored knockout approach was first demonstrated in [6] and has been employed in a number of other cases [10,14,40].

The spectra in Figure 3 show the resulting $\gamma$-ray spectra for this mirrored reaction: Figure 3a shows the $^{49}\mathrm{V}-1p \rightarrow ^{48}\mathrm{Ti}$ reaction and Figure 3b the mirrored $^{49}\mathrm{Fe}-1n \rightarrow ^{48}\mathrm{Fe}$ reaction. One can see very similar population distribution from the spectra. The spectra, as expected, are dominated by the decays from the $2^+_{1,2}$ states (labelled with blue squares), the $4^+_{1,2}$ states (green diamonds) and the $6^+_{1,2}$ states (red stars) [8,41]. The spectra also show the clear benefit of using a Ge $\gamma$-ray tracking array (i.e., GRETINA) for in-beam spectroscopy with relativistic beams. The position-sensitivity afforded by the pulse-shape-analysis approach allowed for accurate Doppler reconstruction (e.g., [38]), reducing the otherwise huge impact of Doppler broadening at these high fragment velocities.

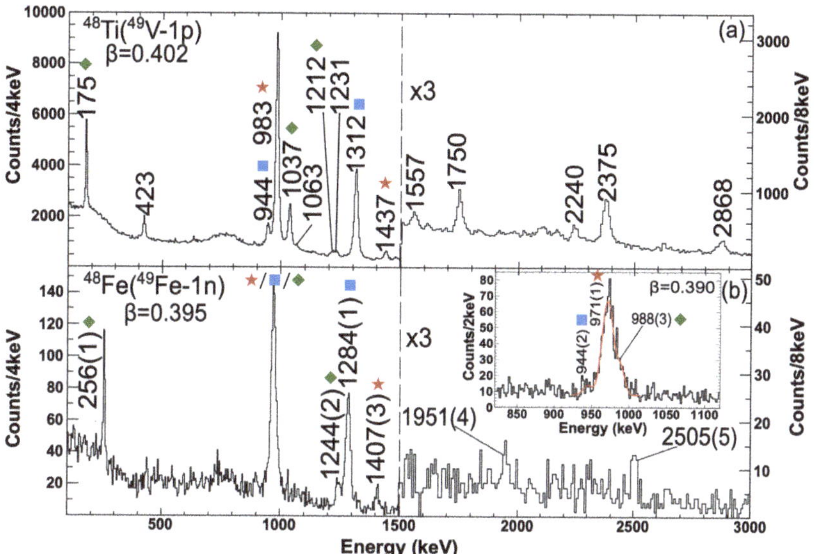


**Figure 3.** The $\gamma$-ray spectra observed in the case study [8]. The spectra are measured with the GRETINA array following the identification and selection of the relevant incoming and outgoing fragment beams. Panel (**a**) shows the $^{49}\mathrm{V}-1p \rightarrow ^{48}\mathrm{Ti}$ reaction and (**b**) the mirrored $^{49}\mathrm{Fe}-1n \rightarrow ^{48}\mathrm{Fe}$ reaction. The peaks are labelled by the $\gamma$-ray energy and the symbols refer to the angular momentum/parity, $J^\pi$, of the states from which these decays proceed. Decays from the $2^+_{1,2}$ states are labelled with blue squares, the $4^+_{1,2}$ states with green diamonds and the $6^+_{1,2}$ states with red stars. The insert in (**b**) shows how the peak around 970 keV comprises three $\gamma$ rays. Adapted from [41].

The use of knockout reactions, and the mirrored-knockout technique, has provided a wealth of data on MED in the upper $f_{\frac{7}{2}}$ region which has, in turn, helped shed light on the role of isospin-non-conserving interactions in the shell-model analysis; see Section 4.1. The $^{56}$Zn case has also yielded information on how occupation of specific shell-model orbitals have a shape-driving effect; see Section 4.2.

## 3. Shell Model Approach for Energy Differences between Excited Analogue States

Without a reliable model to describe MED and TED as a function of $J$, the experimental observations of the variation of MED and TED with $J$ cannot be interpreted in any physically meaningful sense. The shell-model approach to modelling MED and TED has transformed this field of research, allowing interpretation in terms of detailed nuclear structure phenomena including particle alignments and changes in nuclear shape/radii.

Indeed, the happy coincidence that occurred around 20 years ago was that exceptionally powerful large-scale shell-model calculations were becoming available (e.g., [42,43]) in exactly the region where major experimental advances in the spectroscopy of mirror nuclei were taking place—i.e., the lower part of the $pf$ shell.

If perfect isospin symmetry between analogue states is assumed, and that the contributions to MED and TED are entirely related to electromagnetic effects, there are a number of effects that can contribute to MED/TED, and their variation with excitation energy/$J$, which can in principle be calculated in the shell-model approach. The key factor is the multipole effect of re-coupling the angular momentum of pairs of protons, resulting in a decrease in spatial overlap of the protons, with increasing coupled $J$, and hence a reduction in the Coulomb energy. This is straightforward to model in large-scale shell-model calculations through the application of Coulomb matrix elements, calculated in the usual harmonic oscillator (HO) basis, in addition to the nuclear effective interaction. Initial attempts to model MED, using just this approach, were only partially successful (e.g., [44,45]) and it was concluded from that analysis that additional ingredients (including of multipole origin) were missing in the model. Indeed, better agreement was obtained using "empirical" effective $f_{\frac{7}{2}}$ Coulomb matrix elements, extracted from the $A = 42$ mirror nuclei (e.g., [46]) or sets of ad hoc Coulomb matrix elements derived from fits to the data in the centre of the $f_{\frac{7}{2}}$ shell [44].

It was clearly important to develop a consistent shell-model approach for prediction of MED and rooted correctly in the physics. The breakthrough came with the seminal work of Zuker et al. [16], in which multipole and monopole effects were treated together in the same shell-model prescription. The model was developed and tested using MED measured in the centre of the $f_{\frac{7}{2}}$ shell, with shell-model calculations performed with the `ANTOINE` code [42,43] in the full $pf$ space, using the mass-dependent effective interaction for the $pf$-shell, KB3G [47]. This model has formed the basis of the large-scale shell-model approach to MED and TED ever since; see, e.g., [3] for an earlier review. In this approach, the energy differences between analogue states within the shell model can be separated into four components, which can be calculated individually, so that the impact of each can be evaluated.

The first and last terms below are multipole terms. These can be calculated by determining the appropriate matrix elements of the interactions and calculating expectation values through first-order perturbation theory using a set of wave functions calculated in an isoscalar basis. The remaining two components are monopole terms, associated with bulk Coulomb effects and EM-induced shifts in single-particle energies. The four components are as follows.

### 3.1. Coulomb Multipole Interaction: $V_{CM}$

This multipole term accounts for the contribution of the two-body Coulomb interaction to the MED. The contribution to the MED or TED arises due to protons re-coupling angular momentum, with the resulting change in Coulomb energy, and the different numbers of active $pp$ pairs between the isospin-symmetric configurations of the isobaric analogue states. It is accounted for in the shell-model simply through the application of Coulomb matrix elements, calculated in a HO basis.

### 3.2. Single-Particle Contributions: $V_{ll}$ and $V_{ls}$

It was recognised by Zuker et al. [16] that the single-particle splitting between neutron and proton orbitals, induced by the Coulomb interaction, should be accounted for. In the shell model, this can be achieved through introducing shifts between the neutron and proton single particle levels before diagonalisation. The required Coulomb shifts ($V_{ll}$) can be determined through the formalism derived by Duflo and Zuker [48]. Since MED are normalised to the ground state, this term will only become significant where configurations change along the yrast line, and where there are different orbital occupancies between protons and neutrons. This term was eventually neglected by Zuker et al. [16], since they showed that other monopole effects (see Section 3.3) dominate in the specific region being tested. However, this will not always be the case, and the term is routinely included in MED calculations.

A second single particle term, not originally included by Zuker et al. [16], is the electromagnetic spin-orbit effect ($V_{ls}$). This is a purely electromagnetic effect, affecting both proton and neutron levels, associated with the spin magnetic moment of the nucleon interacting with the Coulomb field of the nucleus. The formalism was introduced by Nolen and Schiffer [49] in their description of Coulomb displacement energies. This effect, which has opposite signs for protons and neutrons, can be significant for MED, especially where occupancy of orbitals with $j = l + s$ and $j' = l' - s$ (where $j$, $l$ and $s$ are the total, orbital and spin angular momentum quantum numbers) are both changing (e.g., [50]).

In calculations of MED, both these effects are routinely included. Generally, however, these contributions are expected to cancel in TED calculations, due to the double-difference method of determining TED.

### 3.3. *Radial Contribution:* $V_{Cr}$

A major innovation, introduced by Lenzi et al. [51], and included in the prescription of Ref. [16], is the recognition that the nuclear radius may change along the yrast line with increasing excitation energy, resulting in a change in the bulk Coulomb energy. This, in turn, will contribute to the MED through the difference in $Z$ between the mirror pair. It was recognised [16,51] that orbital radii depend on $l$ and that, in the $f_{\frac{7}{2}}$ region, it was the changing occupancy of the low-$l$ orbitals $p_{\frac{3}{2}}$ and $p_{\frac{1}{2}}$ which would drive the nucleus to larger radii.

Unlike the $V_{ll}$ and $V_{ls}$ terms above, for which the MED will depend on the difference between proton and neutron orbital occupancies, the $V_{Cr}$ term will depend on the average (proton plus neutron) occupancy of the two $p$ orbits. The MED contribution due to the $V_{Cr}$ term is then calculated using

$$\text{MED}_{V_{Cr}}(J) = n\alpha\left[\left(\frac{m_\pi(gs) + m_\nu(gs)}{2}\right) - \left(\frac{m_\pi(J) + m_\nu(J)}{2}\right)\right] \tag{3}$$

where $m(J)$ is the total occupancy of the $p_{\frac{3}{2}}$ and $p_{\frac{1}{2}}$ orbitals for neutrons ($\nu$) and protons ($\pi$), and $n = 2|T_z|$ accounts for the difference in $Z$ between the mirror nuclei. The coefficient $\alpha$ was estimated in Ref. [3] as 200 keV, based on the $A = 41$ mirror nuclei, and this number has been used extensively in the region. In the $f_{\frac{7}{2}}$ region, the occupancy of $p_{\frac{1}{2}}$ is often negligibly small, and so is neglected in the MED calculation. However, above $Z, N \sim 28$ it should be included (e.g., [9]). In the *sd*-shell the same formalism has been used (e.g., [11,52,53]), but instead tracking the occupancy of the $s_{\frac{1}{2}}$ orbital.

Again, as with the previous term, this effect cancels in the calculation of TED.

### 3.4. *Isospin Non-Conserving (INC) Interaction:* $V_B$

Zuker et al. [16] recognised that an additional multipole component is required, in the model, to account for the experimental MED and TED observed in the region (the inclusion of the HO Coulomb matrix elements were shown to be insufficient). Zuker et al. [16] extracted an additional effective INC interaction through comparing the HO Coulomb matrix elements with the MED and TED for the $A = 42$ isospin triplet. An isovector matrix element, $V_B^{(1)}$ was derived for $f_{\frac{7}{2}}$ orbital the from the $A = 42$, $T = 1$ mirror nuclei and an isotensor $f_{\frac{7}{2}}$ matrix element $V_B^{(2)}$ extracted from the TED for $T = 1$ triplet. These matrix elements were derived as a function of angular-momentum coupling $J$, and it was observed that the dominant components appeared to be at $J = 2$ for $V_B^{(1)}$ and $J = 0$ for $V_B^{(2)}$ and both of the order of +100 keV. It was observed [16] that these additional interactions, when included in the shell-model calculations, along with the first and third terms above, allowed for a very good description of the data available at the time. It was later noted [17], once more data became available, that an isovector INC matrix element of the order of $-100$ keV at $J = 0$ gives essentially very similar results to the original value of +100 keV at $J = 2$.

Whatever the origin of this effect, the inclusion of these additional effective isovector and isotensor interactions appeared to be an essential inclusion in the modelling of MED and TED, respectively, at least in the $f_{\frac{7}{2}}$ region. The importance of inclusion of such INC isovector and/or isotensor interactions has also been investigated in the *sd* shell [11,52] and in the upper *pf* shell (e.g., [18,21]).

## 4. Recent Advances Based on Shell-Model Analysis

The shell-model approach described in Section 3 has formed the backbone of this field of study over the last two decades. The strength of the shell-model prescription is the simultaneous inclusion of the multipole and monopole effects, since the relative scale of the contributions of the four components described in Section 3 changes from case to case. Examples of this are the $T = 1$ and $T = 2$ mirror nuclei with $A = 48$ [8,54], one of which is a case study in Section 2.3. Each of these pairs of mirror nuclei, which lie in the exact centre of the $f_{\frac{7}{2}}$ shell, would also be "cross conjugate" nuclei in the assumption of a single isolated $f_{\frac{7}{2}}$ shell. In this extreme assumption, which is not bad for the $f_{\frac{7}{2}}$ shell, all multipole MED would be zero since the number of protons in one nucleus is the same as the number of proton holes in the mirror partner. In the case of the $T = 1$, $A = 48$ mirrors [54] this appears to be the case, and the experimental MED is largely accounted for by the monopole $V_{Cr}$ term. This nicely demonstrates the power of the approach in accounting for a range of phenomena.

In this Section, some of the latest developments in this field, specifically relating to shell-model analysis, are discussed, focussing in particular on the results from the case studies presented in Section 2.

### 4.1. Isospin-Non-Conserving Interactions

One of the key areas of study has been to map out the influence of the additional effective INC interactions (see Section 3.4). In the $f_{\frac{7}{2}}$ shell, a large amount of data have become available which has enabled a more complete numerical evaluation of the influence of these INC effects. In Refs. [17,18], all available MED and TED data in the $f_{\frac{7}{2}}$ shell (at that time) were gathered and modelled using a consistent shell-model approach. The shell-model MED and TED were then fitted to the experimental data, allowing the magnitude of the $J$-dependent INC terms $V_B^{(1)}$ and $V_B^{(2)}$ to vary freely; the former (isovector) term was derived from the MED and the latter (isotensor) term from the TED. The key results are shown in Table 1. The results of two types of fit are presented: one where a single $T = 1$ $V_B$ matrix element is considered (at a coupling of $J = 0$) and the second where all four $T = 1$ matrix elements $J = 0, 2, 4, 6$ were allowed to be non-zero. Only $f_{\frac{7}{2}}$ matrix elements were considered. See Refs. [17,18] for a full discussion of the analysis.

**Table 1.** Data collected from Refs. [17,18]. The isovector $V_B^{(1)}(J)$ and isotensor $V_B^{(2)}(J)$ INC matrix elements, for $f_{7/2}$ pairs, extracted from fits across the whole $f_{7/2}$ shell (see text for details). For the full fits, a monopole centroid has been subtracted as part of the fitting process to allow the $J$-dependence to be fully evaluated. The numbers in the parentheses are the errors on the fitted values.

| **EXTRACTED $V_B^{(k)}$ PARAMETERS FOR THE $f_{\frac{7}{2}}$ ORBITAL** | | | | | | | |
|---|---|---|---|---|---|---|---|
| **Matrix elements $V_B^{(1)}$ (keV)** | | | | **Matrix elements $V_B^{(2)}$ (keV)** | | | |
| $J = 0$ | $J = 2$ | $J = 4$ | $J = 6$ | $J = 0$ | $J = 2$ | $J = 4$ | $J = 6$ |
| **One-parameter fit** | | | | | | | |
| −79(6) | - | - | - | 98(11) | - | - | - |
| **Full fits: centroid-subtracted** | | | | | | | |
| −72(7) | 32(6) | 8(6) | −12(4) | 113(18) | 23(29) | 5(24) | −21(22) |

Two key results emerged from the analysis. Firstly, for the purpose of MED and TED, it is having the correct $J$-dependence of these matrix elements that is crucial in the determination of the theoretical MED and TED; the extracted results do indeed have a

strong $J$-dependence. Secondly, it was shown that a single matrix element at $J = 0$ gives, essentially, as good a fit as allowing all four matrix elements to vary. Hence a prescription in which a single $J = 0$ INC matrix element ($V_B^{(1)}$ or $V_B^{(2)}$) is included is now the commonly used approach for modelling MED and TED (e.g., [8–11,18,52]). The results in Table 1 suggest that an isovector $J = 0$ matrix element of the order of $-100$ keV is required for MED and an isotensor $J = 0$ matrix element of the order of $+100$ keV is required for TED. These conclusions are essentially consistent with the original study of Zuker et al. [16]. Indeed, the fits in Ref. [17] indicates that a positive isovector matrix element at $J = 2$ (as was originally extracted in [16]) has a similar effect as a negative matrix element at $J = 0$. Again, the key contributor is the $J$-dependence, not the absolute magnitude, of these matrix elements.

Figure 4 shows experimental and shell-model MED in the $f_{\frac{7}{2}}$ shell. The solid blue lines contain the full shell-model calculation, performed exactly as described in Section 3, with a single isovector $J = 0$ matrix element of $V_B = -79$ keV, the figure extracted from the fits [17] (see Table 1). The red dashed lines show the calculations without $V_B$ included. It is clear from data like these the crucial role that this effective INC interaction has, especially at low $J$, in the description of MED in the $f_{\frac{7}{2}}$ region.

It is certainly of interest to understand the importance of this effect in other mass regions. In general, this is more challenging in regions where there are more orbitals in play and where the influence of the monopole contributions may be large. In the *sd* shell, inclusion of the INC $V_B$ term also appears to be necessary, with matrix elements of the same order as described above [11,52]. In the upper $fp$ shell, it has been challenging to find a consistent picture for MED, and this remains an open question, e.g., [21]. However, a very recent analysis of the $A = 58$, $T = 1$ mirror nuclei has been performed [9], using the same modelling as that presented for the $A = 56$, $T = 2$ mirrors later in Section 4.2. In the $A = 58$ mirror nuclei, the $f_{\frac{7}{2}}$ shell is expected to be almost fully filled and so the MED will be insensitive to the inclusion of $V_B$ in the $f_{\frac{7}{2}}$ orbital, but will be sensitive to its inclusion in the other $pf$ orbitals. The analysis indicated that a much better match to the experimental MED was obtained when a negative $J = 0$ matrix element for $V_B$ was included for *all* the $fp$ orbitals.

The physical origin of the isovector INC interaction in the modelling of MED remains unclear, and the analysis presented in Reference [17] suggests that the matrix elements and their $J$ dependence (see Table 1) cannot be reconciled easily with the properties of known nuclear charge-symmetry breaking interaction. This therefore points to other electromagnetic contributions missing in the model; see Ref. [17] for a discussion.

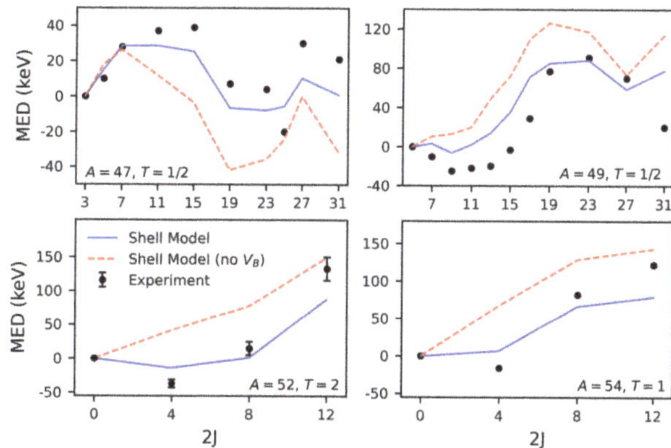


**Figure 4.** Experimental and shell-model MED in the $f_{\frac{7}{2}}$ shell. The solid blue lines contain the full shell-model calculation, including a single isovector $J = 0$ INC matrix element of $V_B = -79$ keV. See text for details. The red dashed lines show the calculations without $V_B$ included. Data for "Shell Model (no $V_B$)" originally presented in Ref. [17]. Where error bars are not visible, they are smaller than the data markers.

Turning to TED, and the impact of the isotensor INC interaction $V_B^{(2)}$, the case study of $^{66}$Se (Section 2) and the $A = 66$, $T = 1$ isobaric triplet provides a practical example. The successful spectroscopy of $^{66}$Se [12] up to $J^\pi = 6^+$ completed the $T = 1$ isobaric triplet and allowed for TED to be determined to $J^\pi = 6^+$. The experimental data are presented in Figure 5. The large negative TED observed are typical of all $T = 1$ triplets; see, e.g., [18]. Figure 5 also contains the result of the shell-model calculation performed following the prescription in Section 3 and using the JUN45 interaction [55]; see black line. The calculations were originally performed in Ref. [18], and updated for this review. The shell-model calculation does not contain calculations related to the two monopole components ($V_{Cr}$ and $V_{ll,ls}$) since these effectively cancel to zero due to the double-difference method of calculating the TED. Hence, only the two multipole interactions, $V_{CM}$ (Coulomb) and $V_B$ (INC), are relevant for this calculation. The shell-model results are plotted in Figure 5, for different strengths of the INC parameter $V_B^{(2)}$, for $J = 0$ couplings. The blue dotted line shows $V_B = 0$ (i.e., just $V_{CM}$ contributes), the black solid line has $V_B = +100$ keV and the red dashed line shows $V_B = +200$ keV. The calculation with $V_B = +100$ keV (black line) is consistent with the prescription in [16] and with the data in Table 1. The $V_B$ interaction was applied equally to all orbits in the $p_{\frac{3}{2}} f_{\frac{5}{2}} p_{\frac{1}{2}} g_{\frac{9}{2}}$ valence space although, in this case, it is the contribution from the $f_{\frac{5}{2}}$ that dominates [18].

Two conclusions can be drawn from the comparison with the shell-model results when $V_B = +100$ keV is applied. The first is that the agreement with experimental TED would fail badly without the inclusion of this additional effective isotensor INC term. Secondly, it can be shown from this analysis [18] that the two components, $V_{CM}$ (Coulomb) and $V_B$ (INC), have approximately the same magnitude when $V_B = +100$ is applied. This is not that surprising since, as noted above, it is the $J$-dependence of the matrix elements that influences the TED, and the Coulomb matrix elements generally vary by around 100 keV from $J = 0$ to $J_{\text{max}}$. The key point is that the contribution of the isotensor INC term, to the TED, is as large as that of the Coulomb two-body interaction.

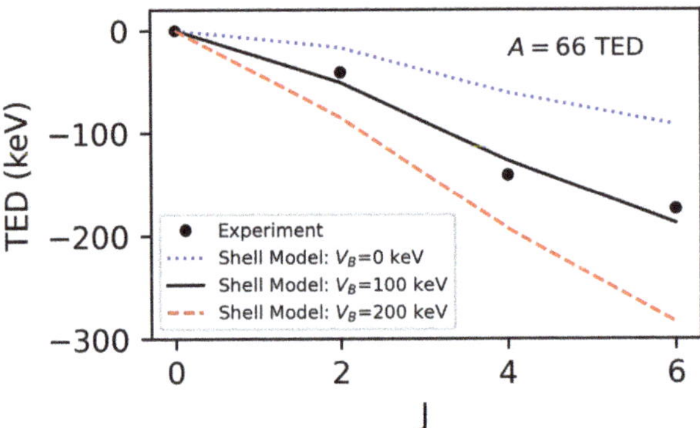


**Figure 5.** Experimental and shell-model TED for the $A = 66$, $T = 1$ isobaric triplet. The black line shows the full shell-model calculation, including a single isotensor $J = 0$ INC matrix element of $V_B = +100$ keV in all orbitals in the valence space. The other lines show the shell-model results using different strengths of the INC parameter, $V_B$. See text for details. The calculations presented are based on the approach of Ref. [18]. The error bars on the data points are smaller than the data markers.

Lenzi et al. [18] performed a similar analysis for all $T = 1$ triplets between $A = 22$ and $A = 66$, using four different interactions, as appropriate to the valence space being used, and applying an isotensor $J = 0$ matrix element of $V_B = +100$ keV in all orbitals. A remarkably consistent picture emerged, with observations very similar to that for $A = 66$; i.e., that the $V_B$ contribution is significant, and required, across the full range of triplets studied. An important point to note is that, for TED, we have seen that the monopole

terms of the shell-model prescription do not contribute significantly. Therefore, the TED is essentially only sensitive to multipole effects and thus represents an observable that can shed light on the nature of effective isospin-non conserving interactions. Since the size of the required $V_B$ interaction appears to be largely independent of mass region, orbital or shell-model interaction, it is natural to examine whether or not the true charge dependence of the nuclear interaction [1] could be the origin. It was shown by Ormand and Brown [56] that nucleon scattering data suggests that the $np$ nuclear interaction is approximately 2–3% stronger than the $pp$ and $nn$ interactions. The analysis of Reference [18] indeed indicated that the scale, and sign, of the effective isotensor interaction $V_B$ interaction appear to be approximately consistent with that estimate for the charge-dependence of the NN interaction. This indeed highlights the power of using energy differences, coupled to a reliable shell-model calculation, to probe effective nucleon interactions.

### 4.2. Nuclear Radii and Neutron Skins

In Section 3, it was demonstrated how occupation of low-$l$ orbitals can contribute to MED and that this can be accounted for in the shell model through tracking of the total (proton plus neutron) occupation of low-$l$ orbits: in the $f_{\frac{7}{2}}$ region this would be the occupancy of the $p_{\frac{3}{2}}, p_{\frac{1}{2}}$ orbitals. This provides the first indication that MED can yield real physical insight into changes in nuclear radii.

Recently, Bonnard et al. [19] have investigated the role of the occupation of low-$l$ "halo" orbitals in driving radii and on their influence in the development of neutron skins. They have been able to show that the effect on the total radius of occupation of one of the low-$l$ orbitals is strongly dependent on the extent of the occupation of that orbital. For example, in the $f_{\frac{7}{2}}$ shell, the occupancy of the $p$ orbits is generally expected to be low (the shell-model occupancies are $\ll 1$). Moreover, the parameterisation of the $V_{Cr}$ term (see Section 3) has been optimised for that region. However, in heavier nuclei, once the $f_{\frac{7}{2}}$ shell is full, the occupancies of the $p$ orbits will increase significantly, and the work of Bonnard et al. [19] suggests that the radial-driving effect of the $p$ orbit will be significantly smaller in this circumstance.

This has been investigated in the $A = 56$, $T = 2$ mirror nuclei following the spectroscopy of $^{56}$Zn [9], discussed as a case study in Section 2.3. Figure 6 shows the experimental MED compared with the shell-model calculations. These calculations have been performed with a modified KB3G interaction, KB3GR (Caurier, E.; Poves, A. *Unpublished work*) which has been optimised for this region. The calculation using the standard parameterisation for the radial term ($\alpha = 200$ keV, see Equation (3)) is shown by the red dashed line. However, in this case, protons in $^{56}$Zn (and neutrons in its mirror, $^{56}$Fe) are already occupying the $p_{\frac{3}{2}}$ orbital, and the results of Reference [19] therefore imply that the radial term $V_{Cr}$ is likely to be overestimated. Therefore, in the analysis of the $A = 56$ mirrors, Fernández et al. [9] reduced the $\alpha$ parameter (see Equation (3)) for the $p_{\frac{3}{2}}$ occupancies, from the standard value of 200 keV. The $\alpha$ parameter for the $p_{\frac{1}{2}}$, which remains largely unoccupied, was left unchanged. The results can be seen in Figure 6 where a smaller value of $\alpha = 50$ keV is applied for the $p_{\frac{3}{2}}$ orbital; see solid blue line. This gives a much better description, in qualitative agreement with the results of Bonnard et al. [19]. It is also noteworthy that the multipole contributions to the MED for this mirror pair turn out to be small, due to particle-hole symmetry; both nuclei have two particles and two holes with respect to $^{56}$Ni. This makes this mirror pair sensitive to the remaining significant monopole contribution, $V_{Cr}$, making this an ideal test case to examine radial effects.

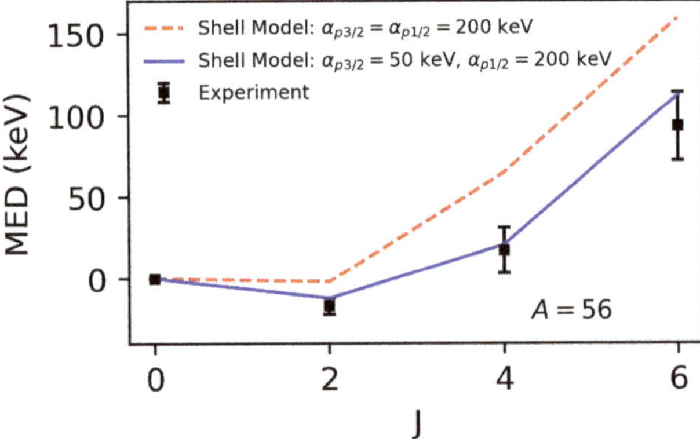


**Figure 6.** Results from [9]. The experimental MED for the $A = 56$, $T = 2$ mirror nuclei compared with the results of shell-model calculations performed with the KB3GR interaction. The model uses the standard parameterisation, but with a varying value of the scaling parameter, $\alpha$ (Equation (3)), used in the determination of the radial contribution to the MED due to the occupation of the $p_{\frac{3}{2}}$ orbital. See text for details.

As well as the *total* nuclear radius having an impact on the Coulomb energy, and hence MED, for a mirror pair, any *difference* between the neutron and proton radii (i.e., neutron skin) could also have an effect on MED if, as isospin symmetry would suggest, the neutron radius of one member of a mirror pair is equal to the proton radius of the other. This idea, also inspired by the study in Ref. [19], was pursued in the analysis of the $A = 23$ mirror nuclei by Boso et al. [11], work that was made possible by the spectroscopy of $^{23}$Mg, our remaining case study (see Section 2.1). The analysis was undertaken using a no-core shell-model approach based on the monopole-corrected interaction (MCI) [57], which contains all the necessary Coulomb and charge-symmetry breaking terms. The MCI matrix elements were computed using different size parameters for neutrons and protons (i.e., allowing for the possibility of different neutron and proton radii). Whilst the proton radius of $^{23}$Na is experimentally known, its neutron radius is not. The neutron radius (and hence neutron skin) was then determined following the method of Duflo and Zuker [48] by adjusting the neutron radius until the experimental ground state mirror displacement energy (MDE) is reproduced by the model. The method was then repeated state by state, in order to reproduce the MED, allowing for the variation of the skin thickness as a function of $J$ for the excited states.

Full details can be found in Ref. [11] but the key results are shown in Figure 7. The neutron skin thickness parameter, $\zeta$, is plotted using the blue circles. $\zeta$, in the paramaterisation of Duflo and Zuker [48], is proportional to difference between the neutron and proton rms (root mean square) radii and is defined as $\zeta = \Delta r_{\nu\pi} A/(T_z e^{g/A})$, where the exponential factor is a correction term, applied for light nuclei [48]. These results show that the neutron skin, as derived from the MED, varies significantly from state to state. This, in turn, implies neutron skin sizes, and their variation with excitation energy/$J$, can influence the MED and, if so, it is an effect currently not included in the MED models. Another key observation is that the skin thickness, has a correlation with the difference between the neutron and proton occupancies of the $s_{\frac{1}{2}}$ orbit. This difference is plotted as $\Delta_{\nu\pi}$ in Figure 7 (red squares). This analysis was repeated for a range of other odd-$A$ mirror nuclei in the *sd* shell, and similar variations of neutron-skin thickness with $J$ were suggested by that analysis; see Ref. [11] for the full results and discussion. Inclusion of effects of this kind in the calculation of MED is clearly an exciting future topic for investigation.

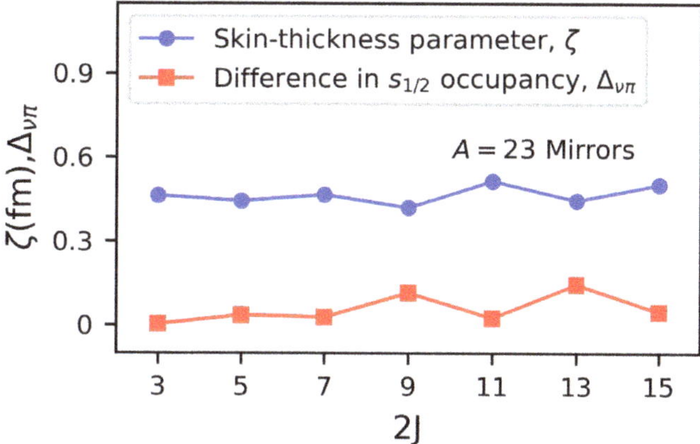


**Figure 7.** Data from Ref. [11]. Blue circles: the neutron skin thickness parameter, $\zeta$, defined in Ref. [48], which is proportional to difference between the neutron and proton rms radii. This parameter has been extracted through fitting to the measured MED. Red squares: $\Delta_{\nu\pi}$, the difference between the neutron and proton occupancies of the $s_{\frac{1}{2}}$ orbit, for each state. See text and Ref. [11] for details.

## 5. Summary and Outlook

In this short review, some of the latest experimental advances have been presented. The advent of the new radioactive beam facilities will allow some of these techniques to be applied to allow spectroscopy of the most exotic proton-rich systems, or to perform high precision tests of the predictions that come from the isospin formalism. The high-intensity intermediate-energy fragmentation beams available at the upcoming FRIB (Facility for Rare Isotope Beams, East Lansing, MI, USA) and FAIR (Facility for Antiproton and Ion Research, Darmstadt, Germany) facilities are expected to have particular impact. Techniques such as those described Section 2.3, can be applied to access nuclei with large proton excess and pursue spectroscopy of mirror nuclei in the upper half of the $fpg$ region. The high-velocity beams also allow for a range of lifetime-measurement techniques to be applied, allowing for precision tests of the isospin-dependence of transition strengths. From a theoretical perspective, it will be especially important to develop a better understanding of the origin of the effective isovector isospin non-conserving (INC) interactions (see Section 4.1). Moreover, the link between mirror energy differences (MED) and radii/neutron skin is especially exciting and should be developed further in future shell-model work. As the study of energy splitting between isobaric multiplets develops in the future, the exciting developments in the shell-model, some of which have been discussed, will have crucial role to play.

**Funding:** This research was funded by the UKRI (UK Research and Innovation) Science and Technology Facilities Council under grant number ST/V001108/1.

**Data Availability Statement:** No new data are presented in this work. The data used can be found in the corresponding references.

**Conflicts of Interest:** The author declares no conflict of interest.

## References

1. Machleidt, R.; Slaus, I. The nucleon-nucleon interaction. *J. Phys. Nucl. G Part. Phys.* **2001**, *27*, R69–R108. [CrossRef]
2. Wigner, E. On the consequences of the symmetry of the nuclear Hamiltonian on the spectroscopy of nuclei. *Phys. Rev.* **1937**, *51*, 106–119. [CrossRef]
3. Bentley, M.A.; Lenzi, S.M. Coulomb energy differences between high-spin states in isobaric multiplets. *Prog. Part. Nucl. Phys.* **2007**, *59*, 497–561. [CrossRef]

4. Ekman, J.; Fahlander, C.; Rudolph, D. Mirror symmetry in the upper $fp$ shell. *Mod. Phys. Lett. A* **2005**, *20*, 2977–2992. [CrossRef]
5. Davies, P.J.; Bentley, M.A.; Henry, T.W.; Simpson, E.C.; Gade, A.; Lenzi, S.M.; Baugher, T.; Bazin, D.; Berryman, J.S.; Bruce, A.M.; et al. Mirror energy differences at large isospin studied through direct two-nucleon knockout. *Phys. Rev. Lett.* **2013**, *111*, 072501. [CrossRef]
6. Milne, S.A.; Bentley, M.A.; Simpson, E.C.; Dodsworth, P.; Baugher, T.; Bazin, D.; Berryman, J.S.; Bruce, A.M.; Davies, P.J.; Diget, C.A.; et al. Mirrored one-nucleon knockout reactions to the $T_z = \pm\frac{3}{2}$ $A = 53$ mirror nuclei. *Phys. Rev. C* **2016**, *93*, 024318. [CrossRef]
7. Milne, S.A.; Bentley, M.A.; Simpson, E.C.; Baugher, T.; Bazin, D.; Berryman, J.S.; Bruce, A.M.; Davies, P.J.; Diget, C.A.; Gade, A.; et al. Isospin symmetry at high spin studied via nucleon knockout from isomeric states. *Phys. Rev. Lett.* **2016**, *117*, 082502. [CrossRef]
8. Yajzey, R.; Bentley, M.A.; Simpson, E.C.; Haylett, T.; Uthayakumaar, S.; Bazin, D.; Belarge, J.; Bender, P.C.; Davies, P.J.; Elman, B.; et al. Spectroscopy of the $T_z = \pm 2$ mirror nuclei $^{48}$Fe/$^{48}$Ti using mirrored knockout reactions. *Phys. Lett. B* **2021**, *823*, 136757. [CrossRef]
9. Fernández, A.; Jungclaus, A.; Doornenbal, P.; Bentley, M.A.; Lenzi, S.M.; Rudolph, D.; Browne, F.; Cortés, M.L.; Koiwai, T.; Taniuchi, R.; et al. Mirror energy differences above the $0f_{7/2}$ shell: First $\gamma$-ray spectroscopy of the $T_z = -2$ nucleus $^{56}$Zn. *Phys. Lett. B* **2021**, *823*, 136784. [CrossRef]
10. Uthayakumaar, S.; Bentley, M.A.; Simpson, E.C.; Haylett, T.; Yajzey, R.; Lenzi, S.M.; Satuła, W.; Bazin, D.; Belarge, J.; Bender, P.C.; et al. Spectroscopy of the $T = \frac{3}{2} A = 47$ and $A = 45$ mirror nuclei via one- and two-nucleon knockout reactions. *Phys. Rev. C* **2022**, *106*, 024327. [CrossRef]
11. Boso, A.; Lenzi, S.M.; Recchia, F.; Bonnard, J.; Zuker, A.P.; Aydin, S.; Bentley, M.A.; Cederwall, B.; Clement, E.; de France, G.; et al. Neutron skin effects in mirror energy differences: The case of $^{23}$Mg$-^{23}$Na. *Phys. Rev. Lett.* **2018**, *121*, 032502. [CrossRef]
12. Ruotsalainen, P.; Jenkins, D.G.; Bentley, M.A.; Wadsworth, R.; Scholey, C.; Auranen, K.; Davies, P.J.; Grahn, T.; Greenlees, P.T.; Henderson, J.; et al. Spectroscopy of proton-rich $^{66}$Se up to $J^\pi = 6^+$: Isospin-breaking effect in the $A = 66$ isobaric triplet. *Phys. Rev. C* **2013**, *88*, 041308. [CrossRef]
13. Debenham, D.M.; Bentley, M.A.; Davies, P.J.; Haylett, T.; Jenkins, D.G.; Joshi, P.; Sinclair, L.F.; Wadsworth, R.; Ruotsalainen, P.; Henderson, J.; et al. Spectroscopy of $^{70}$Kr and isospin symmetry in the $T = 1fpg$ shell nuclei. *Phys. Rev. C* **2016**, *94*, 054311. [CrossRef]
14. Wimmer, K.; Korten, W.; Arici, T.; Doornenbal, P.; Aguilera, P.; Algora, A.; Ando, T.; Baba, H.; Blank, B.; Boso, A.; et al. Shape coexistence and isospin symmetry in $A = 70$ nuclei: Spectroscopy of the $T_z = -1$ nucleus 70Kr. *Phys. Lett. B* **2018**, *785*, 441–446. [CrossRef]
15. Henderson, J.; Jenkins, D.G.; Kaneko, K.; Ruotsalainen, P.; Sarriguren, P.; Auranen, K.; Bentley, M.A.; Davies, P.J.; Görgen, A.; Grahn, T.; et al. Spectroscopy on the proton drip-line: Probing the structure dependence of isospin nonconserving interactions. *Phys. Rev. C* **2014**, *90*, 051303. [CrossRef]
16. Zuker, A.P.; Lenzi, S.M.; Martínez-Pinedo, G.; Poves, A. Isobaric multiplet yrast energies and isospin nonconserving forces. *Phys. Rev. Lett.* **2002**, *89*, 142502. [CrossRef]
17. Bentley, M.A.; Lenzi, S.M.; Simpson, S.A.; Diget, C.A. Isospin-breaking interactions studied through mirror energy differences. *Phys. Rev. C.* **2015**, *92*, 024310. [CrossRef]
18. Lenzi, S.M.; Bentley, M.A.; Lau, R.; Diget, C.A. Isospin-symmetry breaking corrections for the description of triplet energy differences. *Phys. Rev. C.* **2018**, *98*, 054322. [CrossRef]
19. Bonnard, J.; Lenzi, S.M.; Zuker, A.P. Neutron skins and halo orbits in the *sd* and $pf$ shells. *Phys. Rev. Lett.* **2016**, *116*, 212501. [CrossRef]
20. Kaneko, K.; Sun, Y.; Mizusaki, T.; Tazaki, S. Variation in displacement energies Due to isospin-nonconserving forces. *Phys. Rev. Lett.* **2013**, *110*, 172505. [CrossRef]
21. Kaneko, K.; Sun, Y.; Mizusaki, T.; Tazaki, S. Isospin nonconserving interaction in the $T = 1$ analogue states of the mass-70 region. *Phys. Rev. C* **2014**, *89*, 031302. [CrossRef]
22. Azaiez, F. EXOGAM: A $\gamma$-ray spectrometer for radioactive beams. *Nucl. Phys. A* **1999**, *654*, 1003c–1008c. [CrossRef]
23. Skeppstedt, O.; Roth, H.A.; Lindström, L.; Wadsworth, R.; Hibbert, I.; Kelsall, N.; Jenkins, D.G.; Grawe, H.; Górska, M.; Moszyński, M.; et al. The EUROBALL neutron wall – design and performance tests of neutron detectors. *Nucl. Instrum. Meth. Phys. Res. A* **1999**, *421*, 531–541. [CrossRef]
24. Scheurer, J.N.; Aiche, M.; Aleonard, M.M.; Barreau, G.; Bourgine, F.; Boivin, D.; Cabaussel, D.; Chemin, J.F.; Doan, T.P.; Goudour, J.P.; et al. Improvements in the in-beam $\gamma$-ray spectroscopy provided by an ancillary detector coupled to a Ge $\gamma$-spectrometer: The DIAMANT-EUROGAM II example. *Nucl. Instrum. Meth. Phys. Res. A* **1997**, *385*, 501–510. [CrossRef]
25. Valiente-Dobón, J.J.; Jaworski, G.; Goasduff, A.; Egea, F.J.; Modamio, V.; Hüyük, T.; Triossi, A.; Jastrząb, M.; Söderström, P.A.; Di Nitto, A.; et al. NEDA—NEutron Detector Array. *Nucl. Instrum. Meth. Phys. Res. A* **2019**, *927*, 81–86. [CrossRef]
26. Akkoyun, S.; Algora, A.; Alikhani, B.; Ameil, F.; de Angelis, G.; Arnold, L.; Astier, A.; Ataç, A.; Aubert, Y.; Aufranc, C.; et al. AGATA—Advanced GAmma Tracking Array. *Nucl. Instrum. Meth. Phys. Res. A* **2012**, *668*, 26–58. [CrossRef]
27. Korten, W.; Atac, A.; Beaumel, D.; Bednarczyk, P.; Bentley, M.A.; Benzoni, G.; Boston, A.; Bracco, A.; Cederkäll, J.; Cederwall, B.; et al. Physics opportunities with the Advanced Gamma Tracking Array: AGATA. *Eur. Phys. J. A* **2020**, *56*, 137. [CrossRef]

28. Cederwall, B.; Liu, X.; Aktas, O.; Ertoprak, A.; Zhang, W.; Qi, C.; Clément, E.; de France, G.; Ralet, D.; Gadea, A.; et al. Isospin Properties of Nuclear Pair Correlations from the Level Structure of the Self-Conjugate Nucleus $^{88}$Ru. *Phys. Rev. Lett.* **2020**, *124*, 062501. [CrossRef]
29. Steer, A.; Jenkins, D.; Glover, R.; Bondili, S.N.; Pattabiraman, N.; Wadsworth, R.; Eeckhaudt, S.; Grahn, T.; Greenlees, P.; Jones, P.; et al. Recoil-beta tagging: A novel technique for studying proton-drip-line nuclei. *Nucl. Instrum. Meth. Phys. Res. A* **2006**, *565*, 630–636. [CrossRef]
30. Nara Singh, B.S.; Steer, A.N.; Jenkins, D.G.; Wadsworth, R.; Bentley, M.A.; Davies, P.J.; Glover, R.; Pattabiraman, N.S.; Lister, C.J.; Grahn, T.; et al. Coulomb shifts and shape changes in the mass 70 region. *Phys. Rev. C* **2007**, *75*, 061301. [CrossRef]
31. Leino, M.; Äystö, J.; Enqvist, T.; Heikkinen, P.; Jokinen, A.; Nurmia, M.; Ostrowski, A.; Trzaska, W.; Uusitalo, J.; Eskola, K.; et al. Gas-filled recoil separator for studies of heavy elements. *Nucl. Instrum. Meth. Phys. Res. B* **1995**, *99*, 653–656. [CrossRef]
32. Sarén, J.; Uusitalo, J.; Leino, M.; Sorri, J. Absolute transmission and separation properties of the gas-filled recoil separator RITU. *Nucl. Instrum. Meth. Phys. Res. A* **2011**, *654*, 508–521. [CrossRef]
33. Henderson, J.; Ruotsalainen, P.; Jenkins, D.G.; Scholey, C.; Auranen, K.; Davies, P.J.; Grahn, T.; Greenlees, P.T.; Henry, T.W.; Herzáň, A.; et al. Enhancing the sensitivity of recoil-beta tagging. *J. Instrum.* **2013**, *8*, P04025. [CrossRef]
34. Sarén, J.; Uusitalo, J.; Leino, M.; Greenlees, P.; Jakobsson, U.; Jones, P.; Julin, R.; Juutinen, S.; Ketelhut, S.; Nyman, M.; et al. The new vacuum-mode recoil separator MARA at JYFL. *Nucl. Instrum. Meth. Phys. Res. B* **2008**, *266*, 4196–4200. [CrossRef]
35. Kubo, T.; Kameda, D.; Suzuki, H.; Fukuda, N.; Takeda, H.; Yanagisawa, Y.; Ohtake, M.; Kusaka, K.; Yoshida, K.; Inabe, N.; et al. BigRIPS separator and ZeroDegree spectrometer at RIKEN RI Beam Factory. *Prog. Theor. Exp. Phys.* **2012**, *2012*. [CrossRef]
36. Takeuchi, S.; Motobayashi, T.; Togano, Y.; Matsushita, M.; Aoi, N.; Demichi, K.; Hasegawa, H.; Murakami, H. DALI2: A NaI(Tl) detector array for measurements of $\gamma$ rays from fast nuclei. *Nucl. Instrum. Meth. Phys. Res. A* **2014**, *763*, 596–603. [CrossRef]
37. Morrissey, D.J.; Sherrill, B.M.; Steiner, M.; Stolz, A.; Wiedenhoever, I. Commissioning the A1900 projectile fragment separator. *Nucl. Instrum. Meth. Phys. Res. B* **2003**, *204*, 90–96. [CrossRef]
38. Weisshaar, D.; Bazin, D.; Bender, P.C.; Campbell, C.M.; Recchia, F.; Bader, V.; Baugher, T.; Belarge, J.; Carpenter, M.P.; Crawford, H.L.; et al. The performance of the $\gamma$-ray tracking array GRETINA for $\gamma$-ray spectroscopy with fast beams of rare isotopes. *Nucl. Instrum. Meth. Phys. Res. A* **2017**, *847*, 187–198. [CrossRef]
39. Bazin, D.; Caggiano, J.A.; Sherrill, B.M.; Yurkon, J.; Zeller, A. The S800 spectrograph. *Nucl. Instrum. Meth. Phys. Res. B* **2003**, *204*, 629–633. [CrossRef]
40. Spieker, M.; Gade, A.; Weisshaar, D.; Brown, B.A.; Tostevin, J.A.; Longfellow, B.; Adrich, P.; Bazin, D.; Bentley, M.A.; Brown, J.R.; et al. One-proton and one-neutron knockout reactions from $N = Z = 28$ $^{56}$Ni to the $A = 55$ mirror pair $^{55}$Co and $^{55}$Ni. *Phys. Rev. C* **2019**, *99*, 051304. [CrossRef]
41. Yajzey, R. The Isospin Symmetry of the $A = 48$, $T = 2$ Mirror Nuclei Studied Through the Mirrored Knockout Technique. Ph.D. Thesis, University of York, York, UK, 2022. Available online: https://etheses.whiterose.ac.uk/30822/ (accessed on 6 July 2022).
42. Caurier, E.; Nowacki, F. Present Status of Shell Model Techniques. *Acta Phys. Pol. B* **1999**, *30*, 705–714. Available online: https://www.actaphys.uj.edu.pl/R/30/3/705 (accessed on 3 August 2022).
43. Caurier, E.; Martínez-Pinedo, G.; Nowacki, F.; Poves, A.; Zuker, A.P. The shell model as a unified view of nuclear structure. *Rev. Mod. Phys.* **2005**, *77*, 427–488. [CrossRef]
44. Bentley, M.A.; O'Leary, C.D.; Poves, A.; Martinez-Pinedo, G.; Appelbe, D.E.; Bark, R.A.; Cullen, D.M.; Ertürk, S.; Maj, A. Mirror and valence symmetries at the centre of the $f_{7/2}$ shell. *Phys. Lett. B* **1998**, *437*, 243–248. [CrossRef]
45. Williams, S.J.; Bentley, M.A.; Warner, D.D.; Bruce, A.M.; Cameron, J.A.; Carpenter, M.P.; Fallon, P.; Frankland, L.; Gelletly, W.; Janssens, R.V.F.; et al. Anomalous Coulomb matrix elements in the $f_{7/2}$ shell. *Phys. Rev. C* **2003**, *68*, 011301. [CrossRef]
46. Bentley, M.A.; Williams, S.J.; Joss, D.T.; O'Leary, C.D.; Bruce, A.M.; Cameron, J.A.; Carpenter, M.P.; Fallon, P.; Frankland, L.; Gelletly, W.; et al. Mirror symmetry at high spin in $^{51}$Fe and $^{51}$Mn. *Phys. Rev. C* **2000**, *62*, 051303. [CrossRef]
47. Poves, A.; Sánchez-Solano, J.; Caurier, E.; Nowacki, F. Shell model study of the isobaric chains $A = 50$, $A = 51$ and $A = 52$. *Nucl. Phys. A* **2001**, *694*, 157–198. [CrossRef]
48. Duflo, J.; Zuker, A.P. Mirror displacement energies and neutron skins. *Phys. Rev. C* **2002**, *66*, 051304. [CrossRef]
49. Nolen, J.A.; Schiffer, J.P. Coulomb Energies. *Annu. Rev. Nucl. Sci.* **1969**, *19*, 471–526. [CrossRef]
50. Ekman, J.; Rudolph, D.; Fahlander, C.; Zuker, A.P.; Bentley, M.A.; Lenzi, S.M.; Andreoiu, C.; Axiotis, M.; de Angelis, G.; Farnea, E.; et al. Unusual isospin-breaking and isospin-mixing effects in the $A = 35$ mirror nuclei. *Phys. Rev. Lett.* **2004**, *92*, 132502. [CrossRef]
51. Lenzi, S.M.; Mărginean, N.; Napoli, D.R.; Ur, C.A.; Zuker, A.P.; de Angelis, G.; Algora, A.; Axiotis, M.; Bazzacco, D.; Belcari, N.; et al. Coulomb energy differences in $T = 1$ mirror rotational bands in $^{50}Fe$ and $^{50}Cr$. *Phys. Rev. Lett.* **2001**, *87*, 122501. [CrossRef]
52. Lenzi, S.M.; Lau, R. A systematic study of mirror and triplet energy differences. *J. Phys. Conf. Ser.* **2015**, *580*, 012028. [CrossRef]
53. Testov, D.A.; Boso, A.; Lenzi, S.M.; Nowacki, F.; Recchia, F.; de Angelis, G.; Bazzacco, D.; Colucci, G.; Cottini, M.; Galtarossa, F.; et al. High-spin intruder states in the mirror nuclei $^{31}$S and $^{31}$P. *Phys. Rev. C* **2021**, *104*, 024309. [CrossRef]
54. Bentley, M.A.; Chandler, C.; Taylor, M.J.; Brown, J.R.; Carpenter, M.P.; Davids, C.; Ekman, J.; Freeman, S.J.; Garrett, P.E.; Hammond, G.; et al. Isospin symmetry of odd-odd mirror nuclei: Identification of excited states in $N = Z - 2$ $^{48}$Mn. *Phys. Rev. Lett.* **2006**, *97*, 132501. [CrossRef] [PubMed]

55. Honma, M.; Otsuka, T.; Mizusaki, T.; Hjorth-Jensen, M. New effective interaction for $f_5pg_9$-shell nuclei. *Phys. Rev. C* **2009**, *80*, 064323. [CrossRef]
56. Ormand, W.E.; Brown, B.A. Empirical isospin-nonconserving hamiltonians for shell-model calculations. *Nucl. Phys. A* **1989**, *491*, 1–23. [CrossRef]
57. Bonnard, J.; Zuker, A.P. Radii in the *sd* shell and the $1s_{1/2}$ "halo" orbit: A game changer. *J. Phys. Conf. Ser.* **2018**, *1023*, 012016. [CrossRef]

*Review*

# Isospin-Symmetry Breaking within the Nuclear Shell Model: Present Status and Developments

**Nadezda A. Smirnova**

Laboratoire de Physique des Deux Infinis Bordeaux (LP2IB), Centre National de la Recherche Scientifique, Institut National de Physique Nucléaire et de Physique des Particules (CNRS/IN2P3), Université de Bordeaux, 33175 Gradignan, France; nadezda.smirnova@u-bordeaux.fr

**Abstract:** The paper reviews the recent progress in the description of isospin-symmetry breaking within the nuclear shell model and applications to actual problems related to the structure and decay of exotic neutron-deficient nuclei and nuclei along the $N = Z$ line, where $N$ is the neutron number and $Z$ the atomic number. The review recalls the fundamentals of the isospin formalism for two-nucleon and many-nucleon systems, including quantum numbers, the spectrum's structure and selection rules for weak and electromagnetic transitions; and at the end, summarizes experimental signatures of isospin-symmetry breaking effects, which motivated efforts towards the creation of a relevant theoretical framework to describe those phenomena. The main approaches to construct accurate isospin-nonconserving Hamiltonians within the shell model are briefly described and recent advances in the description of the structure and (isospin-forbidden) decay modes of neutron-deficient nuclei are highlighted. The paper reviews major implications of the developed theoretical tools to (i) the fundamental interaction studies on nuclear decays and (ii) the estimation of the rates of nuclear reactions that are important for nuclear astrophysics. The shell model is shown to be one of the most suitable approaches to describing isospin-symmetry breaking in nuclear states at low energies. Further efforts in extending and refining the description to larger model spaces, and in developing first-principle theories to deal with isospin-symmetry breaking in many-nucleon systems, seem to be indispensable steps towards our better understanding of nuclear properties in the precision era.

**Keywords:** nuclear shell model; isospin symmetry and its breaking; structure of neutron-deficient nuclei; superallowed Fermi beta decay; fundamental interactions; astrophysical *rp*-process

**Citation:** Smirnova, N.A. Isospin-Symmetry Breaking within the Nuclear Shell Model: Present Status and Developments. *Physics* **2023**, *5*, 352–380. https://doi.org/10.3390/physics5020026

Received: 2 February 2023
Revised: 5 March 2023
Accepted: 7 March 2023
Published: 31 March 2023

**Copyright:** © 2023 by the author. Licensee MDPI, Basel, Switzerland. This article is an open access article distributed under the terms and conditions of the Creative Commons Attribution (CC BY) license (https://creativecommons.org/licenses/by/4.0/).

## 1. Introduction

### 1.1. Isospin Symmetry in Nuclear Structure

Atomic nuclei are unique quantum many-body systems composed of two sorts of fermions—protons and neutrons, which are known to have similar masses and possess similar properties with respect to the strong interactions. It was Heisenberg [1] (see English translation in Ref. [2]) who soon after the discovery of the neutron, introduced an *isospin formalism* similar to the ordinary spin formalism as an elegant mathematical tool for dealing with protons and neutrons. Nucleons are considered to be isospin $t = 1/2$ particles and represented by two-component spinors spanning an abstract vector space where the isospin operator, $\hat{\mathbf{t}}$, acts. The neutron and the proton are two eigenstates of $\hat{t}_3$ (the third component of the isospin operator):

$$\psi_n(\vec{r}) = \psi(\vec{r})\begin{pmatrix}1\\0\end{pmatrix},\ \psi_p(\vec{r}) = \psi(\vec{r})\begin{pmatrix}0\\1\end{pmatrix},$$

with eigenvalues $m_t = \pm 1/2$, respectively, and $\vec{r}$ the radius vector. The three components of the isospin operator, analogues of the Cartesian components, generate an isospin SU(2) algebra:

$$[\hat{t}_j, \hat{t}_k] = \mathrm{i}\epsilon_{jkl}\hat{t}_l\,, \tag{1}$$

where $j, k, l = 1, 2, 3$, $\epsilon_{jkl}$ is the Levi-Civita symbol, and the square of the isospin operator,

$$\hat{\mathbf{t}}^2 = \hat{t}_1^2 + \hat{t}_2^2 + \hat{t}_3^2 , \tag{2}$$

commutes with each of the components: $[\hat{\mathbf{t}}^2, \hat{t}_j] = 0$.

Operators corresponding to various physical observables can be conveniently expressed using isospin formalism. For example, the third component of the isospin operator $\hat{t}_3$ allows one to express the nucleon charge operator,

$$\hat{q} = \left(\frac{1}{2} - \hat{t}_3\right)\mathrm{e} ,$$

and the ladder operators $\hat{t}_\pm$,

$$\hat{t}_\pm = \hat{t}_1 \pm i\hat{t}_2 , \tag{3}$$

transforming a proton into a neutron and vice versa, can be useful to formulate nuclear $\beta$ decay. Here, "e" denotes the elementary charge.

Nowadays, isospin symmetry is an important concept in particle physics describing a symmetry between $u$ and $d$ quarks with respect to the strong interaction and their similarly light masses as compared to the other known quarks. The isospin character of nucleons, and of other hadrons composed from $u$ and/or $d$ quarks, is a consequence of isospin coupling.

Based on the conservation of charge and the approximate charge-independence of the nuclear forces, Wigner [3] introduced the total isospin operator for an $A$-nucleon system arising from the coupling of the individual isospin operators:

$$\hat{\mathbf{T}} = \sum_{k=1}^{A} \hat{\mathbf{t}}(k) ,$$

or for the components:

$$\hat{T}_\pm = \sum_{k=1}^{A} \hat{t}_\pm(k) , \quad \hat{T}_3 = \sum_{k=1}^{A} \hat{t}_3(k) , \tag{4}$$

with $T(T+1)$ and $M_T = (N-Z)/2$ being eigenvalues of $\hat{\mathbf{T}}^2$ and $\hat{T}_3$, respectively, $N$ the neutron number, and $Z$ the atomic number. A charge-independent nuclear Hamiltonian would commute with $\hat{\mathbf{T}}$,

$$[\hat{H}_{\mathrm{nucl}}, \hat{\mathbf{T}}] = 0 ,$$

or

$$[\hat{H}_{\mathrm{nucl}}, \hat{T}_\pm] = [\hat{H}_{\mathrm{nucl}}, \hat{T}_3] = 0 .$$

An additional isospin quantum number $T$ appears to label $A$-nucleon states besides the total angular momentum, $J$, and parity, $\pi$. The spectrum of $H_{\mathrm{nucl}}$ thus consists of degenerate isobaric multiplets, which can be labeled by $(J^\pi, T)$ in nuclei with the same mass number $A$ and $M_T = -T, \ldots, T$, called isobaric analogue states (IAS).

It was realized long ago that electromagnetic interactions destroy this degeneracy. However, as it was shown by Wigner [4], this leads mainly to dynamical breaking of the isospin SU(2) symmetry. Indeed, the Coulomb interaction between protons, which is the main source of the isospin-symmetry breaking on the nuclear level, can be represented

as a linear combination of an isoscalar ($\hat{V}^{(0)}$), an isovector ($\hat{V}^{(1)}$) and an isotensor ($\hat{V}^{(2)}$) operator:

$$\hat{V}_{\text{Coul}} = \sum_{i<k}^{A} \left(\frac{1}{2} - \hat{t}_3(i)\right)\left(\frac{1}{2} - \hat{t}_3(k)\right) \frac{e^2}{|\vec{r}(i) - \vec{r}(k)|}$$
$$= \sum_{i<k}^{A} \left\{ \underbrace{\left[\frac{1}{4} + \frac{1}{3}\hat{\mathbf{t}}(i)\hat{\mathbf{t}}(k)\right]}_{V^{(0)}} - \underbrace{\frac{1}{2}(\hat{t}_3(i) + \hat{t}_3(k))}_{V^{(1)}} + \underbrace{\left[\hat{t}_3(i)\hat{t}_3(k) - \frac{1}{3}\hat{\mathbf{t}}(i)\hat{\mathbf{t}}(k)\right]}_{V^{(2)}} \right\} \frac{e^2}{|\vec{r}(i) - \vec{r}(k)|}. \quad (5)$$

By estimating the effect of this charge-dependent operator on the isobaric multiplets within the lowest order perturbation theory (due to its expectation value within the states of a given isospin, $T$) and applying the Wigner–Eckart theorem in the isospace, one gets an expression quadratic in $M_T$:

$$\begin{aligned} \langle \eta T M_T | \hat{V}_{\text{Coul}} | \eta T M_T \rangle &= \frac{(T M_T 00 | T M_T)}{\sqrt{2T+1}} \langle \eta T || \hat{V}^{(0)} || \eta T \rangle \\ &+ \frac{(T M_T 10 | T M_T)}{\sqrt{2T+1}} \langle \eta T || \hat{V}^{(1)} || \eta T \rangle \\ &+ \frac{(T M_T 20 | T M_T)}{\sqrt{2T+1}} \langle \eta T || \hat{V}^{(2)} || \eta T \rangle, \end{aligned} \quad (6)$$

where double bar denotes reduction in the isospin space; $(T M_T \lambda \mu | T M_T)$ are the Clebsch–Gordan coefficients; and $\eta$ refers to other quantum numbers characterizing an isobaric multiplet: $\eta = (A, J^\pi, \ldots)$. By inserting Clebsch–Gordan coefficients, one gets:

$$\langle \eta T M_T | V_{\text{Coul}} | \eta T M_T \rangle = E^{(0)}(\eta, T) + E^{(1)}(\eta, T) M_T + E^{(2)}(\eta, T) \left[ 3M_T^2 - T(T+1) \right], \quad (7)$$

where $E^{(\lambda)}(\eta, T)$ are related to the reduced in isospace matrix elements of isotensors, as seen from Equation (6). This expression remains valid if leading-order terms of charge-dependent forces of nuclear origin are included, as discussed in Section 1.2. Such a dependence, re-written for nuclear masses, is known as the isobaric-multiplet mass equation (IMME) [4],

$$\mathcal{M}(\eta, T, M_T) = a(\eta, T) + b(\eta, T) M_T + c(\eta, T) M_T^2, \quad (8)$$

with $\mathcal{M}$ being an atomic mass excess. Experimental $a$, $b$ and $c$ coefficients can be deduced from available data on nuclear masses and spectra of up to about $A = 71$ [5,6].

Interestingly, Equation (8) holds exceptionally well, even for isobaric multiplets with more than three members ($T > 1$). This makes the IMME a powerful tool for predicting the nuclear masses of nuclei along the $N = Z$ line, as illustrated in Section 3. Deviations from the quadratic form are rare and small. They are specifically searched for in experiments, as they can bring important information on the presence of charge-dependent many-body forces or witness strong isospin mixing.

From a group-theoretical point of view [7], Equation (7), or equivalently, Equation (8), expresses a reduction of the isospin SU(2) group to its SO(2) subgroup. The eigenstates of the full Hamiltonian, $\hat{H}_{\text{nucl}} + \hat{V}_{\text{Coul}}$, can still be characterized by the isospin quantum number $T$, but the $(2T+1)$-fold degeneracy inherent to the isotopic multiplets is now removed. This effect is analogous to a Zeeman splitting of atomic levels in the presence of a magnetic field.

As every symmetry, isospin symmetry proposes a number of selection rules for various transition operators, on the basis of their tensorial character with respect to the SU(2) group in isospace. For example, allowed $\beta$-decay, governed by the vector or axial-vector weak

currents, is described by Fermi (F) or Gamow–Teller (GT) operators, respectively. In the impulse approximation, these operators read

$$\hat{O}_{\mathrm{F}}(\beta^{\pm}) = \sum_{k=1}^{A} \hat{t}_{\pm}(k), \quad \hat{O}_{\mathrm{GT}}(\beta^{\pm}) = \sum_{k=1}^{A} \hat{\boldsymbol{\sigma}}(k)\hat{t}_{\pm}(k). \tag{9}$$

Both operators are seen to be isovector components. The Fermi operator is a scalar, and the Gamow–Teller operator is a vector in the ordinary spin space ($\hat{\boldsymbol{\sigma}}$ is the Pauli spin operator). The Wigner–Eckart theorem establishes angular momentum parity, and isospin selection rules can be established for transitions between an initial state $(J_i^{\pi_i}, T_i)$ and a final state $(J_f^{\pi_f}, T_f)$. For Fermi transitions, one has:

$$\Delta J = 0,\ \Delta T = 0,\ \Delta\pi = 0,$$

and for Gamow–Teller transitions, one has:

$$\begin{array}{l} \Delta J = 0,1,\ \Delta T = 0,1,\ \Delta\pi = 0 \\ (\text{no } J_i = 0 \to J_f = 0). \end{array}$$

From this one can conclude that $J_i = 0 \to J_f = 0$ decay can be only by the Fermi type.

A similar analysis can be performed for electromagnetic operators. Assuming a one-body structure of nucleonic convection and spin currents and point-like nucleons, electromagnetic operators can be shown to be a linear combination of an isoscalar and an isovector operator [8], e.g., for an operator of multipolarity $L$, one has $\hat{O}_{LM} = \hat{O}^{(0)}_{LM} + \hat{O}^{(1)}_{LM}$, where $M = -L, \ldots, L$. Therefore, their matrix elements between states of given isospin can be expressed as

$$\begin{aligned} \langle J_f M_f; T_f M_T | \hat{O}_{LM} | J_i M_i; T_i M_T \rangle &= \delta_{T_i T_f} \langle J_f M_f | \hat{O}^{(0)}_{LM} | J_i M_i \rangle \\ &+ \frac{(T_i M_T 1 0 | T_f M_T)}{\sqrt{2T_f + 1}} \langle J_f M_f; T_f || \hat{O}^{(1)}_{LM} || J_i M_i; T_i \rangle, \end{aligned} \tag{10}$$

where $\delta_{T_i T_f}$ is the Kronecker delta.

From Equation (10) one immediately gets the isospin selection rules for electromagnetic transitions [8].

- For $\Delta T = 1$ transitions ($T_f = T_i \pm 1$), the (reduced) matrix elements of analogue transitions in mirror nuclei or between respective analogue states should be identical, since they are governed only by the isovector term.
- In transitions between the states of the same isospin ($T_i{=}T_f{=}T$), both isoscalar and isovector terms contribute, and the matrix element for analogue transitions within an isobaric multiplet exhibits a linear trend as a function of $M_T$:

$$\begin{aligned} \langle J_f M_f; T M_T | \hat{O}_{LM} | J_i M_i; T M_T \rangle &= \langle J_f M_f | \hat{O}^{(0)}_{LM} | J_i M_i \rangle \\ &+ \frac{M_T}{\sqrt{T(T+1)(2T+1)}} \langle J_f M_f; T || \hat{O}^{(1)}_{LM} || J_i M_i; T \rangle. \end{aligned} \tag{11}$$

- Another specific rule can be established for electric dipole operator. In the lowest order of the long-wavelength approximation, the electric-dipole ($E1$) operator is an isovector operator:

$$\hat{O}(E1) = \sum_{k=1}^{A} \mathrm{e}(k)\vec{r}(k) = \sum_{k=1}^{A} \left(\frac{1}{2} - \hat{t}_3(k)\right) \mathrm{e}\vec{r}(k). \tag{12}$$

Hence, $E1$ transitions between the states of the same isospin ($T_i = T_f = T$) in $N = Z$ nuclei are forbidden by the isospin symmetry because of the vanishing Clebsch–Gordan coefficient, $(T\,0\,1\,0|\,T\,0\,) = 0$ (see Equation (11)).

Finally, isospin selection rules govern also nuclear reactions (see also, e.g., Refs. [9–11], for specific topics). Restricting ourselves to nuclear decays, only nucleon, two-nucleon and $\alpha$-particle emission are mentioned here: for example, for isospin-allowed proton emission, the difference in isospin between the initial and final states is $\Delta T = 1/2$; for two-proton emission, it is $\Delta T = 1$; $\alpha$ emission should be consistent with $\Delta T = 0$.

Observation of isospin-forbidden decay modes indicates explicit isospin-symmetry breaking and the presence of isospin mixing in nuclear states.

### 1.2. Isospin-Symmetry Breaking

Although isospin symmetry proved to be quite a useful concept in nuclear and particle physics, which helps to simplify theoretical modeling of the nucleon–nucleon interaction and provides an efficient framework for the nuclear many-body problem, experimental evidence has been accumulated on the breaking of isospin symmetry.

First, it is known that isobaric multiplets are not degenerate. The differences in energy between states forming an isobaric multiplet are called *Coulomb displacement energies*, since the Coulomb interaction is the main contributor to the effect. Such splittings can be explained within *dynamical breaking* of isospin symmetry, as was pointed out in Section 1.1. However, observation of isospin-forbidden decays, i.e., decays which break isospin selection rules, indicates that isospin is not a good quantum number, and there is a certain amount of isospin mixing in nuclear states. To describe such phenomena, one must introduce an explicit breaking of isospin symmetry within a nuclear structure model. Development of microscopic approaches for an accurate description of isospin-symmetry breaking is important not only for understanding the structure and decay of proton-rich nuclei, but also for the evaluation of nuclear-structure corrections to weak processes in nuclei. Taking isospin-symmetry breaking into account may also help to improve our knowledge of certain reactions involving proton-rich nuclei, which are crucial for nuclear astrophysics.

At the nuclear level, isospin symmetry is broken mainly due to the Coulomb interaction among protons (a long-range component of the electromagnetic interaction between protons), and to a minor extent by the proton and neutron mass difference and the presence of the charge-dependent forces of nuclear origin (short-range). At the quark level, these causes can be rooted to the $u$ and $d$ quark mass difference and electromagnetic interactions between the quarks. The need for charge-dependent forces of nuclear origin was established long ago from the analysis of the nucleon–nucleon ($NN$) scattering data. For example, it is known that there are differences in the neutron–neutron ($a_{nn}$), proton–proton ($a_{pp}$, with electromagnetic effects being subtracted) and neutron–proton ($a_{np}$) ${}^1S_0$ (a $T = 1$ channel) scattering lengths [12,13]. Namely, the difference of $a_{nn}$ and $a_{pp}$,

$$a_{nn} - a_{pp} = 1.6 \pm 0.6 \text{ fm}, \tag{13}$$

is a signature of *charge-symmetry breaking* of the strong $NN$ force; and the even larger difference between $a_{np}$ and the average of $a_{nn}$ and $a_{pp}$,

$$\frac{1}{2}(a_{nn} + a_{pp}) - a_{np} = 5.64 \pm 0.40 \text{ fm}, \tag{14}$$

is known as the *charge-independence breaking* property.

Moreover, still long ago, Nolen and Schiffer [14] noticed that the Coulomb force alone cannot satisfactorily explain the binding energy differences in mirror nuclei if one requires the model to reproduce nuclear charge radii and vise versa (the so-called Nolen–Schiffer anomaly). The insufficiency of the two-body Coulomb interaction in reproduction of splittings of isobaric multiplets was also demonstrated in more refined shell-model

calculations (e.g., Refs. [15–17]). Many-body approaches must therefore, take into account short-range charge-dependent components of the nucleon–nucleon interaction.

Henley and Miller [18] proposed to divide two-nucleon forces into four classes according to their isospin characters, namely,

- class I ($V_I$) are charge-independent forces $\{1, \hat{\mathbf{t}}(1)\cdot\hat{\mathbf{t}}(2)\}$;
- class II ($V_{II}$) are forces which break the charge independence, but preserve the charge symmetry of the two-nucleon system, $\{\hat{t}_3(1)\hat{t}_3(2)\}$;
- class III ($V_{III}$) are charge-symmetry breaking forces, which vanish in the neutron-proton system, $\{\hat{t}_3(1)+\hat{t}_3(2)\}$;
- class IV ($V_{IV}$) are forces which do not conserve the isospin of the two-nucleon system: $\{\hat{\mathbf{t}}(1) \times \hat{\mathbf{t}}(2), \hat{t}_3(1) - \hat{t}_3(2)\}$.

If, as an example the two-body Coulomb interaction, acting between protons, is considered, one may notice that it comprises terms of classes I, II and III, as seen in Equation (5). It is important to note that although class II and class III forces commute with the two-nucleon isospin operator, such forces do violate the isospin symmetry in an $A$-nucleon system with $A > 2$.

Isospin-symmetry breaking two-nucleon forces have been constructed and explored in earlier meson-exchange models [12,19] and within the modern chiral effective field theory $\chi$EFT) [13,20–22]. The details of various contributions from hadronic mass splittings and electromagnetic processes can be found in the above-given references. From $\chi$EFT, the following hierarchy was deduced [20]: $V_I > V_{II} > V_{III} > V_{IV}$. In addition, charge-dependent three-nucleon ($3N$) forces have been constructed within $\chi$EFT see, e.g., the review [13] and references therein). Those may contribute to possible deviations of the IMME from its quadratic form, as discussed in Section 3.1 below.

Although charge-dependent realistic inter-nucleon interactions are frequently used in many-body calculations, in particular, in ab initio approaches, there have been few studies specifically focused on the degree of isospin-symmetry breaking. Nevertheless, ab initio Green's function Monte Carlo calculations with charge-dependent forces from the realistic Argonne $v_{18}$ $NN$ + Illinois-7 $3N$ potential supplemented by more refined charge-dependent terms have been performed [23]. Quite good reproduction of the binding-energy differences in a few pairs of light mirror nuclei and the expected amount of isospin-mixing in $^8$Be were reported. A significant feature of those calculations is that they introduced and demonstrated the role of class IV forces. Charge-dependent $NN$+$3N$ forces from $\chi$EFT are used in state-of-the-art no-core shell model calculations for light nuclei [24,25], and the validity of isospin symmetry in electric quadrupole moments of mirror nuclei has been probed within the same theoretical approach in Ref. [26].

This review is devoted rather to the description of isospin-nonconserving phenomena in spectra and decays of heavier nuclei, for which a solution of the nuclear many-body problem needs an approach requiring effective charge-dependent interactions. Various theoretical frameworks aimed at a reliable description of isospin-symmetry breaking have been developed to deal with the problem. Among them are state-of-the-art shell-model calculations [15–17,27–35], including its no-core realization [36] and continuum-coupling extension [37], mean-field approaches and beyond (e.g., [38–47]) and others. Earlier comprehensive reviews on isospin symmetry and its breaking can be found in Refs. [9,48–50].

The present paper focuses rather on a particular theoretical approach to the problem, namely, on the nuclear shell model (e.g., see books [51–54]). Indeed, the shell model conserves all fundamental symmetries of atomic nuclei (such as angular momentum and particle number) and describes quite accurately individual states and transitions at low energies. This makes it an adequate approach for searching for tiny isospin-symmetry breaking effects. In the following sections, we highlight recent progress achieved by the isospin nonconserving shell model. A short summary of selected results has already been published in the proceedings of EuNPC2018 [55].

## 2. Formalism

The starting point of the shell model is a non-relativistic Hamiltonian for point-like nucleons containing nucleon kinetic energies and effective $NN$ interactions (only two-body interactions are considered here):

$$\hat{H} = \sum_{k=1}^{A} \hat{T}_{\rm kin}(k) + \sum_{k<l=1}^{A} \hat{V}_{\rm nucl}(k,l)\,. \tag{15}$$

By adding and subtracting a one-body spherically symmetric potential (e.g., a harmonic-oscillator potential), one can rewrite the Hamiltonian as a sum of an independent-particle Hamiltonian ($\hat{H}_0$) and a residual interaction ($\hat{V}$):

$$\hat{H} = \sum_{k=1}^{A} [\hat{T}_{\rm kin}(k) + \hat{U}(k)] + \left[ \sum_{k<l=1}^{A} \hat{V}_{\rm nucl}(k,l) - \sum_{k=1}^{A} \hat{U}(k) \right] = \hat{H}^{(0)} + \hat{V}\,. \tag{16}$$

The eigenstates of $\hat{H}$ ($\hat{H}\Psi_m = E_m\Psi_m$) are searched for in terms of a complete orthonormal set of eigenfunctions of $\hat{H}_0$ ($\hat{H}_0\Phi_m = E_{0m}\Phi_m$):

$$\Psi_m = \sum_{m'} C_{mm'}\Phi_{m'}\,.$$

Using this expansion, the eigenproblem is reduced for $\hat{H}$ to the diagonalization of the Hamiltonian matrix, $\langle\Phi_{m'}|\hat{H}|\Phi_m\rangle$, computed from single-particle energies of valence-space orbitals, $\varepsilon_{p,n}(a)$, and two-body matrix elements (TBMEs) of the residual interaction, $\langle ab; JMTM_T|\hat{V}|cd; JMT'M_T\rangle$ ($a,b,c,d$ run over valence-space orbitals in a spherically symmetric mean field, i.e., $a = (n_a l_a j_a)$ and so on). As a result, one gets eigenvalues $E_m$ and the corresponding sets of expansion coefficients $\{C_{mm'}\}$. If the nuclear Hamiltonian, which is rotational invariant, is also taken to be charge-independent (the proton and neutron single-particle energies are identical and TBMEs are independent from $M_T$ with $T = T'$), its eigenstates are characterized by the angular momentum and isospin quantum numbers $(JMTM_T)$, thereby forming degenerate spin (isospin) multiplets.

Since the model's space dimensions grow quickly as the number of particles increases, only for light nuclei can the shell model problem be solved for all nucleons considered in a model space comprised of many harmonic-oscillator shells. When using realistic internucleon interaction, the approach is referred to as an ab initio no-core shell model [24]. For heavier nuclei, the shell-model problem is formulated for valence nucleons only, occupying a model space consisting of one or two oscillator shells beyond a closed shell core. This restriction of the model space has been proved to be sufficient for low-energy nuclear structures. However, because of a severely truncated model space, one needs to derive a so-called *effective interaction*.

In this context, the isospin formalism helps to reduce the number of parameters. Nevertheless, construction of robust valence-space effective Hamiltonians remains a challenging and a long-standing problem of nuclear theory. Microscopic effective interactions have been constructed, for example, within the many-body perturbation theory, starting from the pioneering work in 60s [56,57] and continuing on into recent times (for reviews, see Refs. [58–60]). In spite of important advances, microscopic interactions are known to be less successful than more phenomenological parametrizations, based on the adjustment of TBMEs to selected data on nuclear spectra from a given model space. In particular, with two-nucleon forces only, the resulting effective interaction suffers from serious deficiencies in their monopole component [61]. This feature was ascribed to missing $3N$ forces. In addition, a number of theoretical issues in application of many-body perturbation theory to nuclear effective interaction problem have been raised regarding convergence of the expansion [62], which have not convincingly been answered yet.

In the last decade, new non-perturbative approaches to the construction of effective valence-space Hamiltonians have been put forward, based on unitary transformation

techniques—the in-medium similarity-renormalization group approach (IMSRG) [60,63] and the Okubo–Lee–Suzuki transformations of no-core shell-model solutions [64,65]. In addition, similar ideas have been implemented within the coupled-cluster method [66–68]. Moreover, some of these approaches, including modern many-body perturbation theory [69, 70], have successfully incorporated three-nucleon forces in their frameworks, producing state-of-the-art microscopic effective valence-space interactions from first principles.

In spite of all these developments, phenomenological effective interactions still remain a benchmark. Therefore, let us start the discussion of isospin-nonconserving (INC) Hamiltonians from a phenomenological perspective.

### 2.1. Phenomenological Approaches

Phenomenological effective Hamiltonians are typically isospin-conserving; therefore, the Coulomb contribution is usually evaluated and subtracted from the data before it is used in a fit. The resulting interactions are called realistic, and they can provide high accuracy in the description of nuclear excited states and transitions at low energies for a large set of nuclei (ideally, all nuclei) from a given model space. The most famous examples are the Cohen–Kurath Hamiltonians [71] in the $p$ shell; the universal *sd* shell (USD) family of Hamiltonians [72,73], and Kuo-Brown modified KB3G [74] and GXPF1A [75] Hamiltonians in the $pf$ shell.

An attractive option to construct an accurate INC Hamiltonian is thus to adopt a well-established charge-independent Hamiltonian as a lowest-order approximation and to add an INC term. The latter must contain the two-body Coulomb interaction and effective charge-dependent $NN$ forces ($\hat{V}_{\rm CD}$), at least of classes II and III (no class IV forces are discussed here, but eventually, the framework can be extended to include them as well). Such an operator is a sum of an isoscalar, an isovector and an isotensor term:

$$\hat{V}_{\rm INC} = \hat{V}_{\rm Coul} + \hat{V}_{\rm CD} = \sum_{\lambda=0,1,2} \hat{V}^{(\lambda)}_{\rm INC}, \quad \text{where} \quad \begin{cases} \hat{V}^{(0)}_{\rm INC} = (v_{pp} + v_{nn} + v^{T=1}_{np})/3\,, \\ \hat{V}^{(1)}_{\rm INC} = v_{pp} - v_{nn}\,, \\ \hat{V}^{(2)}_{\rm INC} = (v_{pp} + v_{nn})/2 - v^{T=1}_{np}\,. \end{cases}$$

To describe the Coulomb effects of the core, an isovector one-body term is added which gives rise to the so-called *isovector single-particle energies*, $\tilde{\varepsilon}(a){=}\varepsilon_p(a){-}\varepsilon_n(a)$, where $a$ runs over model-space orbitals. In lowest-order perturbation theory, the splitting of the isobaric multiplets is due to the expectation value of this operator; therefore, it is expressed by a quadratic polynomial in $M_T$, similarly to Equation (7):

$$\langle \Psi_{TM_T} | \hat{V}_{\rm INC} | \Psi_{TM_T} \rangle = E^{(0)}(\eta, T) + E^{(1)}(\eta, T) M_T + E^{(2)}(\eta, T) \Big[ 3M_T^2 - T(T+1) \Big].$$

In order to find the best set of parameters of $\hat{V}_{\rm INC}$ and isovector single-particle energies $\tilde{\varepsilon}_a$, one can perform a fit requiring that theoretical isovector and isotensor components allow one to reproduce experimentally deduced $b$ and $c$ IMME coefficients for a wide selection of lowest and excited isobaric multiplets with $T = 1/2, 1, 3/2, \ldots$. This procedure was first proposed in Ref. [15] and was used in the later work related to the *sd*-shell [16,27] and $pf$-shell and heavier nuclei [28]. Among various possible forms of $\hat{V}_{\rm CD}$, modelization of that term either by a $\rho$-exchange Yukawa-type potential (with a scaled meson mass) or by the $T = 1$ term of the isospin-conserving Hamiltonian in the isovector and isotensor channels resulted in similar quality fits [15,27]. At the same time, the use of the $\pi$-exchange potential was found to require much stronger renormalization of the two-body Coulomb force, and therefore, it was not retained.

Figure 1 shows the $b$ coefficients for the lowest doublets and $c$ coefficients for the lowest triplets obtained from such phenomenological interactions for *sd*-shell and $pf$-shell nuclei, in comparison with the experimental values. It is evident that the agreement between theory and experiment is remarkable. The root-mean-square (rms) deviations between theory and experiment represented in Figure 1 are 30 keV (95 keV) for $b$ coefficients in the

*sd* ($pf$) shell and around 9 keV (25 keV) for the $c$-coefficients in the *sd* ($pf$) shell. One can observe that the description of the $pf$-shell $b$ coefficients worsens towards the middle of the shell. By excluding data for $A = 59, 61, 63$, the rms deviation reduces to 55 keV. This problem seems to be linked to the difficulty in the description of nuclei from the upper part of the $pf$ shell because of large dimensions involved, and may not be related to the form of the INC terms. Note also that more realistic forms of $\hat{V}_{\rm CD}$ did not help to improve the fit [16].

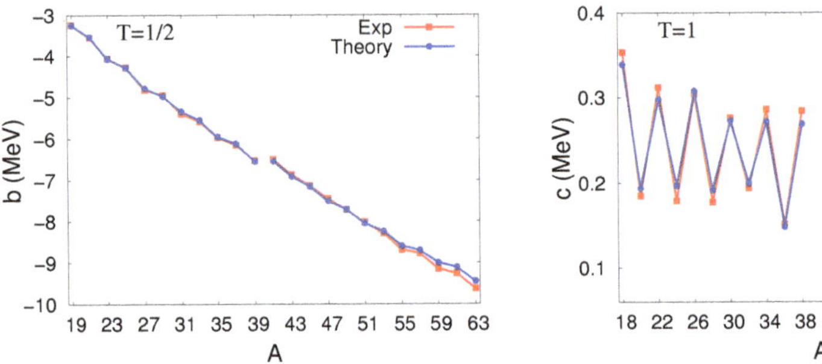


**Figure 1.** Experimental ("Exp") [5,76] and theoretical ("Theory") IMME $b$ coefficients for the lowest doublets (**left**) and $c$ coefficients for the lowest triplets (**right**) in the *sd* and $pf$ shells. The *sd*-shell results were quoted from Ref. [27], and $pf$-shell calculations were performed with GX1Acd interaction [77]. See text for details.

As seen in Figure 1, the shell model well reproduces both the general trends and the fine structure of $b$ and $c$ coefficients. The latter considers the staggering $c$ coefficients as a function of $A$, as well visible in Figure 1 (right): the $c$ coefficients in $A = 4n + 2$ multiplets are systematically larger than those in $A = 4n$ ($n$ being a positive integer). Similarly, the $b$ coefficients in doublets and quartets form two families for $A = 4n + 1$ and $A = 4n + 3$, with opposite phases, however (for doublets, $b$ coefficients are largest in $A = 4n + 1$ nuclei, and for quartets, they are largest for $A = 4n + 3$ nuclei). To amplify the effect, in Figure 2, the differences in $b$ coefficients between $A$ and $A - 2$ nuclei are plotted.

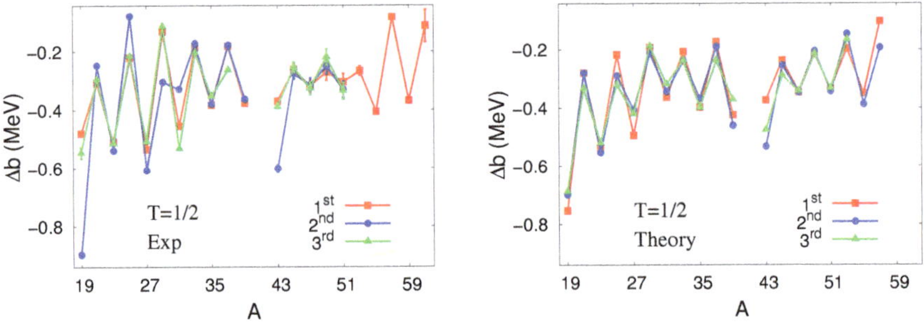


**Figure 2.** Experimental [5,76] (**left**) and theoretical (**right**) differences in IMME $b$ coefficients ($\Delta_b$) for the ground-state, first-excited and second-excited natural-parity $T = 1/2$ multiplets in the *sd* and $pf$ shells. The *sd*-shell results were obtained with the interaction from Ref. [27], and $pf$-shell calculations were performed with GX1Acd interaction [77].

The staggering was noticed long ago and explained by the interplay between the Coulomb force and the pairing TBMEs [78]. It should be visibly present in $b$ coefficients of multiplets with half-integer $T$ and $c$ coefficients of multiplets with integer values of $T$. The same conclusions have been reached [79] within a simpler macroscopic approach

supplemented by different proton and neutron average pairing gaps, which made it possible to grasp the main features of staggering.

Modern microscopic approaches [27,28,34,44,45] using realistic interactions well reproduce the effect. The main advantage of the shell-model type approaches is that they can describe $b$ and $c$ coefficients of excited states as well. Figures 2 and 3 show the differences among $b$ coefficients, $\Delta_b$ and $c$ coefficients for the three lowest positive-parity multiplets in doublets and triplets, respectively. Interestingly, that the amplitudes of oscillations diminish with excitation energy. This hints that the pairing effect gradually weakens as systems become more and more excited.

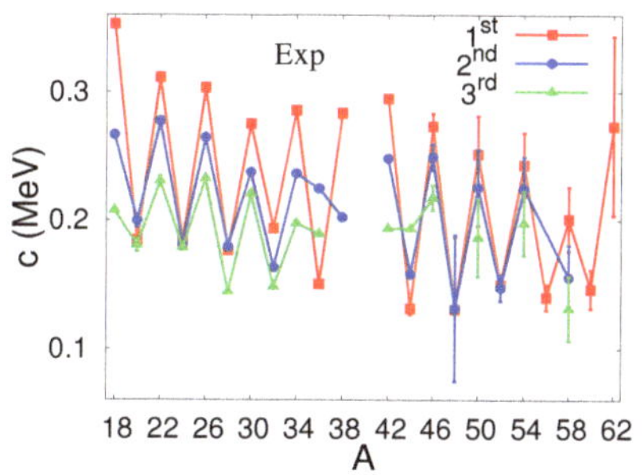

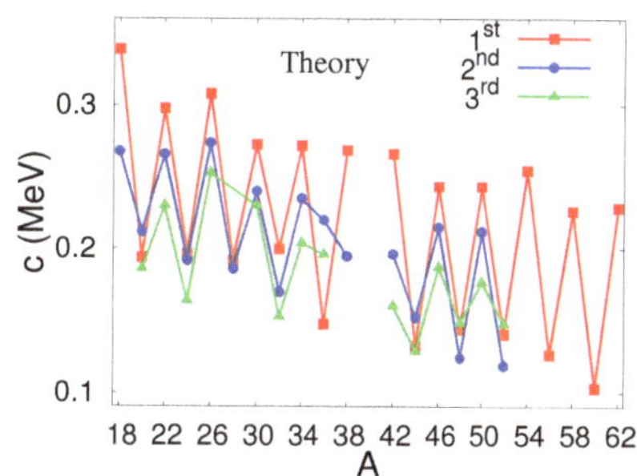


**Figure 3.** Experimental [5,76] (**left**) and theoretical (**right**) IMME $c$ coefficients for the lowest, first-excited and second-excited $T = 1$ multiplets in the $sd$ and $pf$ shells. The $sd$-shell results were obtained with the interaction from Ref. [27], and $pf$-shell calculations were performed with GX1Acd interaction [77]. For $A = 42$, the data are given for $J^\pi = 0^+, 2^+, 4^+$ states. See text for details.

The approach described above can rather well reproduce an extended set of $b$ and $c$ coefficients and provides an attractive tool with which to predict binding energies and states in mirror systems using a method of Coulomb energy differences, described in Section 3.1 below. At the same time, a few drawbacks exist—namely, that it (i) does not allow one to predict nuclear masses on purely theoretical grounds, (ii) does not account for the so-called Thomas–Ehrman shift and (iii) it does not provide enough accuracy in the description of the differences in excitation energies of analogue states, usually referred to as Coulomb energy differences.

Another strategy was put forward by Zuker, Lenzi and collaborators in a series of papers starting from [17] (see also Refs. [80–82] for a recent review). The idea consists in modeling charge-dependent forces of nuclear origin with a few TBMEs, adjusted to reproduce the differences in excitation energies of isobaric multiplets relative to the lowest in energy multiplet. Those quantities are known as mirror energy differences (MEDs) and triplet energy differences (TEDs) in $T = 1$ multiplets, and they are related to the differences in $b$ or $c$ coefficients between the lowest multiplet and an excited one. For example, for triplets,

$$\begin{aligned} \mathrm{MED}(J) &= -2(b(J) - b_0)\,, \\ \mathrm{TED}(J) &= 2(c(J) - c_0)\,, \end{aligned}$$

where $b_0$ ($c_0$) is a $b$ ($c$) coefficient of the lowest triplet. Considered as a function of $J$ along an excitation band (a pattern of excited states linked by pronounced electromagnetic transitions), MEDs and TEDs can bring pertinent information on nuclear structure effects. A vary accurate description has been achieved [17,80] of the $pf$ shell by a phenomenological parameterization of various physical effects, such as changes in nuclear radius (or deformation) and electromagnetic corrections to the single-particle energies, with $\hat{V}_{\mathrm{CD}}$ being modeled by a few $J = 0$ TBMEs in isovector and isotensor channels.

In Ref. [28], it was shown that modelization of $V_{\mathrm{CD}}$ by two $J = 0$, $T = 1$ TBMEs in the $f_{7/2}$ orbital and theoretically calculated single-particle energies was sufficient to reproduce the staggering behavior of $b$ and $c$ coefficients. This may not be surprising, since

we understand the staggering effect is due to the Coulomb contribution to the pairing-type matrix elements.

Later on, the approach was generalized to other model spaces using a more extended form of a charge-dependent term of nuclear origin as a number of TBMEs in Refs. [31,32]. MED and TED appear to be sensitive tools to unveil the structure of excited states, and in particular, TEDs and MEDs can shed light onto pair alignment process or on the shape evolution. Detailed study of the heavy $N = Z$ region allowed researchers to understand co-existing shapes and other effects in $A = 66, 70, 74, 78$ (e.g., [29,83]).

Moreover, MEDs have been shown [84] to depend linearly on the difference between neutron and proton radii, known as "neutron skin", and that they strongly correlate with the $s_{1/2}$-orbital occupation. In general, low-$l$ orbitals, especially $s_{1/2}$ orbitals, are characterized by an extended radius and play thus a special role in nuclear structure. In particular, it was noted that MEDs of states having higher occupation of $s_{1/2}$ are unusually large. It turns out that states in proton-rich nuclei having high occupation of such an orbital experience a stronger shift with respect to their mirror states in neutron-rich partners. This is the essence of the so-called Thomas–Erhman effect [85,86]. Parameterizations of the charge-dependent forces mentioned above do not necessarily include this effect, which thus requires special care. In order to account for the Thomas–Ehrman shift, several approaches have been developed. For example, one can vary the energy of the proton $\varepsilon(s_{1/2})$ single-particle orbital (e.g., Ref. [87]) or quench TBMEs which involve $s_{1/2}$ orbitals [88]. Recently, a direct construction of TBMEs based on a simultaneous fit of isoscalar, isovector and isotensor terms has been undertaken, which lead to a few new types of USD interactions [34], aiming at consistent description of proton-rich and neutron-rich nuclei on similar grounds.

### 2.2. Semi-Phenomenological Approaches

A first step towards a more theoretical framework was to use a more realistic form of $\hat{V}_{\mathrm{CD}}$ on top of phenomenological wave functions. This was introduced in Ref. [16] for the $sd$ shell but found to be less successful than a purely phenomenological charge-dependent term. More recently, in Ref. [33], microscopic charge-independence breaking $pf$-shell Hamiltonians have been constructed from the two-body CD-Bonn, Argonne $v_{18}$ and chiral $\mathrm{N}^3$LO (next-to-next-to-next-to-leading order) potentials on top of the phenomenological GXPF1A interaction. The authors compared theoretical IMME $c$ coefficients as a function of the angular momentum in selected $pf$-shell nuclei with experimental data and conclude that the theory indicates too-strong of a contribution of the charge-independence breaking terms of nuclear origin.

### 2.3. Microscopic Approaches

A recent breakthrough in the construction of the $NN$ interaction from effective field theories and advances in nuclear many-body methods led to the appearance of the first semi-microscopic and fully microscopic effective charge-dependent Hamiltonians. In particular, large-scale calculations for proton-rich nuclei in the extended $sdf_{7/2}p_{3/2}$ and $pfg_{9/2}$ model spaces with effective Hamiltonians, derived within many-body perturbation theory from $\chi$EFT $NN$+3$N$ interactions, have been reported in Ref. [30].

Later on, valence space Hamiltonians were constructed within the IMSRG approach [35] based on two forces obtained within $\chi$EFT. The author tested the ability of their fully ab initio methods to reproduce the experimental IMME $b$ and $c$ coefficients for a large selection of nuclei of interest for superallowed $\beta$-decay applications with $A$ between 10 and 74. Their conclusion is that although the major trend comes out correctly, their results are interaction-dependent and not precise enough to get the fine details.

Numerous modern theoretical investigations of nuclear properties are performed nowadays within ab initio approaches using charge-dependent realistic interactions (for example, those from $\chi$EFT). We believe that specific issues of isospin-symmetry breaking will definitely be addressed in forthcoming studies.

## 3. Structure and Decay of Neutron-Deficient Nuclei

Development of charge-dependent Hamiltonians has its ultimate goal of providing an accurate description of nuclei along the $N = Z$ line and proton-rich nuclei, making it possible to describe the signatures of isospin-symmetry breaking. This Section gives examples of how theoretical IMME coefficients can serve to predict nuclear masses and excited states in mirror nuclei, and it summarizes the progress in the description of isospin-forbidden transitions. The latter provide important tests of isospin mixing in nuclear wave functions to validate theoretical models.

### *3.1. IMME Coefficients for Masses and Excitation Spectra of Proton-Rich Nuclei*

It was recognized long ago that the quadratic IMME, Equation (8), been successful throughout the nuclear chart, can provide a powerful method to determine masses, called the method of Coulomb displacement energies [28,79,89–92]. Namely, the mass excess of a proton-rich nucleus (with $M_T = -T$) on the basis of an experimental mass excess of its neutron-rich mirror (with $M_T = T$) and the theoretical $b$ coefficient as

$$\mathcal{M}(\eta, T, M_T = -T) = \mathcal{M}(\eta, T, M_T = T) - 2\, b(\eta, T)\, T\,. \qquad (17)$$

If theoretically Coulomb displacement energies are calculated, then they can be used straight instead of $2bT$ in Equation (18), as is done in Ref. [28,92]. Since the IMME is also applicable to describe excited multiplets, the method can be used to predict the positions of excited states in proton-rich nuclei.

Even more precise determination of the energy-level position is possible in triplets if two of three members of an isobaric multiplet are known experimentally:

$$\mathcal{M}(\eta, T, M_T = -1) = 2\mathcal{M}(\eta, T, M_T = 0) - \mathcal{M}(\eta, T, M_T = 1) + 2c(\eta, T)\,. \qquad (18)$$

Since the rms (root-mean-square) deviation for $c$ coefficients is typically smaller than that for $b$ coefficients, one would expect to have a smaller theoretical uncertainty value. These methods can be advantageous for determination of the level in proton-rich nuclei of astrophysical interest (e.g., [93]), as pointed out in Section 5.

The methods described above rely on the quadratic IMME given by Equation (8). Indeed, for isobaric mutliplets with $T > 1$, which involve more than three members, deviations from the quadratic law can be expected. An extended IMME equation would include terms proportional to $M_T^3$ and $M_T^4$, i.e.,

$$\mathcal{M}(\eta, T, M_T) = a(\eta, T) + b(\eta, T)M_T + c(\eta, T)M_T^2 + d(\eta, T)M_T^3 + e(\eta, T)M_T^4\,, \qquad (19)$$

which can be tested on quartets and quintets. Up till now, very few cases of non-zero $d$ or $e$ coefficients have been reported [5,6,94]; see also Refs. [95,96] and references therein. Typical values reach tens of keV.

Theoretically [94,97], deviations from a quadratic IMME are possible due to the presence of charge-dependent three-nucleon forces and/or due to isospin-mixing with nearby states. It is worth noting that the diagonalization of an INC shell-model Hamiltonian can generate an extended IMME, and several calculations have been reported [96,98]. To understand the challenge of getting reliable estimations of cubic and quartic terms on purely theoretical grounds, it is sufficient to notice that the rms errors of $b$ and $c$ coefficients are of the same order of magnitude or even larger than possible non-zero values of $d$ and $e$ coefficients. To avoid these ambiguities, a dedicated analysis constraining theory by available experimental information on $A = 32$ quintet have recently been performed [99]. Further efforts towards required precision will be crucial to advance our understanding of the origin of the IMME beyond its quadratic form.

### 3.2. Isospin-Forbidden Decays

Observation of transitions violating isospin selection rules, pointed to in the Introduction, signifies that the states are not pure in isospin. To predict theoretically the magnitudes of isospin impurities based on a fully microscopic calculation represents quite a complicated task. This can be understood as follows. In the shell model discussed here, isospin impurities arise from mixing of states of the same spin and parity but different isospin, if charge-dependent forces are present. Let us consider the simplest case of just two eigenstates of a charge-independent Hamiltonian, $|J^\pi, T\rangle$ and $|J^\pi, T'\rangle$, of isospin $T$ and $T'$, respectively. Inclusion of a charge-dependent interaction will result in new eigenstates, being linear combinations of unperturbed states, as

$$\begin{aligned} |a, J^\pi\rangle &= \sqrt{1-x^2}|J^\pi, T\rangle + x|J^\pi, T'\rangle \\ |b, J^\pi\rangle &= \sqrt{1-x^2}|J^\pi, T'\rangle - x|J^\pi, T\rangle \,. \end{aligned}$$

The mixing amplitude, $x$, in the first order is given by the ratio of the isospin-mixing matrix element and the energy difference between the two states:

$$x \sim \langle J^\pi, T|V_{\rm INC}|J^\pi, T'\rangle / \Delta E \,.$$

It is known that it is difficult to predict theoretically the energy difference between states, especially for an odd–odd nucleus. Uncertainties of a few hundred keV may result in huge uncertainty on the mixing probability. However, we would like to require from theory to robustly predict the value of the mixing matrix element, $\langle V_{\rm INC}\rangle$. In practice, there could be many-state mixing, and the theory should able to deal with such a problem.

Mixing matrix elements depend strongly on the structure of the states considered, and therefore require in each case a dedicated calculation. Systematic calculations of $\langle V_{\rm INC}\rangle$, and distinction between its long-range (Coulomb) and short-range contributions, may bring interesting information, especially when compared to available experimental data (see Refs. [16,49] for an earlier study). From various specific calculations, it seems that typical values of $\langle V_{\rm INC}\rangle$ are between a few keV to a few tens of keV. Maximum values are 150 keV in $p$-shell nuclei [23,100], around 100 keV for *sd*-shell nuclei [16] and about 50 keV in the $pf$ shell [101]. These estimations are in agreement with the largest observed values reported until now: $-145(20)$ keV for $^{8}$Be in the $p$ shell [100], 106(40) keV for $^{24}$Mg [102] in the *sd* shell and 40(23) keV for $^{56}$Cu [103] in the $pf$ shell. Although theoretical uncertainty on energy differences between admixed states hampers direct predictions of isospin impurities from theory, it often turns out that combining calculations with experimental data may be sufficient to constrain predictions, as illustrated in Section 3.2.3 below.

#### 3.2.1. Isospin-Forbidden $\beta$-Decay

To shed light on possible isospin impurities in nuclear states, one must appeal to isospin-forbidden transition probabilities. Let us remark that the only model-independent way to extract the amount of isospin-mixing from experiment is provided by Fermi $\beta$-decay [49]. Since the Fermi operator (9) is given by the isospin ladder operators $\hat{T}_\pm$, its matrix element between members of an isobaric multiplet can be expressed as

$$|M_{\rm F}^0| = |\langle T, M_T \pm 1|\hat{T}_\pm|TM_T\rangle| = \sqrt{(T+M_T)(T-M_T+1)} \,. \tag{20}$$

In isospin-symmetry limit, the whole strength would feed the IAS. A measured depletion of the Fermi strength from the IAS or observation of Fermi transitions to non-analogue states can bring information on the amount of isospin mixing in the IAS. In addition, if a $M_T > 0$ nucleus $\beta^+$ ($\beta^-$) decays, then the mixing is dominantly present in the parent (daughter) nucleus, and inversely for a $M_T < 0$ nucleus. Then, the isospin-forbidden Fermi-matrix element in a non-analogue state can be estimated as $|M_{\rm F}^{IF}| = |x||M_{\rm F}^0|$.

Special cases of purely Fermi, non-analogue $0^+ \to 0^+$ transitions are known, and they bring important information for the tests of the weak interaction in nuclear decays [104].

Distribution of the non-analogue Fermi strength, as experimentally measured recently in $^{62}$Ga [105], can shed light on the mixing matrix elements to cross-check the theory.

Transitions between states of the same (but non-zero) angular momentum, $J^\pi \to J^\pi$ ($J \neq 0$), are governed by both Fermi and Gamow–Teller components of the $\beta$-decay operator. A separation of the Gamow–Teller matrix element is an experimental challenge, bringing, however, direct knowledge on on isospin impurities, as elaborated in Refs. [49,106,107].

#### 3.2.2. Signatures of Isospin-Symmetry Breaking from Electromagnetic Transitions

Observation of other isospin-forbidden decays requires theoretical calculations of corresponding nuclear processes for extraction of the mixing probability. For example, electromagnetic transitions which violate isospin selection rules propose another possibility to test the degree of isospin-symmetry breaking.

Electric dipole transitions play a special role in these explorations due to a specific isovector character of the operator; see Equation (12). In particular, in Section 1.1, it was mentioned that $E1$ transitions between the states of the same isospin in self-conjugate ($N = Z$) nuclei are forbidden by isospin symmetry. A few cases of observation of weak $E1$ transitions in $N = Z$ nuclei between states of the same isospin have been reported [108,109]. This indicates breaking of isospin symmetry in the states involved in the decay. The shell-model calculation of individual $E1$ transition rates is hampered by the fact that the model space should contain orbitals of different parities, which could also lead to a center-of-mass motion. Given that the center-of-mass separation is only approximate, it is a challenge to give a precise estimation of the $E1$ strength. Observed enhancements of $E1$ rates in $N = Z$ nuclei and enhanced asymmetries of mirror $E1$ transitions can be related to the giant isovector monopole resonance [109].

An original idea of using $E2/M1$ (electric quadrupole/magnetic-dipole) mixing ratio of decays in a self-conjugate nucleus $^{54}$Co has been elaborated in Ref. [110] to pin down isospin impurities in a $4^+$ doublet.

Electromagnetic transitions between isobaric analogue states provide other possibilities to test isospin selection rules. For example, linear dependence of the $E2$ matrix elements on $M_T$ in $\Delta T = 1$ analogue transitions have been explored experimentally in a number of triplets (see Refs. [111,112] and references therein), and tests of equality of isovector matrix elements in mirror systems have been carried out [113,114].

An interesting idea to extract the amount of isospin mixing from $E1$ transition rates in mirror nuclei has been proposed and explored in Ref. [115].

Other possibilities to deduce isospin mixing in nuclear states from electromagnetic responses have been explored, e.g., in electron-scattering experiments [116] or via excitation of giant dipole resonance in $N = Z$ nuclei, e.g., in Refs. [117–119].

#### 3.2.3. $\beta$-Delayed Proton, Diproton or $\alpha$ Emission

Nucleon(s) emission may serve as a stringent test of isospin purity [120]. Interesting cases are provided by $\beta$-delayed proton (or two-proton, $\alpha$) emission when an IAS, populated in the $\beta$-decay, is located beyond the corresponding particle separation threshold [121,122]. As follows from a typical energy balance, in this case the proton (two-proton, alpha) emission from the IAS ($J^\pi, T$), populated in the $\beta$-decay of a $M_T < 0$ precursor, is forbidden by isospin symmetry (see Figure 4). Observation of such processes evidences the presence of isospin mixing, mainly, in the IAS which is surrounded by states of another isospin, ($J^\pi, T - 1$). A large amount of mixing can be deduced from the missing Fermi strength. However, small amounts may be hidden by experimental uncertainties.

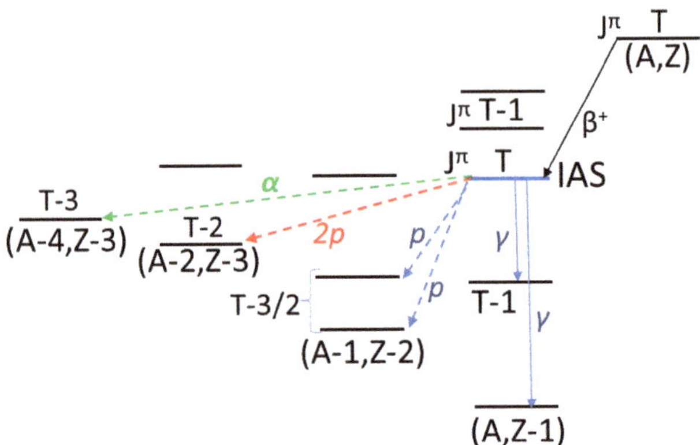


**Figure 4.** Schematic picture of $\beta$-delayed $p$, $\gamma$, $2p$ and $\alpha$ emission from an IAS. See text for details.

To deduce spectroscopic factors from isospin-forbidden proton emission on purely theoretical grounds is challenging [120,123]. Nevertheless, recently, it has been shown that one can deduce isospin mixing using experimental proton-$\gamma$branching ratios in the case of $\beta$-delayed $p\gamma$-emission [101,124] (two-proton or $\alpha$ emission were supposed to be absent or negligible in that study). Since the proton to $\gamma$-decay branching ratio for the IAS, $I_p^{\rm IAS}/I_\gamma^{\rm IAS}$, equals to the ratio of the corresponding widths, with the help of the theoretical electromagnetic width, $\Gamma_\gamma^{\rm IAS}$, one can extract the proton width of the IAS as

$$\Gamma_p^{\rm IAS} = \Gamma_\gamma^{\rm IAS} \frac{I_p^{\rm IAS}}{I_\gamma^{\rm IAS}} . \tag{21}$$

Generally, the shell model provides a relatively robust description of electromagnetic widths, if experimental energies are used. Deduced proton widths are important in astrophysics applications. For example, radiative proton capture is an inverse process, where a nucleus capturing a proton gets excited to a specific level and is de-excited by $\gamma$ emission. Proton and electromagnetic widths are thus essential ingredients with which to estimate the contribution of resonant capture.

In addition, if the angular momentum, $l$, of the proton is unambiguously determined (as in the decay from $0^+$ state), from Equation (21) one can deduce the spectroscopic factor for an isospin-forbidden proton emission from the IAS. To this end, one can estimate theoretical single-particle proton width, $\Gamma_{sp}^{\rm IAS}$, of the IAS and express the spectroscopic factor as

$$S_p^{\rm IAS} = \frac{\Gamma_\gamma^{\rm IAS}}{\Gamma_{sp}^{\rm IAS}} \frac{I_p^{\rm IAS}}{I_\gamma^{\rm IAS}} . \tag{22}$$

Let us remark that this estimation does not rely on the isospin mixing in the IAS, which would depend on energies of admixed states, but only on the experimental ratio of proton and gamma intensities and on the calculated width.

If a two-state mixing hypothesis approximately holds, for example, when the IAS is mostly mixed with a single nearby non-analogue state ($J^\pi, T-1$), then one can approximately estimate the amount of isospin mixing in the IAS. Namely, using the shell-model value for the spectroscopic factor of the admixed state, $S_p^{\rm T-1}$, the probability of mixing can be expressed as $x^2 = S_p^{\rm IAS}/ S_p^{\rm T-1}$. This procedure can be generalized to include isospin-forbidden $2p$ or $\alpha$-particle emission from the IAS.

For several measured proton branches that form the IAS, one can apply the above formalism to each of them separately, since relation (21) holds:

$$\Gamma_{p,i}^{\rm IAS} = \Gamma_{\gamma}^{\rm IAS} \frac{I_{p,i}^{\rm IAS}}{I_{\gamma}^{\rm IAS}} . \tag{23}$$

Proceed to extract spectroscopic factors and isospin mixing, if a two-level mixing model is applicable. This proposes an interesting possibility to cross-check the values of $x^2$ deduced from various branches. Such cases of $\beta$-delayed $p\gamma$ emission from an odd $A$ precursor have also been reported (see, e.g., Refs. [121,122,125]).

Actually, one can also determine an approximate value of isospin mixing in the IAS in a two-level mixing case, even if the set of quantum numbers $(nlj)$ characterizing the emitted proton is not unique. In this case, the proton width is a sum of contributing partial widths corresponding to all allowed orbitals from a given model space, e.g., $\Gamma_p^{\rm IAS} = \sum_{nlj} S_p^{\rm IAS}(nlj)\Gamma_{sp}^{\rm IAS}(nlj)$. Therefore, providing shell-model values of isospin-allowed spectroscopic factors, $S_p^{T-1}(nlj)$, one can estimate the amount of isospin impurity of the IAS to be

$$x^2 = \frac{\Gamma_p^{\rm IAS}}{\sum_{nlj} S_p^{T-1}(nlj)\Gamma_{sp}^{\rm IAS}(nlj)} , \tag{24}$$

where $\Gamma_p^{\rm IAS}$ is deduced as in Equation (21) and the denominator is evaluated theoretically. Finally, individual spectroscopic factors (for each $nlj$ channel) for isospin-forbidden proton emission can be obtained as $S_p^{\rm IAS}(nlj) = x^2 S_p^{T-1}(nlj)$ The uncertainty of theoretical estimation increases in this case, since a few theoretical quantities have to be used. In general, one should also remember that small spectroscopic factors (below 0.1) carry a significant theoretical uncertainty and this may prohibit extraction of the detailed information according to the proposed method.

## 4. Theoretical Isospin-Symmetry Breaking Corrections to Weak Processes in Nuclei

At present, many-body calculations for nuclear structure are required to link experimental information on weak processes involving nuclei to the underlying theories of fundamental interactions. In this context, the nuclear shell model is among the most favorite tools to provide nuclear matrix elements necessary for the tests of the symmetries of the standard model and for the searches for physics beyond it. Those can be probed in nuclear $\beta$-decay, but also in charge–exchange reactions or, eventually, in muon capture experiments. Calculations allowing to account for isospin-symmetry breaking may become vital in studies of individual transitions involving proton-rich nuclei and nuclei along $N = Z$ line.

The discussion below focuses on two activities related to the study of beta decay, which can be described by an effective axial-vecor and vector, $V$–$A$, interaction Hamiltonian,

$$\hat{H}_{V-A} = \frac{G_V}{\sqrt{2}} \hat{J}_\mu^\dagger \hat{j}^\mu + \text{h.c.} , \tag{25}$$

where $J_\mu^\dagger$ (($j^\mu$) is hadronic (leptonic) weak current, the index $\mu$ represents the space-time 4-vector index and takes valuse 0 (time),1, 2, and 3 (space), "h.c." stays for "hermitian conjugate", and $G_V$ is the weak-interaction coupling constant responsible for this semi-

leptonic decay. The most general form of a Lorentz-covariant form of the vector and axial-vector nucleon currents read

$$\hat{J}^{\dagger}_{\mu} = \hat{V}_{\mu} + \hat{A}_{\mu} \quad (26)$$

$$\hat{V}_{\mu} = i\overline{\psi}_p \left( g_V(k^2)\gamma_{\mu} + \frac{g_W(k^2)}{2m_N}\sigma_{\mu\nu}k_{\nu} + ig_S(k^2)k_{\mu} \right)\psi_n \quad (27)$$

$$\hat{A}_{\mu} = i\overline{\psi}_p \left( g_A(k^2)\gamma_{\mu} + \frac{g_T(k^2)}{2m_N}\sigma_{\mu\nu}k_{\nu} + ig_P(k^2)k_{\mu} \right)\gamma_5\psi_n \quad (28)$$

where $\psi_p$ and $\psi_n$ are nucleon field operators, $m_N$ is the nucleon mass; $k_{\mu}$ is the 4-momentum transferred from hadrons to leptons; $\sigma_{\mu\nu} = [\gamma_{\mu}, \gamma_{\nu}]/2i$ and $\gamma_{\mu}$ are Dirac matrices. The six form-factors are arbitrary real functions of Lorentz invariants of $k^2$, to be compatible with time-reversal invariance. At low momentum transfer, they are known as the vector ($g_V$), weak magnetism ($g_W$), scalar ($g_S$), axial-vector ($g_A$), tensor ($g_T$) and pseudo-scalar ($g_P$) coupling constants.

The six terms have definite properties under the $\hat{G}$-parity transformation, which is a combination of charge-conjugation ($\hat{C}$) and rotation in isospin space over 180 degrees about the 2-axis ($\hat{G} = \hat{C}\exp(i\pi\hat{T}_2)$). Those which transform as leading-order vector and axial-vector terms are called first-class currents, and those which have opposite transformation properties are called second-class currents. Of the latter type are the induced scalar term in the vector current and induced tensor term in the axial-vector current.

Various constraints on these coupling constants come from the symmetries underlying the standard model [126]. The most stringent condition is provided by the conserved vector current (CVC) hypothesis, which is based on the similarity in structure of the vector weak current and the isovector electromagnetic current. From CVC, it follows that the vector and weak-magnetism form factors are related to their electromagnetic counterparts, in particular, $g_V(k^2 \to 0) = 1$. This symmetry also implies that the induced scalar term vanishes ($g_S = 0$).

For the axial-vector current, only a partially conserved axial-vector current hypothesis exists, and it is less restrictive: it allows one to relate the main axial-vector coupling constant to the pion–nucleon coupling constant by famous Goldhaber–Trieman relations.

Nuclear $\beta$-decay experiments provide an excellent ground to test the structure of these currents and experimentally determine the magnitude of the coupling constants (see extensive reviews [127,128]). Two particular domains are described below, when theoretical calculation of nuclear matrix elements is required, along with an accurate treatment of isospin-symmetry breaking.

### 4.1. Superallowed Fermi $\beta$-Decay

The most prominent application of the theoretical formalism exposed just above is the calculation of realistic Fermi-matrix elements for $\beta$-decay between $0^+$ states or between the mirror states in $T = 1/2$ nuclei [129].

Indeed, Fermi type $\beta$-decay is governed uniquely by the vector part of the weak current. According to the CVC hypothesis, the absolute $Ft$ values of such transitions in various emitters with a given isospin $T$ should be the same. If this feature holds, from $Ft$ one can deduce the vector's coupling constant, $G_V$, that is responsible for this semi-leptonic decay ($u \to de^+\nu_e$). Combining $G_V$ with the value of fundamental weak coupling constant $G_F$ obtained from a purely leptonic muon decay ($\mu^+ \to e^+\nu_e\overline{\nu}_{\mu}$), one can determine the absolute value of the $|V_{ud}| = G_V/G_F$ matrix element of the Cabibbo–Kobayasi–Maskawa (CKM) quark-mixing matrix:

$$V_{\rm CKM} = \begin{pmatrix} V_{ud} & V_{us} & V_{ub} \\ V_{cd} & V_{cs} & V_{cb} \\ V_{td} & V_{ts} & V_{tb} \end{pmatrix}.$$

Numerical values of CKM matrix elements are important for the unitarity tests, such as the normalization condition for its first row: $|V_{ud}|^2 + |V_{us}|^2 + |V_{ub}|^2 = 1$.

The absolute $Ft$ value is obtained from the experimentally deduced $ft$ value after incorporation of a few non-negligible theoretical corrections [130] as defined by the following equation:

$$Ft^{0^+\to 0^+} \equiv ft^{0^+\to 0^+}(1+\delta_R')(1+\delta_{NS}-\delta_C) = \frac{K}{|M_{\rm F}^0|^2 G_V^2 (1+\Delta_R)}. \tag{29}$$

Here, $f$ is the statistical rate function calculated from the decay energy, $t$ is the partial half-life of the transitions, $K = 2\pi^3\hbar \ln 2(\hbar c)^6/(m_e c^2)^5$, $|M_{\rm F}^0|$ is the Fermi-matrix element in the isospin-symmetry limit (20), $\hbar$ is the reduced Planck's constant, $c$ is the speed of light, and $\Delta_R$, $\delta_R'$ and $\delta_{NS}$ are transition-independent, transition-dependent and nuclear-structure-dependent radiative corrections; and $\delta_C$ is the isospin-symmetry breaking correction due to the lost analogue symmetry between the parent and the daughter nuclear states. The detailed description of the electroweak corrections and the current status in the field can be found in the latest survey by Hardy and Towner [130]. The most prominent feature discussed in recent years is an updated value of the transition independent term, $\Delta_R$, which was re-evaluated using the formalism of the effective field theory, and this brought fragility to the unitarity tests [131].

The present discussion focuses on the isospin-symmetry breaking correction, $\delta_C$. This correction is defined as a deviation of the squared realistic Fermi-matrix element from its isospin-symmetry value: $|M_{\rm F}|^2 = |M_{\rm F}^0|^2(1-\delta_C)$. Therefore, the estimation of $\delta_C$ requires an accurate calculation within a nuclear-structure model which can account for the broken isospin symmetry.

There have been lots of efforts within various theoretical approaches during a few decades already. Figure 5 summarizes predictions from different calculations for the 13 best known transitions (by now, the decay of $^{26}$Si has been added to this dataset).

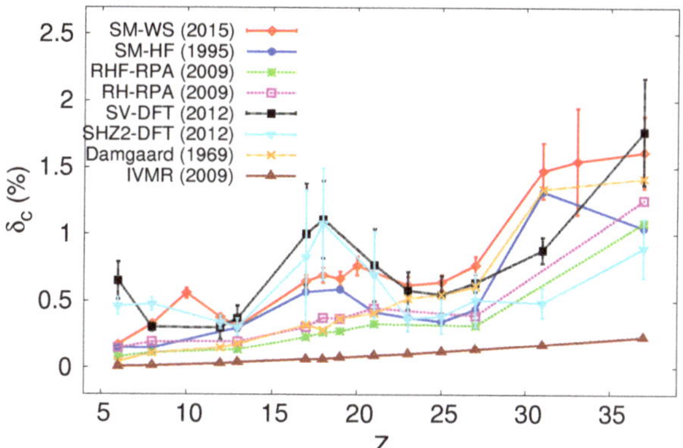


**Figure 5.** Isospin-symmetry breaking correction, $\delta_C$, from various theoretical approaches: SM-WS(2015) [130], SM-HF(1995) [132], RHF-RPA(2009) [39], RH-RPA(2009) [39], SV-DFT(2012) [42], SHZ2-DFT(2012) [42], Damgaard(1969) [133], IVMR(2009) [134]. Figure is adapted from Ref. [135].

The values obviously diverge. In addition, to note is that theoretical approaches assign important uncertainties to their values (those which present associated uncertainties). Currently, evaluation of $\delta_C$ provides the largest contribution to the $Ft$-value uncertainty.

As has been demonstrated in Section 2 above, the INC shell model represents a well-suited tool for the $\delta_C$ calculation. By expressing the Fermi-matrix elements in the second quantization, one gets:

$$M_F = \langle \Psi_f | \hat{T}_+ | \Psi_i \rangle = \sum_\alpha \langle f | \hat{c}^\dagger_{\alpha_n} \hat{c}_{\alpha_p} | i \rangle \langle \alpha_n | \hat{t}_+ | \alpha_p \rangle \,, \tag{30}$$

where $\hat{c}^\dagger_\alpha$ and $\hat{c}_\alpha$ are nucleon creation (destruction) operators; $\alpha$ denotes a full set of spherical quantum numbers, $\alpha = (n_a, l_a, j_a, m_a) \equiv (a, m_a)$ and the two ingredients of Equation (30) are (i) one body-transition densities:

$$\langle f | \hat{c}^\dagger_{\alpha_n} \hat{c}_{\alpha_p} | i \rangle \equiv \rho_\alpha \,, \tag{31}$$

and isospin single particle matrix elements, given by overlap integrals:

$$\langle \alpha_n | \hat{t}_+ | \alpha_p \rangle = \int_0^\infty R_{\alpha_n}(r) R_{\alpha_p}(r) r^2 dr \equiv \Omega_\alpha . \tag{32}$$

Here, $R_\alpha$ denotes the radial part of the single-particle wave function.

It has been pointed out by Miller and Schwenk [136,137] that the use of the exact isospin operator in Equation (30) would involve terms where the radial quantum number, $n_\alpha$, for of a proton state, $\alpha_p$, is different from that of a neutron state $\alpha_n$. Up till now, all shell-model work [132,135,138,139] has been done within an approximation that allows one to express the radial overlaps by Equation (32).

Calculation of realistic Fermi-matrix elements implies the use of one-body transition densities computed using many-body states obtained from the diagonalization of an INC Hamiltonian, and the use of radial wave functions, obtained from a realistic spherical single-particle potential, such as Wood–Saxon (WS) or Hartree–Fock (HF) potential, instead of the harmonic oscillator. Therefore,

$$M_F = \sum_\alpha \rho_\alpha \Omega_\alpha \,,$$

and the model-independent value (20) can be obtained from one-body transitions densities in the isospin limit ($\rho^0_\alpha$) and harmonic-oscillator radial overlaps ($\Omega^0_\alpha = 1$):

$$M^0_F = \sum_\alpha \rho^0_\alpha \Omega^0_\alpha = \sum_\alpha \rho^0_\alpha$$

(the superscript "0" indicates that those quantities were calculated in the isospin limit). Therefore, there are two sources of isospin-symmetry breaking in the Fermi-matrix element: first comes from the difference in configuration mixing of the parent and daughter nuclei as obtained from the shell-model diagonalization of an INC Hamiltonian. The other is due to the deviation of the radial overlaps from unity, when calculated with realistic single-particle wave functions instead of the harmonic-oscillator ones. These deviations of one-body transitions densities and radial overlaps from their isospin-symmetry values are typically small. Keeping only linear terms in small quantities, one can express $|M_F|^2$ as

$$|M_F|^2 \approx |M^0_F|^2 \Big[ 1 - \underbrace{\frac{2}{M^0_F} \sum_\alpha \left( \rho^0_\alpha - \rho_\alpha \right)}_{\delta_{C1}} - \underbrace{\frac{2}{M^0_F} \sum_\alpha \rho^0_\alpha (1 - \Omega_\alpha)}_{\delta_{C2}} \Big] ,$$

From this expression, it is seen that the correction splits into two terms according to the two sources of isospin-symmetry breaking mentioned above:

$$\delta_C \approx \delta_{C1} + \delta_{C2} \,.$$

To get $\delta_{C1}$, it is sufficient to perform calculations with INC interactions. As has been discussed in Section 3, the theoretical value for a depletion of the Fermi strength in the IAS is due to the mixing of the IAS with non-analogue states (see Figure 6 (left)). Therefore, the position of those states is vital.

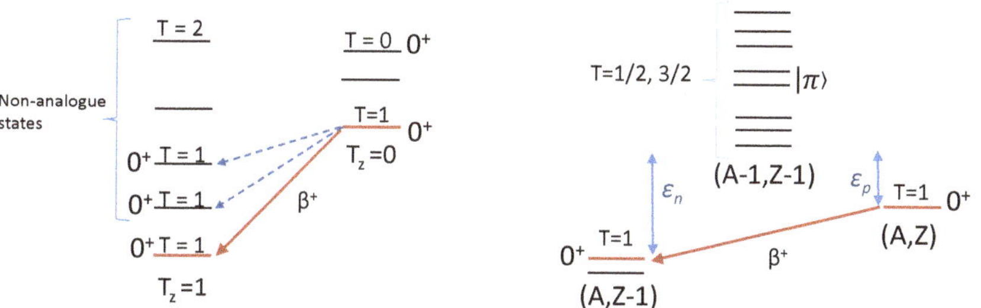


**Figure 6.** Schematic picture of the Fermi strength distribution in the daughter nucleus due to the isospin-symmetry breaking effects, as can be viewed from the shell-model's perspective. **Left**: depletion of the Fermi strength from an IAS because of non-analogue transitions. **Right**: insertion of the intermediate states to better constrain the radial part of the single-particle wave functions.

To avoid this uncertainty, one may scale the strengths of individual transitions to non-analogue states with the energy difference between those states and the IAS [138]:

$$\delta_{C1} = \delta_{C1}^{\text{th}} \left( \frac{\Delta E_{\text{th}}}{\Delta E_{\text{exp}}} \right)^2 .$$

Existing shell-model studies use various parametrizations of the INC Hamiltonian, ranging from realistic phenomenological fits [132,135,140] to individual parametrization of charge-dependent terms to each isobaric multiplet presented in Ref. [138,139]. Since this part of the correction is small, the results of both approaches are within typical uncertainties.

In addition to the isospin-symmetry breaking inside the model space, one has to replace harmonic-oscillator radial wave functions with realistic spherically symmetric wave functions from a WS or a HF potential, including Coulomb. This is the largest part of the correction; see Ref. [138] and references therein. A parametrization of a single-particle potential is crucial for the value of the correction. Due to this reason, potential parameters are adjusted to reproduce proton and neutron separation energies and nuclear charge radii. To achieve this goal, a calculation has to be done beyond the closure approximation. This means instead of inserts, a complete sum of intermediate nucleus states ($\{\pi\}$) in the Fermi-matrix element. Then, the radial-overlap correction can be expressed as

$$\delta_{C2} = \frac{2}{M_{\text{F}}^0} \sum_{\alpha,\pi} \langle f | \hat{c}_{\alpha_n}^{\dagger} | \pi \rangle^0 \langle \pi | \hat{c}_{\alpha_p} | i \rangle^0 (1 - \Omega_\alpha^\pi) .$$

The two ingredients are the spectroscopic amplitudes, $\langle f | \hat{c}_{\alpha_n}^{\dagger} | \pi \rangle^0$, obtained within the isospin-symmetry limit, and the radial-overlap integrals

$$\Omega_\alpha^\pi = \int_0^\infty R_{\alpha_n}^\pi(r) R_{\alpha_p}^\pi(r) r^2 dr ,$$

which now contain dependence on the excitation energy of the intermediate states $\pi$; see Figure 6 (right).

This opportunity to constrain theoretical calculations by experimental observables greatly helps to reduce uncertainty in the potential parameters and guaranties consistency of the results, as has been discussed in detail in Ref. [135]. In particular, the largest

contribution to theoretical uncertainty on $\delta_{C2}$ is because of the experimental uncertainty on the nuclear charge radii.

Up till now, systematic calculations with the WS potential are the only ones who produce corrections consistent with the CVC hypothesis within a non-zero confidence limit [141]. The use of the HF wave functions, pioneered by Ormand and Brown [132,140,142], has been explored by a few other groups as well [138,143,144]. Self-consistent HF potentials are not immediately appropriate for calculations and have to also be adjusted to give rise to experimental proton and neutron separation energies. The procedures exploited by various authors are somewhat different, and, in general, led to smaller corrections than those obtained from a WS potential. This issue has recently been explored in detail in Ref. [144]. In particular, the authors examined the role of previously neglected effects, taking care of the approximate elimination of spurious isospin-mixing, two-body center-of-mass corrections, exact treatment of the exchange Coulomb term and many others. Moreover, INC terms have been added to the energy-density functional. Those corrective terms indeed explain some of the difference between the HF and WS results, allowing to suppose that the remaining part of the difference is due to the need for correlations beyond the HF approximation. Further efforts towards more sophisticated theories should be addressed in future studies.

In spite of these challenges in the computation of theoretical corrections, nuclear $0^+ \to 0^+$ $\beta$-decay provides the best opportunity to test the CVC and to extract the $V_{ud}$ value, among other ways (mirror transitions, neutron or pion beta decay) [130]. Therefore, it is reasonable to persist with efforts in improving theoretical modelization of the isospin-symmetry breaking correction.

### 4.2. *β-Decay between Mirror $T = 1/2$ States*

It was pointed also that $\beta$-decay between mirror states with $T = 1/2$, which are governed by both Fermi and Gamow–Teller operators, can also serve for the tests of the CVC hypothesis and extraction of $V_{ud}$, once the GT component is eliminated [130,145]. To this end, an additional correlation coefficient has to be measured. Similarly to $0^+ \to 0^+$ decay, the experimentally determined $ft$ value has to be corrected for radiative effects and for isospin-symmetry breaking in decaying states. The shell-model framework relies on a similar expression of the realistic Fermi-matrix element, as discussed, with an intermediate state summation which involves a larger number of different states because of non-zero values of angular momenta involved. Currently achieved results are summarized in Ref. [129].

### 4.3. *Gamow–Teller Transitions in Mirror Nuclei*

Another long-standing application is related to the asymmetry of Gamow–Teller $\beta$-decay rates in mirror nuclei, defined as

$$\delta = \left| \frac{M^{+}_{\rm GT}}{M^{-}_{\rm GT}} \right|^2 - 1\,, \tag{33}$$

where $M^{\pm}_{\rm GT}$ are reduced matrix elements for mirror GT $\beta^{\pm}$ transitions. The initial interest in the topic was due to the fact that the contribution to that asymmetry may be due to the presence of the induced tensor term ($g_T$) in the axial-vector current; see Equation (28).

To pin down a possible manifestation of the induced tensor term, an accurate calculation of GT matrix elements, including isospin-symmetry breaking, is required. In the second-quantization formalism, the reduced matrix elements of the GT operator can be expressed as follows:

$$M^{\pm}_{\rm GT} = \langle \Psi_f || \hat{O}_{\rm GT}(\beta^{\pm}) || \Psi_i \rangle = \frac{1}{\sqrt{3}} \sum_{a,b} \langle J_f || [\hat{c}^{\dagger}_a \, \hat{c}_b]^{(1)} || J_i \rangle \langle a, m_{t_a} || \hat{\sigma} \hat{t}_{+} || b, m_{t_b} \rangle\,, \tag{34}$$

where double bars denote reduction in angular momentum, and $\hat{\tilde{c}}_{a,m_a} = (-1)^{j_a+m_a}\hat{c}_{a,-m_a}$. Again, realistic calculations should be ensured beyond the closure approximation, thereby inserting a complete sum over intermediate nucleus states. Several theoretical investigations have been performed from the 60s up to the present, without any indication on a possible application of the analysis to the weak-interaction problem because of high theoretical uncertainty of the nuclear wave functions (see Ref. [146,147] and references therein). Although experimental measurements of mirror transitions main an active field, the main impact of the results is on the structural aspects of the states involved in the decay. In this context, alternative constraints on the induced tensor current from $\beta$-$\alpha$ and $\beta$-$\gamma$ angular correlation experiments tend to be much more advantageous [148,149].

## 5. Astrophysical Applications

One of the greatest motivations to explore the properties of nuclei is their need for nuclear astrophysics. Nuclear masses, half-lives, level densities, and nuclear, electromagnetic and weak-interaction reaction rates represent crucial ingredients for simulations and understanding of astrophysical processes [150]. In particular, the structure of neutron-deficient nuclei is important for comprehension of nucleosynthesis during stellar explosive hydrogen burning. Among the possible sites are X-ray bursts and novae outbursts.

Novae are understood as a result of thermonuclear runaway at the surface of a white dwarf within a binary star system. At high temperatures ($\sim 10^8$ K) and densities in O-Ne type novae, the break-out of the hot CNO (carbon-nitrogen-oxygen) cycle leads to nucleosynthesis of heavier elements $A \geq 20$ by mainly $(p,\gamma)$, in competition with $(\alpha,p)$ and inverse reactions, with the end point around Ca [151]. In X-ray bursts [152,153], based on a neutron star accreting hydrogen matter within a binary system, the temperatures are even higher (up to $2 \times 10^9$ K), and radiative proton capture reactions involve proton-rich nuclei towards the proton-drip line, being the most important reaction type in nucleosynthesis with up to roughly $A \sim 100$ ($rp$ process). Simulations of X-ray bursts exploit a huge set of nuclear reactions which have to be constrained.

For stable nuclei, the proton-capture reaction $Q$-values are relatively high, and the reaction rate may be approximated by a statistical model. For unstable (proton-rich) nuclei, $Q$ values become small (in the order of a few MeV or less), and hence, the reaction rate is dominated by a few isolated resonances above the proton-emission threshold, together with a non-resonant reaction contribution in the energy range within a Gamow peak. In this case, accurate knowledge of resonance energies and decay widths is required.

Current state-of-the-art simulations are based on experimentally deduced information when it is available. If no data exist yet, then one can either deduce the missing information from mirror systems, assuming the isospin symmetry, or appeal to theory. Therefore, higher-precision theoretical calculations are important to reduce uncertainties.

Shell modeling is one of the approaches which can provide detailed information on nuclear states and transitions at low energies. The resonant part of a thermonuclear $(p,\gamma)$ reaction rate for a single resonance can be expressed [154] as

$$N_A\langle\sigma v s.\rangle_r = 1.540 \times 10^{11}\,(\mu T_9)^{-3/2}\,\omega\,\gamma\,\exp\left(\frac{-11.605\,E_r}{T_9}\right)\,\mathrm{cm}^3\,\mathrm{s}^{-1}\,\mathrm{mol}^{-1}, \qquad (35)$$

where $\mu = A_pA/(A_p + A)$ is the reduced atomic mass number and $E_r$ is the resonance energy above the proton-emission threshold (in MeV), $T_9$ is the temperature in GK. The resonance strength $\omega\,\gamma$ (in MeV) depends on the spins of initial $J_i$ and final (resonance) $J_f$ states, the partial proton width $\Gamma_p$ for the entrance channel and gamma widths $\Gamma_\gamma$ for the exit channel:

$$\omega\gamma = \frac{2J_f + 1}{2(2J_i + 1)}\frac{\Gamma_p\Gamma_\gamma}{\Gamma_{tot}}, \qquad (36)$$

with $\Gamma_{tot} = \Gamma_p + \Gamma_\gamma$. The proton decay width depends on the resonance energy via the proton width, which could be estimated from a single-particle potential model and the

shell model's spectroscopic factor. In case of a few resonances, the resonant reaction rate represents a sum of single-resonance rates (35) over contributing final states.

The non-resonant part of $(p,\gamma)$, the reaction rate is given by direct capture transitions to the ground or low-level states of the final nucleus.

A number of radiative proton-capture reaction rates have been evaluated with a good precision for *sd*-shell and $pf$-shell nuclei [155,156], since the shell model provides missing information on resonance states, proton and electromagnetic widths. The INC formalism is of particular interest for such problems. First, using theoretical IMME $b$ coefficients, one can not only provide nuclear masses of proton-rich nuclei [28,91,92], but also determine positions of unknown resonances in a proton-rich nucleus, if the level scheme of a neutron-rich mirror nucleus is known experimentally. The necessity to account for isospin-symmetry breaking to get more accurate results was demonstrated first in Ref. [155] and followed in numerous studies. A use of theoretical $c$ coefficients may even be more advantageous for an $M_T = -1$ nuclei if experimental information on $M_T = 0, 1$ exists (see, e.g., [93,157,158].) The cross-shell $p$-$sd$-$pf$ model space is necessary for the description of negative parity resonances in *sd*-shell nuclei (see, e.g., Refs. [159,160]).

A particularly interesting result was obtained a few years ago, indicating that the Thomas–Ehrman effect may significantly change values of theoretical spectroscopic factors [87]. More attention therefore has to be paid to small values of spectroscopic factors. This also indicates that results on spectroscopic factors from mirror systems should be accepted with caution.

## 6. Conclusions and Perspectives

The nuclear shell model provides a powerful formalism with which to deal with tiny breaking of isospin symmetry in nuclear states. Currently, the most accurate results are due to phenomenological treatment of nuclear wave functions and parametrization of the INC terms of the Hamiltonian. Although more work is required to have a better handle on large model spaces, extended applications to structure and decay proton-rich nuclei and nuclei along the $N = Z$ line support experimental investigations. Important applications of that formalism exist, such as the calculation of isospin-symmetry breaking corrections for Fermi-matrix elements required to test the symmetries underlying the standard model. Finally, isospin-symmetry breaking is nowadays taken into account in the evaluation of thermonuclear reaction rates in proton-rich nuclei, which plays an important role in astrophysical simulations.

While phenomenological approaches still have to be pursued to assure solid support to experimental investigations, the eventual goal of nuclear theorists is to develop fundamental ab initio frameworks for many-body calculations towards a higher precision level that will be relevant for the isospin-symmetry breaking domain.

**Funding:** This research was funded by Master Projects *Isospin-symmetry breaking* (2017–2020) and *Exotic Nuclei, Fundamental Interactions and Astrophysics (ENFIA)* (2020–2023).

**Data Availability Statement:** The data presented can be found in the references cited.

**Acknowledgments:** The author acknowledges collaboration with N. Benouaret, B. Blank, B. A. Brown, Y. H. Lam, W. A. Richter, C. Volpe, and L. Xayavong on different topics related to isospin-symmetry breaking. Large-scale calculations have been performed at MCIA, University of Bordeaux.

**Conflicts of Interest:** The author declares no conflict of interest.

## Abbreviations

The following abbreviations are used in this manuscript

| | |
|---|---|
| CD | charge-dependent |
| CKM | Cabbibo-Kobayashi-Moskawa |
| CNO | carbon-nitrogen-oxygen |
| CVC | conserved vector current |
| $E1$, $E2$ | electric-dipole, electric-quadrupole |
| EFT | effective field theory |
| F | Fermi |
| GT | Gamow-Teller |
| h.c. | hermitian congugate |
| HF | Hartree–Fock |
| IAS | isobaric analogue state |
| IMME | isobaric-multiplet mass equation |
| IMSRG | in-medium similarity-renormalization group |
| INC | Isospin-nonconserving |
| $M1$ | magnetic-dipole |
| MED | mirror energy difference |
| $N^3LO$ | next-to-next-to-next-to-leading |
| $NN$ | nucleon–nucleon |
| rms | root mean square |
| TED | triplet energy difference |
| TMBE | two-body matrix element |
| $V$–$A$ | vector–axial vector |
| WS | Wood-Saxon |
| USD | universal *sd* shell |
| $\chi$EFT | chiral effective field theory |
| $3N$ | three-nucleon |

## References


1. Heisenberg, W. Über den Bau der Atomkerne. I. *Z. Phys.* **1932**, *77*, 1–11. [CrossRef]
2. Hesenberg, W. On the structure of atomic nuclei. I. In *Nuclear Forces*; Brink, D.M., Ed.; Pergamon Press: Oxford, UK, 1965; p. 214.
3. Wigner, E. On the consequences of the symmetry of the nuclear Hamiltonian on the spectroscopy of nuclei. *Phys. Rev.* **1937**, *51*, 106–119. [CrossRef]
4. Wigner, E. Isotopic spin—A quantum number for nuclei. In *Proceedings of the Robert A. Welch Foundation Conference on Chemical Research*; Milligan, W.O., Ed.; Welch Foundation: Houston, TX, USA, 1957; Volume 1, pp. 67–91.
5. Lam, Y.H.; Blank, B.; Smirnova, N.A.; Antony, M.S.; Bueb, J. The isobaric multiplet mass equation for $A \leq 71$ revisited. *At. Data Nucl. Data Tables* **2013**, *99*, 680–703. [CrossRef]
6. MacCormick, M.; Audi, G. Evaluated experimental isobaric analogue states from $T = 1/2$ to $T = 3$ and associated IMME coefficients. *Nucl. Phys. A* **2014**, *925*, 61–95. [CrossRef]
7. Frank, A.; Jolie, A.; Van Isacker, P. *Symmetries in Atomic Nuclei*; Springer Science+Business Media, LLC: New York, NY, USA, 2009. [CrossRef]
8. Warburton, E.K.; Weneser, J. The role of isospin in electromagnetic transitions. In *Isospin in Nuclear Physics*; Wilkinson, D.H., Ed.; North-Holland Publishing Co.: Amsterdam, The Netherlands, 1969; pp. 152–228.
9. Wilkinson, D.H. (Ed.) *Isospin in Nuclear Physics*; North-Holland Publishing Co.: Amsterdam, The Netherlands, 1969.
10. Harney, H.L.; Richter, A.; Weidenmüller, H.A. Breaking of isospin symmetry in compound-nucleus reactions. *Rev. Mod. Phys.* **1986**, *58*, 607–645. [CrossRef]
11. Fujita, Y.; Rubio, B.; Gelletly, W. Spin–isospin excitations probed by strong, weak and electro-magnetic interactions. *Prog. Part. Nucl. Phys.* **2011**, *66*, 549–606. [CrossRef]
12. Machleidt, R. The meson theory of nuclear forces and nuclear structure. In *Advances in Nuclear Physics. Volume 19*; Negele, J.W., Vogt, E., Eds.; Plenum Press: New York, NY, USA, 1989; Chapter 2. [CrossRef]
13. Epelbaum, E.; Hammer, H.-W. ; Meißner, U.-G. Modern theory of nuclear forces. *Rev. Mod. Phys.* **2009**, *81*, 1773–1825. [CrossRef]
14. Nolen, J.A.; Schiffer, J.P. Coulomb energies. *Ann. Rev. Nucl. Sci.* **1969**, *19*, 471–526. [CrossRef]
15. Ormand, W.E.; Brown, B.A. Empirical isospin-nonconserving Hamiltonians for shell-model calculations. *Nucl. Phys. A* **1989**, *491*, 1–23. [CrossRef]
16. Nakamura, S.; Muto, K. ; Oda, T. Isospin-forbidden beta decays in *ls0d*-shell nuclei. *Nucl. Phys. A* **1994**, *575*, 1–45. [CrossRef]
17. Zuker, A.P.; Lenzi, S.M.; Martinez-Pinedo, G.; Poves, A. Isobaric multiplet yrast energies and isospin nonconserving forces. *Phys. Rev. Lett.* **2002**, *89*, 142502. [CrossRef] [PubMed]

18. Henley, E.M.; Miller, G.A. Meson theory of charge-dependent nuclear forces. In *Mesons in Nuclei. Volume 1*; Rho, M., Wilkinson, D.H., Eds.; North-Holland Publishing Co.: Amsterdam, The Netherlands, 1979; pp. 405–434.
19. Miller, G.A.; Nefkens, B.M.K.; Slaus, I. Charge symmetry, quarks and mesons. *Phys. Rep.* **1990**, *194*, 1–116. [CrossRef]
20. van Kolck, U.L. Soft Physics: Applications of Effective Chiral Lagrangians to Nuclear Physics and Quark Models. Ph.D. Thesis, The University of Texas at Austn, Austin, TX, USA, 1993. Available online: https://www.proquest.com/openview/b885fad2126b5b81a16dca7d226f854a/ (accessed on 7 March 2023).
21. Epelbaum, E. Few-nucleon forces and systems in chiral effective field theory. *Prog. Part. Nucl. Phys.* **2006**, *57*, 654–741. [CrossRef]
22. Machleidt, R.; Entem, D.R. Chiral effective field theory and nuclear forces. *Phys. Rep.* **2011**, *503*, 024001. [CrossRef]
23. Wiringa, R.B.; Pastore, S.; Pieper, S.C.; Miller, G.A. Charge-symmetry breaking forces and isospin mixing in $^8$Be. *Phys. Rev. C* **2013**, *88*, 044333. [CrossRef]
24. Barrett, B.R.; Navrátil, P.; Vary, J.P. *Ab initio* no core shell model. *Prog. Part. Nucl. Phys.* **2013**, *57*, 654–741. [CrossRef]
25. Maris, P.; Epelbaum, E.; Furnstahl, R.J.; Golak, J.; Hebeler, K.; Hüther, T.; Kamada, H.; Krebs, H.; Meißner, U.-G.; Melendez, J.A.; et al. Light nuclei with semilocal momentum-space regularized chiral interactions up to third order. *Phys. Rev. C* **2021**, *103*, 054001. [CrossRef]
26. Caprio, M.A.; Fasano, P.J.; Maris, P.; McCoy, A.E. Quadrupole moments and proton-neutron structure in $p$-shell mirror nuclei. *Phys. Rev. C* **2021**, *104*, 034319. [CrossRef]
27. Lam, Y.H.; Smirnova, N.A.; Caurier, E. Isospin nonconservation in $sd$-shell nuclei. *Phys. Rev. C* **2013**, *87*, 054304. [CrossRef]
28. Kaneko, K.; Sun, Y.; Mizusaki, T.; Tazaki, S. Variation in displacement energies due to isospin-nonconserving forces. *Phys. Rev. Lett.* **2013**, *110*, 172505. [CrossRef]
29. Kaneko, K.; Sun, Y.; Mizusaki, T.; Tazaki, S. Isospin-nonconserving interaction in the $T = 1$ analogue states of the mass-70 region. *Phys. Rev. C* **2014**, *89*, 031302. [CrossRef]
30. Holt, J.D.; Menendez, J.; Schwenk, A. Three-body forces and proton-rich nuclei. *Phys. Rev. Lett.* **2013**, *110*, 022502. [CrossRef] [PubMed]
31. Bentley, M.; Lenzi, S.M.; Simpson, S.A.; Diget, C.A. Isospin-breaking interactions studied through mirror energy differences. *Phys. Rev. C* **2015**, *92*, 024310. [CrossRef]
32. Lenzi, S.M.; Bentley, M.; Lau, R.; Diget, C.A. Isospin-symmetry breaking corrections for the description of triplet energy differences. *Phys. Rev. C* **2018**, *98*, 054322. [CrossRef]
33. Ormand, W.E.; Brown, B.A.; Hjorth-Jensen, M. Realistic calculations for $c$ coefficients of the isobaric mass multiplet equation in $1p0f$ shell nuclei. *Phys. Rev. C* **2017**, *96*, 024323. [CrossRef]
34. Magilligan, A.; Brown, B.A. New isospin-breaking "USD" Hamiltonians for the $sd$ shell. *Phys. Rev. C* **2020**, *101*, 064312. [CrossRef]
35. Martin, M.S.; Stroberg, S.R.; Holt, J.D.; Leach, K.G. Testing isospin symmetry breaking in *ab initio* nuclear theory. *Phys. Rev. C* **2021**, *104*, 014324. [CrossRef]
36. Caurier, E.; Navrátil, P.; Ormand, W.E.; Vary, J.P. *Ab initio* shell model for $A = 10$ nuclei. *Phys. Rev. C* **2002**, *66*, 024314. [CrossRef]
37. Michel, N.; Nazarewicz, W.; Płoszajczak, M. Isospin mixing and the continuum coupling in weakly bound nuclei. *Phys. Rev. C* **2010**, *82*, 044315. [CrossRef]
38. Sagawa, H.; Van Giai, N.; Suzuki, T. Effect of isospin mixing on superallowed Fermi $\beta$ decay. *Phys. Rev. C* **1996**, *53*, 2163–2170. [CrossRef]
39. Liang, H.; Van Giai, N.; Meng, J. Isospin corrections for superallowed Fermi $\beta$ decay in self-consistent relativistic random-phase approximation approaches. *Phys. Rev. C* **2009**, *79*, 064316. [CrossRef]
40. Petrovici, A. Isospin-symmetry breaking and shape coexistence in $A \approx 70$ analogs. *Phys. Rev. C* **2015**, *91*, 014302. [CrossRef]
41. Satuła, W.; Dobaczewski, J.; Nazarewicz, W.; Rafalski, M. Microscopic calculations of isospin-breaking corrections to superallowed beta decay. *Phys. Rev. Lett.* **2011**, *106*, 132502. [PubMed]
42. Satula, W.; Dobaczewski, J.; Nazarewicz, W.; Rafalski, M. Isospin-breaking corrections to superallowed Fermi $\beta$ decay in isospin- and angular-momentum-projected nuclear density functional theory. *Phys. Rev. C* **2012**, *86*, 054316. [CrossRef]
43. Satuła, W.; Bączyk, P.; Dobaczewski, J.; Konieczka, M. No-core configuration-interaction model for the isospin- and angular-momentum-projected states. *Phys. Rev. C* **2016**, *94*, 024306. [CrossRef]
44. Bączyk, P.; Dobaczewski, J.; Konieczka, M.; Nakatsukasa, T.; Sato, K.; Satula, W. Isospin-symmetry breaking in masses of $N \approx Z$ nuclei. *Phys. Lett. B* **2018**, *778*, 178–183. [CrossRef]
45. Baczyk, P.; Satula, W.; Dobaczewski, J.; Konieczka, M. Isobaric multiplet mass equation within nuclear density functional theory. *J. Phys. G* **2019**, *46*, 03LT01. [CrossRef]
46. Roca-Maza, X.; Colò, G.; Sagawa, H. Nuclear symmetry energy and the breaking of the isospin symmetry: How do they reconcile with each other? *Phys. Rev. Lett.* **2018**, *120*, 202501. [CrossRef]
47. Naito, T.; Colò, G.; Liang, H.; Roca-Maza, X.; Sagawa, H. Toward *ab initio charge* symmetry breaking in nuclear energy density functionals. *Phys. Rev. C* **2022**, *105*, L021304. [CrossRef]
48. Bertsch, G.F.; Mekjian, A. Isospin impurities in nuclei. *Ann. Rev. Nucl. Sci.* **1972**, *22*, 25–64. [CrossRef]
49. Raman, S.; Walkiewicz, T.A.; Behrens, H. Superallowed $0^+ \to 0^+$ and isospin-forbidden $J^\pi \to J^\pi$ Fermi transitions. *At. Data Nucl. Data Tables* **1975**, *16*, 451–494. [CrossRef]
50. Auerbach, N. Coulomb effects in nuclear structure. *Phys. Rep.* **1983**, *98*, 273–341. [CrossRef]

51. Brussaard P.J.; Glaudemans P.W.M. *Shell-Model Applications in Nuclear Spectroscopy*; North-Holland Publishing Co.: Amsterdam, The Netherlands, 1977.
52. Heyde, K.L.G. *The Nuclear Shell Model*; CRC Press/Taylor & Francis Group: Boca Raton, FL, USA, 2004. [CrossRef]
53. Suhonen, J. *From Nucleons to Nucleus*; Springer: Heidelberg/Berlin, Germany, 2007. [CrossRef]
54. Caurier, E.; Martínez-Pinedo, G.; Nowacki, F.; Poves, A.; Zuker, A.P. The shell model as a unified view of nuclear structure. *Rev. Mod. Phys.* **2005**, *77*, 427-488. [CrossRef]
55. Smirnova, N.A. Isospin-symmetry breaking in nuclear structure. *Nuovo Cim. C* **2019**, *42*, 54. [CrossRef]
56. Bertsch, G.F. Role of core polarization in two-body interaction. *Nucl. Phys.* **1965**, *74*, 234–240. [CrossRef]
57. Kuo, T.T.S.; Brown, G.E. Structure of finite nuclei and the free nucleon-nucleon interaction. An application to $^{18}$O and $^{18}$F. *Nucl. Phys.* **1966**, *85*, 40–86. [CrossRef]
58. Hjorth-Jensen, M.; Kuo, T.T.S.; Osnes, E. Realitic effective interactions for nuclear systems. *Phys. Rep.* **1995**, *261*, 125–270. [CrossRef]
59. Coraggio, A.; Covello, A.; Gargano, A.; Itaco, N.; Kuo, T.T.S. Shell-model calculations and realistic effective interactions. *Prog. Part. Nucl. Phys.* **2009**, *62*, 135–182. [CrossRef]
60. Stroberg, S.R.; Hergert, H.; Bogner, S.; Holt, J.D. Nonempirical interactions for the nuclear shell model: An update. *Ann. Rev. Nucl. Part. Sci.* **2019**, *69*, 307–362. [CrossRef]
61. Poves, A.; Zuker, A.P. Theoretical spectroscopy and the $fp$ shell. *Phys. Rep.* **1981**, *70*, 235–314. [CrossRef]
62. Barrett, B.R. Theoretical approaches to many-body perturbation theory and challenges. *J. Phys. G: Nucl. Part. Phys.* **2005**, *31*, S1349–S1355. [CrossRef]
63. Stroberg, S.R.; Calci, A.; Hergert, H.; Holt, J.D.; Bogner, S.; Roth, R.; Schwenk, A. Nucleus-dependent valence-space approach to nuclear structure. *Phys. Rev. Lett.* **2017**, *118*, 032502. [CrossRef] [PubMed]
64. Dikmen, E.; Lisetskiy, A.F.; Barrett, B.R.; Maris, P.; Shirokov, A.M.; Vary, J.P. *Ab initio* effective interactions for *sd*-shell valence nucleons. *Phys. Rev. C* **2015**, *91*, 064301. [CrossRef]
65. Smirnova, N.A.; Barrett, B.R.; Kim, Y.; Shin, I.J.; Shirokov, A.M.; Dikmen, E.; Maris, P.; Vary, J.P. Effective interactions in the *sd* shell. *Phys. Rev. C* **2019**, *100*, 054329. [CrossRef]
66. Jansen, G.R.; Engel, J.; Hagen, G.; Navrátil, P.; Signoracci, A. *Ab initio* coupled-cluster effective interactions for the shell model: Application to neutron-rich oxygen and carbon isotopes. *Phys. Rev. Lett.* **2014**, *113*, 142502. [CrossRef]
67. Jansen, G.R.; Schuster, M.D.; Signoracci, A.; Hagen, G.; Navrátil, P. Open *sd*-shell nuclei from first principles. *Phys. Rev. C* **2016**, *94*, 011301. [CrossRef]
68. Sun, Z.H.; Morris, T.D.; Hagen, G.; Jansen, G.R.; Papenbrock, T. Shell-model coupled-cluster method for open-shell nuclei. *Phys. Rev. C* **2018**, *98*, 054320. [CrossRef]
69. Fukui, T.; De Angelis, L.; Ma, Y.Z.; Coraggio, A.; Gargano, A.; Itaco, N.; Xu, F. Realistic shell-model calculations for *p*-shell nuclei including contributions of a chiral three-body force. *Phys. Rev. C* **2018**, *98*, 044305. [CrossRef]
70. Ma, Y.Z.; Coraggio, A.; De Angelis, L.; Fukui, T.; Gargano, A.; Itaco, N.; Xu, F. Contribution of chiral three-body forces to the monopole component of the effective shell-model Hamiltonian. *Phys. Rev. C* **2019**, *100*, 034324. [CrossRef]
71. Cohen, S.; Kurath, D. Effective interactions for the 1p shell. *Nucl. Phys.* **1965**, *73*, 1–24. [CrossRef]
72. Wildenthal, B.H. Empirical strengths of spin operators in nuclei. *Prog. Part. Nucl. Phys.* **1984**, *11*, 5–51. . [CrossRef]
73. Richter, W.A.; Brown, B.A. New "USD" Hamiltonians for the *sd* shell. *Phys. Rev. C* **2006**, *85*, 045806. [CrossRef]
74. Poves, A.; Sanchez-Solano, J.; Caurier, E.; Nowacki, F. Shell model study of the isobaric chains $A = 50$, $A = 51$ and $A = 52$. *Nucl. Phys. A* **2001**, *694*, 157–198. [CrossRef]
75. Honma, M.; Otsuka, T.; Brown, B.A.; Mizusaki, T. New effective interaction for $pf$-shell nuclei and its implications for the stability of the $N = Z = 28$ closed core. *Phys. Rev. C* **2004**, *69*, 034335. [CrossRef]
76. Zhang, Y.H.; Zhang, P.; Zhou, X.H.; Wang, M.; Litvinov, Yu.A.; Xu, H.S.; Xu, X.; Shuai, P.; Lam, Y.H.; Chen, R.J.; et al. Isochronous mass measurements of $T\_z = -1$ $fp$-shell nuclei from projectile fragmentation of $^{58}$Ni. *Phys. Rev. C* **2018**, *98*, 014319. [CrossRef]
77. Brown, B.A.; Rae, W.D.M. The shell-model code NuShellX. *Nucl. Data Sheets* **2014**, *120*, 115–118. [CrossRef]
78. Jänecke, J. Vector and tensor Coulomb energies. *Phys. Rev. C* **1966**, *147*, 735–742. [CrossRef]
79. Klochko, O.; Smirnova, N.A. Isobaric-multiplet mass equation in a macroscopic-microscopic approach. *Phys. Rev. C* **2021**, *103*, 024316. [CrossRef]
80. Bentley, M.A.; Lenzi, S.M. Coulomb energy differences between high-spin states in isobaric multiplets. *Prog. Part. Nucl. Phys.* **2007**, *59*, 497–561. [CrossRef]
81. Warner, D.D.; Van Isacker, P.; Bentley, M.A. The role of isospin symmetry in collective nuclear structure. *Nat. Phys.* **2006**, *2*, 311–318. [CrossRef]
82. Bentley, M.A. Excited states in isobaric multiplets—Experimental advances and the shell-model approach. *Physics* **2022**, *4*, 995–1011. [CrossRef]
83. Lenzi, S.M.; Poves, A.; Macchiavelli, A.O. Isospin symmetry breaking in the mirror pair $^{73}$Sr - $^{73}$Br. *Phys. Rev. C* **2020**, *102*, 031302. [CrossRef]
84. Boso, A.; Lenzi, S.M.; Recchia, F.; Bonnard, J.; Zuker, A.P.; Aydin, S.; Bentley, M.A.; Cederwall, B.; Clement, E.; de France, G.; et al. Neutron skin effects in mirror energy differences: The case of $^{23}$Mg–$^{23}$Na. *Phys. Rev. Lett.* **2018**, *121*, 032502. [CrossRef] [PubMed]
85. Thomas, R.G. An analysis of the energy levels of the mirror nuclei, $C^{13}$ and $N^{13}$. *Phys. Rev.* **1952**, *88*, 1109–1125. [CrossRef]

86. Ehrman, J.B. On the displacement of corresponding energy levels of $C^{13}$ and $N^{13}$. *Phys. Rev.* **1951**, *81*, 412–416. [CrossRef]
87. Longfellow, B.; Gade, A.; Brown, B.A.; Richter, W.A.; Bazin, D.; Bender, P.C.; Bowry, M.; Elman, B.; Lunderberg, B.E.; Weisshaar, D.; et al. Measurement of key resonances for the $^{24}$Al$(p,\gamma)^{25}$Si reaction rate using in-beam $\gamma$-ray spectroscopy. *Phys. Rev. C* **2018**, *97*, 054307. [CrossRef]
88. Cenxi, Y.; Qi, C.; Xu, F.; Suzuki, T.; Otsuka, T. Mirror energy difference and the structure of loosely bound proton-rich nuclei around $A = 20$. *Phys. Rev. C* **2014**, *89*, 044327. [CrossRef]
89. Pape, A.; Antony, M.S. Masses of proton-rich $T_z < 0$ nuclei with isobaric mass equation. *At. Data Nucl. Data Tables* **1988**, *39*, 201–203. [CrossRef]
90. Brown, B.A. Diproton decay of nuclei on the proton drip line. *Phys. Rev. C* **1991**, *43*, 1513(R)–1517(R). [CrossRef]
91. Ormand, W.E. Mapping the proton drip line up to $A = 70$. *Phys. Rev. C* **1997**, *55*, 2407–2417. [CrossRef]
92. Brown, B.A.; Clement, R.R.C.; Schatz, H.; Volya, A.; Richter, W.A. Proton drip-line calculations and the rp process. *Phys. Rev. C* **2002**, *65*, 045802. [CrossRef]
93. Richter, W.A.; Brown, B.A.; Signoracci, A.; Wiescher, M. Properties of $^{26}$Mg and $^{26}$Si in the *sd* shell model and the determination of the $^{26}$Al$(p,\gamma)^{26}$Si reaction rate. *Phys. Rev. C* **2011**, *83*, 065803. [CrossRef]
94. Benenson, W.; Kashy, E. Isobaric quartests in nuclei. *Rev. Mod. Phys.* **1979**, *51*, 527–540. [CrossRef]
95. Zhang, Y.H.; Xu, H.S.; Litvinov, Yu.A.; Tu, X.L.; Yan, X.L.; Typel, S.; Blaum, K.; Wang, M.; Zhou, X.H.; Sun, Y.; et al. Mass measurements of the neutron-deficient $^{41}$Ti, $^{45}$Cr, $^{49}$Fe, and $^{53}$Ni nuclides: First test of the isobaric multiplet mass equation in $fp$-shell nuclei. *Phys. Rev. Lett* **2012**, *109*, 102501. [CrossRef] [PubMed]
96. Brodeur, M.; Kwiatkowski, A.A.; Drozdowski, O.M.; Andreoiu, C.; Burdette, D.; Chaudhuri, A.; Chowdhury, U.; Gallant, A.T.; Grossheim, A.; Gwinner, G.; et al. Precision mass measurements of magnesium isotopes and implications for the validity of the isobaric mass multiplet equation. *Phys. Rev. C* **2017**, *96*, 034316. [CrossRef]
97. Bertsch, G.F.; Kahana, S. $T_z^3$ term in the isobaric multiplet equation. *Phys. Lett. B* **1970**, *33*, 193–194. [CrossRef]
98. Signoracci, A.; Brown, B.A. Effects of isospin mixing in the $A = 32$ quintet. *Phys. Rev. C* **2011**, *84*, 031301. [CrossRef]
99. Kamil, M.; Triambak, S.; Magilligan, A.; García, A.; Brown, B.A.; Adsley, P.; Bildstein, V.; Burbadge, C.; Diaz Varela, A.; Faestermann, T.; et al. Isospin mixing and the cubic isobaric multiplet mass equation in the lowest $T = 2$, $A = 32$ quintet. *Phys. Rev. C* **2022**, *104*, L061303. [CrossRef]
100. Barker, F.C. Intermediate coupling shell-model calculations for light nuclei. *Nucl. Phys.* **1966**, *83*, 418–448. [CrossRef]
101. Smirnova, N.A.; Blank, B.; Brown, B.A.; Richter, W.A.; Benouaret, N.; Lam, Y.H. Theoretical analysis of isospin mixing with the $\beta$ decay of $^{56}$Zn. *Phys. Rev. C* **2016**, *93*, 044305. [CrossRef]
102. Hoyle, C.D.; Adelberger, E.G.; Blair, J.S.; Snover, K.A.; Swanson, H.E.; Von Lintig, R.D. Isospin mixing in $^{24}$Mg. *Phys. Rev. C* **1983**, *27*, 1244–1259. [CrossRef]
103. Orrigo, S.; Rubio, B.; Fujita, Y.; Blank, B.; Gelletly, W.; Agramunt, J.; Algora, A.; Ascher, P.; Bilgier, B.; Cáceres, L.; et al. Observation of the $\beta$-delayed $\gamma$-proton decay of $^{56}$Zn and its impact on the Gamow-Teller strength evaluation. *Phys. Rev. Lett.* **2014**, *112*, 222501. [CrossRef] [PubMed]
104. Hagberg, E.; Koslowsky, V.T.; Hardy, J.C.; Towner, I.S.; Hykawy, J.G.; Savard, G.; Shinozuka, T. Tests of isospin mixing corrections in superallowed $0^+ \to 0^+$ $\beta$ decays. *Phys. Rev. Lett* **1994**, *73*, 396–399. [CrossRef] [PubMed]
105. MacLean, A.D.; Laffoley, A.T.; Svensson, C.E.; Ball, G.C.; Leslie, J.T.; Andreoiu, C.; Babu, A.; Bhattacharjee, S.S.; Bidaman, H.; Bildstein, V.; et al. High-precision branching ratio measurement and spin assignment implications for $^{62}$Ga superallowed $\beta$ decay. *Phys. Rev. C* **2020**, *102*, 054325. [CrossRef]
106. Schuurmans, P.; Camps, J.; Phalet, T.; Severijns, N.; Vereecke, B.; Versyck, S. Isospin mixing in the ground state of $^{52}$Mn. *Nucl. Phys. A* **2000**, *672*, 89–98. [CrossRef]
107. Severijns, N.; Vénos, D.; Schuurmans, P.; Phalet, T.; Honusek, M.; Srnka, D.; Vereecke, B.; Versyck, S.; Zákoucký, D.; Köster, U.; et al. Isospin mixing in the $T = 5/2$ ground state of $^{71}$As. *Phys. Rev. C* **2005**, *71*, 064310. [CrossRef]
108. Farnea, E.; de Angelis, G.; Gadea, A.; Bizzeti, P.G.; Dewald, A.; Eberth, J.; Algora, A.; Axiotis, M.; Bazzacco, D.; Bizzeti-Sona, A.M.; et al. Isospin mixing in the $N = Z$ nucleus $^{64}$Ge. *Phys. Lett. B* **2004**, *551*, 56–62. [CrossRef]
109. Bizzeti, P.G.; de Angelis, G.; Lenzi, S.M.; Orlandi, R. Isospin symmetry violation in mirror $E1$ transitions: Coherent contributions from the giant isovector monopole resonance in the $^{67}$As–$^{67}$Se doublet. *Phys. Rev. C* **2012**, *86*, 044311. [CrossRef]
110. Lisetskiy, A.F.; Schmidt, A.; Schneider, I.; Friessner, C.; Pietralla, N.; von Brentano, P. Isospin mixing between low-lying states of the odd-odd $N = Z$ nucleus $^{54}$Co. *Phys. Rev. Lett.* **2002**, *89*, 012502. [CrossRef]
111. Prados-Estevez, F.M.; Bruce, A.M.; Taylor, M.J.; Amro, H.; Beausang, C.W.; Casten, R.F.; Ressler, J.J.; Barton, C.J.; Chandler, C.; Hammond, G. Isospin purity of $T = 1$ states in the $A = 38$ nuclei studied via lifetime measurements in $^{38}$K. *Phys. Rev. C* **2007**, *75*, 014309. [CrossRef]
112. Giles, M.M.; Nara Singh, B.S.; Barber, L.; Cullen, D.M.; Mallaburn, M.J.; Beckers, M.; Blazhev, A.; Braunroth, T.; Dewald, A.; Fransen, C.; et al. Probing isospin symmetry in the ($^{50}$Fe, $^{50}$Mn, $^{50}$Cr) isobaric triplet via electromagnetic transition rates. *Phys. Rev. C* **2019**, *99*, 044317. [CrossRef]
113. Bizzeti, P.G.; Bizetti-Sona, A.M.; Cambi, A.; Mandò, M.; Maurenzig, P.R.; Signorini, C. Strength of analogue $E2$ transitions in $^{30}$Si and $^{30}$P. *Lett. Nouvo Cim.* **1969**, *16*, 775. [CrossRef]

114. Ekman, J.; Rudolph, D.; Fahlander, C.; Zuker, A.P.; Bentley, M.A.; Lenzi, S.M.; Andreoiu, C.; Axiotis, M.; de Angelis, G.; Farnea, E.; et al. Unusual isospin-breaking and isospin-mixing effects in the $A = 35$ mirror nuclei. *Phys. Rev. Lett.* **2004**, *92*, 132502. [CrossRef] [PubMed]
115. Pattabiraman, N.S.; Jenkins, D.G.; Bentley, M.A.; Wadsworth, R.; Lister, C.J.; Carpenter, M.P.; Janssens, R.V.F.; Khoo, T.L.; Lauritsen, T.; Seweryniak, D.; et al. Analog $E1$ transitions and isospin mixing. *Phys. Rev. C* **2008**, *78*, 024301. [CrossRef]
116. von Neumann-Cosel, P.; Gräf, H.-D.; Krämer, U.; Richter, A.; Spamer, E. Electroexcitation of isoscalar and isovector magnetic dipole transitions in $^{12}$C and isospin mixing. *Nucl. Phys. A* **2000**, *669*, 3–13. [CrossRef]
117. Corsi, A.; Wieland, O.; Barlini, S.; Bracco, A.; Camera, F.; Kravchuk, V.L.; Baiocco, G.; Bardelli, L.; Benzoni, G.; Bini, M.; et al. Measurement of isospin mixing at a finite temperature in $^{80}$Zr via giant dipole resonance decay. *Phys. Rev. C* **2011**, *84*, 041304(R). [CrossRef]
118. Ceruti, S.; Camera, F.; Bracco, A.; Avigo, R.; Benzoni, G.; Blasi, N.; Bocchi, G.; Bottoni, S.; Brambilla, S.; Crespi, F.C.L.; et al. Isospin mixing in $^{80}$Zr: From finite to zero temperature. *Phys. Rev. Lett.* **2016**, *115*, 222502. [CrossRef]
119. Gosta, G.; Mentana, A.; Camera, F.; Bracco, A.; Ceruti, S.; Benzoni, G.; Blasi, N.; Brambilla, S.; Capra, S.; Crespi, F.C.L.; et al. Probing isospin mixing with the giant dipole resonance in the $^{60}$Zn compound nucleus. *Phys. Rev. C* **2021**, *103*, L041302. [CrossRef]
120. Brown, B.A. Isospin-forbidden $\beta$-delayed proton emission. *Phys. Rev. Lett.* **1990**, *65*, 2753–2756. [CrossRef]
121. Dossat, C.; Adimi, F.; Aksouh, F.; Becker, F.; Bey, A.; Blank, B.; Borcea, C.; Borcea, R.; Boston, A.; Caamano, M.; et al. The decay of proton-rich nuclei in the mass $A = 36 - 56$ region. *Nucl. Phys. A* **2005**, *792*, 18–86. [CrossRef]
122. Blank, B.; Borge, M.J.G. Nuclear structure at the proton drip line: Advances with nuclear decay studies. *Prog. Part. Nucl. Phys.* **2008**, *60*, 403–483. [CrossRef]
123. Ormand, W.E.; Brown, B.A. Isospin-forbidden proton and neutron emission in 1*s*-0*d* shell nuclei. *Phys. Lett. B* **1986**, *174*, 128–132. [CrossRef]
124. Smirnova, N.A.; Blank, B.; Richter, W.A.; Brown, B.A.; Benouaret, N.; Lam, Y.H. Isospin mixing from $\beta$-delayed proton emission. *Phys. Rev. C* **2017**, *95*, 054301. [CrossRef]
125. Saxena, M.; Ong, W.-J.; Meisel, A.; Hoff, D.E.M.; Smirnova, N.; Bender, P.C.; Burcher, S.P.; Carpenter, M.P.; Carroll, J.J.; Chester, A.; et al. $^{57}$Zn $\beta$-delayed proton emission establishes the $^{56}$Ni *rp*-process waiting point bypass. *Phys. Lett. B* **2022**, *829*, 137059. [CrossRef]
126. Towner, I.S.; Hardy, J.C. Currents and their couplings in the weak sector of the Standard Model. In *Symmetries and Fundamental Interactions in Nuclei*; Henley, E.M., Haxton, W.C., Eds.; World Scientific: Singapore, 1995; pp. 183–249. [CrossRef]
127. Severijns, N.; Beck, M.; Naviliat-Cuncic, O. Tests of the standard electroweak model in nuclear beta decay. *Rev. Mod. Phys.* **2006**, *78*, 991–1040. [CrossRef]
128. González-Alonso, M.; Naviliat-Cuncic, O.; Severijns, N. New physics searches in nuclear and neutron $\beta$-decay. *Prog. Part. Nucl. Phys.* **2019**, *104*, 165–223. [CrossRef]
129. Towner, I.S.; Hardy, J.C. The evaluation of $V_{ud}$ and its impact on the unitarity of the Cabibbo–Kobayashi– Maskawa quark-mixing matrix. *Rep. Prog. Phys.* **2010**, *73*, 046301. [CrossRef]
130. Hardy, J.C.; Towner, I.S. Superallowed $0^+ \to 0^+$ nuclear $\beta$ decays: 2020 critical survey, with implications for $V_{ud}$ and CKM unitarity. *Phys. Rev.* **2020**, *102*, 045501. [CrossRef]
131. Seng, C.-Y.; Gorchtein, M.; Patel, H.H.; Ramsey-Musolf, M.J. Reduced Hadronic Uncertainty in the Determination of $V_{ud}$. *Phys. Rev. Lett.* **2018**, *121*, 241804. [CrossRef]
132. Ormand, W.E.; Brown, B.A. Isospin-mixing corrections for $fp$-shell Fermi transitions. *Phys. Rev. C* **1995**, *52*, 2455–2460. [CrossRef]
133. Damgaard, J. Corrections to the $ft$-values of $0^+ \to 0^+$ superallowed $\beta$-decays. *Nucl. Phys. A* **1969**, *130*, 233–240. [CrossRef]
134. Auerbach, N. Coulomb corrections to superallowed $\beta$ decay in nuclei. *Phys. Rev. C* **2009**, *79*, 035502. [CrossRef]
135. Xayavong, L.; Smirnova, N.A. Radial overlap correction to superallowed $0^+ \to 0^+$ $\beta$ decay reexamined. *Phys. Rev. C* **2018**, *97*, 024324. [CrossRef]
136. Miller, G.A.; Schwenk, A. Isospin-symmetry-breaking corrections to superallowed Fermi $\beta$ decay. Formalism and schematic models. *Phys. Rev. C* **2008**, *78*, 035501. [CrossRef]
137. Miller, G.A.; Schwenk, A. Isospin-symmetry-breaking corrections to superallowed Fermi $\beta$ decay: Radial excitations. *Phys. Rev. C* **2009**, *80*, 064319. [CrossRef]
138. Towner, I.S.; Hardy, J.C. Improved calculations of isospin-symmetry breaking corrections to superallowed Fermi $\beta$ decay. *Phys. Rev. C* **2008**, *77*, 025501. [CrossRef]
139. Hardy, J.C.; Towner, I.S. Superallowed $0^+ \to 0^+$ nuclear $\beta$ decays: 2014 critical survey, with precise results for $V_{ud}$ and CKM unitarity. *Phys. Rev.* **2015**, *91*, 025501. [CrossRef]
140. Ormand, W.E.; Brown, B.A. Corrections to the Fermi matrix element for superallowed $\beta$ decay. *Phys. Rev. Lett.* **1989**, *62*, 866–869. [CrossRef]
141. Towner, I.S.; Hardy, J.C. Comparative tests of isospin-symmetry breaking corrections to superallowed $0^+ \to 0^+$ nuclear $\beta$ decay. *Phys. Rev. C* **2010**, *82*, 065501. [CrossRef]
142. Ormand, W.E.; Brown, B.A. Calculated isospin-mixing corrections to Fermi $\beta$-decays in 1*s*0*d*-shell nuclei with emphasis on $A = 34$. *Nucl. Phys. A* **1985**, *440*, 274–300. [CrossRef]

143. Xayavong, L.; Smirnova, N.; Bender, M.; Bennaceur, K. Shell-model calculation of isospin-symmetry breaking correction to super-allowed Fermi beta decay. *Acta Phys. Pol. B. Proc. Supp.* **2017**, *10*, 285–290. [CrossRef]
144. Xayavong, L.; Smirnova, N.A. Radial overlap correction to superallowed $0^+ \to 0^+$ nuclear $\beta$ decays using the shell model with Hartree-Fock radial wave functions. *Phys. Rev. C* **2022**, *105*, 044308. [CrossRef]
145. Naviliat-Cuncic, O.; Severijns, N. Test of the conserved vector current hypothesis in $T = 1/2$ mirror transitions and new determination of $V_{ud}$. *Phys. Rev. Lett.* **2009**, *102*, 142302. [CrossRef] [PubMed]
146. Towner, I.S. Mirror asymmetry in allowed Gamow-Teller $\beta$-decay. *Nucl. Phys. A* **1973**, *216*, 589–602. [CrossRef]
147. Smirnova, N.A.; Volpe, M.C. On the asymmetry of Gamow-Teller $\beta$-decay rates in mirror nuclei in relation with second-class currents. *Nucl. Phys. A* **2003**, *714*, 441–462. [CrossRef]
148. Grenacs, L. Induced weak currents in nuclei. *Ann. Rev. Nucl. Part. Sci.* **1985**, *35*, 455–499. [CrossRef]
149. Minamisono, K.; Nagatomo, T.; Matsuta, K.; Levy, C.D.P.; Tagishi, Y.; Ogura, M.; Yamaguchi, M.; Ota, H.; Behr, J.A.; Jackson, K.P.; et al. Low-energy test of second-class current in $\beta$ decays of spin-aligned $^{20}$F and $^{20}$Na. *Phys. Rev. C* **2011**, *84*, 055501. [CrossRef]
150. Langanke, K.; Martinez-Pinedo, G. Nuclear weak-interaction processes in stars. *Rev. Mod. Phys.* **2003**, *75*, 812–862. [CrossRef]
151. Jose, J.; Hernanz, M.; Iliadis, C. Nucleosynthesis in classical novae. *Nucl. Phys. A* **2006**, *777*, 550–578. [CrossRef]
152. Wallace, R.K.; Woosley, S.E. Explosive hydrogen burning. *Astrophys. J. Supp. Ser.* **1981**, *45*, 389–420. [CrossRef]
153. Schatz, H.; Aprahamian, A.; Görres, J.; Wiescher, M.; Rauscher, T.; Rembges, J.F.; Thielemann, F.K.; Pfeiffer, B.; Möller, P.; Kratz, K.-L.; et al. rp-process nucleosynthesis at extreme temperature and density conditions. *Phys. Rep.* **1998**, *294*, 167–263. [CrossRef]
154. Fowler, W.A.; Hoyle, F. Neutrino processes and pair formation in massive stars and supernovae. *Astrophys. J. Supp.* **1964**, *9*, 201–319. [CrossRef]
155. Herndl, H.; Görres, J.; Wiescher, M.; Brown, B.A.; Van Wormer, L. Proton capture reaction rates in the *rp* process. *Phys. Rev. C* **1995**, *52*, 1078–1094. [CrossRef] [PubMed]
156. Fisker, J.L.; Barnard, V.; Görres, J.; Langanke, K.; Martinez-Pinedo, G.; Wiescher, M. Shell-model based reaction rates for $rp$-process nuclei in the mass range $A = 44 - 63$. *At. Data Nucl. Data Tables* **2001**, *79*, 241–292. [CrossRef]
157. Richter, W.A.; Brown, B.A. Shell-model studies of the *rp* reaction $^{35}\mathrm{Ar}(p,\gamma)^{36}$K. *Phys. Rev. C* **2012**, *85*, 045806. [CrossRef]
158. Lam, Y.H.; Herger, A.; Lu, N.; Jacobs, A.M.; Smirnova, N.A.; Kurtukian-Nieto, T.; Johnston, T.; Kubono, S. The regulated NiCu cycles with the new $^{57}\mathrm{Cu}(p,\gamma)^{58}$Zn reaction rate and its influence on type I X-ray bursts: The GS 1826-24 clocked burster. *Astrophys. J.* **2022**, *929*, 73–88. [CrossRef]
159. Brown, B.A.; Richter, W.A.; Wrede, C. Shell-model studies of the astrophysical rapid-proton-capture reaction $^{30}\mathrm{P}(p,\gamma)^{31}$S. *Phys. Rev. C* **2014**, *89*, 062801. [CrossRef]
160. Richter, W.A.; Brown, B.A.; Longland, R.; Wrede, C.; Denissenkov, P.; Fry, C.; Herwig, F.; Kurtulgil, D.; Pignatari, M.; Reifarth, R. Shell-model studies of the astrophysical $rp$-process reactions $^{34}\mathrm{S}(p,\gamma)\ ^{35}$Cl and $^{34g,m}\mathrm{Cl}(p,\gamma)\ ^{35}$Ar. *Phys. Rev. C* **2020**, *102*, 025801. [CrossRef]


**Disclaimer/Publisher's Note:** The statements, opinions and data contained in all publications are solely those of the individual author(s) and contributor(s) and not of MDPI and/or the editor(s). MDPI and/or the editor(s) disclaim responsibility for any injury to people or property resulting from any ideas, methods, instructions or products referred to in the content.

*Review*

# Double Beta Decay: A Shell Model Approach

**Mihai Horoi**

Department of Physics, Central Michigan University, Mount Pleasant, MI 48859, USA; mihai.horoi@cmich.edu

**Abstract:** Studies of weak interaction in nuclei are important tools for testing different aspects of the fundamental symmetries of the Standard Model. Neutrinoless double beta decay offers an unique venue of investigating the possibility that neutrinos are Majorana fermions and that the lepton number conservation law is violated. Here, I use a shell model approach to calculate the nuclear matrix elements needed to extract the lepton-number-violating parameters of a few nuclei of experimental interest from the latest experimental lower limits of neutrinoless double beta decay half-lives. The analysis presented here could reveal valuable information regarding the dominant neutrinoless double beta decay mechanism if experimental half-life data become available for different isotopes. A complementary shell model analysis of the two-neutrino double beta decay nuclear matrix elements and half-lives is also presented.

**Keywords:** neutrino properties; double beta decay; nuclear shell model; many-body methods

**Citation:** Horoi, M. Double Beta Decay: A Shell Model Approach. *Physics* **2022**, *4*, 1135–1149. https://doi.org/10.3390/physics4040074

Received: 6 June 2022
Accepted: 7 September 2022
Published: 26 September 2022

**Publisher's Note:** MDPI stays neutral with regard to jurisdictional claims in published maps and institutional affiliations.

**Copyright:** © 2022 by the author. Licensee MDPI, Basel, Switzerland. This article is an open access article distributed under the terms and conditions of the Creative Commons Attribution (CC BY) license (https://creativecommons.org/licenses/by/4.0/).

## 1. Introduction

The recent experimental discovery of neutrino oscillations [1,2] proved that neutrinos have mass, and this discovery was awarded a Nobel prize in 2015 [3,4]. Neutrino oscillation experiments can only provide information about the squared mass differences, while other properties of neutrinos, such as their mass hierarchy, their absolute masses, or their fermionic signatures, Dirac or Majorana, remain to be determined. However, this new information coming from the neutrino oscillations experiments has led to new interest in neutrino physics and in particular in their nature as Dirac or Majorana fermions that may be unraveled by neutrinoless double beta decay investigations.

Neutrinoless double beta decay ($0\nu\beta\beta$) is one of the best experimental approaches for identifying processes that violate the lepton number conservation, thus signaling beyond the Standard Model (BSM) physics. If neutrinoless double beta transitions occur, then the lepton number conservation is violated by two units, and the black-box theorems [5–8] indicate that the light left-handed neutrinos are Majorana fermions. As a consequence, the BSM extension of the Standard Model Lagrangian would be significantly different from that where neutrinos are Dirac fermions. Theoretical investigations of $0\nu\beta\beta$ decay combine lepton number violation (LNV) amplitudes with leptonic phase-space factors (PSFs) and nuclear matrix elements (NMEs). The NMEs are computed using a large variety of nuclear structure methods and specific models. Among the LNV models considered, the left-right symmetric model [9–13] is among the most popular, and its predictions are currently investigated at the Large Hadron Collider [14]. In some recent papers [15–17], I have investigated observables that could identify the contributions of different left-right symmetric model mechanisms to the $0\nu\beta\beta$ decay rate, such as the angular distribution and the energy distribution of the two outgoing electrons that could be measured. A more general approach is effective field theory (EFT), which considers an expansion of the BSM Lagrangian consistent with the Standard Model symmetries and including LNV and neutrino mass mechanisms. This approach has the advantage of being independent of specific models, and it can be used to describe in a unified manner BSM-sensitive observables, including those related to $0\nu\beta\beta$ decay. One can then use the existing data/limits from

different experiments to evaluate the energy scales up to which the effective field operators are not broken and limits for effective low-energy couplings.

The theoretical analysis of the $0\nu\beta\beta$ decay process has many steps, including the nuclear structure calculation of the NMEs. However, in the first step, the weak interaction of quarks and leptons described by the BSM EFT Lagrangian is considered in the lowest order (see the diagram in Figure 1). In the next step, the hadronization process to nucleons and exchanging pions is considered, leading to the diagram in Figure 2. Furthermore, the nucleons are treated in the impulse approximation leading to free space $0\nu\beta\beta$ transition operators, and the nucleon dynamics inside the nuclei are treated using nonperturbative nuclear wave functions, which are later used to obtain the nuclear matrix elements needed to calculate the $0\nu\beta\beta$ observables, such as half-lives and two-electron angular and energy distributions [15]. A modern approach that can be used to make the transition from quarks and gluons to nucleons and pions is based on the chiral effective field theory of pions and nucleons [18,19]. This approach introduces a number of effective low-energy couplings, which in principle can be calculated from the underlying theory of strong interaction using lattice QCD techniques [18] or may be extracted within some approximation from the known experimental data [19]. These couplings may have new complex phases, and they could include effective contributions from the exchange of heavier mesons. The lattice QCD approach is in progress (see, e.g., Ref. [20]), but it has proven to be difficult for extracting some of the necessary weak nucleon coupllowestngs, even the known $g_A$ [20].

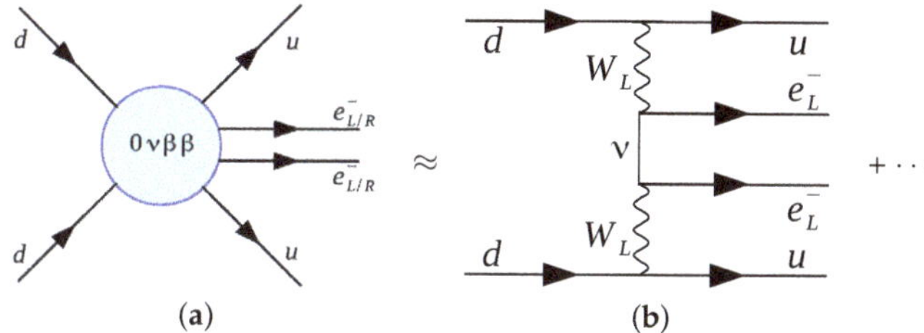


**Figure 1.** The $0\nu\beta\beta$ decay process diagrams: (**a**) typical $0\nu\beta\beta$ decay diagram at the quark ($u$ and $d$) level presents the generic description of the process and (**b**) light left-handed ($L$) neutrino exchange diagram shows the most studied case in the literature, that of the light left-handed neutrino exchange. Here, "..." stands for other diagrams involving left- and right-handed ($R$) leptons (see, e.g., Figure 1 of Ref. [17] for model diagrams).

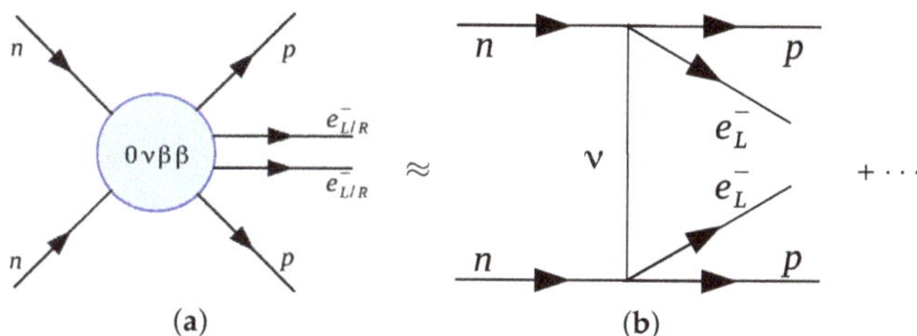


**Figure 2.** Similar to Figure 1, the nucleon-level diagrams of $0\nu\beta\beta$ decay process: (**a**) the typical $0\nu\beta\beta$ decay process nucleon-level diagram presents the generic description of the process and (**b**) the light left-handed neutrino exchange diagram shows the light left-handed neutrino exchange. Here, "..." stands for other higher-order effective field theory (EFT) diagrams (see Figure 2 of Ref. [17]).

Here, as in Ref. [17], I use the formalism of Refs. [21–24] that provides a general EFT approach to the BSM Lagrangian. It also provides a somewhat older hadronization scheme, which is needed to obtain the neutrinoless double beta decay transition operators. To extract new limits for the effective Majorana mass and for the low-energy EFT couplings from the current experiment for the isotopes listed in Table 1 below, I use the assumption

that only one single coupling in the BSM Lagrangian may dominate the $0\nu\beta\beta$ amplitude. In the analysis, about 20 nuclear matrix elements and nine phase-space factors are needed. I use the existing neutrinoless double beta decay data to extract the limits for the BSM EFT couplings and limits of validity for the energy scale of the BSM Lagrangian. In addition, the calculated ratio of half-lives for different isotopes could be useful in guiding the experimental effort, in estimating their scales and costs, in fine-tuning the experimental searches for the $0\nu\beta\beta$ transition mechanism, and also in providing a better view and comparison of the status of various experimental efforts. Our analysis suggests that the experimental confirmation of $0\nu\beta\beta$ decay rates for several isotopes could possibly help in identifying the dominant mechanism responsible for the transition.

**Table 1.** The $Q_{\beta\beta}$ values (in MeV), the experimental half-lives $T^{2\nu}_{1/2}$ [25,26] and $T^{0\nu}_{1/2}$ limits (in years), and the calculated PSFs $G_{2\nu}$ [27] and $G_{01}$ ($G_{02}$–$G_{09}$ can be found elsewhere [17]) (in years$^{-1}$) for all five isotopes currently under investigation.

| | $^{48}$Ca | $^{76}$Ge | $^{82}$Se | $^{124}$Sn | $^{130}$Te | $^{136}$Xe |
|---|---|---|---|---|---|---|
| $Q_{\beta\beta}$ [28] | 4.268 | 2.039 | 2.998 | 2.291 | 2.528 | 2.458 |
| $G_{01} \cdot 10^{14}$ | 2.45 | 0.23 | 1.00 | 0.887 | 1.41 | 1.45 |
| $G_{2\nu} \cdot 10^{20}$ | 1480 | 4.51 | 150.3 | 51.45 | 142.7 | 133.7 |
| $T^{2\nu}_{1/2} \cdot 10^{-20}$ | 0.53 | 19 | 0.87 | >0.1 | 7.9 | 22 |
| $T^{0\nu}_{1/2} \cdot 10^{-23} >$ | 0.2 [29] | 800 [30] | 2.5 [31] | >0.01 [26] | 40 [32] | 1100 [33] |

One important step in describing the $0\nu\beta\beta$ decay observables is obtaining the appropriate NMEs. The nuclear structure methods used for NME calculations are the interacting shell model [34–52], proton-neutron quasi random phase approximation (pnQRPA) [21–24,53–57], interacting moson model [58–61], projected Hartree–Fock–Bogoliubov [62], energy density functional [63], and relativistic energy density functional method [64]. The NMEs calculated with different methods and by different groups show sometimes large variations by a factor of 3–5 [65,66]. Most references only provide NMEs for the light left-handed Majorana neutrino exchange mechanism, but some provide results for the right/left heavy neutrino exchange and some more exotic mechanisms. Ref. [50] provides tables and plots that compare results for the light left-handed neutrino exchange and for the heavy right-handed neutrino exchange, while Ref. [17] provides tables with all NMEs necessary for the EFT approach. I calculate the NMEs using shell model techniques [36,41–51] and a preferred set of effective Hamiltonians that were tested for a wide set of nuclei. The shell model calculations of NMEs use a relatively small single-particle model space, but they are better suited and more reliable for $0\nu\beta\beta$ decay calculations because they take into account all the correlations around the Fermi surface, respect all nuclear many-body problem symmetries, and can take into account the effects of the missing single particle space via many-body perturbation theory (the effects were shown to be small [67]). In addition, it was shown [68,69] that the QRPA approaches using the same model spaces and effective Hamiltonian as in the shell model produce NMEs within 25% of the shell model results. Furthermore, I test the shell model methods and the effective Hamiltonians by comparing the calculations of spectroscopic observables for the nuclei involved in the transition to the experimental data, as presented in Refs. [41,50,70]. I do not consider any quenching for the bare $0\nu\beta\beta$ operator in these calculations. Such a choice is different from that for the simple Gamow–Teller operator used in the single beta and two-neutrino double beta decay ($2\nu\beta\beta$), where a quenching factor of about 0.7 is necessary [69]. For the PSFs, I use an effective theory based on the formalism of Ref. [71], but fine-tuned as to take into account the effects of a Coulomb-field-distorting finite-size proton distribution in the daughter nuclei. Table 1 provides information relevant for the main nuclei that can be calculated using shell model techniques (see Equations (1) and (13) below for a precise definition of the PSFs used).

In this paper, I mostly review the shell model techniques needed to accomplish the plan outline above. The numerical results and their analysis are available in different papers that are appropriately cited below. Although most material described below reviews results already published, some new results can be found at the end of Section 3.2 and in Section 4. The paper is organized as follows: Section 2 analyzes the contributions of several BSM mechanisms to neutrinoless double beta decay, and it presents the framework of effective field theory for neutrinoless double beta decay; Section 3 presents an analysis of the $0\nu\beta\beta$ nuclear matrix elements in the shell model approach; Section 4 presents an analysis of the $2\nu\beta\beta$ nuclear matrix elements in the shell model approach; Section 5 is dedicated to conclusions.

## 2. Neutrinoless Double Beta Decay And Neutrino Physics

The main mechanism considered to be responsible for neutrinoless double beta decay is the mass mechanism that assumes that the neutrinos are Majorana fermions and relies on the assumption that the light left-handed neutrinos have mass. However, the possibility that right-handed currents could contribute to neutrinoless double beta decay ($0\nu\beta\beta$) has been already considered for some time [71,72]. Recently, $0\nu\beta\beta$ studies [13,73] have adopted the left-right symmetric model [11,74] for the inclusion of right-handed currents at quark level. In addition, the $R$-parity-violating ($\mathcal{R}_p$) supersymmetric (SUSY) model can also contribute to the neutrinoless double beta decay process [75–77].

### 2.1. LNV Models Contributing to $0\nu\beta\beta$

In the framework that includes the left-right symmetric model and $R$-parity-violating SUSY model, after hadronization, the $0\nu\beta\beta$ half-life can be written as a sum of products of PSFs, BSM LNV parameters, and their corresponding NMEs [15]:

$$\left[T^{0\nu}_{1/2}\right]^{-1} = G_{01} g_A^4 \Big| \eta_{0\nu} M_{0\nu} + \left(\eta^L_{N_R} + \eta^R_{N_R}\right) M_{0N} + \eta_{\tilde{q}} M_{\tilde{q}} + \eta_{\lambda'} M_{\lambda'} + \eta_\lambda X_\lambda + \eta_\eta X_\eta \Big|^2. \quad (1)$$

Here, $G_{01}$ is a phase-space factor that can be calculated with good precision for most cases [27,28,78,79], $g_A$ is the axial vector coupling constant, $\eta_{0\nu} = \langle m_{\beta\beta} \rangle / m_e$, effective Majorana neutrino mass (see Equation (3)), and $m_e$ is the electron mass. $\eta^L_{N_R}$ and $\eta^R_{N_R}$ are the heavy neutrino parameters with left-handed and right-handed currents, respectively [13,36], $\eta_{\tilde{q}}$ and $\eta_{\lambda'}$ are $\mathcal{R}_p$ SUSY LNV parameters [80], and $\eta_\lambda$ and $\eta_\eta$ are parameters for the so-called "$\lambda$–" and "$\eta$–mechanisms", respectively [13]. $M_{0\nu}$ and $M_{0N}$ are the light and the heavy neutrino exchange NMEs, $M_{\tilde{q}}$ and $M_{\lambda'}$ are the $\mathcal{R}_p$ SUSY NMEs, and $X_\lambda$ and $X_\eta$ denote the combinations of NMEs and other PSFs ($G_{02}$–$G_{09}$) corresponding to the the $\lambda$–mechanism involving right-handed leptonic and right-handed hadronic currents and the $\eta$–mechanism with right-handed leptonic and left-handed hadronic currents, respectively [15]. Assuming a seesaw type I dominance [81], the term $\eta^L_{N_R}$ is considered negligible if the heavy mass eigenstates are larger than 1 GeV [52], and I ignore it here. For consistency with the literature, the remaining term $\eta^R_{N_R}$ is labeled as $\eta_{0N}$.

Here, I exclusively describe transitions from the spin/parity $J^\pi = 0^+$ ground state (g.s.) of the parent nucleus to the final $J^\pi = 0^+$ ground state of the daughter nucleus. There is also the possibility of $0\nu\beta\beta$ decay to the excited states of the daughter, such as the first $J^\pi = 2^+$. This alternative is rarely considered in the literature, mainly because besides a significant reduction in the effective Q-values for most isotopes, thus reducing the corresponding phase space factors, it has also been known for some time that based on a general analysis the NMEs for this transition are suppressed for the mass mechanism [72]. In addition, the initial numerical estimates of the NMEs corresponding to the $\eta_\eta$ and $\eta_\lambda$ in Equation (1) showed that they were also suppressed [82]. Recently, it was found that more up-to-date QRPA calculations of these right currents' contributions could lead to a significant increase in the matrix elements for the lambda mechanism that might compete

with the transition to the $J^{\pi} = 0^{+}$ ground state, at least for case of $^{136}$Xe [83,84]. These new findings are clearly interesting, and I plan to investigate them using shell model techniques similar to the ones described below and report them in future publications.

Table 1 presents the $Q_{\beta\beta}$ values, the most recent experimental half-life limits, and the nine PSFs for the $0\nu\beta\beta$ transitions to the ground states of the daughter nucleus for five isotopes considered in this investigation. The PSFs were calculated using a new effective method described in detail in Ref. [27]. $G_{01}$ values were calculated with a screening factor ($s_f$) of 94.5, while for $G_{02}$–$G_{09}$ I used $s_f = 92.0$, which was shown to provide results close to those of the more accurate approach described in Ref. [85].

As indicated in Equation (1), the main observable related to $0\nu\beta\beta$ decay is the half-life of the process. It is unlikely that this unique observable, even if measured for several isotopes, could provide enough information to identify different mechanisms that may contribute to this process. In Ref. [15], I investigated other observables that could be used to disentangle contributions from different mechanisms, such as the two-electron angular and energy distributions, in addition to the half-life data from several isotopes. I considered the case where one mechanism dominates, i.e., there is one single term in the decay amplitude of Equation (1). Table 2 of Ref. [17] shows the shell model values of the NMEs that enter Equation (1). Details regarding the definitions of specific NMEs can be found in Refs. [17,49]. All NMEs were calculated using the interacting shell model (ISM) approach [36,43–46,49,52] (see also Ref. [49] for a review) and included short-range correlation effects based on the CD-Bonn parametrization [41], finite-size effects [80], and, when appropriate, optimal closure energies [70] (see Section 3 for more details). Table 2 of Ref. [17] also presents the upper limits for the corresponding LNV parameters extracted from the lower limits of the half-lives under the assumption of one-mechanism dominance. However, less general analyses are available based on QRPA [71,80,85–87], NMEs, and other interactive shell model NMEs [34–37].

If only the main diagram in Figure 2b is considered, the associated mechanism is known as the light neutrino exchange mechanism and the half-life of Equation (1) becomes

$$\left[T_{1/2}^{0\nu}\right]^{-1} = G_{01} g_A^4 \frac{|\langle m_{\beta\beta}\rangle|^2}{m_e^2} M_{0\nu}^2, \tag{2}$$

with the effective neutrino mass given by following sum over the light mass eigenstates:

$$|\langle m_{\beta\beta}\rangle| = \left| \sum_{i \in light} U_{ei}^2 m_i \right|, \tag{3}$$

where $U_{ei}$ are the complex matrix elements of the first row in the Pontecorvo–Maki–Nakagawa–Sakata (PMNS) neutrino mixing matrix. This quantity is very often used in the literature as an example of how one could potentially extract additional information about neutrino physics parameters, such as neutrino mass ordering and the mass of the lowest mass eigenstate, from the experimental value of $T_{1/2}^{0\nu}$ [88].

### 2.2. EFT Approach to $0\nu\beta\beta$ Decay

As mentioned in the introduction, a more general approach could be constructed based on the effective field theory extension of the Standard Model. Such an EFT analysis is preferable because it does not rely on specific models and the parameters could be constrained by the existing $0\nu\beta\beta$ data and by data from the Large Hadron Collider and other experiments. In addition, the models considered in Equation (1) always lead to a subset of terms in the low-energy ($\sim$200 MeV) effective field theory Lagrangian. EFT considers all terms in the BSM Lagrangian allowed by the symmetries, some of them corresponding to the model terms incorporated in Equation (1), but the couplings might have a wider meaning. Other terms in the EFT Lagrangian are new, not directly identifiable with those originating from specific models.

At the quark level, Figure 1 shows the generic $0\nu\beta\beta$ Feynman diagrams contributing to the $0\nu\beta\beta$ process. I consider contributions coming from the light left-handed Majorana neutrino (Figure 1b) and a long-range part coming from the low-energy four-fermion charged-current interaction (see Ref. [17] for details). After hadronization (see Figure 2), the extra terms in the Lagrangian require the knowledge of about 20 individual NMEs [22–24,75,80,89]. One can write the half-life in a factorized compact form:

$$\left[T_{1/2}^{0\nu}\right]^{-1} = g_A^4 \left[\sum_i |\mathcal{E}_i|^2 \mathcal{M}_i^2 + \mathrm{Re}\left(\sum_{i\neq j} \mathcal{E}_i \mathcal{E}_j \mathcal{M}_{ij}\right)\right]. \tag{4}$$

Here, the $\mathcal{E}_i$ contain the neutrino physics parameters, $\mathcal{E}_1 = \eta_{0\nu}$ represent the exchange of light left-handed neutrinos, $\mathcal{E}_{2\text{–}6} = \{\epsilon_{V-A}^{V+A}, \epsilon_{V+A}^{V+A}, \epsilon_{S\pm P}^{S+P}, \epsilon_{TR}^{TR}, \eta_{\pi\nu}\}$ are the long-range LNV parameters, and $\mathcal{E}_{7\text{–}14} = \{\varepsilon_1, \varepsilon_2, \varepsilon_3^{LLz(RRz)}, \varepsilon_3^{LRz(RLz)}, \varepsilon_4, \varepsilon_5, \eta_{1\pi}, \eta_{2\pi}\}$ denote the short-range LNV parameters at the quark level (see Ref. [17] for definitions of notations and details). The contributions of pion-exchange diagrams are also included in the so-called "higher-order term in nucleon currents" [80]. However, they are constrained by partial conservation of axial current (PCAC) and are only included in the light neutrino exchange contribution in Figure 2a. This contribution changes the associated NMEs by only 20%, and one concludes that it does not represent a serious double counting issue.

In Equation (4), $\mathcal{M}_i^2$ and $\mathcal{M}_{ij}$ are combinations of NMEs and integrated PSFs [27] denoted with $G_{01}$–$G_{09}$ (see Ref. [17] for definitions and details). In some cases, the interference terms $\mathcal{E}_i\mathcal{E}_j\mathcal{M}_{ij}$ are small [90] and can be neglected, but not all of them [91]. In Ref. [15], I analyzed a subset of terms contributing to the half-life formula, with Equation (1) originating from the left-right symmetric model. In that restrictive case, I showed that one can disentangle different contributions to the $0\nu\beta\beta$ decay process using two-electron angular and energy distributions as well as the half-lives of two selected isotopes.

## 3. Neutrinoless Double Beta Decay Nuclear Matrix Elements

From previous Sections, one can conclude that the analysis of main experimental data regarding the $0\nu\beta\beta$, the half-lives of multiple isotopes, and the two-electron angular and energy distributions [15] require a set of nuclear matrix elements. In this Section, I describe different techniques for calculating NMEs, starting with the direct summation on the states in the intermediate nucleus $(Z-1, N-1)$, where $Z$ denotes the atomic number and $N$ the number of neutrons in a nucleus, and continuing with the often used closure approximation. An alternative method that performs a summation on the intermediate states in the $(Z-2, N)$ or $(Z, N-2)$ nuclei is described in Ref. [46].

### 3.1. The Anatomy of the $0\nu\beta\beta$ NMEs

The nuclear matrix elements needed in Equations (1)–(4) describe the transition from an initial nucleus $|i\rangle = |0_i^+\rangle$ to a final nucleus $|f\rangle = |0_f^+\rangle$, and the matrix elements can also be presented as a sum over intermediate nuclear states $|\kappa\rangle = |J_\kappa^\pi\rangle$ with certain angular momentum $J_\kappa$, parity $\pi$, and energy $E_\kappa$:

$$M_\alpha^{0\nu} = \sum_\kappa \sum_{1234} \langle 13|\mathcal{O}_\alpha|24\rangle \langle f|\hat{c}_3^\dagger \hat{c}_4|\kappa\rangle\langle\kappa|\hat{c}_1^\dagger \hat{c}_2|i\rangle, \tag{5}$$

where operators $\mathcal{O}_\alpha$—with $\alpha$ denoting Gamow-Teller (GT), Fermi (F), tensor (T), etc. operators—contain neutrino potentials, spin and isospin operators, and explicit dependence on the intermediate state energy $E_\kappa$. The most common of the operators can be found in Refs. [17,43], and they include vector and axial nucleon form-factors that take into account nucleon size effects. The calculation details for two-body matrix elements, $\langle 13|\mathcal{O}_\alpha|24\rangle$, are discussed in Appendix D of Ref. [43]. Let us note that the two-body wave functions in the matrix elements (5) are not antisymmetrized, as one would expect for nuclear two-body matrix elements. The wave functions should be understood as

$|24\rangle = |2\rangle \cdot |4\rangle$ and $|13\rangle = |1\rangle \cdot |3\rangle$, where 1, 2, 3, and 4 represent single-nucleon quantum numbers, e.g., $1 = \{\tau_{1z}, n_1, l_1, j_1, \mu_1\}$

Calculations using a summation on intermediate states is very time-consuming, due to the need for obtaining a large number of intermediate states $\kappa$ and the associated one-body transition densities $\langle f|\hat{c}_3^\dagger \hat{c}_4|\kappa\rangle$ and $\langle \kappa|\hat{c}_1^\dagger \hat{c}_2|i\rangle$ in Equation (5), which can only be conducted efficiently in J-scheme codes such as NuShellX code [92]. The results and analyses for most of the nuclei in Table 1 can be found in Refs. [16,43,45,48,70].

Although time-consuming, this method has the advantage of being applicable for a large class of effective nuclear Hamiltonian and transition operators. For example, it can be used for isospin-breaking nuclear Hamiltonians and with transition operators that are treating asymmetrically the initial neutron single particle (s.p.) states vs. the final proton s.p. states, such as the in-medium similarity renormalization group and realistic shell model methods. This method is always applicable for transitions to the $2^+$ states in the daughter nucleus, even in cases when the transition operator is not a rotational scalar anymore [83,84].

If one replaces the energies of the intermediate states in the form-factors by an average constant value, one obtains the closure approximation. The operators $\mathcal{O}_\alpha \to \tilde{\mathcal{O}}_\alpha \equiv \mathcal{O}_\alpha(\langle E\rangle)$ become energy-independent and the sum over the intermediate states in the nuclear matrix element (5) can be taken explicitly using the completeness relation:

$$\sum_\kappa \langle f|\hat{c}_3^\dagger \hat{c}_4|\kappa\rangle\langle \kappa|\hat{c}_1^\dagger \hat{c}_2|i\rangle = \langle f|\hat{c}_3^\dagger \hat{c}_4 \hat{c}_1^\dagger \hat{c}_2|i\rangle. \tag{6}$$

The advantage of this approximation is significant because it eliminates the need for calculating a very large number of states in the intermediate nucleus, which could be computationally challenging, especially for heavy systems. One needs only to calculate the two-body transition densities (see Section 3.2) between the initial and final nuclear states. This approximation is very good due to the fact that the values of $q$ that dominate the matrix elements are of the order of 100–200 MeV, while the relevant excitation energies are only of the order of 10 MeV. The obvious difficulty related to this approach is that I have to find a reasonable value for this average energy, $\langle E\rangle$, which can effectively represent the contribution of all the intermediate states. This average energy needs to account also for the symmetric part of the two-body matrix elements $\langle 13|\mathcal{O}_\alpha|24\rangle$ in Equation (7) below. Indeed, the two-body wave functions $|13\rangle$ and $|24\rangle$ are not antisymmetric; by replacing the energies of the intermediate states with a constant, only the antisymmetric parts of these matrix elements are taken into account.

Most reported calculations are using closure approximation with some closure energies taken from Ref. [93]. By comparing the closure and the summation method results for different isotopes in different model spaces, I find [48,70] the optimal closure energies for a given model space and effective Hamiltonian (see end of Section 3.2 for examples). The optimal closure energies for a given model space and effective Hamiltonian can then be found by performing a calculation for a (fictitious) $0\nu\beta\beta$ NME of lower complexity.

### 3.2. The $0\nu\beta\beta$ NME in Closure Approximation

In the closure approximation, the $0\nu\beta\beta$ NME can be reduced to a sum of the products of two-body transition densities (TBTD), defined by the right-hand side of Equation (6), and antisymmetrized two-body matrix elements,

$$\begin{aligned} M_\alpha^{0\nu} &= \sum_{j_p j_{p'} j_n j_{n'} J_\pi} \mathrm{TBTD}\Big(j_p j_{p'}, j_n j_{n'}; J_\pi\Big) \\ &\quad \Big\langle j_p j_{p'}; J^\pi T \mid \tau_{-1}\tau_{-2} O_{12}^\alpha \mid j_n j_{n'}; J^\pi T\Big\rangle, \end{aligned} \tag{7}$$

where $O_{12}^\alpha$ are the two-body operators corresponding to different transitions (here denoted by $\alpha = F, GT, T, Fq, GTq, \ldots$ [17]) contributing to some of the diagrams of the $0\nu\beta\beta$ process

in Figure 2 (see Ref. [17] for details). One should not confuse the isospin $T$ in the two-body matrix elements with the tensor operator notation for $\alpha$.

Having the two-body matrix elements ready, one can calculate the NME in Equation (7) if two-body transition densities TBTD$\left(j_p j_{p'}, j_n j_{n'}; J_\pi\right)$ are known. Most of the shell model codes do not provide two-body transition densities. One alternative approach is to take advantage of the isospin symmetry that most of the effective interactions have, which creates wave functions with good isospin. The approach described below works also when the proton and neutron are in different shells. If the above conditions are satisfied, one can transform the two-body matrix elements of a change in isospin $\Delta T = 2$ operator using the Wigner–Eckart theorem, from a change in isopspin projection $\Delta T_z = -2$ to $\Delta T_z = 0$, which can be further used to describe transitions between states in the same nucleus.

Then the transformed matrix elements preserve spherical symmetry and they can be used as a piece of a Hamiltonian, $H^{\alpha}_{\beta\beta}$, which violates isospin symmetry, but it is a scalar with respect to rotational group. One can then lower by two units the isospin projection of the g.s. of the parent nucleus that has the higher isospin $T_>$, e.g., that of $^{48}$Ca, thus becoming an isobar analog state of the daughter nucleus that has isospin $T_< = T_> - 2$, e.g., in $^{48}$Ti. Denoting by $\mid 0^+_{i<}\ T_> >$ the transformed state, one can now calculate the many-body matrix elements of the transformed $0\nu\beta\beta$ operator,

$$M^{0\nu}_{\alpha}(T_z = T_<) = \left\langle 0^+_f\ T_< \mid H^{\alpha}_{\beta\beta} \mid 0^+_{i<}\ T_> \right\rangle. \tag{8}$$

Choosing $\mid 0^+_{i<}\ T_> >$ as a starting Lanczos vector and performing one Lanczos iteration with $H^{\alpha}_{\beta\beta}$, one obtains

$$H^{\alpha}_{\beta\beta} \mid 0^+_{i<}\ T_> >= a_1 \mid 0^+_{i<}\ T_> > + b_1 \mid L_1 >, \tag{9}$$

where $\mid L_1 >$ is the new Lanczos vector. Then, one can calculate the matrix elements in Equation (8):

$$M^{0\nu}_{\alpha}(T_z = T_<) = b_1 \left\langle 0^+_f\ T_< \mid L_1 \right\rangle. \tag{10}$$

The transition matrix elements in Equation (7) can then be recovered using again the Wigner–Eckart theorem,

$$M^{0\nu}_{\alpha} = M^{0\nu}_{\alpha}(T_z = T_<) \times C^{T_>\ 2\ T_<}_{T_>\ -2\ T_<} / C^{T_>\ 2\ T_<}_{T_<\ 0\ T_<}, \tag{11}$$

where $C^{T_1\ T_2\ T}_{T_{z1}\ T_{z2}\ T_z}$ are isospin Clebsch-Gordan coupling coefficients.

This procedure can be implemented in most nuclear shell model codes. The transformation of the g.s. of the parent to an analog state of the daughter can be performed very quickly, and one Lanczos iteration represents a small load as compared with the calculation of the g.s. of the daughter. The additional calculations described in Equations (9)–(11) require smaller resources than those necessary to calculate the TBTDs.

The form of the NME described in Equation (7) assumes that the underlying many-body Hamiltonian and the resulting wave functions have good isospin symmetry. That might not be the case when the Coulomb interaction is included or/and ab initio approaches to obtain the effective shell model Hamiltonian, such as the in-medium similarity renormalization group [94] or realistic shell model [95], are used. In that case, one could project the parent and daughter wave functions on good isospin components and extend the above procedure for each pair of isospin components considering the appropriate jump in isospin (which might be a difference of 2). In practice, the contributions from the main isospin components described in the above procedure dominate.

There are also some limitations to this method. For example, the in-medium similarity renormalization group and realistic shell model methods, as well as the G-matrix-like re-

summation approach (see, e.g., [96] and references therein) also provide effective operators, which breaks the symmetry of the two-body matrix elements,

$$\left\langle j_p j_{p'}; J^\pi T \mid \tau_{-1}\tau_{-2}O^\alpha_{12} \mid j_n j_{n'}; J^\pi T \right\rangle$$

between the $nn$ and $pp$ two-body states of the initial and final nucleus. In that case, one could consider the average of the corresponding two-body matrix elements [96]. This method is not always applicable for transitions to the $2^+$ states in the daughter nucleus because if one uses the contributions from the right-handed currents, such as those of the $\lambda$ and $\eta$ mechanisms (see Section 2.1), then the transition operator is not a rotational scalar anymore [83,84]. In those cases, the $H^\alpha_{\beta\beta}$ cannot be defined and the TBTDs are needed.

It would be interesting to compare the half-lives of two isotopes to identify the dominant mechanisms contributing to $0\nu\beta\beta$ decays. It is often the case that even within the shell model approach, using two different effective Hamiltonians leads to significantly different results. Given that the conclusion of two-isotope analysis is sensitive to the accuracy of NMEs, it is important to consider at least two sets of effective Hamiltonians. In addition, for consistency, I use the optimal closure energies [43,45,48], with $\langle E\rangle$ corresponding to each Hamiltonian and model space. One set of NMEs is obtained using the Hamiltonians preferred by our CMU (Central Michigan University) group: for $^{48}$Ca in the $pf$ model space $(0f_{7/2}, 1p_{3/2}, 0f_{5/2}, 1p_{1/2})$, I use GXPF1A effective Hamiltonian [97] with $\langle E\rangle = 0.5$ MeV; for $^{76}$Ge and $^{82}$Se in the $jj44$ model space $(0f_{5/2}, 1p_{3/2}, 1p_{1/2}, 0g_{9/2})$, I choose JUN45 [98] with $\langle E\rangle = 3.4$ MeV; and for $^{124}$Sn, $^{130}$Te, and $^{136}$Xe in the $jj55$ model space $(0g_{7/2}, 1d_{5/2}, 1d_{3/2}, 1s_{1/2}, 0h_{11/2})$, I use SVD effective Hamiltonian [99] with $\langle E\rangle = 3.5$ MeV. The second set of NMEs I calculate using the Hamiltonians preferred by the Strasbourg–Madrid group: in this case, for $^{48}$Ca I use KB3G [100] with $\langle E\rangle = 2.5$ MeV, for $^{76}$Ge I use $^{82}$Se GCN.28-50 with $\langle E\rangle = 10$ MeV, and for $^{130}$Te and $^{136}$Xe I use GCN.50-82 with $\langle E\rangle = 12$ MeV [101] (see Section 3.1 for the definition of the optimal closure energies $\langle E\rangle$). The numerical analysis is given in Ref. [17], where I find that using the ratio of experimental half-lives one could identify if a selected few mechanisms may be dominant.

## 4. Two-Neutrino Double Beta Decay

Two-neutrino double beta decay ($2\nu\beta\beta$) is an associated process allowed by the Standard Model, which was observed experimentally for about 10 isotopes, including most in Table 1. Here, I describe an improved spectra-function technique for calculating associated NMEs in very large model spaces in which a direct summation on intermediate states is not practical. For the $2\nu\beta\beta$ mode, the relevant NMEs are of Gamow–Teller type, and have the following expression for decays to states in the grand-daughter that have the angular momentum $J = 0$ [93]:

$$M_{2\nu} = \sum_k \frac{\langle 0^+_f||\sigma\tau^-||1^+_k\rangle\langle 1^+_k||\sigma\tau^-||0^+_i\rangle}{(E_k + E_0)}. \tag{12}$$

Here, $E_k$ is the excitation energy of the $1^+_k$ state of the intermediate odd-odd nucleus and $E_0 = \frac{1}{2}Q_{\beta\beta} + \Delta M$. $Q_{\beta\beta}$ is the Q-value corresponding to the $\beta\beta$ decay to the final $0^+_f$ state of the grand-daughter nucleus, and $\Delta M$ is the mass difference between the parent (e.g., $^{48}$Ca) and the intermediate nucleus (e.g., $^{48}$Sc). The most common case is the decay to the $0^+_1$ g.s. of the grand-daughter, but decays to the first excited $0^+_2$ state were also observed [80].

The $2\nu\beta\beta$ decay half-life is given by

$$\left[T^{2\nu}_{1/2}\right] = G_{2\nu} \cdot g_A^4 \cdot \left(m_e c^2 \cdot M_{2\nu}\right)^2 \tag{13}$$

In Ref. [51], I fully diagonalized 250 $1^+$ states of the intermediate nucleus $^{48}$Sc in the $pf$ valence space to calculate the $2\nu\beta\beta$ NME for $^{48}$Ca. This method of the direct diagonalization of a large number of states can be used for somewhat heavier nuclei using the J-scheme shell model code NuShellX [92], but for large-dimension cases one needs an alternative method. In particular, the m-scheme dimensions needed for the $^{48}$Ca NME calculations when taking into account up to $2\hbar\omega$ excitations $sd$-$pf$ valence space are larger than 1 billion (716 million for $^{48}$Sc). These large dimensions also require a method more efficient than direct diagonalization. The pioneering work on $^{48}$Ca [102] used a strength-function approach that converges after a small number of Lanczos iterations, but it requires large-scale shell model diagonalization when one wants to check the convergence. Ref. [103] proposed an alternative method which converges very quickly, but it did not provide full recipes for all its ingredients, and it was never used in practical calculations. Here, I propose a simple numerical scheme to calculate all coefficients of the expansion proposed in Ref. [103]. Following Ref. [103], I choose as a starting Lanczos vector $L_1^{\pm}$ either the initial or final states in the decay (only $0^+$ to $0^+$ transitions are considered here), on which is applied the Gamow–Teller operator,

$$|\sigma\tau^- 0_i^+ >= c_-|dw_- >\equiv c_-|L_1^- >, \tag{14}$$
$$|\sigma\tau^+ 0_f^+ >= c_+|dw_+ >\equiv c_+|L_1^+ > . \tag{15}$$

The results are the "door-way" states $|dw_{\pm}>$ multiplied by the constants $c_{\pm}$, which represent the square roots of the respective Gamow–Teller sum rule. Ref. [103] showed that the matrix element in Equation (12) could be calculated using one of the following two equations:

$$M_{2\nu}(0^+) \approx 3c_+c_- \sum_m g_m^- < dw_+|L_m^- >\equiv M_{2\nu}^{\mathrm{GT}-} , \tag{16}$$
$$M_{2\nu}(0^+) \approx 3c_+c_- \sum_m g_m^+ < L_m^+|dw_- >\equiv M_{2\nu}^{\mathrm{GT}+} . \tag{17}$$

Here, the sum is over the Lanczos vectors $L_m$. One can show that the $g_m^{\pm}$ factors can be calculated with the following formula after $N$ Lanczos iterations:

$$g_m^{\pm} = \sum_{k=1}^{N} \frac{V_{1k}^{\pm} V_{mk}^{\pm}}{E_L^N(1_k^+) - E_{g.s.} + E_0} . \tag{18}$$

Here, $V_{mk}$ are the eigenvectors of the N-order Lanczos matrix corresponding to eigenvalue $E_L^N(1_k^+)$. The advantage of using Equations (14)–(18) is that in order to check the convergence at each iteration one only needs the Lanczos vectors, which have to be stored anyway, and not the eigenvectors of the many-body Hamiltonian. The $g_m^{\pm}$ can be calculated very quickly, and only the last overlap in the sum of Equation (16) or Equation (17) needs to be calculated at each iteration. This algorithm can provide a gain in efficiency by a factor of about two as compared with the strength-function approach of Ref. [102].

Another advantage of this method is that it can be used with both M-scheme and J-scheme shell model codes, while a direct summation in Equation (12) on the $1^+$ states in the intermediate nucleus can only be performed using J-scheme codes. The method described here requires about 20 Lanczos iterations for convergence. I estimate (see, e.g., [51]) that good convergence for the direct summation in Equation (12) requires about 300–500 $1^+$ that usually can be achieved with about 5000–10,000 iterations. Given the input/output burden associated with so many iterations, I estimate computational speed improvement by a factor of about 1000 in the present method as compared with the direct summation method.

It is known that a good comparison of the shell model results with experimental data requires a multiplicative quenching factor for the Gamow–Teller operator. This numerical analysis when compared with the experimental data [45,49–51,70] indicates that for the selected effective Hamiltonians, only quenching factors between 0.6 and 0.74 are needed.

## 5. Conclusions

In this paper, I provide an overview of the double beta decay process and describe in some detail the shell model approach for the calculation of the nuclear matrix elements necessary for the analysis of experimental data.

Analyzing the physics related to the neutrinoless double beta decay process, one observes that it would entail physics beyond the Standard Model, namely the lepton number violation, which may lead to the conclusion that neutrinos may be the only known Standard Model fermions that are of Majorana type. This information may be crucial for properly extending the Standard Model Lagrangian to describe the observed neutrino masses and other LNV processes.

I describe the $0\nu\beta\beta$ decay half-life using BSM mechanisms induced by new particles such as the left-right symmetric model or SUSY and also use a more general EFT approach that includes the most general LNV addition to the Standard Model Lagrangian. Both approaches lead to similar numbers of NMEs associated with either model-specific or EFT-linked LNV couplings.

The largest part of the paper is dedicated to the techniques for calculating the needed NMEs within the shell model approach. For $0\nu\beta\beta$, I analyze the different scenarios under which the NMEs can be calculated in the closure approximation that is good within 10%. I also describe how to calculate the same NMEs beyond closure and identify optimal closure energies which can minimize the error of the less time-consuming closure approximation.

Two-neutrino double beta decay is an associated process allowed by the Standard Model, which was observed experimentally for about 10 isotopes. Here, I describe an improved spectra-function technique of calculating the associated NMEs in very large model spaces for which the dwerect summation on intermediate states is impractical.

Finally, although most of the paper reviews results already published, some new results regarding techniques for calculating $0\nu\beta\beta$ and $2\nu\beta\beta$ NMEs in extreme situations can be found at the end of Sections 3.2 and 4.

**Funding:** This research received support from the US Department of Energy grant DE-SC0022538 "Nuclear Astrophysics and Fundamental Symmetries". The author acknowledges a Central Michigan University Faculty Research and Creative Endeavors (FRCE) type A grant for travel support.

**Conflicts of Interest:** The author declares no conflict of interest.

## Abbreviations

The following abbreviations are used in this manuscript:

| | |
|---|---|
| $0\nu\beta\beta$ | neutrinoless double beta decay |
| $2\nu\beta\beta$ | two-neutrino double beta decay |
| BSM | beyond the Standard Model |
| CMU | Central Michigan University |
| EFT | effective field theory |
| g.s. | ground state |
| s.p. | single particle |
| LNV | lepton number violation |
| NME | nuclear matrix element |
| PCAC | Partial Conservation of Axial Current |
| PMNS | Pontecorvo–Maki–Nakagawa–Sakata |
| PSF | phase space factor |
| QCD | Quantum Chromodynamics |
| QRPA | quasiparticle random phase approximation |
| SUSY | super symmetry |
| TBTD | two-body transition densities |

## References


1. Fukuda, Y. et al. [Super-Kamiokande Collaboration] Evidence for oscillation of atmospheric neutrinos. *Phys. Rev. Lett.* **1998**, *81*, 1562–1567. [CrossRef]
2. Ahmad, Q.R. et al. [SNO Collaboration] Measurement of the rate of $\nu_e + d \rightarrow p + p + e^-$ interactions produced by $^8B$ solar neutrinos at the Sudbury Neutrino Observatory. *Phys. Rev. Lett.* **2001**, *87*, 071301. [CrossRef]
3. Kajita, T. Nobel Lecture: Discovery of atmospheric neutrino oscillations. *Rev. Mod. Phys.* **2016**, *88*, 03051. [CrossRef]
4. McDonald, A.B. Nobel Lecture: The Sudbury Neutrino Observatory: Observation of flavor change for solar neutrinos. *Rev. Mod. Phys.* **2016**, *88*, 030502. [CrossRef]
5. Schechter, J.; Valle, J.W.F. Neutrinoless doubl-$\beta$ decay in SU(2)$\times$U(1) theories. *Phys. Rev. D* **1982**, *25*, 2951–2954. [CrossRef]
6. Nieves, J. Dirac and pseudo-Dirac neutrinos and neutrinoless double beta decay. *Phys. Lett. B* **1984**, *147*, 375–379. [CrossRef]
7. Takasugi, E. Can the neutrinoless double beta decay take place in the case of Dirac neutrinos? *Phys. Lett. B* **1984**, *149*, 372–376. [CrossRef]
8. Hirsch, M.; Kovalenko, S.; Schmidt, I. Extended black box theorem for lepton number and flavor violating processes. *Phys. Lett. B* **2006**, *642*, 106–110. [CrossRef]
9. Pati, J.C.; Salam, A. Lepton number as the fourth "color". *Phys. Rev. D* **1974**, *10*, 275–289. [CrossRef]
10. Mohapatra, R.N.; Pati, J.C. "Natural" left-right symmetry. *Phys. Rev. D* **1975**, *11*, 2558–2561. [CrossRef]
11. Senjanovic, G.; Mohapatra, R.N. Exact left-right symmetry and spontaneous violation of parity. *Phys. Rev. D* **1975**, *12*, 1502–1505. [CrossRef]
12. Keung, W.-Y.; Senjanovic, G. Majorana neutrinos and the production of the right-handed charged gauge boson. *Phys. Rev. Lett.* **1983**, *50*, 1427–1430. [CrossRef]
13. Barry, J.; Rodejohann, W. Lepton number and flavor violation in TeV-scale left-right symmetric theories with large left-right mixing. *J. High Energy Phys.* **2013**, *2013*, 153. [CrossRef]
14. Khachatryan, V. et al. [CMS Collaboration] Search for heavy neutrinos and W bosons with right-handed couplings in proton-proton collisions at $\sqrt{s}$ = 8 TeV. *Eur. Phys. J. C* **2014**, *74*, 3149. [CrossRef]
15. Horoi, M.; Neacsu, A. Analysis of mechanisms that could contribute to neutrinoless double-beta decay. *Phys. Rev. D* **2016**, *93*, 113014. [CrossRef]
16. Neacsu, A.; Horoi, M. Shell Model studies of competing mechanisms to the neutrinoless double-beta decay in $^{124}$Sn, $^{130}$Te, and $^{136}$Xe. *Adv. High Energy Phys.* **2016**, *2016*, 1903767. [CrossRef]
17. Horoi, M.; Neacsu, A. Shell model study of using an effective field theory for disentangling several contributions to neutrinoless double-$\beta$ decay. *Phys. Rev. C* **2018**, *98*, 035502. [CrossRef]
18. Cirigliano, V.; Dekens, W.; de Vries, J.; Graesser, M.L.; Mereghetti, E. Neutrinoless double beta decay in chiral effective field theory: Lepton number violation at dimension seven. *J. High Energy Phys.* **2017**, *2017*, 82. [CrossRef]
19. Cirigliano, V.; Dekens, W.; Graesser, M.; Mereghetti, E. Neutrinoless double beta decay and chiral $SU(3)$. *Phys. Lett. B* **2017**, *769*, 460–464. [CrossRef]
20. Berkowitz, E.; Brantley, D.; Bouchard, C.; Chang, C.C.; Clark, M.A.; Garron, N.; Joo, B.; Kurth, T.; Monahan, C.; Monge-Camacho, H.; et al. An accurate calculation of the nucleon axial charge with lattice QCD. *arXiv* **2017**, arXiv:1704.01114. [CrossRef]
21. Hirsch, M.; Klapdor-Kleingrothaus, H.V.; Kovalenko, S.G. On the SUSY accompanied neutrino exchange mechanism of neutrinoless double beta decay. *Phys. Lett. B* **1996**, *372*, 181–186. [CrossRef]
22. Päs, H.; Hirsch, M.; Klapdor-Kleingrothaus, H.V.; Kovalenko, S.G. Towards a superformula for neutrinoless double beta decay. *Phys. Lett. B* **1999**, *453*, 194–198. [CrossRef]
23. Päs, H.; Hirsch, M.; Klapdor-Kleingrothaus, H.V.; Kovalenko, S.G. A Superformula for neutrinoless double beta decay. II. The short range part. *Phys. Lett. B* **2001**, *498*, 35–39. [CrossRef]
24. Deppisch, F.F.; Hirsch, M.; Päs, H. Neutrinoless double-beta decay and physics beyond the standard model. *J. Phys. G: Nucl. Part. Phys.* **2012**, *39*, 124007. [CrossRef]
25. Barabash, A.S. Precise half-life values for two-neutrino double-$\beta$ decay: 2020 review. *Universe* **2020**, *6*, 159. [CrossRef]
26. Barabash, A.S.; Hubert, P.; Nachab, A.; Konovalov, S.I.; Vanyushin, I.A.; Umatov, V. Search for $\beta^+$ EC and ECEC processes in $^{112}$Sn and $\beta^-\beta^-$ decay of $^{124}$Sn to the excited states of $^{124}$Te. *Nucl. Phys. A* **2008**, *807*, 269–289. [CrossRef]
27. Neacsu, A.; Horoi, M. An effective method to accurately calculate the phase space factors for $\beta^-\beta^-$ decay. *Adv. High Energy Phys.* **2016**, *2016*, 7486712. [CrossRef]
28. Stoica, S.; Mirea, M. New calculations for phase space factors involved in double-$\beta$ decay. *Phys. Rev. C* **2013**, *88*, 037303. [CrossRef]
29. Arnold, R. et al. [NEMO-3 Collaboration] Measurement of the double-beta decay half-life and search for the neutrinoless double-beta decay of $^{48}$Ca with the NEMO-3 detector. *Phys. Rev. D* **2016**, *93*, 112008. [CrossRef]
30. Agostini, M. et al. [GERDA Collaboration] Final results of GERDA on the search for neutrinoless Double-$\beta$ decay. *Phys. Rev. Lett.* **2020**, *125*, 252502. [CrossRef]

31. Waters, D. et al. [NEMO-3 and SuperNEMO Collaborations] Latest Results from NEMO-3 and Status of the SuperNEMO Experiment. Talk at the *Neutrino 2016:* XXVII Intern. Conf. on Neutrino Physics and Astrophysics, London, UK, 4–9 July 2016. Available online: http://neutrino2016.iopconfs.org/IOP/media/uploaded/EVIOP/event_948/10.25__5__waters.pdf (accessed on 1 September 2022).
32. Adams, D.Q. et al. [CUORE Collaboration] Improved limit on neutrinoless double-beta decay in $^{130}$Te with CUORE. *Phys. Rev. Lett.* **2020**, *124*, 122501. [CrossRef] [PubMed]
33. Gando, A. et al. [KamLAND-Zen Collaboration] Search for Majorana neutrinos near the inverted mass hierarchy region with KamLAND-Zen. *Phys. Rev. Lett.* **2016**, *117*, 082503. [CrossRef]
34. Retamosa, J.; Caurier, E.; Nowacki, F. Neutrinoless double beta decay of $^{48}$Ca. *Phys. Rev. C* **1995**, *51*, 371–378. [CrossRef] [PubMed]
35. Caurier, E.; Nowacki, F.; Poves, A.; Retamosa, J. Shell model studies of the double beta decays of $^{76}$Ge, $^{82}$Se, and $^{136}$Xe. *Phys. Rev. Lett.* **1996**, *77*, 1954–1957. [CrossRef]
36. Horoi, M. Shell model analysis of competing contributions to the double-$\beta$ decay of $^{48}$Ca. *Phys. Rev. C* **2013**, *87*, 014320. [CrossRef]
37. Neacsu, A.; Stoica, S. Constraints on heavy neutrino and SUSY parameters derived from the study of neutrinoless double beta decay. *Adv. High Energy Phys.* **2014**, *2014*, 724315. [CrossRef]
38. Caurier, E.; Menendez, J.; Nowacki, F.; Poves, A. Influence of pairing on the nuclear matrix elements of the neutrinoless $\beta\beta$ decays. *Phys. Rev. Lett.* **2008**, *100*, 052503. [CrossRef] [PubMed]
39. Menendez, J.; Poves, A.; Caurier, E.; Nowacki, F. Disassembling the nuclear matrix elements of the neutrinoless $\beta\beta$ decay. *Nucl. Phys. A* **2009**, *818*, 139–151. [CrossRef]
40. Caurier, E.; Martinez-Pinedo, G.; Nowacki, F.; Poves, A.; Zuker, A.P. The shell model as a unified view of nuclear structure. *Rev. Mod. Phys.* **2005**, *77*, 427–487. [CrossRef]
41. Horoi, M.; Stoica, S. Shell model analysis of the neutrinoless double-$\beta$ decay of $^{48}$Ca. *Phys. Rev. C* **2010**, *81*, 024321. [CrossRef]
42. Neacsu, A.; Stoica, S.; Horoi, M. Fast, efficient calculations of the two-body matrix elements of the transition operators for neutrinoless double-$\beta$ decay. *Phys. Rev. C* **2012**, *86*, 067304. [CrossRef]
43. Sen'kov, R.A.; Horoi, M. Neutrinoless double-$\beta$ decay of $^{48}$Ca in the shell model: Closure versus nonclosure approximation. *Phys. Rev. C* **2013**, *88*, 064312. [CrossRef]
44. Horoi, M.; Brown, B.A. Shell-model analysis of the $^{136}$Xe double beta decay nuclear matrix elements. *Phys. Rev. Lett.* **2013**, *110*, 222502. [CrossRef] [PubMed]
45. Sen'kov, R.A.; Horoi, M.; Brown, B.A. Neutrinoless double-$\beta$ decay of $^{82}$Se in the shell model: Beyond the closure approximation. *Phys. Rev. C* **2014**, *89*, 054304. [CrossRef]
46. Brown, B.A.; Horoi, M.; Sen'kov, R.A. Nuclear structure aspects of neutrinoless double-$\beta$ decay. *Phys. Rev. Lett.* **2014**, *113*, 262501. [CrossRef] [PubMed]
47. Neacsu, A.; Stoica, S. Study of nuclear effects in the computation of the $0\nu\beta\beta$ decay matrix elements. *J. Phys. G: Nucl. Part. Phys.* **2014**, *41*, 015201. [CrossRef]
48. Sen'kov, R.A.; Horoi, M. Accurate shell-model nuclear matrix elements for neutrinoless double-$\beta$ decay. *Phys. Rev. C* **2014**, *90*, 051301(R). [CrossRef]
49. Neacsu, A.; Horoi, M. Shell model studies of the $^{130}$Te neutrinoless double-$\beta$ decay. *Phys. Rev. C* **2015**, *91*, 024309. [CrossRef]
50. Horoi, M.; Neacsu, A. Shell model predictions for $^{124}$Sn double-$\beta$ decay. *Phys. Rev. C* **2016**, *93*, 024308. [CrossRef]
51. Horoi, M.; Stoica, S.; Brown, B.A. Shell-model calculations of two-neutrino double–$\beta$ decay rates of $^{48}$Ca with the GXPF1A interaction. *Phys. Rev. C* **2007**, *75*, 034303. [CrossRef]
52. Blennow, M.; Fernandez-Martinez, E.; Lopéz-Pavon, J.; Menendez, J. Neutrinoless double beta decay in seesaw models. *J. High Energy Phys.* **2010**, *2010*, 096. [CrossRef]
53. Šimkovic, F.; Pantis, G.; Vergados, J.D.; Faessler, A. Additional nucleon current contributions to neutrinoless double–$\beta$ decay. *Phys. Rev. C* **1999**, *60*, 055502. [CrossRef]
54. Suhonen, J.; Civitarese, O. Effects of orbital occupancies and spin-orbit partners on $0\nu\beta\beta$-decay rates. *Nucl. Phys. A* **2010**, *847*, 207–232. [CrossRef]
55. Faessler, A.; Meroni, A.; Petcov, S.T.; Šimkovic, F.; Vergados, J. Uncovering multiple *CP*-nonconserving mechanisms of $(\beta\beta)_{0\nu}$ decay. *Phys. Rev. D* **2011**, *83*, 113003. [CrossRef]
56. Mustonen, M.T.; Engel, J. Large-scale calculations of the double-$\beta$ decay of $^{76}$Ge, $^{130}$Te, $^{136}$Xe, and $^{150}$Nd in the deformed self-consistent Skyrme quasiparticle random-phase approximation. *Phys. Rev. C* **2013**, *87*, 064302. [CrossRef]
57. Faessler, A.; Gonzalez, M.; Kovalenko, S.; Šimkovic, F. Arbitrary mass Majorana neutrinos in neutrinoless double beta decay. *Phys. Rev. D* **2014**, *90*, 096010. [CrossRef]
58. Barea, J.; Iachello, F. Neutrinoless double-$\beta$ decay in the microscopic interacting boson model. *Phys. Rev. C* **2009**, *79*, 044301. [CrossRef]
59. Barea, J.; Kotila, J.; Iachello, F. Limits on neutrino masses from neutrinoless double-$\beta$ decay. *Phys. Rev. Lett.* **2012**, *109*, 042501. [CrossRef]
60. Barea, J.; Kotila, J.; Iachello, F. Nuclear matrix elements for double-$\beta$ decay. *Phys. Rev. C* **2013**, *87*, 014315. [CrossRef]

61. Barea, J.; Kotila, J.; Iachello, F. $0\nu\beta\beta$ and $2\nu\beta\beta$ nuclear matrix elements in the interacting boson model with isospin restoration. *Phys. Rev. C* **2015**, *91*, 034304. [CrossRef]
62. Rath, P.K.; Chandra, R.; Chaturvedi, K.; Lohani, P.; Raina, P.K.; Hirsch, J.G. Neutrinoless $\beta\beta$ decay transition matrix elements within mechanisms involving light Majorana neutrinos, classical Majorons, and sterile neutrinos. *Phys. Rev. C* **2013**, *88*, 064322. [CrossRef]
63. Rodriguez, T.R.; Martinez-Pinedo, G. Energy density functional study of nuclear matrix elements for neutrinoless $\beta\beta$ decay. *Phys. Rev. Lett.* **2010**, *105*, 252503. [CrossRef] [PubMed]
64. Song, L.S.; Yao, J.M.; Ring, P.; Meng, J. Relativistic description of nuclear matrix elements in neutrinoless double-$\beta$ decay. *Phys. Rev. C* **2014**, *90*, 054309. [CrossRef]
65. Faessler, A.; Rodin, V.; Šimkovic, F. Nuclear matrix elements for neutrinoless double-beta decay and double-electron capture. *J. Phys. G: Nucl. Part. Phys.* **2012**, *39*, 124006. [CrossRef]
66. Vogel, P. Nuclear structure and double beta decay. *J. Phys. G: Nucl. Part. Phys.* **2012**, *39*, 124002. [CrossRef]
67. Holt, J.D.; Engel, J. Effective double-$\beta$-decay operator for $^{76}$Ge and $^{82}$Se. *Phys. Rev. C* **2013**, *87*, 064315. [CrossRef]
68. Jiao, C.F.; Horoi, M.; Neacsu, A. Neutrinoless double-$\beta$ decay of $^{14}$Sn, $^{130}$Te, and $^{136}$Xe in the Hamiltonian-based generator-coordinate method. *Phys. Rev. C* **2018**, *98*, 064324. [CrossRef]
69. Brown, B.A.; Fang, D.L.; Horoi, M. Evaluation of the theoretical nuclear matrix elements for $\beta\beta$ decay of $^{76}$Ge. *Phys. Rev. C* **2015**, *92*, 041301(R). [CrossRef]
70. Sen'kov, R.A.; Horoi, M. Shell-model calculation of neutrinoless double-$\beta$ decay of $^{76}$Ge. *Phys. Rev. C* **2016**, *93*, 044334. [CrossRef]
71. Doi, M.; Kotani, T.; Takasugi, E. Double-beta decay and Majorana neutrino. *Prog. Theor. Phys. Suppl.* **1985**, *83*, 1–175. [CrossRef]
72. Doi, M.; Kotani, T.; Nishiura, H.; Takasugi, E. Double beta decay. *Progr. Theor. Exp. Phys.* **1983**, *69*, 602–635. [CrossRef]
73. Rodejohann, W. Neutrinoless double-beta decay and neutrino physics. *J. Phys. G: Nucl. Part. Phys.* **2012**, *39*, 124008. [CrossRef]
74. Mohapatra, R.N.; Pati, J.C. Left-right gauge symmetry and the "isoconjugate" model of *CP* violation. *Phys. Rev. D* **1975**, *11*, 566–571. [CrossRef]
75. Hirsch, M.; Klapdor-Kleingrothaus, H.łV.; Kovalenko, S.G. Supersymmetry and neutrinoless double $\beta$ decay. *Phys. Rev. D* **1996**, *53*, 1329–1348. [CrossRef]
76. Kolb, S.; Hirsch, M.; Klapdor-Kleingrothaus, H.V. Squark mixing and its consequences for $\not{R}_p$ minimal supersymmetric standard model couplings. *Phys. Rev. D* **1997**, *56*, 4161–4165. [CrossRef]
77. Faessler, A.; Gutsche, T.; Kovalenko, S.; Šimkovic, F. Pion dominance in *R*-parity violating supersymmetry induced neutrinoless double beta decay. *Phys. Rev. D* **2008**, *77*, 113012. [CrossRef]
78. Suhonen, J.; Civitarese, O. Weak-interaction and nuclear-structure aspects of nuclear double beta decay. *Phys. Rep.* **1998**, *300*, 123–214. [CrossRef]
79. Kotila, J.; Iachello, F. Phase-space factors for double-$\beta$ decay. *Phys. Rev. C* **2012**, *85*, 034316. [CrossRef]
80. Vergados, J.D.; Ejiri, H.; Šimkovic, F. Theory of neutrinoless double-beta decay. *Rep. Prog. Phys.* **2012**, *75*, 106301. [CrossRef]
81. Bhupal Dev, P.S.; Goswami, S.; Mitra, M. TeV-scale left-right symmetry and large mixing effects in neutrinoless double beta decay. *Phys. Rev. D* **2015**, *91*, 113004. [CrossRef]
82. Tomoda, T. $0^+ \to 2^+$ neutrinoless $\beta\beta$ decay of $^{76}$Ge. *Nucl. Phys. A* **1988**, *484*, 635–646. [CrossRef]
83. Fang, D.-L.; Faessler, A. Nuclear matrix elements for the $0\nu\beta\beta(0^+ \to 2^+)$ decay of $^{76}$Ge within the two-nucleon mechanism. *Phys. Rev. C* **2021**, *103*, 045501. [CrossRef]
84. Fang, D.-L.; Faessler, A. $0\nu\beta\beta$ to the first 2+ state with two-nucleon mechanism for L-R symmetric model. *arXiv* **2022**, arXiv:2208.08595. [CrossRef]
85. Štefanik, D.; Dvornicky, R.; Šimkovic, F.; Vogel, P. Reexamining the light neutrino exchange mechanism of the $0\nu\beta\beta$ decay with left- and right-handed leptonic and hadronic currents. *Phys. Rev. C* **2015**, *92*, 055502, [CrossRef]
86. Muto, K.; Bender, E.; Klapdor, H.V. Nuclear structure effects on the neutrinoless double beta decay. *Z. Phys. A* **1989**, *334*, 187–194. [CrossRef]
87. Faessler, A.; Šimkovic, F. Double beta decay. *J. Phys. G Nucl. Part. Phys.* **1998**, *24*, 2139–2178. [CrossRef]
88. Avignone, F.T., III; Elliott, S.R.; Engel, J. Double beta decay, Majorana neutrinos, and neutrino mass. *Rev. Mod. Phys.* **2008**, *80*, 481. [CrossRef]
89. Deppisch, F.F.; Harz, J.; Huang, W.-C.; Hirsch, M.; Päs, H. Falsifying high-scale baryogenesis with neutrinoless double beta decay and lepton flavor violation. *Phys. Rev. D* **2015**, *92*, 036005. [CrossRef]
90. Ahmed, F.; Neacsu, A.; Horoi, M. Interference between light and heavy neutrinos for $0\nu\beta\beta$ decay in the left–right symmetric model. *Phys. Lett. B* **2017**, *769*, 299–304. [CrossRef]
91. Ahmed, F.; Horoi, M. Interference effects for $0\nu\beta\beta$ decay in the left-right symmetric model. *Phys. Rev. C* **2020**, *101*, 035504. [CrossRef]
92. Nuclear Structure Resources. Available online: https://people.nscl.msu.edu/~brown/resources/resources.html (accessed on 1 September 2022).
93. Tomoda, T. Double beta decay. *Rep. Prog. Phys.* **1991**, *54*, 53–126. [CrossRef]
94. Belley, A.; Payne, C.G.; Stroberg, S.R.; Miyagi, T.; Holt, J.D. Ab initio neutrinoless double-Bbta decay matrix elements for $^{48}$Ca, $^{76}$Ge, and $^{82}$Se. *Phys. Rev. Lett.* **2021**, *126*, 042502. [CrossRef]

95. Coraggio, L.; Gargano, A.; Itaco, N.; Mancino, R.; Nowacki, F. Calculation of the neutrinoless double-$\beta$ decay matrix element within the realistic shell model. *Phys. Rev. C* **2020**, *101*, 044315. [CrossRef]
96. Kwiatkowski, A.A.; Brunner, T.; Holt, J.D.; Chaudhuri, A.; Chowdhury, U.; Eibach, M.; Engel, J.; Gallant, A.T.; Grossheim, A.; Horoi, M.; et al. New determination of double-$\beta$-decay properties in $^{48}$Ca: High-precision $Q_{\beta\beta}$-value measurement and improved nuclear matrix element calculations. *Phys. Rev. C* **2014**, *89*, 045502. [CrossRef]
97. Honma, M.; Otsuka, T.; Brown, B.A.; Mizusaki, T. New effective interaction for $pf$-shell nuclei and its implications for the stability of the $N = Z = 28$ closed core. *Phys. Rev. C* **2004**, *69*, 034335. [CrossRef]
98. Honma, M.; Otsuka, T.; Mizusaki, T.; Hjorth-Jensen, M. New effective interaction for $f_5pg_9$-shell nuclei. *Phys. Rev. C* **2009**, *80*, 064323. [CrossRef]
99. Qi, C.; Xu, Z.X. Monopole-optimized effective interaction for tin isotopes. *Phys. Rev. C* **2010**, *86*, 044323. [CrossRef]
100. Poves, A.; Sánchez-Solano, J.; Caurier, E.; Nowacki, F. Shell model study of the isobaric chains $A = 50$, $A = 51$ and $A = 52$. *Nucl. Phys. A* **2001**, *694*, 157–198. [CrossRef]
101. Caurier, E.; Nowacki, F.; Poves, A.; Sieja, K. Collectivity in the light xenon isotopes: A shell model study. *Phys. Rev. C* **2010**, *82*, 064304. [CrossRef]
102. Caurier, E.; Poves, A.; Zuker, A.P. A full $0\hbar\omega$ description of the $2\nu\beta\beta$ decay of $^{48}$Ca. *Phys. Lett. B* **1990**, *252*, 13–17. [CrossRef]
103. Engel, J.; Haxton, H.; Vogel, P. Effective summation over intermediate states in double-beta decay. *Phys. Rev. C* **1992**, *46*, 2153(R). [CrossRef]

 *physics* 

*Review*

# Shell Model Applications in Nuclear Astrophysics †

**Gabriel Martínez-Pinedo [1,2,3,*] and Karlheinz Langanke [1,2]**

1 GSI Helmholtzzentrum für Schwerionenforschung, Planckstraße 1, 64291 Darmstadt, Germany; k.langanke@gsi.de
2 Institut für Kernphysik (Theoriezentrum), Fachbereich Physik, Technische Universität Darmstadt, Schlossgartenstraße 2, 64298 Darmstadt, Germany
3 Helmholtz Forschungsakademie Hessen für FAIR, GSI Helmholtzzentrum für Schwerionenforschung, Planckstraße 1, 64291 Darmstadt, Germany
* Correspondence: g.martinez@gsi.de

† This paper is an extended and updated version of the contribution to the 11th International Spring Seminar on Nuclear Physics: Shell Model and Nuclear Structure—Achievements of the Past Two Decades, Ischia, Italy, 12–16 May 2014.

**Abstract:** In recent years, shell model studies have significantly contributed in improving the nuclear input, required in simulations of the dynamics of astrophysical objects and their associated nucleosynthesis. This review highlights a few examples such as electron capture rates and neutrino-nucleus cross sections, important for the evolution and nucleosynthesis of supernovae. For simulations of rapid neutron-capture (r-process) nucleosynthesis, shell model studies have contributed to an improved understanding of half lives of neutron-rich nuclei with magic neutron numbers and of the nuclear level densities and $\gamma$-strength functions that are both relevant for neutron capture rates.

**Keywords:** shell model; core-collapse supernova; r-process nucleosynthesis; neutrino–nucleus reactions; electron capture

**Citation:** Martínez-Pinedo, G.; Langanke, K. Shell Model Applications in Nuclear Astrophysics. *Physics* **2022**, *4*, 677–689. https://doi.org/10.3390/physics4020046

Received: 7 April 2022
Accepted: 17 May 2022
Published: 17 June 2022

**Publisher's Note:** MDPI stays neutral with regard to jurisdictional claims in published maps and institutional affiliations.

**Copyright:** © 2022 by the authors. Licensee MDPI, Basel, Switzerland. This article is an open access article distributed under the terms and conditions of the Creative Commons Attribution (CC BY) license (https://creativecommons.org/licenses/by/4.0/).

## 1. Introduction

The interacting shell model, which takes in account correlations beyond mean field in a valence space, is generally considered as the method of choice to describe medium-mass nuclei [1–3]. Such nuclei play crucial roles for the dynamics of astrophysical objects and their associated nucleosynthesis. Unfortunately, a direct experimental determination of the required input is often prohibited due to the extreme conditions of the astrophysical environment in terms of temperature, density and also proton-to-neutron ratio; hence, the information has to be modeled. Here, the shell model has led to decisive progress in many cases in recent years, mainly due to its ability to account for the relevant correlations among nucleons and to accurately reproduce low-energy spectra and electromagnetic transitions [2,4,5].

This paper summarizes some of the progress achieved on the basis of shell model studies. Here, two different versions of the interacting shell model have been exploited: the diagonalization shell model [2] and the Shell Model Monte Carlo (SMMC) approach [6,7]. Diagonalization shell model calculations, which in contrast to SMMC allow for detailed spectroscopy, have been performed to derive rates for weak interaction processes of nuclei up to the iron-nickel mass range [8–11]. In particular, the shell model rates for electron captures on nuclei have significant impact on the presupernova core evolution of massive stars [12,13], the core evolution at the end of the hydrostatic evolution of medium-mass stars [11,14,15] and on the nucleosynthesis in thermonuclear supernovae [16].

The SMMC approach is based on a statistical description of the nucleus at finite temperature. In contrast to diagonalization, the shell model allows the derivation of nuclear properties at finite temperatures in extremely large model spaces by taking the relevant nuclear correlations into account [6,7]. SMMC has been the basis for deriving

electron capture rates for heavier neutron-rich nuclei for which cross-shell correlations are essential in establishing the capture on nuclei as the main weak interaction process for the dynamics of the core collapse of a massive star [17–19].

Both varieties of the interacting shell model have improved the nuclear input required for simulations of rapid neutron-capture (r-process) nucleosynthesis. Diagonal shell model calculations have been used to derive half lives for neutron-rich nuclei with magic neutron numbers (called waiting points), which are crucial for the mass flow during the nucleosynthesis process [20–22]. Shell model calculations have also been used to study the general behavior of electromagnetic transitions, which are essential for modeling neutron capture rates, where an experimentally observed increase in the dipole's strength function at low energies has drawn attention recently [23,24]. The nuclear level density is another important ingredient in modeling neutron capture rates. Here, SMMC calculations have allowed a microscopic derivation of level densities, also allowing the exploration of parameter dependencies, used in phenomenological approachesl (see, e.g., [25,26]) .

## 2. Weak Interaction Processes in Supernovae

A massive star ends its life in a supernova explosion triggered by the gravitational collapse of its inner core that is no longer supported by energy released in charged-particle reactions [19,27]. Electron captures on nuclei have three important consequences during the collapse [4,28]: (i) electron captures reduce the number of electrons and hence the pressure with which the degenerate (relativistic) electron gas counteracts against the gravitational contraction; (ii) the neutrinos, generated by the capture process, leave the star mainly unhindered, carrying away energy and keeping the entropy in the core low such that heavy nuclei survive during collapse; (iii) electron capture changes a proton in the nucleus into a neutron, driving the core composition to be a more neutron-rich (and heavier) nuclei. In the late stage of the collapse, coherent scattering with nuclei and inelastic scattering with electrons are responsible for neutrinos becoming trapped and thermalized in what is called the homologous core [27]. Other neutrino–nucleus interactions are of minor importance during collapse; however, they play a role in the nucleosynthesis processes following supernova explosions and for the detection of supernova neutrinos.

### *2.1. Electron Capture on Nuclei*

At the stellar conditions early in the collapse at which the core composition (described by nuclear statistical equilibrium) is dominated by nuclei from the iron-nickel mass range ($pf$-shell nuclei), electron capture is dominated by Gamow-Teller ($GT_+$) transitions. The subscript refers to the isospin component in the GT operator such that in $GT_+$ transitions a proton is changed into a neutron, in $GT_-$ transitions, which are relevant for $\beta^-$ decay of nuclei with neutron excess, a neutron is changed into a proton, and the $GT_0$ strength, important for describing low-energy inelastic neutrino-nucleus scattering, refers to transitions between proton states and neutron states. It is now possible to derive converged low-energy spectra and transitions of $pf$-shell nuclei in the respective model space [2]. In fact, it turned out that, in addition to a constant renormalization of the Gamow–Teller operator [29,30], such highly correlated wave functions are required to describe the strong fragmentation and total value of $GT_+$ strength [31], as experimentally determined by charge-exchange experiments [32,33].

The formalism for the calculation of electron capture rates has been introduced in ref. [34,35]. Note that the strong energy dependence of phase space as well as the fact that the electron Fermi energy is of the same order as the Q-value (the energy difference between initial and final nuclear states) of the abundant nuclei under presupernova conditions makes a detailed and accurate description of the $GT_+$ distribution an important requirement for a reliable description of stellar electron capture during this phase of the collapse. That the diagonalization shell model is up to this task and indeed the method of choice to describe stellar weak-interaction rates during presupernova collapse has been demonstrated by Cole and collaborators [36]. In [36], the capture rates, derived from experimental $GT_+$ data

for all $pf$-shell nuclei for which data exist, are compared with rates calculated within the shell model using two different residual interactions. As shown in Figure 1, the agreement is quite satisfactory at the conditions at which these nuclei are abundant and relevant for the core dynamics. A tabulation of shell model capture rates for $pf$-shell nuclei has been made available based on large-scale studies using a variation of the Strasbourg–Madrid KB3 interaction [10]. More recent studies using an improved residual interaction basically confirmed prior calculations. These studies led to slight improvements for selected mid-$pf$-shell nuclei; see Figure 1.

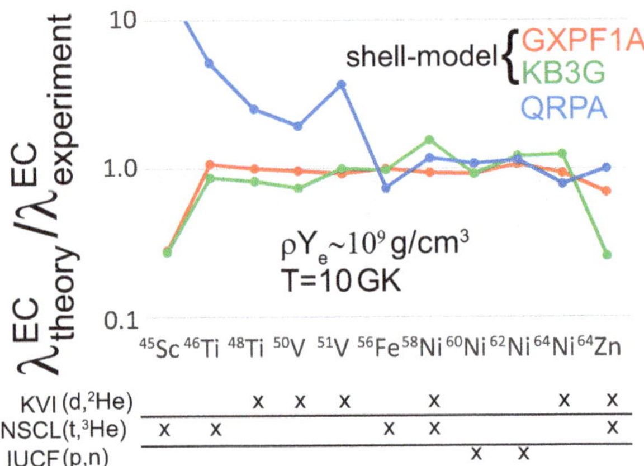


**Figure 1.** Comparison of electron captures rates, calculated from experimental $GT_+$ data and distributions, derived from the large-scale shell model calculations with two different interactions (KB3G [37] and GXPF1 [38]) and from a Quasiparticle Random Phase Approximation (QRPA) approach [36]. See text for details. The conditions correspond to the early stage of the collapse where the capture rates are sensitive to details of the $GT_+$ distribution. The shell model rates have been quenched with typical factor of $(0.74)^2$, as derived in [30]. $\rho Y_e$ and $T$ denote the electron density and temperature, respectively. KVI, UCSL and IUCF stay for the laboratories at which the experiments were performed. Taken from [39] with permission.

Stars in the mass range of 8–12$M_\odot$ ($M_\odot$ denotes the Sun mass) received a lot of attention recently as they fill the gap between low-mass stars, which end their lives as white dwarfs and massive stars which, as discussed above, run through the full circle of hydrostatic burning stages ending finally as core-collapse supernovae. The intermediate mass stars are not massive enough to ignite all advanced hydrostatic burning stages and instead degenerate ONe or ONeMg cores. Electron captures are crucial for the final fate of the stars, where the most abundant nuclei, $^{24}$Mg and $^{20}$Ne, are of key importance together with selected Urca pairs, which reduce the temperature of the core. Shell model rates for *sd*-shell nuclei exist since several years [8]. The important capture rate on $^{24}$Mg has been recently updated, mainly due to improved experimental data [11]. The capture rate on $^{20}$Ne has also been updated with, however, two remarkable highlights. First, it has been pointed out that the rate at the relevant astrophysical conditions could be decisively altered due to the influence of the second forbidden transition between the $^{20}$Ne and $^{20}$F ground states [11]. Such a situation is a novum, as the electron capture process is usually dominated by permitted transitions and (first) forbidden transitions are contributed only in high-temperature, high-density environments. The transition was very recently measured in a dedicated experiment [15], and it was indeed confirmed that it increases the capture rate in the astrophysically relevant range by orders of magnitude (Figure 2). The measured transition strength also agrees with the value calculated within the shell model [40]. Secondly, the electron capture rate on $^{20}$Ne at the astrophysical conditions, relevant for the core evolution of intermediate-mass stars, is now completely determined experimentally [15,40].

The improved $^{20}$Ne electron capture rate has interesting consequences for the final core evolution as the faster electron capture supports the ignition of oxygen burning at slightly smaller densities and off-center. Simulations, exploiting the larger rate, indicate that some intermediate-mass stars might explode as thermonuclear rather than electron capture supernovae [40]. Final conclusions can, however, only been drawn after multidimensional simulations of the core evolution with improved treatments of convection becoming available [41,42]. Other nuclei, for which the electron capture rates are dominated by second forbidden transitions under astrophysical conditions, are $^{24}$Na and $^{27}$Al [43]. The latter is expected to play a minor role on the evolution of ONeMg cores. The former may trigger convectional instabilities that again require multidimensional modeling.

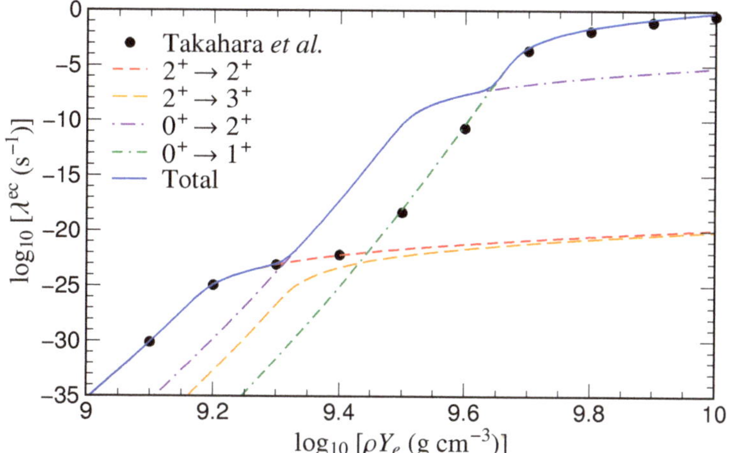


**Figure 2.** Electron capture (ec) rate for $^{20}$Ne as function of density and for a specific temperature ($\log T[\mathrm{K}] = 8.6$) relevant for the core evolution of intermediate-mass stars. The rate is broken down to the individual state-by-state contributions. In the density regime, particularly relevant for core evolution, the rate is dominated by the second-forbidden ground-state-to-ground-state transition. The rates labeled 'Takahara et al.' are derived from allowed transitions calculated in the shell model [44]. Taken from [11].

In the later stage of the collapse of massive stars, the nuclei present in the core composition become heavier and more neutron-rich. The appropriate model space to describe electron capture for such nuclei is too large (requiring two major shells) to allow for shell model diagonalization calculations. The calculations are then based on the SMMC variant of the shell model [7], which allows the determination of nuclear properties at finite temperatures and in large multi-shell model spaces taking the relevant nuclear correlations into account. Such correlations are particularly important for nuclei with proton number below and neutron number, $N$, above an oscillator shell closure (such as $N = 40$). In such states, $GT_+$ transitions would be completely blocked by the Pauli principle in the Independent Particle Model (IPM) [45] suppressing electron capture on nuclei drastically. However, it has been shown in [17,46] that nuclear correlations induced by the residual interaction move nucleons across the shell gap, enabling $GT_+$ transitions and making electron capture on nuclei the dominating weak interaction process during collapse [17,47]. Let us add two remarks. The unblocking of the $GT_+$ strength across the $N = 40$ shell closure has been experimentally confirmed for $^{76}$Se (with 34 protons and 42 neutrons) [48], in agreement with shell model studies [49]. Furthermore, shell model studies certainly show that the description of cross-shell correlations is a rather slowly converging process that requires the consideration of multi-particle multi-hole configurations [49–51].

Recently, it has been pointed out that the $N = 50$ shell closure could serve as a severe obstacle for electron capture on the very neutron-rich nuclei encountered in the later stage of the collapse [52]. This finding was apparently confirmed by measurements of the $GT_+$ strength in the $N = 50$ nuclei $^{86}$Kr and $^{88}$Se, which showed basically the vanishing strength for the ground state [53,54]. However, the situation is decisively different at the high temperaures (about 1 MeV) present in the collapsing core when $N = 50$ nuclei are abundantly present. Here, thermal excitations mix orbitals across the shell gap and unblock the GT transitions in this way. This was confirmed in two independent calculations for neutron-rich $N = 50$ nuclei using a thermal Quasiparticle Random Phase Approximation (QRPA) approach [55,56], in agreement with the earlier results obtained within the SMMC studies [17,57].

Based on the diagonalization shell model and the SMMC results and assuming a nuclear statistical equilibrium distribution for the composition, electron capture rates have been tabulated for the range of astrophysical conditions encountered during collapse of massive stars [57]. These rates consider potential screening effects of the astrophysical surroundings. The rate tabulation of Ref. [57] is now incorporated in many of the leading supernova simulation codes. It turns out that the rates have significant impact on collapse simulations. In the presupernova phase ($\rho < 10^{10}$ g/cm$^3$), the captures proceed slower than assumed before, and for a short period during silicon burning, $\beta$-decays can compete [12,13]. As a consequence, the core is cooler, more massive and less neutron rich before the final collapse. However, for a long time simulations of this final collapse assumed that electron captures on nuclei are prohibited by the Pauli blocking mechanism, as mentioned above (see, e.g., [27]). However, based on the SMMC calculations, it has been shown in [17] that capture on nuclei dominates over capture on free protons. The changes compared to the previous simulations are significant [17–19]. Importantly, the shock is now created at a smaller radius with more infalling material to traverse, but the density, temperature and entropy profiles are also strongly modified [18].

Finally, let us note that the shell model electron capture rates [10,58], which are noticeably slower than the pioneering rates of Fuller et al. (FFN) for $pf$-shell nuclei [34], have important consequences in nucleosynthesis studies for thermonuclear (type Ia) supernovae assuming the single-degenerate scenario as they result in a smaller reduction in the electron-to-nucleon ratio being the burning front [16]. As a consequence, very neutron-rich nuclei such as $^{50}$Ti and $^{54}$Cr are significantly suppressed compared to calculations, which use FFN rates [59]. In fact, in calculations using the shell-model rates, no nuclide is significantly overproduced compared to solar abundances [16].

### 2.2. *Neutrino–Nucleus Scattering*

At sufficiently high densities ($\rho > 4 \times 10^{11}$ g cm$^{-3}$), neutrinos become trapped and thermalized in the collapsing core by coherent scattering on nuclei and inelastic scattering on electrons. It had been suggested that de-excitation of thermally excited nuclei by neutrino pair emission [60] and inelastic neutrino–nucleus scattering [61] might be other modes contributing to neutrino thermalization. Although both processes have been found as rather unimportant cooling mechanisms [47,62], they have interesting impacts elsewhere. Neutrino pair emission has been identified as the major source of neutrino types other than electron neutrinos (produced overwhelmingly by electron capture) [47]. As the consequences of inelastic neutrino scattering are based on shell model calculations, the latter are briefly summarized. The formalism for the calculation of neutrino-nucleus reactions has been introduced in Ref. [63].

Supernova neutrinos have rather low energies (of order 10 MeV). Therefore, inelastic neutrino scattering of such neutrinos is dominated by allowed $GT_0$ transitions. Unfortunately, no data about inelastic neutrino scattering on nuclei exist at such energies. Due to its success in describing $GT_+$ (and $GT_-$) distributions, one can expect that the shell model will also reproduce the $GT_0$ component quite well. Nevertheless, a validation of the shell model approach to inelastic neutrino-nucleus scattering is desired. This can be achieved by

exploiting the fact that the $GT_0$ strength is, in a rather good approximation, proportional to the M1 strength of spherical nuclei [64]. In fact, precision M1 data, obtained by inelastic electron scattering for such nuclei, are well reproduced by shell model calculations [64,65]. The same approaches can also be used to derive $GT_0$ distributions for excited nuclear states, which can be thermally populated at finite supernova conditions [64,66]. At higher neutrino energies, forbidden transitions also contribute to the inelastic scattering cross section, which has been derived by RPA calculations. Supernova simulations that incorporate inelastic neutrino–nucleus scattering indicate that this mode has a noticeable effect on the early neutrino spectra emitted from supernova [62]. Here, nuclei act as obstacles for high-energy neutrinos which are down scattered in energy. This reduces significantly the tail of the neutrino spectra and, hence, also the predicted event rates for the observation of supernova neutrinos by earthbound detectors [62].

Charged-current and neutral-current neutrino–nucleus reactions are key to a specific nucleosynthesis process (called neutrino nucleosynthesis [67]), which are initiated by neutrinos emitted after core bounce in the supernova. Upon passing through the outer layers of the star, these neutrinos excite nuclei above particle thresholds so that the subsequent decay is by particle emission (mainly of protons or neutrons). Neutrino nucleosynthesis has been identified as the main or a strong source for the production of selected isotopes, $^{11}$B and $^{19}$F, from charged- and neutral-current reactions on the abundant isotopes $^{12}$C and $^{20}$Ne; $^{138}$La and $^{180}$Ta mainly by charged-current reactions on Ba and Ta isotopes, which had been previously been produced by the slow neutron-capture process (s-process) [67–71]. The partial neutrino–nucleus cross sections have been obtained by combining shell-model or RPA excitation functions with statistical model decay probabilities [72–74]. A particular interest in neutrino nucleosynthesis arises from the fact that the abundances of the produced nuclides depend on the spectra of those neutrino types ($\nu_\mu, \nu_\tau$ and their antiparticles and $\nu_e$), which have likely not been observed from supernova 1987A.

In principle, neutrino–nucleus reactions also play a role in the $\nu p$ process that operates in the neutrino-driven wind during cooling of the newborn proton–neutron star [75–77]. Simulations, however, show that the main neutrino reaction is the absorption of $\bar{\nu}_e$ on protons that produce a continuous source of free neutrons, which drives the process and allows mass flow through long-lived waiting points such as $^{64}$Ge. The $\nu p$ process is discussed as a potential source of isotopes such as $^{92}$Nb and $^{94,96}$Ru.

## 3. r-Process Nucleosynthesis

The r-process is the astrophysical origin of about half of the elements heavier than iron [78]. It occurs in an astrophysical environment with extreme neutron densities [79,80]. The r-process site has been a mystery for a long time until the observation of the neutron star merger event GW170817 by gravitational waves and its associated electromagnetic signal proved that heavy elements are produced by neutron star mergers [81,82]. The observed electromagnetic transient, called "kilonova," agreed well with prior predictions [83].

r-process simulations show that the reaction path in the nuclear chart runs through nuclei with such large neutron excesses that most of them have yet not been made in the laboratory and their properties have to be modeled. The relevant nuclear properties are masses, half lives, fission rates and yields and neutron capture rates [80]. Shell model calculations improved the determination of half lives for the nuclei with magic neutron numbers, which with their relatively long half-lives, act as obstacles in the r-process flow. They have also demonstrated new methods to calculate electromagnetic strength functions and nuclear level densities, which are both required to calculate neutron capture rates within the framework of the statistical model. For a very recent review of the various astronomical, astrophysical, nuclear, and atomic aspects of r-process nucleosynthesis; see Ref. [84].

r-Process nucleosynthesis proceeds by successive neutron captures and beta decays, which increase the mass and charge numbers, respectively. Nuclear half lives decide the time required to produce the heaviest elements, beginning from free protons and neutrons

that exist in the hot environment of merging neutron stars before matter is ejected and cooled allowing nuclei to form. Hence, nuclear beta decays compete with the time scales of the dynamical evolution of the ejected matter. There has been important progress made by measuring the half lives of some intermediate-mass nuclei on the r-process path [85,86]. However, most half lives still have to be modeled. Global sets of r-process half lives have been determined by QRPA calculations on the basis of phenomenological parametrizations [87,88] and more recently of microscopic Hartree-Fock-Bogoliubov (HFB) or density functional approaches [89–93].

Particularly important for the r-process mass flow are the waiting point nuclei at the magic neutron numbers $N = 50, 82$ and 126, which have rather long half lives due to their closed-shell configurations. For these nuclei, large-scale shell model calculations exist. Importantly, a few of these half lives could also been measured, showing good agreement with the shell model results: for $^{78}$Ni, an experiment done at the National Superconducting Cyclotron Laboratory (NSCL) experiment found a half life of $110 \pm 40$ ms [94], while the shell model predicted 127 ms [4]. Data and shell model results for the $N = 82$ waiting points are compared in Figure 3. Unfortunately, no data exist yet for $N = 126$ waiting points. For these nuclei, two independent shell model calculations have pointed to the importance of forbidden transitions induced by intruder states [21,22]. These forbidden transitions are predicted to shorten the half lives of the $N = 126$ waiting points noticeably and enhanced the mass flow through these waiting points [95]. This implies more r-process material available for fission, thus affecting the abundances of the second r-process peak around atomic mass number, $A = 130$, which for very neutron-rich ejecta is built up by fission yields [95,96]. The enhanced mass flow also increase late-time $\alpha$-decays from the decaying r-process matter, which influence the kilonova signal [97].

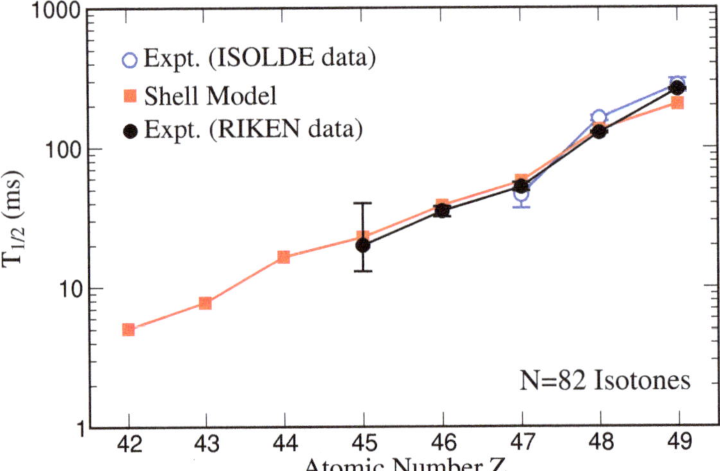


**Figure 3.** Comparison of shell model half lives for neutron number $N = 82$ r-process (rapid neutron-capture process) waiting point nuclei with data [86,98–100]. The GT strengths underlying the shell model results have been quenched with the standard factor of $(0.74)^2$ [30]. Taken from [84].

Neutron capture rates become relevant for r-process nucleosynthesis once the process drops out of $(n,\gamma) \rightleftarrows (\gamma,n)$ equilibrium at temperatures below about 1 GK. Neutron capture rates are traditionally derived within the statistical Hauser–Feshbach model, although this approach might not always be justified for r-process nuclei; see discussion and references in [84]. Important ingredients in the Hauser–Feshbach approach are the nuclear level densities and the $\gamma$-strength functions [80]. Shell model calculations have provided a better understanding of both quantities.

A method has been presented to derive level densities within the SMMC approach by exploiting its ability to describe nuclei in extremely large model spaces and to account for the correlations among nucleons [25,26]. The method has been used to explore the effects of parity, angular-momentum and pairing on the level density [101–103]. Based on SMMC studies, Alhassid et al. [104] presented an approach in which a microscopically derived parity-dependence is incorporated into phenomenological level density formulas. This approach has been used to derive a large set of r-process nuclei by also employing a temperature-dependent parametrization of the pairing parameter modeled after SMMC calculations [105]. These improved level densities are now part of statistical model packages NON-SMOKER and SMARAGD, developed by Rauscher [106–108]. An alternative microscopic approach to level densities, built on the HFB model, has been derived by Goriely and collaborators [109–111].

Experimentally determined dipole $\gamma$-strength functions show an upbend of the strength towards low gamma energies [112,113], which can have important impacts on neutron capture rates [23,24,113–116]. The upbend in the M1 strength has been studied and reproduced in shell model calculations for $pf$-shell and heavier nuclei [117,118]. Similar studies have been used to calculate the M1 contribution to the neutron capture rate in a consistent state-by-state approach [119]. This study found that the rate will be dominated by a single resonance if this state happens to fall into the Gamow window of the reaction. Such a situation is difficult to describe within a statistical approach. The calculation also shows that the M1 scissors mode observed in deformed nuclei [120] can lead to a significant enhancement of the capture rate.

## 4. Summary

Due to the extreme densities, temperatures or neutron excesses encountered in astrophysical environments, the properties of nuclei cannot be measured directly in a laboratory and have to be modeled. If these properties are strongly influenced by nucleon correlations, the diagonalization shell model is the method of choice. In recent years, such studies have been performed to derive the electron capture rates and neutrino-induced cross sections for nuclei in the *sd*- and $pf$-shell advancing our understanding of the core evolution of intermediate-mass and massive stars. Another important application of the diagonalization shell model was the calculation of half-lives for rapid neutron-capture process (r-process) nuclei with magic neutron numbers, which serve as waiting points for the r-process' mass flow. This example also shows the limitation of current shell model applications as such studies would be also very desirable for the other nuclei on the r-process path, but cannot be performed yet as the required model spaces exceed current computational possibilities. These limitations in model space can be overcome within the Shell Model Monte Carlo (SMMC) approach, which is an alternative formulation of the shell model. This approach describes nuclear properties at finite temperature, but is not capable of detailed spectroscopy. Thus, the SMMC cannot be used to calculate r-process half-lives, which need a state-by-state description of transition strength. However, the ability of the SMMC approach to describe nuclear properties at finite temperatures including correlations paves the way to determine electron capture rates of heavier nuclei, which are crucial for the fate of core-collapse supernovae. In particular, SMMC allows the evaluation of how the Pauli blocking of Gamow–Teller strength at closed shells is overcome by correlations. On the basis of these studies, it could be demonstrated that neither $N = 40$ nor $N = 50$ neutron shell closure serve as severe obstacles for electron capture on nuclei. It is now commonly accepted that electron capture proceeds on nuclei throughout the entire collapse.

**Author Contributions:** writing—original draft preparation, K.L. and G.M.-P.; writing—review and editing, K.L. and G.M.-P.; All authors have read and agreed to the published version of the manuscript.

**Funding:** This work was supported by the Deutsche Forschungsgemeinschaft (DFG, German Research Foundation), Project-ID 279384907, SFB 1245 *Nuclei: From Fundamental Interactions to Structure and Stars*, the Helmholtz Forschungsakademie Hessen für FAIR and the European Research Council (ERC) under the European Union's Horizon 2020 research and innovation programme (ERC Advanced Grant KILONOVA No. 885281).

**Data Availability Statement:** Not applicable.

**Acknowledgments:** The authors are grateful for the long-term collaboration with the members of the Strasbourg-Madrid shell model group. We have learnt a lot from Etienne Caurier, Frederic Nowacki, Alfredo Poves and Andres Zuker. Our work has also benefited strongly from collaborations with many astrophysicists, most notably with Raphael Hix, Hans-Thomas Janka and Friedel Thielemann.

**Conflicts of Interest:** The authors declare no conflict of interest.

## References

1. The National Research Council. *Nuclear Physics: Exploring the Heart of Matter*; The National Academies Press: Washington, DC, USA, 2013. Available online: https://nap.nationalacademies.org/read/13438/ (accessed on 15 May 2022).
2. Caurier, E.; Martinez-Pinedo, G.; Nowacki, F.; Poves, A.; Zuker, A.P. The shell model as a unified view of nuclear structure. *Rev. Mod. Phys.* **2005**, *77*, 427–488. [CrossRef]
3. Coraggio, L.; Covello, A.; Gargano, A.; Itaco, N.; Kuo, T.T.S. Shell-model calculations and realistic effective interactions. *Prog. Part. Nucl. Phys.* **2009**, *62*, 135–182. [CrossRef]
4. Langanke, K.; Martínez-Pinedo, G. Nuclear weak-interaction processes in stars. *Rev. Mod. Phys.* **2003**, *75*, 819–862. [CrossRef]
5. Grawe, H.; Langanke, K.; Martinez-Pinedo, G. Nuclear structure and astrophysics. *Rep. Prog. Phys.* **2007**, *70*, 1525–1582. [CrossRef]
6. Johnson, C.W.; Koonin, S.E.; Lang, G.H.; Ormand, W.E. Monte Carlo methods foer the nuclear shell model. *Phys. Rev. Lett.* **1992**, *69*, 3157–3160. [CrossRef]
7. Koonin, S.E.; Dean, D.J.; Langanke, K. Shell model Monte Carlo methods. *Phys. Rep.* **1997**, *278*, 1–77. [CrossRef]
8. Oda, T.; Hino, M.; Muto, K.; Takahara, T.; Sato, K. Rate tables for the weak processes of *sd*-shell nuclei in stellar matter. *At. Data Nucl. Data Tables* **1994**, *56*, 231–403. [CrossRef]
9. Dean, D.J.; Langanke, K.; Chatterjee, L.; Radha, P.B.; Strayer, M.R. Electron capture on iron group nuclei. *Phys. Rev. C* **1998**, *58*, 536–544. [CrossRef]
10. Langanke, K.; Martinez-Pinedo, G. Rate tables for the weak processes of $pf$-shell nuclei in stellar environments. *At. Data Nucl. Data Tables* **2001**, *79*, 1–46. [CrossRef]
11. Martínez-Pinedo, G.; Lam, Y.H.; Langanke, K.; Zegers, R.G.T.; Sullivan, C. Astrophysical weak-interaction rates for selected $A = 20$ and $A = 24$ nuclei. *Phys. Rev. C* **2014**, *89*, 045806. [CrossRef]
12. Heger, A.; Langanke, K.; Martinez-Pinedo, G.; Woosley, S.E. Presupernova collapse models with improved weak-interaction rates. *Phys. Rev. Lett.* **2001**, *86*, 1678–1681. [CrossRef] [PubMed]
13. Heger, A.; Woosley, S.E.; Martinez-Pinedo, G.; Langanke, K. Presupernova evolution with improved rates for weak interactions. *Astrophys. J.* **2001**, *560*, 307–325. [CrossRef]
14. Nomoto, K. Evolution of 8–10 $M_\odot$ stars toward electron capture supernovale: II. Collapse of an O + Ne + Mg core. *Astrophys. J.* **1987**, *322*, 206–214. [CrossRef]
15. Kirsebom, O.S.; Jones, S.; Strömberg, D.F.; Martínez-Pinedo, G.; Langakc, K.; Röpke, F.K.; Brown, B.A.; Eronen, T.; Fynbo, H.O.U.; Hukkanen, M.; et al. Discovery of an exceptionally strong $\beta$-decay transition of $^{20}$F and the fate of intermediate-mass stars. *Phys. Rev. Lett.* **2019**, *123*, 262701. [CrossRef] [PubMed]
16. Brachwitz, F.; Dean, D.J.; Hix, W.R.; Iwamoto, K.; Langanke, K.; Martinez-Pinedo, G.; Nomoto, F.; Strayer, M.R.; Thieleman, F.-K.; Umeda, H. The role of electron captures in Chandrasekhar-mass models for type Ia supernovae. *Astrophys. J.* **2000**, *536*, 934–947. [CrossRef]
17. Langanke, K.; Martínez-Pinedo, G.; Sampaio, J.M.; Dean, D.J.; Hix, W.R.; Messer, O.E.B.; Mezzacappa, A.; Liebendörfer, M.; Janka, H.-T.; Rampp, M. Electron capture rates on nuclei and implications for stellar core collapse. *Phys. Rev. Lett.* **2003**, *90*, 241102. [CrossRef]
18. Hix, W.R.; Messer, O.E.B.; Mezzacappa, A.; Liebendörfer, M.; Sampaio, J.; Langanke, K.; Dean, D.J.; Martínez-Pinedo, G. The consequences of nuclear electron capture in core collapse supernovae. *Phys. Rev. Lett.* **2003**, *91*, 210102. [CrossRef]
19. Janka, H.-T.; Langanke, K.; Marek, A.; Martínez-Pinedo, G.; Mueller, B. Theory of core-collapse supernovae. *Phys. Rep.* **2007**, *442*, 38–74. [CrossRef]
20. Martinez-Pinedo, G.; Langanke, K. Shell-model half-lives for the $N = 82$ nuclei and their implications for the $r$-process. *Phys. Rev. Lett.* **1992**, *83*, 4502–4505. [CrossRef]
21. Suzuki, T.; Yoshida, T.; Kajino, T.; Otsuka, T. $\beta$ decays of isotones with neutron magic number of $N = 126$ and $r$-process nucleosynthesis. *Phys. Rev. C* **2012**, *85*, 015802. [CrossRef]
22. Zhi, Q.; Caurier, E.; Cuenca-García, J.J.; Langanke, K.; Martíz-Pinedo, G.; Sieja, K. Shell-model half-lives including first-forbidden contributions for $r$-process waiting-point nuclei. *Phys. Rev. C* **2013**, *87*, 025803. [CrossRef]

23. Larsen, A.C.; Goriely, S.; Bernstein, L.A.; Bleuel, D.L.; Bracco, A.; Brown, B.A.; Camera, F.; Eriksen, T.K.; Frauendorf, S.; Giacoppo, F.; et al. Upbend and M1 scissors mode in neutron-rich nuclei—Consequences for r-process $(n, \gamma)$ reaction rates. *Acta Phys. Pol. B* **2015**, *46*, 509–512. [CrossRef]
24. Larsen, A.C.; Spyrou, A.; Liddick, S.N.; Guttormsen, M. Novel techniques for constraining neutron-capture rates relevant for $r$-process heavy-element nucleosynthesis. *Prog. Part. Nucl. Phys.* **2019**, *107*, 69–108. [CrossRef]
25. Nakada, H.; Alhassid, Y. Total and parity-projected level densities of iron-region nuclei in the auxiliary fields Monte Carlo shell model. *Phys. Rev. Lett.* **1997**, *79*, 2939–2942. [CrossRef]
26. Langanke, K. Shell model Monte Carlo level densities for nuclei around $A \sim 50$. *Phys. Lett.* **1998**, *438*, 235–241. [CrossRef]
27. Bethe, H.A. Supernova mechanisms. *Rev. Mod. Phys.* **1990**, *62*, 801–867. [CrossRef]
28. Bethe, H.A.; Brown, G.E.; Applegate, J.; Lattimer, J.M. Equation of state in the gravitational collapse of stars. *Nucl. Phys. A* **1979**, *324*, 487–533. [CrossRef]
29. Langanke, K.; Dean, D.J.; Radha, P.B.; Alhassid, Y.; Koonin, S.E. Shell model Monte Carlo studies of $fp$-shell nuclei. *Phys. Rev. C* **1995**, *52*, 718–725. [CrossRef]
30. Martínez-Pinedo, G.; Poves, A.; Caurier, E.; Zuker, A.P. Effective $g_A$ in the $pf$-shell. *Phys. Rev. C* **1996**, *53*, R2602–R2605. [CrossRef]
31. Caurier, E.; Langanke, K.; Martínez-Pinedo, G.; Nowacki, F. Shell-Model calculations of stellar weak interaction rates. I. Gamow-Teller distributions and spectra of nuclei in the mass range $A = 45 - 65$. *Nucl. Phys. A* **1999**, *653*, 439–452. [CrossRef]
32. Vetterli, M.C.; Jackson, K.P.; Celler, A.; Engel, J.; Yen, S. The $^{70,72}$Ge$(n,p)^{70,72}$Ga reactions: Suppression of Gamow–Teller strength near $N = 40$. *Phys. Rev. C* **1992**, *45*, 997–1004. [CrossRef] [PubMed]
33. Frekers, D.; Alanssari, M. Charge-exchange reactions and the quest for resolution. *Eur. Phys. J. A* **2018**, *54*, 177. [CrossRef]
34. Fuller, G.M.; Fowler, W.A.; Newman, M.J. Stellar weak-interaction rates for *sd*-shell nuclei. I. Nuclear matrix element systematics with application to $^{26}$Al and selected nuclei of importance to the supernova problem *Astrophys. J. Suppl. Ser.* **1980**, *42*, 447–473. [CrossRef]
35. Fuller, G.M.; Fowler,W.A.; Newman, M.J. Stellar weak interaction rates for intermediate mass nuclei. III. Rate tables for the free nucleons and nuclei with $A = 21$ to $A = 60$. *Astrophys. J. Suppl. Ser.* **1982**, *48*, 279–320. [CrossRef]
36. Cole, A.L.; Anderson, T.S.; Zegers, R.G.T.; Sam, M.; Austin, B.; Brown, A.; Valdez, L.; Gupta, S.; Hitt, G.W.; Fawwaz, O. Gamow–Teller strengths and electron-capture rates for $pf$-shell nuclei of relevance for late stellar evolution. *Phys. Rev. C.* **2012**, *86*, 015809. [CrossRef]
37. Poves, A.; Sánchez-Solano, J.; Caurier, E.; Nowacki, F. Shell model study of the isobaric chains $A = 50$, $A = 51$ and $A = 52$. *Nucl. Phys. A.* **2001**, *694*, 157–198. [CrossRef]
38. Honma, M.; Otsuka, T.; Brown, B.A.; Mizusaki, T. New effective interaction for $pf$-shell nuclei and its implications for the stability of the $N = Z = 28$ closed core. *Phys. Rev. C* **2004**, *69*, 034335. [CrossRef]
39. Langanke, K.; Martinez-Pinedo, G.; Zegers, R.M.T. Stellar electron capture. *Rep. Prog. Phys.* **2021**, *84*, 066301. [CrossRef]
40. Kirsebom, O.S.; Hukkanen, M.; Kankainen, A.; Trzaska, W.H.; Ströberg, D.F.; Martínz-Pinedo, G.; Andersen, K.; Bodewits, E.; Canete, L.; Cederkäl, J. Measurement of the $2^+ \to 0^+$ ground-state transition in the $\beta$ decay of $^{20}$F. *Phys. Rev. C* **2019**, *100*, 065805. [CrossRef]
41. Jones, S.; Röpke, F.K.; Pakmor, R.; Seitenzahl, I.R.; Ohlmann, S.T.; Edelmann, P.V.F. Do electron-capture supernovae make neutron stars? First multidimensional hydrodynamic simulations of the oxygen deflagration. *Astron. Astrophys.* **2016**, *593*, A72. [CrossRef]
42. Zha, S.; Leung, S.C.; Suzuki, T.; Nomoto, K. Evolution of ONeMg core in super-AGB stars toward electron-capture supernovae: Effects of updated electron-capture rate. *Astrophys. J.* **2019**, *886*, 22. [CrossRef]
43. Strömberg, D.F.; Martínez-Pinedo, G.; Nowacki, F. Forbidden electron capture on $^{24}$Na and $^{27}$Al in degenerate oxygen-neon stellar cores. *Phys. Rev. C* **2022**, *105*, 025803. [CrossRef]
44. Takahara, M.; Hino, M.; Oda, T.; Muto, K.; Wolters, A.A.; Glaudemans, P.W.M.; Sato, K. Microscopic calculation of the rates of electron captures which induce the collapse of the O+Ne+Mg cores. *Nucl. Phys. A* **1989**, *504*, 167–192. [CrossRef]
45. Fuller, G.M. Neutron shell blocking of electron capture during gravitational collapse. *Astrophys. J.* **1982**, *252*, 741–764. [CrossRef]
46. Langanke, K.; Kolbe, E.; Dean, D.J. Unblocking of the Gamow–Teller strength in stellar electron capture on neutron-rich germanium isotopes. *Phys. Rev. C* **2001**, *63*, 032801(R). [CrossRef]
47. Fischer, T.; Langanke, K.; Martínez-Pinedo, G. Neutrino-pair emission from nuclear de-excitation in core-collapse supernova simulations. *Phys. Rev. C* **2013**, *88*, 065804. [CrossRef]
48. Grewe, E.-W.; Bäumer, C.; Dohmann, H.; Frekers, D.; Harakeh, M.N.; Hollstein, S.; Johansson, H.; Popescu, L.; Rakers, S.; Savran D.; et al. The $(d,^2\text{He})$ reaction on $^{76}$Se and the double-$\beta$-decay matrix elements for $A = 76$. *Phys. Rev. C* **2008**, *78*, 044301. [CrossRef]
49. Zhi, Q.; Langanke, K.; Martínez-Pinedo, G.; Nowacki, F.; Sieja, K. The $^{76}$Se Gamow–Teller strength distribution and its importance for stellar electron capture rates. *Nucl. Phys. A* **2011**, *859*, 172–184. [CrossRef]
50. Dean, D.J.; Ressell, M.T.; Hjorth-Jensen, M.; Koonin, S.E.; Langake, K.; Zuker, A.P. Shell-model Monte Carlo studies of neutron-rich nuclei in the 1$s$-0$d$-1$p$-0$f$ shells. *Phys. Rev. C* **1999**, *59*, 2474–2486. [CrossRef]
51. Caurier, E.; Langanke, K.; Martínez-Pinedo, G.; Nowacki, F.; Vogel, P. Shell model description of isotope shifts in calcium. *Phys. Lett. B* **2001**, *522*, 240–244. [CrossRef]
52. Sullivan, C.; Connor, E.O.; Zegers, R.G.T.; Grubb, T.; Austin, S.M. The sensitivity of core-collapse supernovae to nuclear electron capture. *Astrophys. J.* **2016**, *816*, 44. [CrossRef]

53. Titus, R.; Ney, E.M.; Zegers, R.G.T.; Bazin, D.; Belarge, J.; Bender, P.C.; Brown, B.A.; Campbell, C.M.; Elman, B.; Engel, J. et al. Constraints for stellar electron-capture rates on $^{86}$Kr via the $^{86}\mathrm{Kr}(t,^3\mathrm{He}+\gamma)^{86}\mathrm{Br}$ reaction and the implications for core-collapse supernovae. *Phys. Rev. C* **2019**, *100*, 045805. [CrossRef]
54. Zamora, J.C.; Zegers, R.G.T.; Austin, S.M.; Bazin, D.; Brown, B.A.; Bender, P.C.; Crawford, H.L.; Engel, J.; Falduto, A.; Gade, A.; et al. Experimental constraint on stellar electron-capture rates from the $^{88}\mathrm{Sr}(t,^3\mathrm{He}+\gamma)^{88}\mathrm{Rb}$ reaction at 115 MeV/u. *Phys. Rev. C* **2019**, *100*, 032801. [CrossRef]
55. Dzhioev, A.A.; Langanke, K.; Martínez-Pinedo, G.; Vdovin, A.I.; Stoyanov, C. Unblocking of stellar electron captures for neutron-rich $N = 50$ nuclei at finite temperatures. *Phys. Rev. C* **2020**, *101*, 025805. [CrossRef]
56. Litvinova, E.; Robin, C. Impact of complex many-body correlations on electron capture in thermally excited nuclei around $^{78}$Ni. *Phys. Rev. C* **2021**, *103*, 024326. [CrossRef]
57. Juodagalvis, A.; Langanke, K.; Hix, W.R.; Martínez-Pinedo, G.; Sampaio, J.M. Improved estimate of stellar electron capture rates on nuclei. *Nucl. Phys. A* **2010**, *848*, 454–478. [CrossRef]
58. Langanke, K.; Martínez-Pinedo, G. Shell-model calculations of stellar weak interaction rates: II. Weak rates for nuclei in the mass range $A = 45 - 65$ in supernovae environments. *Nucl. Phys. A* **2000**, *673*, 481–508. [CrossRef]
59. Iwamoto, K.; Brachwitz, F.; Nomoto, K.; Kishimoto, N.; Umeda, H.; Hix, W.R.; Thielemann, F.-K. Nucleosynthesis in Chandrasekhar mass models for type Ia supernovae and constraints on progenitor systems and burning-front propagation. *Astrophys. J. Suppl. Ser.* **1999**, *125*, 439–462. [CrossRef]
60. Fuller, G.M.; Meyer, B.S. High-temperature neutrino-nucleus processes in stellar collapse. *Astrophys. J.* **1991**, *376*, 701–716. [CrossRef]
61. Bruenn, S.W.; Haxton, W.C. Neutrino-nucleus interactions in core-collapse supernovae. *Astrophys. J.* **1991**, *376*, 678–700. [CrossRef]
62. Langanke, K.; Martínez-Pinedo, G.; Müller, B.; Janka, H.-T.; Marek, A.; Hix, W.R.; Juodagalvis, A.; Sampaio, J.M. Effects of inelastic neutrino-nucleus scattering on supernova dynamics and radiated neutrino spectra. *Phys. Rev. Lett.* **2008**, *100*, 011101. [CrossRef] [PubMed]
63. Donnelly, T.W.; Peccei, R.D. Neutral current effects in nuclei. *Phys. Rep.* **1979**, *50*, 1–85. [CrossRef]
64. Langanke, K.; Martinez-Pinedo, G.; von Neumann-Cosel, P.; Richter, A. Supernova inelastic neutrino-nucleus cross sections from high-resolution electron scattering experiments and shell-model calculations". *Phys. Rev. Lett.* **2004**, *93*, 202501. [CrossRef] [PubMed]
65. von Neumann-Cosel, P.; Poves, A.; Retamosa, J.; Richter, A. Magnetic dipole response in nuclei at the $N = 28$ shell closure: A new look. *Phys. Lett. B* **1998**, *443*, 1–6. [CrossRef]
66. Juodagalvis, A.; Langanke, K.; Martínez-Pinedo, G.; Hix, W.R.; Dean, D.J.; Sampaioc, J.M. Neutral-current neutrino-nucleus cross sections for $A \sim 50$–65 nuclei. *Nucl. Phys. A* **2005**, *747*, 87–108. [CrossRef]
67. Woosley, S.E.; Hartmann, D.H.; Hofmann, R.D.; Haxton, W.C. The $\nu$-process. *Astrophys. J.* **1990**, *356*, 272–301. [CrossRef]
68. Heger, A.; Kolbe, E.; Haxton, W.C.; Langanke, K.; Martínez-Pinedo, G.; Woosley, S.E. Neutrino nucleosynthesis. *Phys. Lett. B* **2005**, *606*, 258–264. [CrossRef]
69. Byelikov, A.; Adachi, T.; Fujita, H.; Fujita, K.; Fujita, Y.; Hatanaka, K.; Heger, A.; Kalmykov, Y.; Kawase, K.; Langanke, K. Gamov–Teller strength in the exotic, odd-odd nuclei $^{138}$La and $^{180}$Ta and its relevance for neutrino nucleosynthesis. *Phys. Rev. Lett.* **2007**, *98*, 082501. [CrossRef]
70. Sieverding, A.; Martínez-Pinedo, G.; Huther, L.; Langanke, K.; Heger, A. The $\nu$-process in the light of an improved understanding of supernova neutrino spectra. *Astrophys. J.* **2018**, *865*, 143. [CrossRef]
71. Sieverding, A.; Langanke, K.; Martínez-Pinedo, G.; Bollig, R.; Janka, H.-T.; Heger, A. The $\nu$-process with fully time-dependent supernova neutrino emission spectra. *Astrophys. J.* **2019**, *876*, 151. [CrossRef]
72. Kolbe, E.; Langanke, K.; Thielemann, F.-K., Vogel, P. Inclusive $^{12}\mathrm{C}(\nu_\mu,\mu)^{12}\mathrm{N}$ reaction in the continuum random phase approximation. *Phys. Rev C* **1995**, *52*, 3437–3441. [CrossRef] [PubMed]
73. Kolbe, E.; Langanke, K.; Martínez-Pinedo, G.; Vogel, P. Neutrino–nucleus reactions and nuclear structure. *J. Phys. G Nucl. Part. Phys.* **2003**, *29*, 2569–2596. [CrossRef]
74. Balasi, K.G.; Langanke, K.; Martínez-Pinedo, G. Neutrino–nucleus reactions and their role for supernova dynamics and nucleosynthesis. *Prog. Part. Nucl. Phys.* **2015**, *85*, 33–81. [CrossRef]
75. Fröhlich, C.; Martínez-Pinedo, G.; Liebendörfer, M. Thielemann, F.-K.; Bravo, E.; Hix, W.R.; Langanke, K.; Zinner, N.T. Neutrino-induced nucleosynthesis of $A > 64$ nuclei: The $\nu p$-process. *Phys. Rev. Lett.* **2006**, *96*, 142502. [CrossRef] [PubMed]
76. Pruet, J.; Woosley, S.E.; Buras, R.; Janka, H.-T.; Hoffman, R.D. Nucleosynthesis in the hot convective bubble in core-collapse supernovae. *Astrophys. J.* **2006**, *623*, 325–336. [CrossRef]
77. Wanajo, S. The rp-process in neutrino-driven winds. *Astrophys. J.* **2006**, *647*, 1323–1340. [CrossRef]
78. Burbidge, E.M.; Burbidge, G.R.; Fowler, W.A.; Hoyle, F. Synthesis of the elements in stars. *Rev. Mod. Phys.* **1957**, *29*, 547–650. [CrossRef]
79. Thielemann, F.-K.; Arcones, A.; Käppeli, R.; Liebendörfer, M.; Rauscher, T.; Winteler, C.; Fröhlich, C.; Dillmann, I.; Fischer, T.; Martinez-Pinedo, G.; et al. What are the astrophysical sites for the $r$-process and the production of heavy elements? *Prog. Part. Nucl. Phys.* **2011**, *66*, 346–353. [CrossRef]
80. Cowan, J.J.; Thielemann, F.-K.; Truran, J.W. The R-process and nucleochronology. *Phys. Rep.* **1991**, *208*, 267–394. [CrossRef]

81. Abbot, B.P. et al. [LIGO Scientific Collaboration and Virgo Collaboration; Fermi GBM; INTEGRAL; IceCube Collaboration; AstroSat Cadmium ZincTelluride Imager Team; IPN Collaboration; The Insight-HXMT Collaboration; ANTARES Collaboration; The Swift Collaboration; AGILE Team; The 1M2H Team; The Dark Energy Camera GW-EM Collaboration and the DES Collaboration; The DLT40 Collaboration,GRAWITA: GRAvitational Wave Inaf TeAm; The Fermi Large Area Telescope Collaboration; ATCA: Australia Telescope CompactArray; ASKAP: Australian SKA Pathfinder; Las Cumbres Observatory Group; OzGrav; DWF(Deeper, Wider, Faster Program); AST3, and CAASTRO Collaborations; The VINROUGE Collaboration; MASTER Collaboration; J-GEM, GROWTH, JAGWAR, Caltech-NRAO, TTU-NRAO, and NuSTAR Collaborations; Pan-STARRS; TheMAXITeam; TZACConsortium; KU Collaboration; NordicOptical Telescope; ePESSTO; GROND; Texas Tech University; SALT Group; TOROS: Transient Robotic Observatory of the SouthCollaboration; The BOOTES Collaboration; MWA: Murchison Widefield Array; The CALET Collaboration; IKI-GW Follow-upCollaboration; H.E.S.S. Collaboration; LOFAR Collaboration; LWA: Long Wavelength Array; HAWC Collaboration; The Pierre Auger Collaboration; ALMA Collaboration; Euro VLBI Team; Pi of the Sky Collaboration; The Chandra Team at McGill University; DFN:Desert Fireball Network; ATLAS; High Time Resolution Universe Survey; RIMAS and RATIR; SKA South Africa/MeerKAT] Multi-messenger observations of a binary neutron star merger. *Astrophys. J. Lett.* **2017**, *848*, L12. [CrossRef]
82. Cowperthwaite, P.S.; Berger, E.; Villar, V.A.; Metzger, B.D.; Nicholl, M.; Chornock, R.; Blanchard, P.K.; Fong, W.; Margutti, R.; Soares-Santos, M.; et al. The electromagnetic counterpart of the binary neutron star merger LIGO/Virgo GW170817. II. UV, optical, and near-infrared light curves and comparison to kilonova models. *Astrophys. J. Lett.* **2017**, *848*, L17. [CrossRef]
83. Metzger, B.D.; Martńez-Pinedo, G.; Darbha, S.; Quataert, E.; Arcones, A.; Kasen, D.; Thomas, R.; Nugent, P.; Panov, I.V.; Zinner, N.T. Electromagnetic counterparts of compact object mergers powered by the radioactive decay of *r*-process nuclei. *Mon. Not. R. Astron. Soc.* **2010**, *406*, 2650–2662. [CrossRef]
84. Cowan, J.J.; Sneden, C.; Lawler, J.E.; Aprahamian, A.; Wiescher, M.; Langake, K.; Martínez-Pinedo, G.; Thielemann, F.-K. Origin of the heaviest elements: The rapid neutron-capture process. Process. *Rev. Mod. Phys.* **2021**, *93*, 015002. [CrossRef]
85. Wu, J. Nishimura, S.; Lorusso, G.; Möller, P.; Ideguchi, E.; Regan, P.-H.; Simpson, G.S.; Söderström, P.A.; Waller, P.M.; Watanable, H.; et al. 94 $\beta$-decay half-lives of neutron-rich $_{55}$Cs to $_{67}$Ho: Experimental feedback and evaluation of the *r*-process rare-earth peak formation. *Phys. Rev. Lett.* **2017**, *118*, 072701.
86. Lorusso, G.; Nishimura, S.; Xu, Z.Y.; Jungclaus, A.; Shimizu, Y.; Simpson, G.S.; Söderström, P.-A.; Watanabe, H.; Browne, F.; Doornenbal, P.; et al. $\beta$-decay half-lives of 110 neutron-rich nuclei across the $N = 82$ shell gap: Implications for the mechanism and universality of the astrophysical *r* process. *Phys. Rev. Lett.* **2015**, *114*, 192501. [CrossRef]
87. Möller, P.; Nix, J.R.; Kratz, K.-L. Nuclear properties for astrophysical and radioactive-ion-beam applications. *At. Nucl. Data Tables* **1997**, *66*, 131–343. [CrossRef]
88. Borzov, I.N.; Goriely, S. Weak interaction rates of neutron-rich nuclei and the *r*-process nucleosynthesis. *Phys. Rev. C* **2000**, *62*, 035501. [CrossRef]
89. Borzov, I.N. Gamow–Teller and first-forbidden decays near the *r*-process paths at $N = 50$, 82, and 126. *Phys. Rev. C* **2003**, *67*, 025802. [CrossRef]
90. Marketin, T.; Huther, L.; Martinez-Pinedo, G. Large-scale evaluation of $\beta$-decay rates of *r*-process nuclei with the inclusion of first-forbidden transitions. *Phys. Rev. C* **2016**, *93*, 025805. [CrossRef]
91. Mustonen, M.T.; Engel, G. Global description of $\beta^-$ decay in even-even nuclei with the axially-deformed Skyrme finite-amplitude method. *Phys. Rev. C* **2016**, *93*, 014304. [CrossRef]
92. Shafer, T.; Engel, J.; Frölich, C.; Mclaughlin, G.C.; Mumpower, M.; Surman, R. $\beta$ decay of deformed *r*-process nuclei near $A = 80$ and $A = 160$, including odd-$A$ and odd-odd nuclei, with the Skyrme finite-amplitude method. *Phys. Rev. C* **2016**, *94*, 055802. [CrossRef]
93. Ney, E.M.; Engel, J.; Li, T.; Schunck, N. Global description of $\beta^-$ decay with the axially deformed Skyrme finite-amplitude method: Extension to odd-mass and odd-odd nuclei. *Phys. Rev. C* **2020**, *102*, 034326. [CrossRef]
94. Hosmer, P.T.; Schatz, H.; Aprahamian, A.; Arndt, O.; Clement, R.R.C.; Estrade, A.; Kratz, K.-L.; Liddick, S.N.; Mantica, P.F.; Mueller, W.F.; et al. Half-life of the doubly magic *r*-process nucleus $^{78}$Ni. *Phys. Rev. Lett.* **2005**, *94*, 112501. [CrossRef] [PubMed]
95. de Jesús Mendoza-Temis, J.; Wu, M.-R.; Langanke, K.; Martínez-Pinedo, G.; Bauswein, A.; Janka, H.-T. Nuclear robustness of the *r* process in neutron-star mergers. *Phys. Rev. C* **2015**, *92*, 055805. [CrossRef]
96. Petermann, I.; Langanke, K.; Martínez-Pinedo, G.; Panov, I.V.; Reinhard, P.-G.; Thielemann, F.-K. Have superheavy elements been produced in nature? *Eur. Phys. J. A* **2012**, *48*, 122. [CrossRef]
97. Wu, M.-R.; Barnes, J.; Martínez-Pinedo, G.; Metzger, B.D. Fingerprints of heavy-element nucleosynthesis in the late-time lightcurves of kilonovae. *Phys. Rev. Lett.* **2019**, *122*, 062701. [CrossRef]
98. Pfeiffer, B.; Kratz, K.-L.; Thielemann, F.-K.; Walters, W.B. Nuclear structure studies for the astrophysical r-process. *Nucl. Phys. A* **2001**, *693*, 282–324. [CrossRef]
99. Dillmann, I.; Kratz, K.-L.; Wöhr, A.; Arndt, O.; Brown, B.A.; Hoff, P.; Hjorth-Jensen, M.; Köster, U.; Ostrowski, A.N.; Pfeiffer, D.; et al. $N = 82$ shell quenching of the classical *r*-process "waiting-point" nucleus $^{130}$Cd. *Phys. Rev. Lett.* **2003**, *91*, 162503. [CrossRef]
100. Fogelberg, B.; Gausemel, H.; Mezilev, K.A.; Hoff, P.; Mach, H.; Sanchez-Vega, M.; Lindroth, A.; Ramström, E.; Genevey, J.; Pinston, J.A.; et al. Decays of $^{131}$In, $^{131}$Sn, and the position of the $h_{11/2}$ neutron hole state. *Phys. Rev. C* **2004**, *70*, 034312. [CrossRef]
101. Alhassid, Y.; Liu, S.; Nakada, H. Particle-number reprojection in the shell model Monte Carlo method: Application to nuclear level densities. *Phys. Rev. Lett.* **1999**, *83*, 4265–4268. [CrossRef]

102. Alhassid, Y.; Liu, S.; Nakada, H. Spin projection in the shell model Monte Carlo method and the spin distribution of nuclear level densities. *Phys. Rev. Lett.* **2007**, *99*, 162504. [CrossRef] [PubMed]
103. Langanke, K. Shell Model Monte Carlo studies of pairing correlations and level densities in medium-mass nuclei. *Nucl. Phys. A* **2006**, *778*, 233–246. [CrossRef]
104. Alhassid, Y.; Bertsch, G.F.; Liu, S.; Nakada, H. Parity dependence of nuclear level densities. *Phys. Rev. Lett.* **2000**, *84*, 4313–4316. [CrossRef] [PubMed]
105. Mocelj, D.; Rauscher, T.; Martínez-Pinedo, G.; Langanke, K.; Pacearescu, L.; Faessler, A.; Thielemann, F.-K.; Alhassid, Y. Large-scale prediction of the parity distribution in the nuclear level density and application to astrophysical reaction rates. *Phys. Rev. C* **2007**, *75*, 045805. [CrossRef]
106. Rauscher, T.; Thielemann, F.-K. Astrophysical reaction rates from statistical model calculations. *At. Nucl. Data Tables* **2000**, *75*, 1–351. [CrossRef]
107. Rauscher, T.; Thielemann, F.-K. Tables of nuclear cross sections and reaction rates: An addendum to the paper "Astrophysical reaction rates from statistical model calculations". *At. Nucl. Data Tables* **2001**, *79*, 47–64. [CrossRef]
108. Rauscher, T. The path to improved reaction rates for astrophysics. *Int. J. Mod. Phys. E* **2011**, *20*, 1071–1169. [CrossRef]
109. Goriely, S.; Hilaire, S.; Koning, A.J. Improved microscopic nuclear level densities within the Hartree-Fock-Bogoliubov plus combinatorial method. *Phys. Rev. C* **2008**, *78*, 064307. [CrossRef]
110. Koning, A.J.; Hilaire, S.; Goriely, S. Global and local level density models. *Nucl. Phys. A* **2008**, *810*, 13–76. [CrossRef]
111. Goriely, S.; Hilaire, S.; Girod, M. Latest development of the combinatorial model of nuclear level densities. *J. Phys. Conf. Ser.* **2012**, *337*, 012027. [CrossRef]
112. Guttormsen, M.; Chankova, R.; Agvaanluvsan, U.; Algin, E.; Bernstein, L.A.; Ingebretsen, F.; Lönnroth, T.; Messelt, S.; Mitchell, G.E.; Rekstad, J.; et al. Radiative strength functions in $^{93-98}$Mo. *Phys. Rev. C* **2005**, *71*, 044307. [CrossRef]
113. Larsen, A.C.; Chankova, R.; Guttormsen, M.; Ingebretsen, F.; Messelt, S.; Rekstad, J.; Siem, S.; Syed, N.U.H.; Øegård, S.W.; Lönnroth, T.; et al. Microcanonical entropies and radiative strength functions of $^{50,51}$V. *Phys. Rev. C* **2006**, *73*, 064301. [CrossRef]
114. Goriely, S. Radiative neutron captures by neutron-rich nuclei and the r-process nucleosynthesis. *Phys. Lett. B* **1998**, *436*, 10–18. [CrossRef]
115. Larsen, A.C.; Goriely, S. Impact of a low-energy enhancement in the $\gamma$-ray strength function on the neutron-capture cross section. *Phys. Rev. C* **2010**, *82*, 014318. [CrossRef]
116. Litvinova, E.; Ring, P.; Tselyaev, V.; Langanke, K. Relativistic quasiparticle time blocking approximation. II. Pygmy dipole resonance in neutron-rich nuclei. *Phys. Rev. C* **2009**, *79*, 054312. [CrossRef]
117. Sieja, K. Electric and magnetic dipole strength at low energy. *Phys. Rev. Lett.* **2017**, *119*, 052502. [CrossRef]
118. Sieja, K. Shell-model study of the $M1$ dipole strength at low energy in the $A > 100$ nuclei. *Phys. Rev. C* **2018**, *98*, 064312. [CrossRef]
119. Loens, H.P.; Langanke, K.; Martínez-Pinedo, G.; Sieja, K. M1 strength functions from large-scale shell model calculations and their effect on astrophysical neutron capture cross-sections. *Eur. Phys. J. A* **2012**, *48*, 34. [CrossRef]
120. Bohle, D.; Richter, A.; Steffen, W.; Dieperink, A.E.L.; Lo Ludice, N.; Palumbo, F.; Scholten, O. New magnetic dipole excitation mode studied in the heavy deformed nucleus $^{156}$Gd by inelastic electron scattering *Phys. Lett.* **1984**, *137*, 27–31. [CrossRef]